Entomology

Entomology

by

D.N. Roy

The Indian Institue of Medical Research

&

A.W.A. Brown

The University of Western Ontario, London, Canada

BIOTECH

2023

BIOTECH BOOKS

D.N. Roy & A.W.A. Brown

ISBN: 978-81-7622-110-8

Published by : **BIOTECH BOOKS**
4762-63/23, Ansari Road, Darya Ganj,
New Delhi - 110002
Phones : 011 4300 3222
Email : biotechbooks@yahoo.co.in

PRINTED IN INDIA

Part I

ENTOMOLOGY

(Medical & Veterinary)

by

D.N. Roy

PREFACE

The first edition of this book was published in 1946. It was written under the stress of the Great War when India remained almost isolated from the rest of the world. Its shortcomings, therefore, could not be avoided. The book has now been revised and brought up to date. This has been rendered possible by the collaboration of Professor A. W. A. Brown of the University of Western Ontario, Canada, who has contributed the section on Insecticides. Such collaboration of two authors from two distant countries is a unique feature in the authorship of a book of this kind. At the back of it lies our conviction that without the acceptance of the thesis of "One World" in the cultural sphere and possibly in other spheres also, the progress of this complicated world of to-day is bound to be delayed. It is also our earnest desire that the knowledge as to how insects, particularly mosquitoes, flies and fleas carry disease organisms, and of the remarkable advancements made in the control of these insects should spread. These insects are alone responsible for as much loss of human lives as those of all other pathogenic diseases in the tropical countries of the world.

In this book due emphasis has been laid on the characteristics of disease-carrying insects to render their identification with certainty. Their life history and bionomics have also been adequately dealt with on which the control measures are based. Its scope has, however, been widened by including the behaviour of the parasite during the "insect phase" in its life cycle. The Medical Entomologist combines the functions of an entomologist, a biologist and a parasitologist at the same time. The book has, therefore, been written to satisfy the needs of the entomologist in order that he may fulfil these difficult functions without any undue strain on him.

Though this book has been written with the primary object of giving an account of insect life and the behaviour of such insects towards the spread of disease organisms in this country, information on similar lines in respect of other countries has also been incorporated though in a comparatively brief manner. A book giving a total picture of insects of the entire world and their parasitological relationships is a necessity for every entomologist and parasitologist and it should rightly devolve on the World Health Organisation to prepare it. The W.H.O. should start by organising research centres in different parts of the tropics. This will help collaboration of intellectuals in the common cause of humanity and in alleviating human sufferings.

The materials for this book have been collected from various books and periodicals, the principal amongst the latter being the Tropical Diseases Bulletin, Review of Applied Entomology, the Indian Journal of Medical Research and the Indian Journal of Malariology.

I cannot pass over without acknowledging the help I received from my friends abroad and my old colleagues in the School of Tropical Medicine and the Institute of Hygiene, Calcutta. I should also record my appreciation of the troubles taken by Dr. Jyoti Dhar in reading through the proofs. My thanks are also due to artists, late Mr. J. K. Mullick and Mr. Ramen Roy of Calcutta and Mr. A. P. Bendre of Armed Forces Medical College, Poona.

D. N. ROY

The Indian Institute for Medical Research.

PART I

CONTENTS

Pages

Introduction 1-2

Chapter I. Phylum Annelida 3-4

Oligochaeta, Polychaeta, Hirudinea. Anti-leech remedies. Treatment of infestation.

Chapter II. Structure of Insects 5-17

Classification. Arthropoda in relation to diseases and discomforts. Therapeutic uses of Arthropoda and Animals of medical importance. Geographical Regions of the World. Type : the Cockroach. Biological warfare.

Chapter III. Order Diptera 18-21

Classification : Families, Subfamilies
Tribes Anophelini, Culicini and Megarhinini. Their life history and bionomics.

Chapter IV. Mosquito (Subfam. Culicinae) 22-34

Diseases. Tribes Anophelini, Culicini.

Chapter V. Genus *Anopheles*. 35-79

Life history and Binomics. Parasites. Races or Varieties. *Anopheles* species of India.
Malaria. *Anopheles* and Malaria. The use of infected *Anopheles* in therapeutics. Black Spore. Geographical distribution of malaria-carrying *Anopheles*. Malaria and malaria-mosquito survey. Collection and dissection of mosquitoes. Detection of infection in the mosquito. Staining of sporozoites. Spleen-index and Parasite-index. Artificial infection of *Anopheles*. Source of stomach blood of mosquitoes. Malaria control. Identification of female *Anopheles* species of India. Identification of Anopheline larvae of India.

Chapter VI. Tribe Culicini 80-102

Genus *Culex*. *Culex fatigans*. Genus *Mansonia*. Genus *Armigeres*. Mosquitoes and filarial diseases of man. Genus *Aedes*. *Aed. aegypti*. Mosquitoes employed in testing antimalarial drugs. Disease relationship. Control.

Chapter VII. Other Nematocera 103-122

Family Psychodidae. Genus *Phlebotomus*. Disease relationship. Breeding. Recovery of sand-fly larvae from soil. Identification. Dissection of spermatheca and of pharynx and buccal cavity. Collection and preservation. Control. Geographical distribution of sand-flies in India.
Familes: Chironomidae, Ceratopogonidae and Simuliidae. Disease ralationship.

Chapter VIII. Orthorrapha-Brachycera 123-131

Families: Tabanidae and Pangonidae. Genera: *Tabanus*, *Haematopota*, *Pangonia*.
Diptera-Aschiza. Genus *Tubifera*.

Chapter IX. Diptera-Cyclorrapha 132-179

Muscoidea. Diseases. Control. Dissection.
Acalyptrata Muscoidea. *Drosophila*, *Hippelates*, *Siphuncnlina*.
Calyptrata Muscoidea—Muscidae, Calliphoridae, Sarcophagidae, Anthomyidae, Oestridae, Tachinidae.
Blood-sucking Muscidae—*Stomoxys*, *Glossina*. *Lyperosia*.
Myiasis. Maggot therapy. Rearing larvae from cases of myiasis.
Diptera—Pupipara.

Chapter X. Order Siphonaptera 180-200

Characters. Disease relationship. Fleas and plague. Flea index. Identification of living fleas. Detection of plague bacilli in fleas. Classification.
Family Pulicidae. Identification of plague-carrying fleas. Rearing fleas.
Rats. Plague rats of India and their identifications. Chief ectoparasites of rats in India. Examination of rats for plague. Anti-rat and anti-flea measures.
Family Sarcopsyllidae. Chiggerosis.

Chapter XI. Orders Hymenoptera, Coleoptera and Lepidoptera 201-206

Their harmful effects. Scarabiasis.
Dermatitis.

Chapter XII. Order Hemiptera (Rhynchota) 207-213

Families. Genera: *Cimex*, *Triatoma*, *Rhodnius*. Chagas' disease. Control.

Chapter XIII. Order Anoplura 214-227

Suborders. Families. Genera: *Pediculus*, *Phthirus*, *Haematopinus*, *Linognathus*, *Bovicola*, *Polyplax*, *Solenopotes*. Disease relationship. Control. Suborder Mallophaga.

Chapter XIV. Class Arachnida 228-259

Order Araneida. Superfamily Ixodoidea. Suborders. Ixodidae. Argasidae. Genera of hard ticks. Human diseases. Different forms of typhus fever and their vectors. Blood parasites of domestic animals carried by hard ticks. *Argas*. *Ornithodorus*. Disease relationship. Relapsing fever ticks. Collection. Dissection. Examination. Breeding, Control.

Chapter XV. Mites 260-272

Family Trombiculidae. Important genera. Scrub typhus. Rickettsialpox. Collection of mites in the field. Prevention.
Families: Tarsonemidae, Tyroglyphidae, Sarcoptidae. Scabies. Sarcopticidal drugs.
Families: Gamasidae, Demodecidae.

Chapter XVI. Orders Araneida, Scorpionidea & other Arachnida 273-279

Chapter XVII. Class Crustacea & other Arthropoda 280-285

Cyclops. *Daphnia*. Disease relationship.

Chapter XVIII. Notes on Laboratory Technique 286-290

INTRODUCTION

Animals may be classified into two broad groups, chordates and nonchordates. The vertebrates come under the phylum Chordata while the nonchordates include the invertebrates; the latter have several distinct subdivisions or phyla. The invertebrates like the prawn, the scorpion, the centipede, the cockroach, the mosquito, etc., constitute the phylum Arthropoda (*arthron* meaning jointed and *podon* meaning foot).

The distinctive features of the Arthropoda are: (*a*) bilateral symmetry of the body; (*b*) skin hard and rigid; (*c*) metameric segmentation of the body; (*d*) segmentation marked externally: each typical segment bears a pair of jointed appendages; (*e*) central nervous system ventral, the heart dorsal, and the alimentary canal between them; (*f*) muscles striated; (*g*) body cavity is haemocœle which is in free communication with the circulatory system; (*h*) visual organs in the form of compound eyes and simple eyes or ocelli; (*i*) growth by periodical casting of skin; and (*j*) jaws present.

The Arthropoda is subdivided into: (*i*) Trilobita, (*ii*) Chelicerata, and (*iii*) Mandibulata. The Trilobita are extinct and occur as fossils. The Chelicerata include: Eurypterida, Xiphosura (Merostomata) and Arachnida. This group is peculiar; there are no antennæ, and their jaws are called the cheliceræ for their peculiar form and for uncertainty of their homology. The Eurypterida are fossils some of which attained a gigantic size. The Xiphosura include the king crab which is a marine animal and is recognisable by the dome-shaped carapace and by the long postanal spine. The Arachnida include the scorpion, the spider, the mite and the tick. The head carries at the anterior end a pair of stout appendages called the pedipalpi while the thorax bears four pair of walking legs. The abdomen is without any appendage and bears stigmata or respiratory apertures.

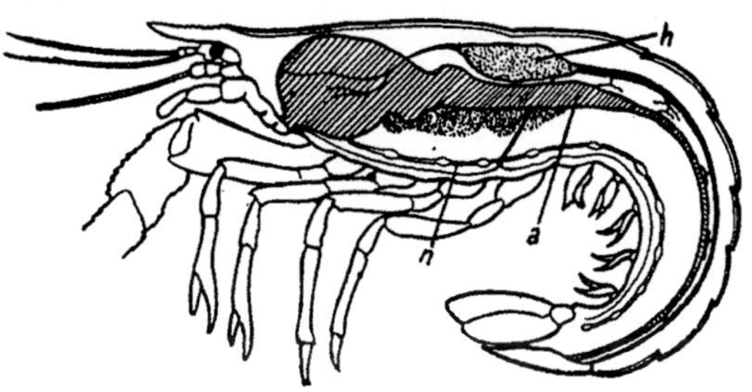

FIG. 1.

Side view of a prawn showing the relative positions of the alimentary canal, heart and nerve chain. *a*, alimentary canal; *h*, heart; *n*, nerve chain.

The group Mandibulata include : the Crustacea, the Myriapoda and the Insecta (Hexapoda). The well-known examples of Crustacea are the water flea, the barnacle, the prawn and the crab. The Crustacea carry on the head two pairs of antennæ. The abdomen bears biramous appendages or forked limbs. The Myriapoda include the millipede and the centipede which carry one pair of antennæ, and their trunk bears numerous legs. In the millipede the genital aperture is situated on the ventral side of the thorax towards the head end, while in the centipede it is at the posterior end of the abdomen. The centipede carries a pair of poison claws on the under surface of the head. The poison claws are appendages of the trunk shifted forward to the head.

The class Insecta has the following distinctive features. The adult body is divisible into three distinct regions: (*i*) the head, (*ii*) the thorax, and (*iii*) the abdomen. The head carries a pair of antennæ or feelers and three pairs of jaws; visual organs exist in the form of compound eyes and the simple eyes or ocelli. The thorax consists of three segments each of which carries a pair of jointed legs. The legs end in claws. The thorax also carries one or two pairs of wings. The wings may be absent. The abdomen consists of eleven segments. This number may be reduced.

Respiratory apertures are called the spiracles or stigmata. They are ten pair in number. Two pair are distributed on the thorax and eight on the abdomen. The spiracles lead air into tracheæ or air tubes, which ramify within the body and end in finer branches. These finer branches called the tracheoles enter within the substance of all the organs to help in respiratory exchange of gases.

The digestive system consists of a tube running from the mouth and ending at the anus. The circulatory system is open and the body cavity is a hæmocoele. Excretory organs occur as malpighian tubules which open into the hind portion of the gut.

The class Onychophora (example, *Peripatus*) occupies a position between the phylum Arthropoda and the phylum Annelida (typical example, the earthworm). Peripatus is a caterpillar-like animal; the head carries a pair of antennæ, a pair of jaws, and has apertures of slime glands. The trunk carries numerous legs, which end in claws, but are not jointed. The skin is soft and wrinkled. The remarkable feature is its excretory organs which resemble nephridia or the typical excretory tubules of Annelida.

CHAPTER I

PHYLUM ANNELIDA

The body is elongated and is divided by a number of rings; the body cavity is a true coelom being lined by layers of mesoderm. This Phylum comprises three different types of worms. 1. Oligochæta (earthworms): the locomotor appendages consist of chitinous setæ or bristles attached in rows to the sides and the ventral surface. 2. Polychæta (bristle worms) : the parapodia are highly developed bearing numerous long setæ. There is a definite head with eyes and tentacles. 3. Hirudinea (leeches) : they possess locomotor and adhesive suckers either at both extremities or posteriorly; bristles on the body are absent.

Oligochaeta or Earthworms. These have been included by Blanchard among the pseudoparasites of man. Heymons (1926) and Muller (1926) have reported freshwater oligochaeta being passed in living states by patients. The species involved was *Pachydrilus lineatus*.

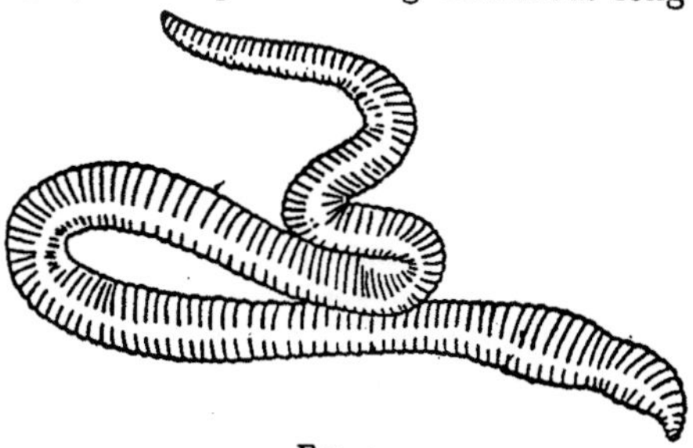

FIG. 2.
An earthworm.

FIG. 3.
A polychaete (after Strickland and Biswas).

Polychaeta or Bristle worms. Though the bristle worms are essentially marine animals, they are very adaptable to changes in their environment, such as the salinity of water in which they may find themselves. There are only two instances where their occurrence in the human body has been reported by Biswas and Strickland, 1927, and Strickland and Roy, 1933, in the *Indian Medical Gazette*.

Hirudinea or Leeches. Leeches belong to the class Hirudinea and are worm-like. They are all hermaphrodites. The eggs are laid in cocoons. They are essentially blood-suckers and attach themselves to any warm-blooded animal they may find. They leave the host after having engorged.

Diagnosis of leeches is based on (*a*) the number and arrangements of eyes and of internal cæca; (*b*) the position of the genital openings; (*c*) the size of the posterior sucker and character of the anterior sucker; and (*d*) the number of teeth on jaws.

The common leeches which concern us are (*a*) *Haemadipsa zeylanica*; (*b*) *H. sylvestris*; (*c*) *Lymnatis nilotica*; (*d*) *Dinobdella ferox*; (*e*) *Hirudo medicinalis* and (*f*) *H. birmanica*. Their distribution and habitat are as follows :

H. zeylanica Moq.-Tand. : This leech has a wide distribution throughout south-eastern continental Asia and the oriental islands. It is the only land leech in Ceylon and is not found on the Indian mainland. They are extremely small and wiry looking. They are an intolerable pest in Ceylon attacking man and domestic animals. The dorsal surface of this leech is reddish yellow or greenish and is generally traversed by four black lines; there are two lateral orange stripes.

H. sylvestris Blanchard : In the living state they are much thinner than they appear in the preserved condition. This is the principal land leech in Assam, Darjeeling, Sikkim and Burma.

L. nilotica Savig. ; They are about three times as strong as the land leech and are distributed in

Palestine, Persia, Europe and North Africa. They live in springs and similar water courses. They are known as "horse leech" on account of their habit of entering the air passages of horse and other domestic animals. The young leech is swallowed with drinking water and usually fastens on the mucous mem-

FIG. 4.
Diagram of a leech.

brane of the mouth, pharynx and larynx. In this situation it may even cause death of the host. Man is occasionally attacked. In the Gallipoli campaign in World War I several cases occurred among soldiers.

D. ferox Blanchard : These are of very large size and are found in tanks, ponds and in marshes. They are essentially cattle leeches and the report of the oozing of blood of domestic animals from the nostril refers to an invasion of the air passages by this species in India. The parasitic infestation is generally of long duration. This species is widely distributed in India, Burma and Ceylon. In the dry zones and also in the deserts they are conspicuously absent.

H. birmanica Blanchard : It is the common medicinal leech of India. It is found in rivers, streams, swamps, tanks or ponds. It is widely distributed in India.

H. medicinalis Lin. : It is the common medicinal leech of Europe and also occurs in North Africa. They are found in marshes, pools and small streams. They live generally on animal's blood. In India this species is represented by *H. birmanica* which are found in tanks. They attack cattle.

Anti-leech remedies. Infusion of tobacco leaves, copper sulphate and common salt are no doubt toxic to leeches but these do not possess any repellent properties. In this respect nearly all volatile oils such as oil of citronella, lemon grass oil, oil cassia, oil ocimum are not only toxic to leeches when applied directly on their bodies, but they also possess considerable repellent properties. Among these oils, those of cassia and citronella seem to be the most powerful, either when prepared with vaseline, a drachm to an ounce, and applied on the skin or mixed with rectified spirit in the same proportion as above and sprayed on socks. Putties when wetted with infusion of tobacco leaves or their midribs do not repel leeches but when they have crawled and stayed on the wet parts for about a minute or so, the animals are stupefied and are soon killed or they drop down in a semidead condition. Washing nut soaked in water is useful so long as it is wet, but as soon as the soap dries, it ceases to act. Infusion of Derris elliptica is also highly toxic to leech. Copper sulphate 1 in 50,000 solution will kill both the young and mature leech; in 1 in 10,000 it is extremely irritating to the leech. Dimethyl phthalate has been proved to be very effective and a durable land-leech repellent. A dose at the rate of 4 c.cm. per sq. ft. applied to cloth in a band 1 inch wide will repel leech for about 6 days. Stammers (1950) has examined a very large number of chemicals for land-leech repellent properties against *H. zeylanica*. DDT is ineffective.

Treatment of infestation of the nasopharynx. For the treatment of infestation of the nasopharynx *Aqua auranti floris* is ideal. When the official preparation (40 per cent strength) is diluted with an equal volume of water and sprayed into the nasopharynx through the nostril in the case of cattle, and preferably through the mouth in the case of human subjects, it at once acts on the leech and causes it to become dislodged. It is extremely toxic to leeches. When *aqua auranti floris* is not available, Eau de Cologne, which contains it, can be used. It is harmless to the mucous membrane.

CHAPTER II

STRUCTURE OF INSECTS

Segmentation. The body of an insect is composed of a number of segments or rings. A segment is defined as a part of the body that not only repeats itself but also contains a pair of cœlomic sacs, a pair of nerve ganglia and a pair of appendages. Segmentation of the body in insects is well marked in the embryo, but in the adult profound changes occur so that in certain regions as the head no sign of segmentation is seen. The appendages are either modified, or disappear as in the anterior portion of the abdomen. The cœlomic cavities are replaced or lost and the body cavity becomes a hæmocoele. The nerve ganglia likewise are displaced or fused together to form complex ganglia as in the head. The difference from annelida is that the internal segmentation as marked by septa or internal partition walls is not visible. Segmentation in insects as in other Arthropoda is said, therefore, to be only external. In the insects the segments fall into three distinct regions, namely, the head, the thorax and the abdomen. In the head region there is no external segmentation, the anterior segments of the body being fused together to form the head capsule. The appendages of the head are one pair of antennæ and three pair of jaws. The minimum number of segments in the head as judged by these visible appendages is, therefore, four; the actual number of segments in the head, however, is more. The pair of ganglia of the antennary segment becomes fused with the ganglia of the eye and other segments that disappeared in the adult so as to form a complex ganglion known as supra-œsophageal ganglion or brain situated above the pharynx or the œsophagus. The ganglia of the mandibular and the maxillary segments are fused to form the subœsophageal ganglion situated ventral to the œsophagus, and is connected with the supra-œsophageal ganglion by a pair of nerve connectives. The supra-œsophageal ganglion sends nerves to the antennæ and the eyes, while the subœsophageal ganglion innervates the jaws.

Notwithstanding disappearance of the sutural lines on the head, the head can be distinguished into different areas. On the dorsal surface the anterior portion is known as the clypeus, to which the labrum or the upper lip overhanging the mouth is sutured. The area between the antennæ and behind the clypeus is the frons. The top of the head is known as the epicranium while the sides are the genae.

The thorax consists of three distinct segments, the prothorax, the mesothorax and the metathorax. Each segment can be distinguished into a dorsal area or the notum, a ventral area or the sternum and two lateral ones or pleurons. The dorsum of the prothorax is called the pronotum, that of the mesothorax, mesonotum and that of the metathorax, metanotum. Similarly the terms prosternum, mesosternum and metasternum are used.

The prothorax bears the first pair or the prothoracic legs. Similarly the mesothorax, the mesothoracic, *i.e.*, the second pair of legs, and the metathorax, the metathoracic legs.

The first pair of wings is mesothoracic and the second pair is metathoracic in position. In Diptera the latter pair is replaced by rod-like balancers, the halteres. In each thoracic segment there is a pair of ganglion which is connected with that of the next segment. Anteriorly the prothoracic ganglion is connected with the subœsophageal ganglion of that side. The thoracic ganglia innervate the legs of the corresponding segments.

The abdominal ganglia are reduced in number and the number also varies. The first abdominal ganglion is often fused with the last thoracic while the last abdominal ganglion is a composite one, being formed of three ganglia of the three abdominal segments. Each abdominal ganglion gives off a pair of nerves to the muscles of that segment.

The abdomen is composed of 11 or less number of segments. In most insects the appendages of the abdomen have disappeared while those of the posterior abdominal segments, namely of the VIII and IX segments, are converted into the external genitalia. In the female the external genitalia in certain insects serve as the ovipositor or become, as in the bee, the sting.

Integument. The integument or the body wall of insects which forms the exoskeleton is hard and rigid, and affords protection to the soft parts of the body. The water proofing properties of the cuticle is due to the lipoids of the epicuticle rather than the hardness of the exocuticle. The integument consists of the cuticle, the epidermis and the basement membrane. The cuticle is secreted by the epidermis, and is composed of protein which is present in no less quantity than the chitin. It is the chitin and protein which form the analogue of mucoprotein, rather than of the keratin in horn. Chitin is insoluble in acids and alkalis. In microscopical sections, the cuticle appears stratified and is distinguished into the epicuticle, the exocuticle and the endocuticle. The epicuticle, the outermost, is an extremely thin membrane and makes the insect's cuticle impervious to water. The exocuticle, the primary cuticula, is the thick layer, dark brown in colour. It makes the cuticle rigid. The endocuticle, the secondary cuticle, made up of several wavy layers of superimposed lamellæ, gives elasticity to the cuticle. The endocuticle is colourless.

There are fine ducts of glands that pass out through the cuticle, for conveyance of secretory substances elaborated by the epidermal glands such as the wax and the scent glands, to the exterior.

The thickness of the cuticle varies. The epicuticle and also the exocuticle may remain soft and flexible in the intersegmental region or is hardened in the region of certain segments into plates called the sclerites.

The cuticle also lines the internal surface of the buccal cavity and other organs such as the tracheæ formed by ectodermal invagination.

The external surface of the cuticle may be drawn out into fine hairs, setæ, spines, or scales.

The cuticle may be actually coloured or may look coloured due to pigments in hypodermal layer or as a result of reflection or interference of light rays on the sculpture of the cuticle.

The epidermis, a single layer of nucleated cells, is the ectoderm; but in the insect, this is spoken of as the hypodermis although the dermal layer of the vertebrates is wanting. The hypodermal cells may be modified to form various integumentary glands, *e.g.*, the wax or the scent glands.

Growth. Since the soft parts of the insect are encased within the rigid armour of the cuticle, the body cannot increase in size with age unless this hard case bursts open and is shed. This happens periodically till the insect attains stage of sexual maturity or the adult stage. The casting of the cuticle is known as moulting or ecdysis. The discarded skin is known as the exuvia. Often its own cast skin is eaten by the insect. Moulting involves all parts of the body derived from the ectoderm having a cuticular lining. After a moult the size of the body of an insect increases, with changes in the structure of the body.

It is believed that a hormone secreted by the glands corpora allata situated in the head region just behind the brain, induces moulting.

The number of moults is usually constant for a species but is altered by temperature and humidity of the environment, and also by food. Scarcity of food reduces the number of moultings and the adult insect is smaller than the average. This confers protection against famine conditions.

After hatching of an insect from the egg, a succession of moults occurs and the span of life of an insect is thus divided into stages called stadia; the form at any particular stadium is known as instar. The last instar is the adult stage of the insect. The adult is sexually mature and is also called the imago.

The transformation from young to adult stage is known as metamorphosis; this may be gradual or sudden. According to the degree of the change, the metamorphosis is calssified into different types, *e.g.*, (*a*) Ametabolous because no marked changes occur as the young passes into the adult stage as in the silver fish of the order Apterygota. The lice which secondarily are wingless, are also ametabolous. (*b*) Paurometabolous where the changes after each moult are gradual as in the cockroach, grasshopper and the locust. The young of these insects resembles the parent except in wings, genitalia and size, and is spoken of as the nymph. (*c*) Hemimetabolous or Direct or Incomplete, as seen in the May fly, the dragon fly, and in bugs where the changes are more marked than in gradual metamorphosis but less so than in the complete metamorphosis. (*d*) Holometabolous or Indirect or Complete as in the moth and the butterfly, the ant and the bee where the changes are very striking. In these examples the larva on hatching from the egg, looks altogether different from the adult. The larva represents the feeding stage and increases in size at successive moults. Finally it then stops feeding and becomes pupa which is the resting or the quiescent stage. The pupa takes no food. At the pupa stage sudden anatomical changes occur and new organs are formed. The breaking down of larval tissues is known as histolysis which is brought about by chemical changes under the influence of a hormone from the corpus allata. The new organs are formed from buds called imaginal buds produced by the old disintegrating organs. The transformed insect is the adult or the imago.

If the insect passes through two different or more types of the larval forms, it is said to undergo Hypermetamorphosis. This is seen in the life cycle of certain Coleoptera, parasitic Hymenoptera, in the Coccidæ of Hemiptera and in a few Diptera like the Tachinid fly.

Appendages. The appendages of the body show adaptive features. The appendages on the head, *i.e.* the cephalic appendages, include the antennæ and the three pair of jaws. These are of different types in different groups of insects. The antennæ are long jointed processes, acting mainly as sensory organs. They vary in form and in number of segments. The proximal part of the antennæ is called the scape or the bulb. The second segment is known as the pedicel. The remaining joints form the flagellum (or clavola). The various forms of antennæ are described as setaceous or bristle like, filiform or thread like, serrate or saw like, pectinate or comb like, clavate or club shaped, capitate or knobbed.

The three pair of jaws which form the mouth-parts are modified according to the nature of food and the feeding habits. The food may be solid like plant bodies or the flesh of animals, or liquid like sap of plants or blood. If the food is chewed as in the case of predacious insects the mouth-parts are adapted for seizing and tearing the food. Among the sucking insects like the butterfly which sips the nectar from the flowers, the mouth-parts are drawn out in the form of a long tube called the proboscis. The mouth-parts may be modified for piercing and sucking as in the mosquito. Notwithstanding the differences the mouth-parts in insects are typically composed of the labrum or the upper lip, a pair of mandibles, a pair of the first maxillæ, a pair of the second maxillæ known as the labium or the lower lip, the unpaired hypopharynx which forms the cuticular lining of the floor of the pharynx and an unpaired epipharynx which lines the roof of the pharynx. The first and second maxillæ are provided with palps. Some of the mouth-parts may atrophy.

The appendages of the thorax are the three pair of legs and the wings. These show considerable modifications in different species of insects.

Each leg consists of the following segments : coxa, trochanter, femur, tibia and tarsus. The tarsus consists of varying number of segments from one to six. This serves as the foot of the insect and bears at the terminal end the claws. It may also carry on the ventral surface cushion-like structure called the pulvillus.

Wings develop as a sac-like fold of the body wall on the dorsal regions of the thorax.

In course of development the upper and lower walls of the embryonic wings become apposed and these appear membraneous. Along certain lines however the walls of the wing remain separate by thickening and hollowing of lines which are termed veins. The veins during development contain blood. Tracheal branches enter the wing veins along the lines of future veins for supplying air to the wing.

The system of veins in a wing is known as venation and is useful for classification.

The texture of the wing also varies, being alternatively hard or membraneous. The anterior or the first pair of wings, if thickened to form the wing covers of the second pair, is termed the elytra. The wings of the second pair rest folded under the elytra and are spread during flight.

Digestive system. The alimentary canal is divisible (according to the mode of development) into : (*i*) the foregut or the stomodæum, (*ii*) the midgut or the mesenteron, (*iii*) the hindgut or the proctodaeum. The foregut and the hindgut are formed from the ectodermal invagination while the midgut develops from the endoderm.

The foregut is differentiated into the pharynx, the œsophagus, the crop, and the proventriculus or the gizzard. The pharynx lies behind the mouth.

The mesenteron is the stomach of the insect and is known as the ventriculus or the chylific stomach, its junction with the stomodæum being guarded by valves. Often blind pouches called the cæca are given off by the ventriculus.

The proctodæum, the junction of which with the ventriculus is marked by the malpighian tubules, is divisible into an anterior portion, the intestine and a posterior portion, the rectum which opens by anus to the exterior. The intestine is subdivisible into the anterior, the ileum and the posterior, the colon.

Associated with the mouth-parts and opening into the anterior end of the gut are the labial glands called the salivary glands. The salivary glands are of various types.

Circulatory system. The circulatory system consists of the body cavity and the heart. The body cavity, the hæmocœle, contains hæmolymph or blood. The heart which is a dorsal tubular organ has apertures on sides called ostia which put the hæmocœle in communication with the heart. Anteriorly the heart is prolonged into a narrow tube called the aorta which discharges its contents into the hæmocœle. There is no vein for draining blood back into the heart and the circulatory system is therefore described as open.

Blood or haemolymph is a colourless liquid often tinged with green or yellow pigment. Floating in the body cavity are the blood cells or hæmocytes. In the hæmocœle also occur cluster of large cells called the pericardial cells in the adjoining region of the heart. They are the nephrocytes and absorb colloidal particles from the blood. Associated with the hæmocœle are other kinds of cells known as the fat body and the œnocytes. Circulation of blood occurs by pulsation of the heart and is aided by respiratory movements of the abdomen.

Respiratory system. The respiratory system consists of the tracheæ, tracheoles and spiracles.

The tracheæ are air tubes strengthened inside by tænidia, the spiral thickenings of the intima, which is continuous with the cuticle of the body. An epithelial layer of nucleated cells called the ectotrachea lies outside the intima. The basement membrane on the outside supports the epithelial cells.

The tracheæ are arranged in the form of a pair of trunks which run in the longitudinal direction of the body. The trunks are connected with the spiracles of the corresponding side by short tracheal branches which follow transverse direction. They are also connected with each other by short branches.

Each trunk and also the connecting ones give out branches which subdivide to ramify over all parts of the body. The ultimate branches are known as tracheoles. These are devoid of tænidia and enter muscles, nerves and other tissues.

The spiracles are the external openings of the tracheal system and are guarded by valves which control inlet and outlet of the respiratory air. There are usually ten pair of spiracles, two on the thorax and eight on the abdomen. Usually the anterior pair is mesothoracic and the second pair is metathoracic in position. There may be variations, as for example, in lice (Anoplura) which has one thoracic and six abdominal pairs.

The position of spiracles on the body also varies. In the larval forms some of the spiracles are closed while others remain open and functional.

Excretory system. The chief excretory organs are the Malpighian tubules which are fine slender tubes often very long and float in the hæmocœle. They open into the gut at the junction of the mesenteron and the intestine. The free end of each tube is closed. Each tube is lined inside by one layer of epithelial cells. These cells extract the nitrogenous waste products, *e.g.*, uric acid, urates, calcium oxalate, by diffusion

2

from the blood and which pass into the cavity of the intestine and pass out as the fæces.

The number of malpighian tubules is constant for the same species of insect but varies in different insects from 2 to 100 or more.

Reproductive system. Sexes are separate. The reproductive system in both sexes consists of a pair of germinal tubes, a pair of ducts and a median terminal tube which opens to the exterior.

In the male, the germinal tube is known as the testis, which is an elongated tube or a lobed organ. The tube is thrown into coils enveloped by connective tissue sheath. The paired ducts are called the vasa deferentia and are dilated in certain regions for storage of mature sperms. The accessory glands which are differentiated into ectodenia and mesodenia according as they originate from the ectoderm or the mesoderm, open into the median duct near its junction with the vas deferens of either side. The median duct is distinguished as the ejaculatory duct. The copulatory apparatus consists of a basal portion and the terminal portion which is called the aedeagus or the penis having chitinous hooks or spines at the terminal end. The basal portion carries a pair of long processes or lobes called the parameres which help to hold the female during coitus.

In the female each ovary is comprised of a number of tubes called the ovarioles. Each ovariole is distinguishable into an apical portion, the germarium and the vitellarium or the follicle. The latter encloses a linear row of chambers for holding developing oocytes.

The wall of the egg chamber is composed of a single layer of epithelial cells. These in the last chamber secrete the chorion which forms the egg shell. The chorion is perforated at the anterior end by fine apertures called the micropyle to admit the stored spermatozoa, flowing out from the spermatheca. The spermatheca is a minute chitinous sac with muscular coatings and contains a gland for supplying nourishment to the sperms stored in the sac.

The fertilized eggs are laid to the exterior singly or in batches at intervals. These are covered by a sticky substance given out by the accessory glands called the colleterial glands. In the cockroach the secretion hardens into a capsule enclosing the eggs. In certain Hemiptera this gives rise to a frothy substance in which the eggs are laid.

In parthenogenetic forms eggs undergo development without being fertilized. Parthenogenesis occurs in Aphids and in a number of insects.

The ovipositor or the egg laying apparatus which is thrust deep into an object for deposition of eggs into it, is made of three pair of elongated hard blades or valves. These fit each other to form a tapering hollow tube for the passage of eggs, and are developed as appendages of the eighth and ninth segments of the abdomen. In the Hymenoptera the ovipositor is modified into a sting.

Nervous system. The central nervous system consists of the brain and a ventral nerve cord. The complex supra-œsophageal ganglion is called the brain. The subœsophageal ganglion is a complex structure and supplies nerves to the mouth-parts. The ventral nerve cord consists of a series of paired ganglia situated on the ventral side of the thoracic and abdominal regions. The thoracic ganglia innervate the musculature of the legs and wings and so these control locomotion. The abdominal ganglia supply nerves to the muscles of the corresponding segments.

Classification

Insects far exceed in number any other types of animals; within limits of their structure they have evolved into varied forms. They are classified into groups.

The present scientific binominal nomenclature was introduced by Linnaeus in 1751 and it has been systematically used for animals from 1758.

Thus the Animal Kingdom is divided into several phyla, each phylum into classes, each class into orders, each order into families, each family into subfamilies, subfamily into genera, genus into species. Thus the common mosquito is known as *Culex fatigans* Weid. The first word is the genus and the second, the species. The name of the author is put after the species. The generic name always starts with a capital letter.

An outline of the classification is given in the following tables.

Phylum Arthropoda. It comprises the following important classes :

Orders :

Myriapoda		Crustacea		Arachnida	Insecta
Chilopoda (Centipede)	Diplopoda (Millipede)	Copepoda (Cyclops)	Daphnia.	(*i*) Acarina—(ticks and mites). (*ii*) Araneida (spiders). (*iii*) Scorpionidea. (*iv*) Pentastomida (tongue worms). (*v*) Vermiformia (follicle mites).	Subclass : (*A*) *Apterygota* (primitive wingless) : Orders : (*i*) *Thysanura* (silver fish). (*ii*) *Protura* (Eosenteron). (*iii*) *Collembola* (springtails). Subclass : (*B*) *Pterygota* (primitive winged insects) : Series : (*a*) *Exopterygota* (Heterometabola)—wings develop internally; metamorphosis complete.

Orders of Exopterygota : (*i*) *Orthoptera* (cockroach, mantids, crickets etc.). (*ii*) *Dermaptera* (earwigs). (*iii*) *Plecoptera* (stone flies). (*iv*) *Isoptera* (white ants or termites). (*v*) *Embioptera* (embia). (*vi*) *Psocoptera* (book lice). (*vii*) *Ephemeroptera* (May flies). (*viii*) *Odonata* (dragon flies). (*ix*) *Thysanoptera* (thrips). (*x*) *Hemiptera* (bugs) : Suborders : (*a*) *Heteroptera* (bed bug, cone-nosed bug *etc.*). (*b*) *Homoptera* (cicads, plant lice *etc.*). (*xi*) *Anoplura* (lice) : Suborders : (*a*) *Siphunculata* (sucking lice). (*b*) *Mallophaga* (biting lice).

Orders of Endopterygota : (*i*) *Neuroptera* (alderflies, antlions). (*ii*) *Mecoptera* (scorpion fly). (*iii*) *Trichoptera* (caddis fly). (*iv*) *Lepidoptera* (moths and butterflies). (*v*) *Coleoptera* (beetles). (*vi*) *Hymenoptera* (ants, wasps *etc.*) (*vii*) *Diptera* : Suborders : (*a*) *Nematocera* (mosquitoes and other midges). (*b*) *Brachycera* (horse fly). (*c*) *Cyclorrapha* (true flies). (*d*) *Pupipara* (tick flies). (*viii*) *Siphonaptera* (fleas.)

Arthropoda Acting as True Carriers of Disease Organisms Affecting Man

Disease	Causative organism	Arthropod vector
Malaria	*Plasmodia* spp.	*Anopheles* spp.
Kala-azar	*Leishmania*	*Phlebotomus argentipes* (India). *P. chinensis* (China). *P. perniciosus* (Mediterranean area). *Phlebotomus* spp. in South America
Oriental sore	*Leishmania*	*P. sergenti*. *P. papatasii*.
Espundia (S. America) (mucocutaneous leishmaniasis)		*Phlebotomus* spp.
Cutaneous form of Oroya fever	*Bartonella bacilliformis*	*Phlebotomus*.
Sleeping sickness	*Trypanosoma*	*Glossina palpalis*. *G. tachinoides*. *G. swynnertoni*. *G. morsitans*.
Chagas' disease	*Trypanosoma cruzi*	Reduviid bugs of *Triatoma* and *Rhodnius* spp.
Epidemic (louse-borne) relapsing fever	*Treponema recurrentis*	*Pediculus humanus*.
Endemic (tick-borne) relapsing fever	*T. duttoni*	*Ornithodorus moubata*. *O. papillipes* and other species of *Ornithodorus*.

Disease	Causative organism	Arthropod vector
Typhus (tick-borne) : Rocky Mountain spotted fever	*Dermacentroxenus rickettsi*	*Dermacentor andersoni.*
Tick typhus (South African)	*D. rickettsi*	*Haemaphysalis leachi.*
Sao Paulo rural typhus	*D. rickettsi*	*Amblyomma cajennense.*
Fievere boutonnense	*D. rickettsi*	*Rhipicephalus sanguineus.*
"Q" fever	*Coxiella burneti*	Different species of hard ticks.
Typhus (louse-borne) : Epidemic or classical	*R. prowazeki*	*Pediculus humanus.*
Murine (endemic) typhus (Brill's disease)	*R. prowazeki* var. *mooseri*	*Xenopsylla cheopis.*
Typhus (mite-borne) : Scrub typhus, Tsutsugamushi disease (Mossman fever of Australia)	*R. orientalis*	Larvae of *Trombicula deliensis* and *T. akamushi.*
Trench fever	*R. quintana*	*P. humanus.*
Rickettsialpox	*R. acari*	The mouse mite *Allodermanyssus sanguinensis.*
Plague	*Pasteurella pestis*	*Xenopsylla cheopis* and other fleas.
Yellow fever	Virus	*Aedes aegypti.*
Dengue fever	Virus	*Aedes aegypti.*
Sand-fly fever	Virus	*Phlebotomus papatasii.*
Tick paralysis	toxin	Hard ticks of *Dermacentor* and *Ixodes* sp.
Tularaemia	*Bacterium tularense*	*Dermacentor venustus.* *Chrysops discalis* (?).
Filariasis	*Wuchereria bancrofti* *W. malayi*	*Culex fatigans.* *Mansonioides* sp. *Anopheles sinensis.*
Calabar swelling (African loaiasis)	*Filaria loa*	*Chrysops dimidiata.* *C. silacea.*
Perstans filariasis	*Acanthocheilonema perstans*	*Culicoides austeni.* *C. grahami.*
Onchocerciasis	*Onchocerca volvulus*	*Simulium damnosum.*
Guinea worm	*Dracunculus medinensis*	*Cyclops leuckerti.* *Cyclops.*
Tape worm	*Diphyllobothrium latum*	*Diaptomus gracilis.*
Gnathostomiasis	*Gnathostoma spinigerum*	*Cyclops* sp.

Arthropoda Acting as Mechanical Carriers of Diseases of Man

Disease	Organism	Carriers
Cholera	*Vibrio cholerae*	Particularly house-flies.
Dysentery	*Entamoeba histolytica* *Shigella dysenteriae*	do
Enteric fever	*Salmonella typhi* *Salmonella paratyphi* *S. enteritides*	House-flies.
Poliomyelitis	Virus	House-flies and Stomoxys.
Anthrax	*Bacillus anthracis*	Biting-and non-biting flies.
Intestinal worms		House-fly; occasionally cockroach.
Yaws	*Treponema pertenue*	House-flies and eye-flies.
Conjunctivitis		Eye-flies.
Trypanosomiasis of sleeping sickness		Sometimes mechanically by *Glossina.*
Tuberculosis	*Mycobacterium tuberculosis*	House flies.

Arthropoda Being Themselves Injurious to Human Being

Myiasis	Several species of Muscoidea being responsible for causing myiasis.
Scarabiasis	Due to the invasion of the alimentary canal by certain types of beetle.
Chiggerosis	Due to burrowing by gravid females of *Tunga penetrans* in the skin.
Lousiness	Due to the presence of lice on the head and body including the pubic region.
Itch	Due to the presence of *Sarcoptes* mite in the skin.
Tongue worms	The condition is generally detected after death.
Other forms of itch	Such as grocer's itch, copra itch *etc.*, due to different species of mites.

Arthropoda in Relation to Diseases and Discomforts of Man. (1) They are obnoxious pests; examples : house-flies, eye-flies *etc.* (2) They bite, producing at times a certain amount of unpleasant after effects : mosquitoes, fleas, ticks, lice *etc.* (3) They produce a kind of diseased condition by their presence either on the skin, in one of the natural orifices of the body or in the intestine : *e.g.*, external and internal myiasis, scarabiasis or beetle infestation *etc.* (4) They produce dermatitis by their irritating effects; as in spider-lick dermatitis, and conjunctivitis and dermatitis as caused by caterpillar hairs. (5) They produce symptoms of local and general toxaemia due to the introduction of a venomous glandular secretion : *e.g.*, sting by centipede, spider, scorpion, wasp *etc.* (6) They carry disease germs mechanically : house-flies and v. cholerae, tabanid flies and trypanosome of surra, blood-sucking flies and anthrax *etc.* (7) They act as vectors in which the development of the pathogenic organisms always takes place : *e.g.*, mosquitoes and plasmodial parasites, avian and mammalian ; mosquitoes and filarial embryos ; sand-flies and leishmania ; tsetse-fly and trypanosomes of sleeping sickness *etc.*

Therapeutic Uses of Arthopoda and Animals of Medical Importance. (1) Use of malaria-infected mosquitoes for inducing malaria in cases of paresis. (2) The treatment of arthritis with bee venom. (3) The application of extracts from certain Meloid beetles as counter-irritants. (4) The use of blow-fly larvae in the treatment of infected wounds. (5) Leeches for letting out blood. (6) Hirudin has well-marked physiological properties. (7) The employment of lice and ticks for preparing specific anti-rickettsia vaccine. (8) For the preparation of anti-venom against spider and scorpion stings, heads of the spider and tails of scorpions are employed.

Geographical Regions of the World. The distribution of land and fresh water animals in six regions as described by Wallace is given below :

(1) *Palaearctic Region.* It includes European continent, Africa north of Sahara and the whole of Asia not included in the tropical region (Oriental region). (2) *Oriental Region.* It includes the tropical parts of Asia, *e.g.*, nearly the whole of India, the part west of the Indus and the Himalayas being excluded, Ceylon, China south of Yangtsekiang, Burma, Borneo, Philippines *etc.* (3) *Nearctic Region.* It includes the northern parts of the United States and further north. (4) *Australian Region.* This includes Australia, New Zealand, New Guinea, Celebes and Polynesia. (5) *Neotropical Region.* Central and South America and West Indies. (6) *Ethiopian Region.* The whole of Africa, south of the Mediterranean and excluding the parts north of the Sahara, southern parts of Arabia and Madagascar.

Type : **The Cockroach**

(*Periplaneta americana*)

In order to understand the structure of an insect, a cockroach, which is a typical representative of the class Insecta, may be selected. Its main features are described below.

Head. The head is deflexed and does not show any indication of segmentation. The head can be separated into the following regions : (*a*) Epicranium covers the dorsal and posterior surfaces of the head. (*b*) Clypeus is a broad plate covering the front of the head below the epicranium. (*c*) Labrum, which is the upper lip. (*d*) Genæ or cheeks covering the sides of the head, behind and below the eyes. (*e*) Eyes are reniform elevations one on each side of the head. They are compound being made up of numerous ommatidia.

The head bears : (1) A pair of many-jointed flagella-like antennæ. To the inner sides of the bases of the antennæ are a pair of small oval patches called the fenestra which represent a highly primitive ocellus. (2) Mouth parts : (*a*) labrum or upper lip overhanging the aperture of the mouth; (*b*) a pair of stout mandibles which lie below the genæ and articulate with the sides of both the epicranium and the clypeus; their inner edges are toothed. (*c*) Behind the mandibles are the first pair of maxillæ (jaws). Each maxilla consists of two segments, cardo and stipes, the latter supporting an outer lobe, the galea and an inner lobe, the lacinia. To the maxilla is attached a 5-jointed palp. While with the help of the powerful mandibles solid food is chewed,

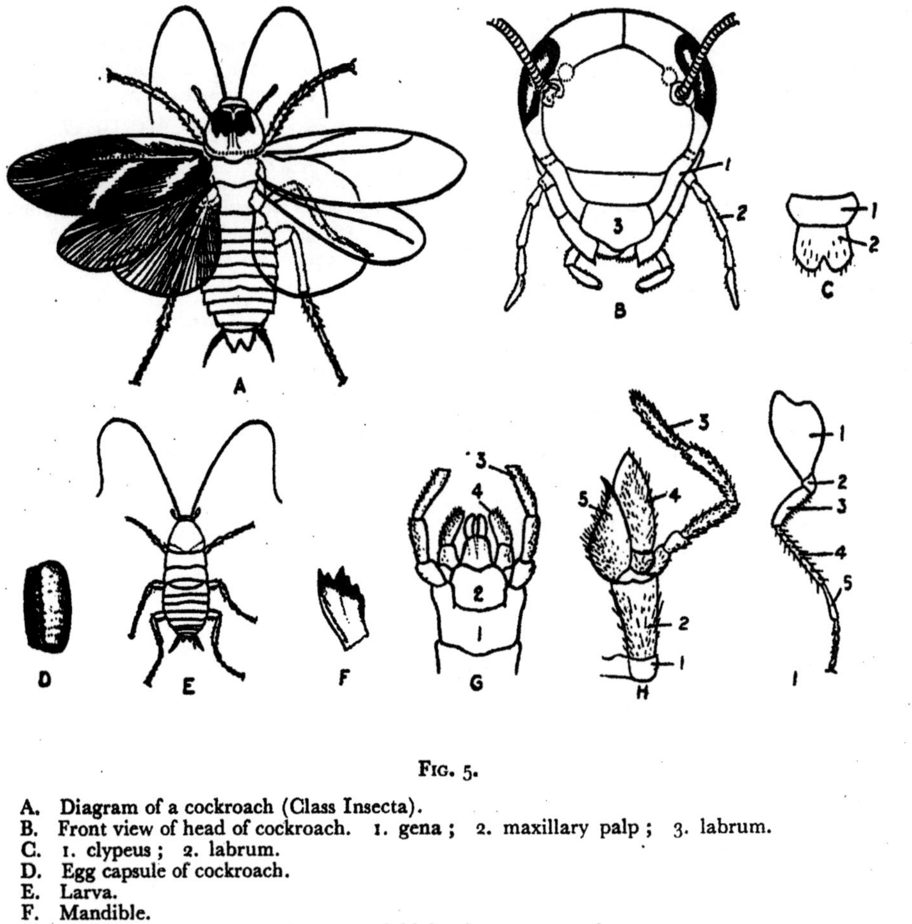

Fig. 5.

A. Diagram of a cockroach (Class Insecta).
B. Front view of head of cockroach. 1. gena; 2. maxillary palp; 3. labrum.
C. 1. clypeus; 2. labrum.
D. Egg capsule of cockroach.
E. Larva.
F. Mandible.
G. 1. submentum; 2. mentum; 3. labial palp; 4. paraglossa.
H. 1. cardo; 2. stipes; 3. maxillary palp; 4. galea; 5. lacinea.
I. Leg of cockroach. 1. coxa; 2. trochanter; 3. femur; 4. tibia; 5. first tarsal segment.

the maxillæ assist in holding and masticating the food. (*d*) Behind the first maxillæ are the second maxillæ which have the two basal segments united together into a single structure which supports the prementum; the latter carries two pairs of distal lobes, an outer pair the paraglossæ, and an inner pair the glossæ; these two structures collectively form the ligula. The united second maxilla is called the labium or lower lip. A pair of 3-jointed labial palps is attached to the prementum. A soft median process, the tongue (lingua or hypopharynx) is attached to the upper surface of the mentum.

Thorax. The thorax consists of three separate segments, prothorax, mesothorax and metathorax. The dorsal plate of each segment is called the tergum, the ventral plate, the sternum, and the plate which joins the tergum and the sternum laterally on each side is the pleuron.

Attached to the anterior border of the tergum of the mesothorax are the anterior wings or elytra. This is highly chitinised and is hard. It takes no part in the flight of the insect and its function is merely to protect the more delicate second pair of wings which are membraneous. When at rest they are kept folded up longitudinally like a fan under the elytra. The venation is of a radiating nature with a network of cross veins. The second pair of wings articulate with the tergum of the metathorax. In the female of *Blattella orientalis* the wings are vestigial.

To the ventral surface of each segment of the thorax is attached a pair of legs. On account of the presence of three pair of legs, insects are also called hexapoda. Each leg is composed of: (1) coxa, (2) trochanter, (3) femur, (4) tibia, (5) five tarsal joints, to the last segment of which are attached the pulvilli and a pair of claws.

Abdomen. The abdomen consists of 10 segments. In the male the 9th segment bears a kind of rudimentary styles. In the female the sternum of the 7th segment is very much more prominent than in the male. The anus lies below the 10th tergum which is produced backwards into a thin flexible plate, notched at the posterior border.

Internal anatomy :--Digestive system : The alimentary system is divided into foregut, midgut and hindgut. The foregut consists of : (*a*) mouth; (*b*) buccal cavity which receives the ducts of the salivary glands. Each gland is divided into lobes. There is a separate salivary receptacle. The duct of the salivary receptacle joins with that of the salivary glands into a common duct and the two common ducts of each side unite to form a single duct; (*c*) pharynx, which is provided with powerful muscles; (*d*) œsophagus; (*e*) crop; (*f*) proventriculus which is lined inside with teeth.

The midgut consists of the stomach proper. To it are attached the hepatic cœca. The stomach with the hepatic cœca is the mesenteron; the regions in front and behind the cœca are the stomodæum and the proctodæum respectively, the two last named portions being lined by a chitinous cuticle. The point of junction between the stomach and the intestine is indicated by the attachment of the Malpighian tubuleswhich are excretory in function. The Malpighian tubules are yellowish, thread-like and extremely long. The intestine consists of two parts, the ileum and the large intestine which ends in the rectum. The latter is provided with papillæ. The rectum opens externally at the anus. Corresponding enzymes are present in the salivary glands and stomach according to the feeding habits of the cockroach which is omnivorous.

The peritrophic membrane is a chitinous membrane which hangs from the proventriculus downwards thus forming an internal loose lining of the mid and hind gut. It originates probably from the secretion of the proventricular glands and while it is intact at its anterior end it is generally considerably damaged towards the posterior part. The posterior end remains loose and is not attached to the inside of the hindgut. Its exact function is not known. It is however present in almost all insects and their larvæ.

Circulatory system. The fluid blood circulates in the hæmocœle and does not flow in any well defined tubular structure except in the heart and aorta. The heart is an elongated tube, closed behind and open in front, running along the middle line of the abdomen and thorax immediately behind the terga. The aorta is a narrow tube which is a continuation of the heart anteriorly. The pulsation of the heart can be seen by placing a cockroach on its ventral surface.

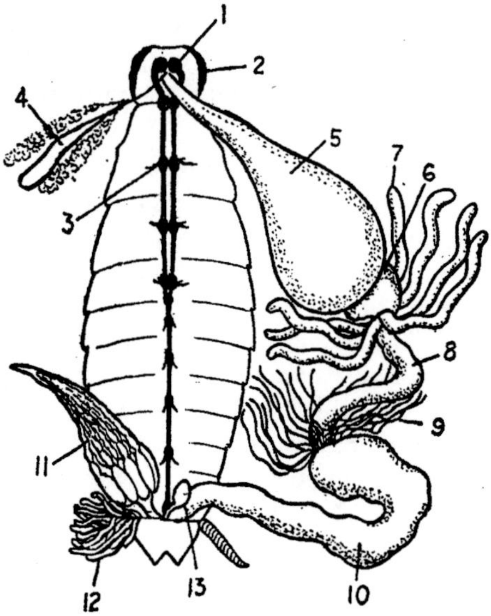

Fig. 6.

Dissection of female cockroach.
1. cerebral ganglion ; 2. eye ; 3. ventral nerve chain with ganglia ; 4. salivary glands with salivary vesicles ; 5. crop ; 6. gizzard ; 7. hepatic caeca ; 8. midgut ; 9. malpighian tubules ; 10. hindgut ; 11. ovary ; 12. arborescent glands ; 13. spermatheca.

Respiration. Respiration takes place through tracheæ which ramify over every organ of the body. They open on the surface at the stigmata. One pair of stigmata lies on the side of the thorax between the prothorax and mesothorax and the second pair between the mesothorax and metathorax. Eight are situated on either side of the abdomen between the terga and sterna of the segments. Each spiracle is guarded by a valve.

Excretion. Solid excreta is given off in crystalline form also in aqueous solution in which the Malpighian tubules play the most important part. Nephrocytes which are special cells arranged in groups are supposed to possess excretory function, as after injection of ammonia carmine into the blood, carmine particles are retained in the cytoplasm of these cells.

Reproduction. *Male* : (*a*) a pair of testes with corresponding vasa deferentia; (*b*) a pair of vesiculæ seminales, two tufts of whitish cœca which together constitute what is known as the "mushroom-shaped gland;" these open into the anterior end of (*c*) the ejaculatory duct which opens between the 9th and 10th sterna.

The characteristic smell of the cockroach is due to large branched tubular glands in the male insects.

Female : (*a*) a pair of ovaries each consisting of 8 ovarioles; (*b*) the two oviducts unite into a single tube which opens in a median aperture on the 8th abdominal sternum; (*c*) a pair of receptaculum seminis or spermatheca which open in the common oviduct; (*d*) collateral glands, arborescent glands, a pair of ramifying glandular tubules which open behind the spermatheca, their calcareous secretion helping to form the egg capsules.

Ommatidium. Ommatidium is a component unit of a compound eye and consists of a complete visual element. The outer layer is the cornea lying in a hexagonal facet and is often thickened to have the shape of a biconvex lens. Beneath the cornea is the corneagen; these cells form new cells for the cornea.

The cornea of adjacent ommatidia forms a continuous sheath. Beneath the corneagen is the crystalline cone which is enclosed in certain large cells. Below the cone lies a rod-like body which has a striated appearance called the rhabdom which is formed from the secretion of the retinal cells. Pigment cells lie between the cone and the rhabdom of the separate ommatidia. The rhabdom is regarded as the perceptive part of the ommatidia. The rhabdom is separated from the nerve cells and fibres by a thin membrane.

The simple eye or ocellus is fundamentally the same as one of the visual elements of a compound eye. It has only one lens and each retinal unit is a single cell, of which the distal part is unpigmented. A simple eye is capable of perceiving light and darkness but for more clear vision, *e.g.*, for perception of motion, a compound eye is essential. There is no focusing mechanism in the eye of an insect.

Biological Warfare

The newest weapons for future warfare under development to-day are sometimes termed ABC which stand for Atomic, Biological and Chemical weapons.

The subject of biological warfare, or warfare by germs has been very much in the public eye since the time the Chinese accused the U.S.A. of insect and germ warfare in Korea.

The use of poisonous germs against an enemy is not really a new idea. In medieval days it was normal practice to sling plague corpses into beleaguered towns. Poisonous plant materials which have been used for centuries in poison-tipped arrows and spears cannot strictly be termed as biological warfare. The ancient Egyptian practice of enclosing live cobras in small sealed earthenware jars and throwing at the enemy perhaps cannot also strictly be called biological weapons of warfare.

The exact nature of the germs and the methods of employment will necessarily vary according to the conditions of the enemy country where these will be used. It is difficult to comprehend the full potentialities of their use in war which are dimly understood even by those who will be the users. Conversely, the small groups of scientists who understand their own particular and specialised subject also understand such potentialities purely from the scientific angle but not from their larger effects on the crippling of war efforts and on the deterioration of the morale of the civilian and military populations caused by such warfare.

Biological warfare, owing to the difficulty of properly controlling the attack, will hardly, if ever, be used in the battlefield proper. Its use will be confined to the saboteurs operating well behind enemy lines or in open country. Dissemination of such organisms will also most probably be by direct human agency though the employment of insects for this purpose cannot be ruled out. It is not known whether the bacteria will survive or to what extent their potency will remain unharmed at the time they reach the targets, when they are used as aerial clouds or in bombs. The insects which are likely to be employed are the louse, the rat-flea, the stegomyia mosquito and the house-fly, and the organisms that may be used for infecting them are *Rickettsia prowazeki*, *Pasteurella pestis*, yellow fever virus and the intestinal organisms such as typhoid, dysentery and cholera bacilli respectively. Regarding their ultimate effects, a number of factors will operate such as the climatic and other environments, density of the population, density of insect population *etc.*

CHAPTER III

ORDER DIPTERA

The members of this order which includes mosquitoes and flies, are of considerable importance to man, especially in the dissemination of diseases. The diseases transmitted by mosquitoes alone are responsible for as much loss of human lives as are those of all other pathogenic diseases.

The order Diptera possesses all the characters of the class Insecta except that the hind pair of wings is replaced by a pair of club-shaped organs used for balancing and called halteres. The metamorphosis in Diptera is complete. This order is divided into two suborders, Orthorrapha and Cyclorrapha. The former represents in addition to mosquitoes and midges other insects like *Tabanus*, *Chrysops*, *etc.* Cyclorrapha includes the true flies. The difference between the two suborders is shown below :—

Orthorrapha	*Cyclorrapha*
Represents the midge group of insects, also gad-flies.	Represents flies, both biting and non-biting, also tick-flies.
(*a*) Larval head is distinct. (*b*) The last larval skin is shed and the pupa is naked, *i.e.*, it is not enclosed in the larval skin. (*c*) The adult escapes through a longitudinal slit that appears on the back of the cephalothorax of the pupa.	(*a*) The larva is maggot-like and the head is not differentiated. (*b*) The last larval skin is not shed; it hardens to form the puparium which encloses the pupa. (*c*) The adult escapes from the puparium through a circular slit that appears at its anterior part.

Orthorrapha is again sub-divided into two groups: (1) Nematocera and (2) Brachycera, and their principal characters are given below:

Orthorrapha

Nematocera (Mosquito and other midges)	*Brachycera* (Gad-flies)
(*a*) Antennae long and filamentous and composed of a large number of segments which are more or less similar in appearance.	(*a*) Antennae much shorter and seldom filamentous; commonly composed of 3 dissimilar segments.
(*b*) Maxillary palps are elongate and flexible and are commonly composed of 4 or 5 joints.	(*b*) The palps are not flexible and usually much shorter; composed of 2 segments.
(*c*) The third longitudinal vein is never forked though several veins are branched.	(*c*) Several longitudinal veins may be forked but never the 2nd.

Cyclorrapha, which includes the true flies, is classified into two groups, Aschiza and Schizophora.

Cyclorrapha

Aschiza (Hover flies).	*Schizophora* (True flies including tick-flies).
Frontal lunule indistinct and frontal suture absent.	Frontal lunule distinct and frontal suture present.

Orthorrapha-Nematocera—may be grouped into numerous families but from our point of view only four need be considered. They are all blood suckers and some of them are well-known carriers of disease organisms.

Important Families of Orthorrapha—Nematocera

Culicidae (Mosquitoes)	Psychodidae (Sandflies)	Chironomidae (Midges)	Ceratopogonidae (Biting midges)	Simuliidae (Black flies)
2nd longitudinal vein forked only once; body and wings may or may not be covered with scales.	2nd longitudinal vein forked twice; body and wings densely hairy.	Short proboscis; mouth parts adapted for sucking only. The first longitudinal vein stops a little beyond the middle of the wing, no scales on wings and body.	Like Chironomidae. Mouth parts adapted for piercing the skin and for sucking blood.	Fly-like with short but formidable proboscis; antennae small, many jointed, and project out like horns of an animal; no scales on wings and body.

The family Culicidæ is further subdivided into three subfamilies Culicinæ, Chaoborinæ and Dixinæ, among them the first two are incapable of piercing the skin and hence are unimportant.

Dixinae	*Chaoborinae*	*Culicinae*
Mouth parts short and unfit for piercing; no scales but only short hairs on wings and body; legs are without scales.	Proboscis short and stiff and not adapted for piercing the skin; no scales on the wing veins; presence of fringe scales along the posterior border of the wing; vein 1 relatively short which ends far before the wing tip and not far from the tip of the subcosta.	Long, slender proboscis; mouth parts of some adapted for piercing and for sucking; scales present on the body, legs and wings.

Dixinae. This group is confined to one genus *Dixa*, the larvæ of which are aquatic. The body of the larva is bent so that the anterior and posterior ends almost meet. Pseudopods armed with hooks are present on the ventral surface of some of the mid-abdominal segments. There is a respiratory cup on the last segment. The larva rests at the surface of water. Whole insect, notably the antennæ (in both sexes) and the wings, almost bare; head and thorax as in Chaoborinæ.

FIG. 7.
Larva of Dixa (after Miall).

Chaoborinae. The Chaoborinæ form a small group of insects which closely resemble mosquitoes. The early stages are aquatic. The larvæ are predacious and feed on other insects or their larvæ. They are found in ponds along with Anopheline larvæ and like the latter they lie horizontally at the surface and often remain motionless for several minutes in which position they are almost invisible on account of their extreme transparence. Hence they are called "phantom gnats." They are devoid of mouth brushes, the antennæ being prehensile. The mandibles possess well developed teeth. Proboscis of the adult not longer than the head, palpi curled downwards and inwards; antennæ of male plumose; thorax projecting somewhat over the head; the whole insect including wings hairy, scales being absent except on the wing fringe.

It contains the genus *Corethrella* which is more or less cosmopolitan. The adults are found in great numbers around lakes and large ponds. So far as is known the adults do not feed.

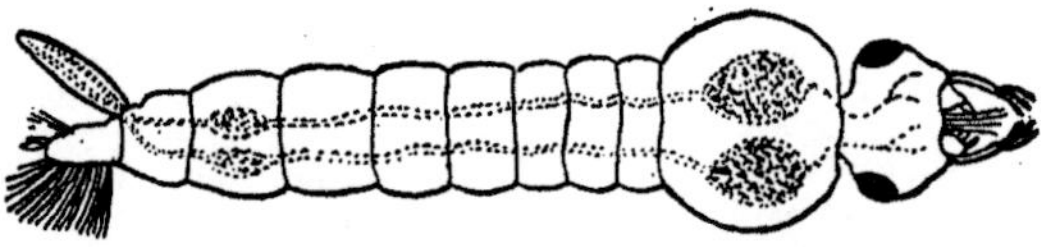

FIG. 8.
Larva of Mochlonyx (after Miall).

Culicinae. The Culicinæ, which include mosquitoes can be easily differentiated from others by the following characters: (1) Long, slender proboscis. (2) Mouth parts of some adapted for piercing and sucking. (3) The presence of scales on wings and a fringe of scales along its posterior border. (4) Characteristic wing venation.

The subfamily Culicinæ is further grouped into tribes; the following only deserve our consideration. (a) Tribe Anophelini, (d) Tribe Culicini, (c) Tribe Megarhinini.

Tribe Anophelini. Free edge of scutellum semilunar; the palpi are as long as proboscis in both male and female; in the male the end of the palpi is clubbed; abdomen never uniformly invested by broad overlapping scales. They generally rest with the back straight, *i.e.*, the head, thorax and abdomen being in a line; some anophelines adopt the culicine attitude. The larva floats on the surface of water.

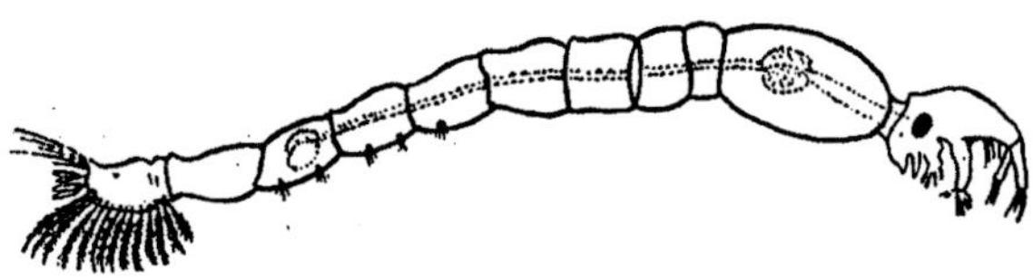

FIG. 9.
Larva of Chaoborus (after Miall).

Tribe Culicini. Free edge of scutellum distinctly trilobed; mesonotum without bristles; the palpi are short in the female and long in the male; abdomen uniformly

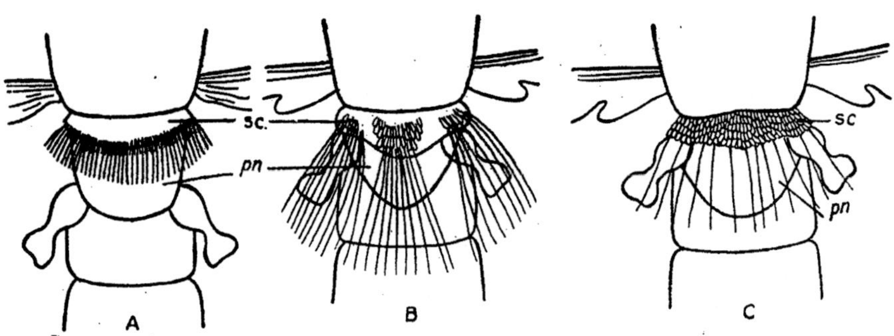

FIG. 10.
Mesoscutellum of (A) *Anopheles*, (B) *Culex* and (C) *Megarhinus*. sc, scutellum ; pn, postnotum.

invested by close overlapping scales; in the resting state they are hunch-backed. The larva hangs from the surface of water. The palpi in the male are long and are more or less of uniform thickness whereas in the female they are very short.

This tribe is again subdivided into two groups;

(1) Aedes group. Eggs laid singly; last segment of female abdomen narrow, usually completely retractile into the penultimate; claws of female at least on the four anterior legs, nearly always toothed.

Genera. *Aedes*, *Janthinosoma*, *Psorophora etc.*

(2) Culex group. Eggs laid in masses; last segment of female abdomen broad, immovable; claws of female never toothed.

Genera. *Culex*, *Taeniorynchus*, *Theobaldia*, *Mansonioides*, *etc.*

Tribe Megarhinini. Free edge of scutellum indistinctly trilobed or semilunar; proboscis distally bent like a hook; abdomen completely invested with scales.

Genus **Megarhinus.** The adult is of large size and is as a rule brightly coloured with green, blue, yellow, purple and white scales which have a bright metallic lusture. The proboscis is very long and has a characteristic shape. It thins out considerably towards the distal part and is bent like a hook. The insect is incapable of piercing the skin. In addition to the characteristic proboscis there are other characters which are helpful in the identification of the tribe. The posterior fork cell of the wing is very short; there is a V-shaped thickening in the wing membrane between the branches of the 5th vein. The sexes are easily differentiated by the antennæ. The larvæ are carnivorous. Their mouth parts are structurally modified to enable them to catch hold on to living objects which come within their reach. Cannibalism is very freely observed among them. They feed on mosquito larvæ and other midges. Eggs are laid in clusters and the larvæ are found in tree-holes, and bamboo stumps during the monsoon. These mosquitoes are all confined to the tropics.

FIG. 11.
Head of *Megarhinus.*

M. splendens Wied: It is widely distributed in the tropics. Its chief characters are: abdomen with conspicuous tufts of hairs on segments 6, 7 and 8; those on 7 black, on 8 orange; first abdominal tergite blue or green in the middle, pale yellow at the sides; fore-tibia dark at apex.

The antennae of the larva are almost bare and the hairs of the mouth brushes are peculiarly bent. The eggs are dropped on the surface of water from a height of 2 or 3 inches. The larvae are carnivorous. The adults feed solely on the surface moisture of plants. On account of its peculiar breeding habits in bamboo stumps, tree-holes, earthen pots *etc.*, the larvae of this mosquito have been used to suppress the breeding of *Aed. scutellaris* in tree-holes in Fiji with questionable results.

M. edwardsi Barraud: Like *M. splendens* it is also a large species. It can be readily distinguished from other allied species by the broad, pale yellow border to the mesonotum. The larvae are found in tree-holes. It has been reported from the Himalayas.

M. minimus Theo.: It can be separated by its size and entirely dark tibiae and tarsi. It is a bamboo breeder and its distribution is confined to South India and Ceylon.

CHAPTER IV

MOSQUITO (Subfam. Culicinæ)

Adult: anatomy. The body of a mosquito is divided into head, thorax and abdomen; a neck connects the head with the thorax.

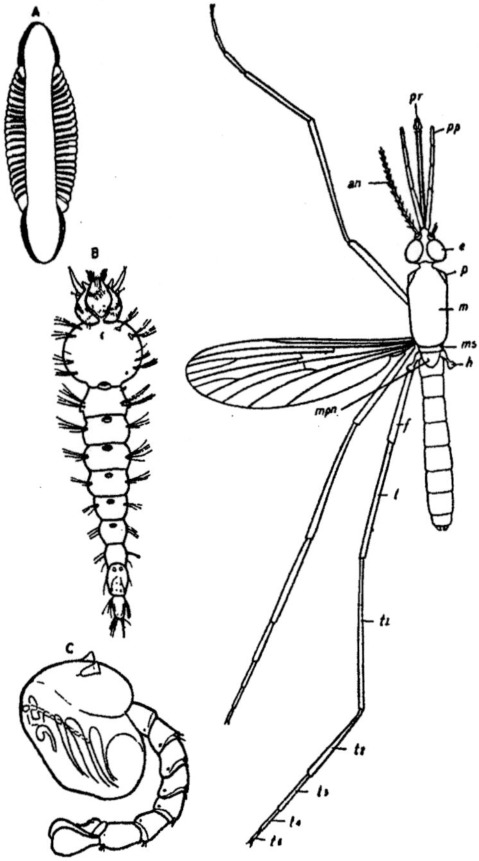

FIGS. 12 & 13.

Showing the complete metamorphosis of a mosquito.

Fig. 12.—A. Anopheline egg. B. Anopheline larva. C. Anopheline pupa.

Fig. 13.—Diagram of a female *Anopheles*. an, antenna ; e, eye ; m, mesonotum ; p, pronotum ; pp, maxillary palp ; pr, proboscis ; f, femur ; h, haltere ; mpn, mesopostnotum ; ms, mesoscutellum ; t, tibia ; t1 to t5, tarsal segments.

Head: The head is spherical and is produced anteriorly into a prominent clypeus. The eyes occupy most of the head; behind them the latter is covered with scales. The eyes are compound, being composed of numerous ommatidia, each of which is a separate visual element; ocelli absent; occiput clothed with scales and hairs.

Each antenna is composed of 14 to 16 segments; the first antennal segment is small and is obscured by the large second segment or torus. The second is large and bulbous. On the second segment lies Johnston's organ, sensory in function. The antenna carries whorls of coarse hairs round the base of each segment, and numerous fine, short hairs are scattered over the rest of the surface. In the male the antenna is plumose, being densely hairy. In the female it is sparsely hairy or pilose. This affords a ready means of identifying the two sexes.

The proboscis is long and appears as a single organ projecting forward from the head. It consists of the labium which is long and slender, and at its distal end are two small triangular lobes the labellæ, said to represent the labial palpi. Between the two labellæ lies a fine membrane called Dutton's membrane. It was once thought that the microfilariæ escaped from the proboscis of the mosquito by piercing the Dutton's membrane and made their way into the wound. It is now known that the filariæ force their way through the tip of the labella.

The labium is clothed with scales and is a flexible hollow structure with a longitudinal groove along the whole length of its dorsal surface in which the actual mouth organs lie concealed. These are : (*a*) a pair of maxillæ; (*b*) a pair of mandibles; (*c*) labrum-epipharynx; and (*d*) hypopharynx.

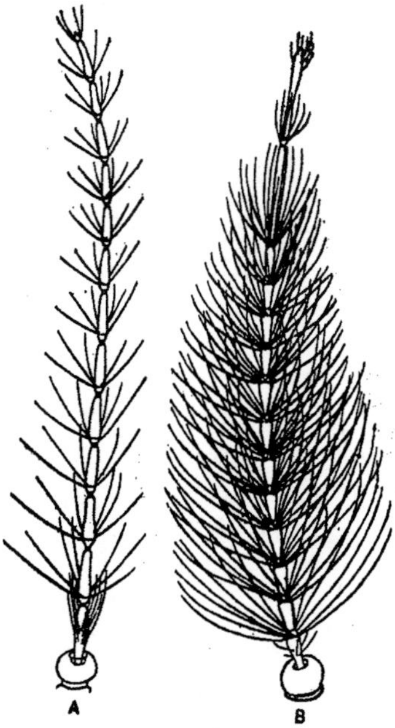

FIG. 14.
Antenna of mosquito : A, female ; B, Male.

The maxilla is thin and blade-like and is provided with a number of teeth at the distal end. The number of teeth varies in different species. It was once thought that the number bore some definite relation with its adaptibility to pierce human or animal skin. This view of Roubaud (1921) has not been corroborated. In males the maxillæ are reduced and the serrations are not so well developed. The mandible is much thinner than the maxilla; its distal end is very finely serrated. In the male the mandibles are very much reduced. In the mosquito the labrum and the epipharynx are fused to form a single structure. The hypopharynx lies on the ventral surface of the labrum-epipharynx. It represents the tongue and arises from the floor of the mouth. It carries the salivary duct. At the time of feeding the lateral margins of the labrum-epipharynx appose with those of the hypo-pharynx to form a complete tube for the passage of food. Vogel (1920) and Robinson (1939) are of the opinion that the food channel is formed by the labrum alone, the hypopharynx playing no essential part in its formation. As the salivary duct runs along inside the hypopharynx, the salivary secretion cannot come in contact with the contents of the food-tube except at the extreme distal end where the salivary duct opens.

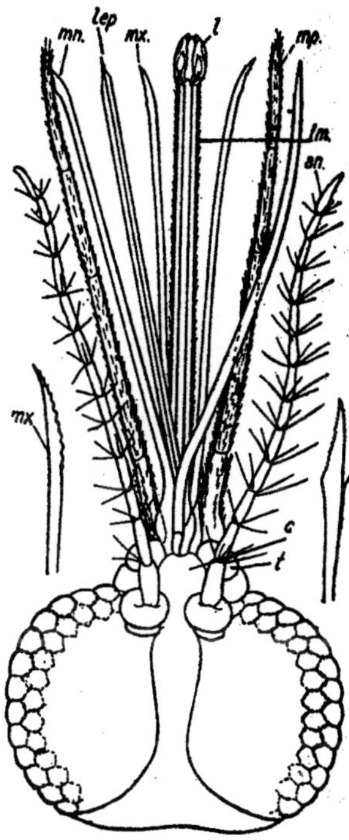

FIG. 15.
Head of a female Anopheles showing the different mouth parts.
an, antenna ; c, clypeus ; l, labella ; lm, labium ; mn, mandible ; mp, maxillary palp ; mx, maxilla ; t, torus or 2nd antennal segment ; lep, labrum-epipharynx.

The female mosquito settles on the skin and with its proboscis feels for a suitable spot for puncturing the skin for biting. The labellæ are applied to the skin and as soon as they are pressed, this causes the labium to bend like an elbow and simultaneously the other mouth organs such as the mandibles, maxillæ, epipharynx and hypopharynx are protruded from the end of the labella on to the skin. The skin is punctured by the maxillæ and mandibles. The maxillæ probably enlarge the wound through which the epipharynx and the hypopharynx are pushed.

The palpi are attached to the base of the maxilla. The difference in the two sexes is often very marked. The palpi are clothed with scales. The number of joints is variable but as a rule they are composed of 4 segments.

Thorax. The thorax is covered with scales and is formed by the fusion of the three segments, prothorax, mesothorax and metathorax. The wings are attached to the mesothorax and the halteres to the metathorax while each segment of the thorax bears a pair of legs.

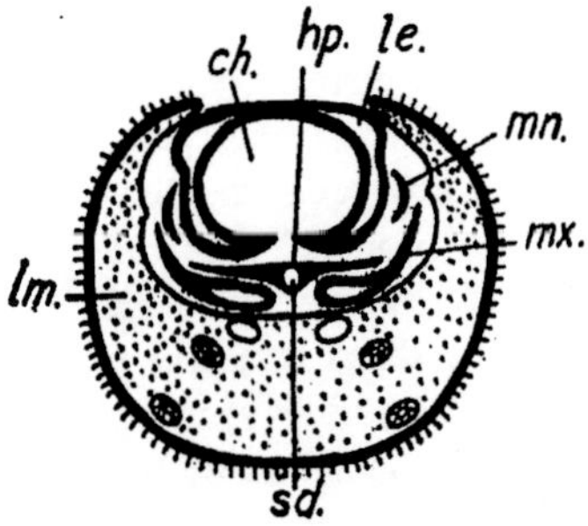

FIG. 16.

Transverse section through the proboscis of a female mosquito. (After Nuttall and Shipley).

ch, food channel ; hp, hypopharynx ; le, labrum-epipharynx ; lm, labium ; mn, mandible ; mx, maxilla ; sd, salivery duct.

Though the thorax is composed of three segments, the middle one or the mesothorax is the largest and occupies practically the whole of the dorsum of the thorax, the other segments being not at all prominent.

The prothoracic lobes lie at the antero-lateral regions of the mesothorax and project on either side of the neck. The pronotum is divided into two parts which are called anterior and posterior pronotal lobes. The coxa of the first pair of legs is attached to the prosternum.

The dorsal part of the mesothorax is domeshaped. The mesonotum or the sclerite which covers the mesothorax dorsally is divided by an incomplete suture. The mesonotum is thus divided into an anterior part called mesoproscutum and a posterior one called mesoscutum. There is a conspicuous transverse ridge at the posterior part of the mesoscutum which separates the mesoscutum from the mesoscutellum. The shape of the posterior margin of the scutellum is important in the identification of the tribe. It is evenly rounded in Anophelini, distinctly trilobed in Culicini, and indistinctly trilobed or even semilunar in Megarhinini. The scutellum overhangs a domeshaped plate which is known as the mesopostnotum.

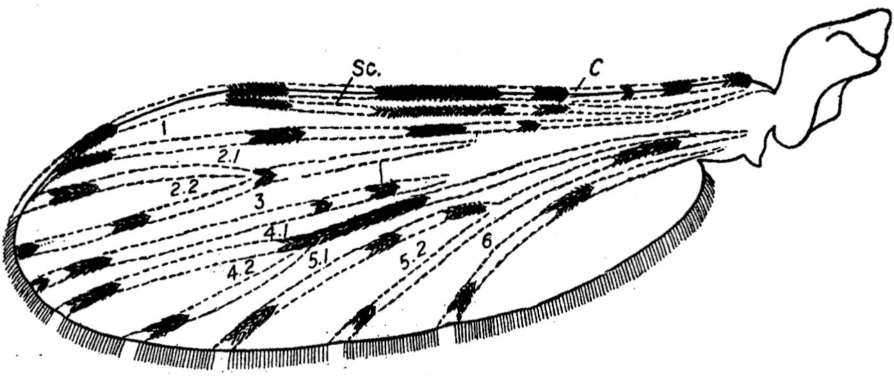

FIG. 17.

Wing of an anopheline mosquito.

C, Costal vein ; Sc, Subcostal vein ; 1—6, longitudinal veins.

The mesopleuron which forms the lateral wall of the mesothorax consists of two plates, mesoepisternum and mesoepimeron. The coxæ of the second pair of legs are attached to the mesosternum.

The metathorax is much reduced. The coxæ of the third pair of legs are attached to the metathorax.

The thorax bears two pairs of spiracles. The anterior pair lies on the mesothorax just posterior to the posterior pronotal lobe. The posterior pair lies on the metathorax above the insertion of the coxa of the third pair of legs. The spiracles are slit-like.

The halteres resemble drum sticks and represent the second pair of wings of insects. They are attached to the sides of the metathorax. They have been assigned different functions by different authors; these are sound production, maintenance of equilibrium or balancing, and the third, hearing.

The wing is attached to the mesothorax. It has a distinct shape and a characteristic venation. All wing veins are clothed with scales. The ornamentation or dappling of wings is entirely due to the arrangement of pale or dark scales on the veins.

On the wing there are eight veins, namely, the costa, subcosta, and six longitudinal veins, the second, fourth, and fifth of which are forked. In addition, the posterior border of the wing is fringed with scales.

A cross vein connects the third with the second and fourth longitudinal veins; another cross vein is linked with the stem of the fourth and the upper branch of the fifth.

Nomenclature of wing veins

Christophers and older authors		Comstock-Needham
Symbol	Name	Name
C	Costal	Costal
SC	Subcostal or auxiliary	Subcostal
R1	First longitudinal	Radius
Rs	Second longitudinal	Radial sector
R 2	Upper or anterior branch of 2nd longitudinal vein	Radius 2
R 3	Lower or posterior branch of 2nd longitudinal vein	Radius 3
R 4+5	Third longitudinal	Radius 4+5
M	Fourth longitudinal	Media
M 1	Upper or anterior branch of 4th longitudinal	Media 1
M 2	Lower or posterior branch of 4th longitudinal	Media 2
Cu	Fifth longitudinal	Cubitus
Cu 1	Upper or anterior branch of 5th longitudinal vein	Cubitus 1
Cu 2	Lower or posterior branch of 5th longitudinal vein	Cubitus 2
An	6th longitudinal	Analis

Nomenclature of cells

Christophers and older authors		Comstock-Needham
Symbol	Name	Name
c	Costal cell	Costal cell
sc	Subcostal cell	Subcostal cell
r 1	Marginal cell	Cell r 1
r 2	First submarginal	Cell r 2
r 3	Second submarginal	Cell r 3
r 5	First posterior	Cell r 5
m 1	Second posterior	Cell m 1
m 2	Third posterior	Cell m 2
cu 1	Fourth posterior	Cell cu 1
cu 2	Auxiliary	Cell cu 2
an	Spurious cell	Anal cell

When the imago emerges from the pupa the wings are small, but they are soon distended and after a short interval, the upper and lower surfaces coalesce, leaving the tracheæ as veins.

The leg is thickly covered with scales. Each segment of the thorax bears a pair of legs. Each leg consists of : (*a*) coxa, (*b*) trochanter, (*c*) femur, (*d*) tibia and (*e*) five tarsal segments, the last one ending in a pair of claws. The claws may be either simple or may contain one or two teeth. A considerable difference is noticed in the arrangement of these teeth not only in the two sexes but also in different tribes. The empodium is hair-like and the pulvilli are developed only in *Culex*, being rudimentary in others. The coxa and the trochanter are very small. The femur is the first visible segment from the dorsum. Both femora and tibiæ are slender and of equal length. The first tarsal segment is as long as the tibia.

Abdomen. The abdomen consists of 9 or 10 segments, the last two being modified for sexual purposes. Within a short time after emergence the tip of the abdomen in the male mosquito undergoes a torsion of 180 degrees, the true dorsal surfaces of these segments being really ventral surfaces and *vice versa*.

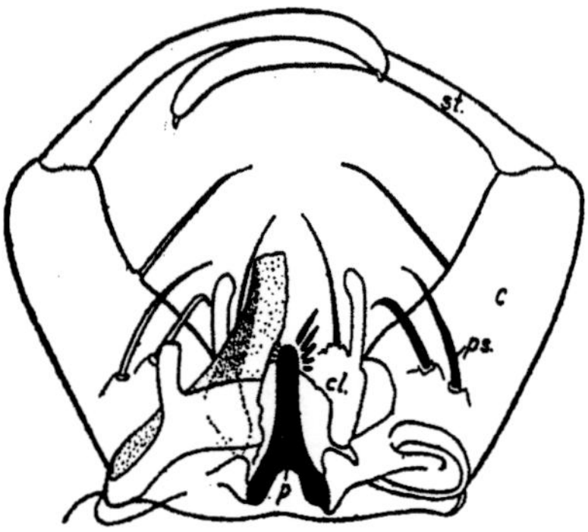

FIG. 18.

Diagram of the male terminalia of an Anopheline imago (dorsal view). (After Christophers).

c, coxite ; cl, claspette ; p, phallosome ; ps, parabasal spine ; st, style.

The male genitalia consist of the 9th segment together with its appendages. The whole of this segment is made up by the tergite as the sternite is very narrow. Its appendages are : (*a*) coxites or side pieces to which is articulated (*b*) the clasper; (*c*) on the inner side of each coxite is a parabasal lobe which carries parabasal spines ; (*d*) claspettes or basal lobes lie at the inner basal portion of the coxites; (*e*) harpago, a lobe-like process lying on the basal lobes; (*f*) phallosome which lies ventrally between the bases of the coxites and carries leaflets.

The anal lobe or the 10th segment lies within 9th tergite ventral to the phallosome.

The female terminalia is of little use. The torsion undergone by the male does not take place in the female. The most conspicuous parts are the cerci. The anus opens at the tip of the 10th segment, above and between the cerci.

Lying in the 8th segment is a round highly chitinised organ, the spermatheca. It is a single organ in *Anopheles* but consists of three round bodies in *Culex*.

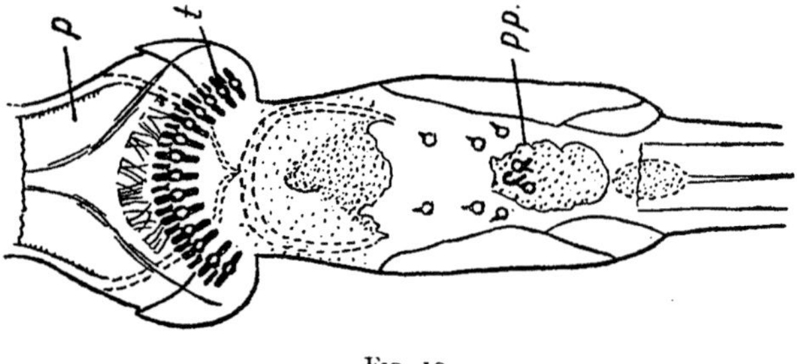

FIG. 19.

Dorsal view of buccal cavity and commencement of pharynx of an anopheline mosquito (after Sinton and Covell).

p, pharynx ; pp, palatal papillae ; t, bucco-pharyngeal armature.

Alimentary canal. The stomodæum and proctodæum develop from invaginations of the ectoderm and the lining of these regions is therefore ectodermal or chitinous in character. The mesenteron is the midgut. The food channel is formed by the apposition of the labrum-epipharynx and hypopharynx. The mouth lies in the region where the mouth parts coalesce and is situated slightly behind the front edge of the clypeus. The pharynx is a large flask-shaped organ, the posterior part being widely dilated. The anterior portion is divisible into two regions, the prepharynx and the midpharynx. The posterior portion may be called the postpharynx and constitutes the pumping organ. The dorsal and the ventral walls of the pharynx are continuous with the epipharynx and the hypopharynx respectively. On the ventral wall of the postpharynx are some chitinised modified setæ in double rows. These are the pharyngeal armature of Sinton and Covell (1927). These have been used to classify *Anopheles* into five groups (Barraud and Covell, 1928). The chitinous armatures are probably homologous with the T-shaped ridges found in the pharynx of saprophagous Diptera larvæ. The pharynx is provided with strong muscles which assist in suction. At the posterior end it becomes continuous with the œsophagus which is much wider. The œsophagus is provided with three diverticulæ, two dorsal and one ventral. The two dorsal ones are small; the ventral sac is very long and extends into the abdomen sometimes even beyond the 5th segment. They all contain

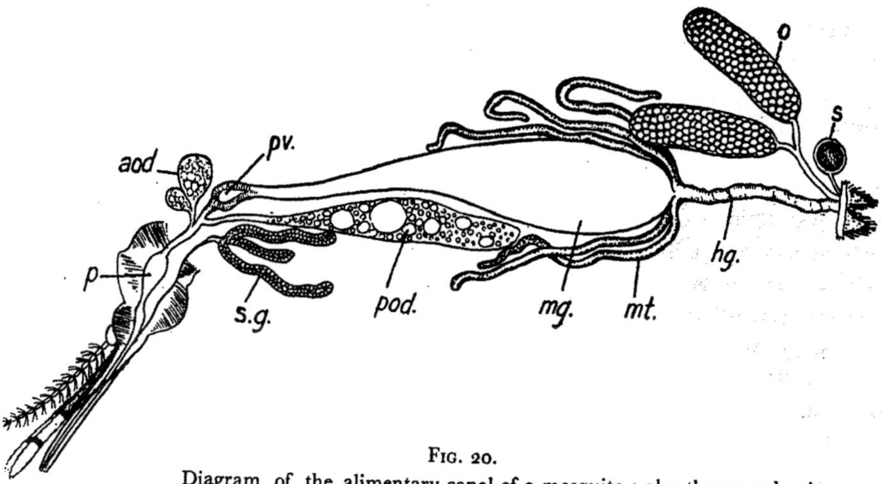

FIG. 20.

Diagram of the alimentary canal of a mosquito ; also the reproductive system of a female mosquito.

aod, anterior oesophageal diverticula ; hg, hindgut ; mg, midgut ; o, ovary ; p, pharynx ; pod, posterior oesophageal diverticula ; pv, proventriculus ; s, spermatheca ; sg, salivary glands ; mt, malpighian tubules.

minute bubbles of gas and can thus be easily distinguished from the salivary glands. No satisfactory explanation has hitherto been put forward regarding the exact function of the diverticulæ. The presence of air-bubbles in these sacs has given rise to the supposition that the diverticula provides these insects with the buoyancy required to enable them to undertake long flights in the air. Such air-sacs are, however, wanting in most other nematocera.

The proventriculus lies at the anterior part of the midgut. It is a muscular organ formed by the invagination of the œsophagus into the stomach. In the mosquito it acts as a muscular valve which prevents the regurgitation of food from the stomach. The midgut or the stomach is elongated, its anterior part being narrow and the posterior part dilated. The mid-intestine is followed by the hind-intestine and at the junction lies Malpighian tubules which are five in number. They are excretory in function. The hind-intestine is smaller and much narrower than the mid-intestine. The rectum is provided with six rectal papillæ. The anus opens on the ventral surface of the last segment.

The salivary gland, one on each side, consists usually of three lobes or acini, the two lateral ones being long and tubular and the middle one considerably shorter. Considerable difference in the structure of the two glands is manifest. The single ducts on each side travel forwards into the neck and head, where they unite to form a common salivary duct. The common duct enters the salivary pump situated below the anterior part of the prepharynx and then opens into the hypopharynx. One of the acini is not infrequently bifid distally and additional or accessory acini are sometimes found.

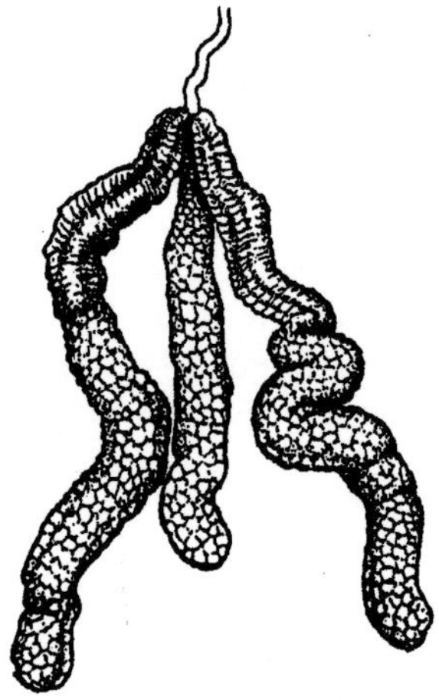

FIG. 21. Salivary glands of one side separated from the head.

The salivary secretion is in part withdrawn into the stomach with the blood when the mosquito feeds. This is borne out by finding sporozoites intermingled with the blood in the stomach and ventral œsophageal diverticulum of infected mosquitoes shortly after feeding.

Cornwall and Patton (1914) found the salivary glands contain a powerful anticoagulin and that the secretion agglutinates the red cells of the blood. This was confirmed by Yorke and Macfie (1924). The injection of salivary secretion is accompanied by a wheal and the redness is due to subsequent scratching (Roy, 1927).

Reproductive system. The reproductive system in the female consists of a pair of large ovaries and when gravid, they contain eggs. The ovary connects with a pair of oviducts which are lined with chitin; they unite to form a common oviduct, its lower part being called the vagina. The number of spermatheca varies in anopheline and culicine mosquitoes. In anophelines only one spermatheca is found, while in culicines they are usually three in number. The accessory gland is very small and is unpaired.

In the male the internal reproductive organs consist of : (*a*) a pair of testes, (*b*) a pair of vasa deferentia, (*c*) a seminal vesicle, (*d*) ejaculatory duct, and (*e*) a pair of accessory glands opening into the distal part of the ejaculatory duct.

Respiratory system. The respiratory exchange is effected by means of a network of air tubes or tracheæ which ramify through the body. They open externally on the skin. A pair of such openings or spiracles lie one on each side of the thorax.

Life history. The following constitutes the general life history of a mosquito considered on a broad basis.

Egg. A mosquito passes through four distinct stages, *i.e.*, egg, larva, pupa and adult, the first three being aquatic. Eggs are small dark bodies and are just visible to the naked eye. They are laid on water, leaves and stems of aquatic plants, mud and even on wet ground where tidal water is likely to reach. The number of eggs laid by different species varies greatly. The full complement of eggs is generally discharged in one sitting though exceptions occur. The shape of the egg varies greatly not only in different tribes but also in different species. *Anopheles* and *Aedes* mosquitoes lay their eggs singly, while *Culex* eggs are laid in clusters. The egg is white in colour and is soft but after it has been discharged it turns black and becomes quite hard. Eggs of some species can withstand a considerable amount of desiccation but a large majority of the eggs die when kept away from water for some time. The egg stage in the tropics under the most favourable conditions does not last more than two days.

Larva. The egg shell is broken by a head breaker found on the head of the first stage larva. The larva is an active wriggler. It has eyes and it feeds on solids like algæ, organic particles, and microscopic living organisms, and chews with its mandibles. It shows little or no discrimination in the choice of food. By means of the mouth brushes a current of water is set up and particles of food are drawn towards the mouth. The mouth brush also acts as a sieve or filter. From time to time the brushes are cleaned with the comb or pecten situated on the sides of the 8th abdominal segment.

What induces the adult to select a special type of water for breeding is difficult to say. It is no doubt true that some species are more attracted by the physico-chemical properties of the water, others in addition, probably to a special type of food existing in the water.

In mosquito larvæ a proteolytic enzyme is present in the gut which acts in alkaline medium. Amylase and invertase are also found.

As it is possible to rear larvæ of *Aedes aegypti* in Berkefeld filtered water, it is apparent that mosquito larvæ are able to utilise materials in solution and in suspension (Hinman, 1932). Howland (1930) found that vital stains like neutral red and methylene blue are absorbed with great rapidity by the gut cells and from this it is surmised that the contents of algæ in solution are taken up in the same way.

Anopheline larvæ float just below the surface of water while those of culicine hang from the surface. When alarmed, the larva generally dives to the bottom and remains motionless feigning death. Mosquito larvæ breathe oxygen from the air and while *Anopheles* larvæ cannot remain submerged in water for a long time, *Culex* and *Aedes* can do so.

The breathing apparatus consists of parallel air tubes in the body of the larva which open by a common orifice at the tail end. In the larvæ of *Culex* and *Aedes* the orifice is situated at the end of the prolongation of the air tube, known as the siphon, which is placed at an angle to the line of the body. In *Anopheles* they open on the 8th segment. Both larvæ of *Anopheles* and *Culex* possess perispiracular glands; these secrete an oily substance which is responsible for the hydrofuge properties of the spiracular region of the mosquito larvæ. The differential wetting of the spiracular region and the action of oil on mosquito larvæ depend on the presence of this oily secretion (Keilin *et. al.*, 1935.) The testes in certain *Anopheles* larvæ are often well seen. The larval life of all mosquitoes is interrupted by four stages; at the end of each stage a skin is cast off. It therefore grows by stages.

Anopheline larvæ are not cannibalistic by nature but become so when they are crowded in a limited space. Some *Culicines*, *e.g.*, *Lutzia*, *Armigeres*, and *Megarhinus* larvæ are predators : they also prey on each other.

In the prepupal state the larva does not feed. The duration of the larval life varies according to the food supply and temperature. In warm climates this stage may be completed within five days.

Pupa. The pupa is also an active swimmer. It sees with its eyes but cannot feed, as the mouth parts are entirely suppressed. On the dorsal surface of the cephalothorax is a pair of "horns" or trumpets which are the external openings of the respiratory system. When the pupa comes to the surface of the water, the trumpets break through the surface film and thus communication is established with the air. Shortly before the adult emerges, it straightens out, and a split appears along the mid-dorsal line of the thorax. The mosquito gradually emerges through this opening, the head first and the legs last. It sits on the empty pupal case for some time and as soon as the body becomes sufficiently hard, it flies away.

The duration of the pupal stage in the different species is somewhat variable. In Calcutta the time taken varies from 26 hours in summer to 48 hours in winter and depends on the seasonal conditions and temperature factor. All the progeny of a given female will not develop at the same rate neither will all the eggs of a progeny develop into the adult stage. The mortality is the heaviest during the larval stage and as it grows, the mortality lessens. Under the most favourable conditions a mosquito may complete the cycle from egg to the adult stage within 9 days and sometimes even less.

Pupæ of anopheline and culicine mosquitoes can be easily recognised by their siphon tube. This is funnel-shaped, *i.e.*, short and broad in anopheline and long and narrow in culicine mosquitoes.

Adult. The posture assumed by a resting imaginal mosquito is to a great extent characteristic of the tribe. Anophelines have a tendency to sit with their head, thorax and abdomen in a straight line, whereas Culicines have hunchbacked appearance. *A. culicifacies*, however, assumes a *Culex*-posture while resting.

The proportion of sexes has been stated by some to be equal, whereas others have noted a preponderance of either males or females.

Very little is really known about the day-time resting places of mosquitoes. Although they are known to take shelter in the day time in dark and often damp places, it is by no means easy to find them in large numbers. In addition to human habitations and cowsheds they hide in cracks, crevices, tree holes, roof thatch, shrubs, and jungles, not necessarily near breeding and feeding places, where they are believed to rest in the day time. It is always difficult to correlate adult catches in the house with the density of larval breeding, thus suggesting that they generally hide in some unknown places in the day time. Their infiltration in the dwelling houses at night is influenced by their natural desire for a blood meal and for this reason the males are found in insignificant proportion in such situation. Mosquitoes as a rule seek dark coloured objects for sitting and hiding.

Mating is generally assumed to take place in the evening or at dusk on the wing in the open at a considerable altitude above the earth. The males which remain near the breeding place collect in swarms and execute what has been described as a nuptial dance. In some species, however, mating can take place in a limited space even in test tubes. Fertilization generally takes place from 12 to 24 hours after emergence of the adult.

A mosquito does not feed within 12 to 24 hours after emergence. Though both males and females can live on vegetable sap and carbohydrate food, a blood meal on the part of the female of most mosquitoes is essential for the maturation of the ovum. For this a satisfactory amount of blood is necessary not only for the maturation of the first batch of eggs but also for subsequent batches. *A. albimanus* can ingest as much as 0.8 mg. of blood and *A. maculipennis* 1.8 mg. (1.73 cmm.), while an *Aedes* generally takes between 0.3 and 0.35 mg. As soon as oviposition has been completed, the mosquito is ready to take another blood feed. The intervening period between the ingestion of blood and oviposition is taken for the digestion of the blood meal. In the most favourable season it takes about four days for the maturation of the eggs inside the ovary. The numerical ratio of egg production of a mosquito does not depend on the nature of blood feed whether of man or of animals, but probably on the amount of blood ingested by it (Roy, 1931, 1936). Not all mosquitoes can be induced to take blood feed in captivity. When a mosquito once starts feeding it generally feeds to repletion. This is borne out from results of precipitation tests. Only a small number of individuals are seen to take both bovine and human blood in one night. Given a substantial blood meal virgin mosquitoes are capable of laying eggs and as these are not fertilized, they will not hatch. The biting propensity of a mosquito has a definite relationship to the intensity of light, temperature, humidity etc. All mosquitoes show some preferential habits for the blood of man or of animals. The blood found in a gorged female in a habitation is not necessarily human blood, since the house may be merely a temporary resting place, the meal having been obtained from an animal in the vicinity. Mosquitoes are primarily attracted to a combination of scent and warmth given off by the human body.

Mosquitoes are known to hibernate, *i.e.*, they pass over from an unfavourable to a favourable season. Hibernation takes place in the winter generally in the adult stage. At this time they do not feed and remain in a semidormant state. It is doubtful if mosquitoes can hibernate in the larval and egg stages; these are really instances of delayed growth rather than of actual hibernation. During hibernation there is an accumulation of fat and an increase of water in the mosquito. As the fat disappears the space that is occupied is partly filled by an increased amount of air in the diverticula, so that the insect's loss of weight is greater than the reduction in size (Buxton, 1935).

They are liable to be dispersed by railway trains, steamers, aeroplanes, and country boats. Wind may also carry them to distant places though they have a particular dread of boisterous weather.

Comparatively few observations have been made on the longevity of mosquitoes in nature. The life of a mosquito is primarily influenced by temperature and humidity. They generally die when the temperature is high and the humidity low. A low temperature is also fatal to them. The male is as a rule shorter lived than the female.

All insects have a tendency to be attracted to very strong light. On this basis various light traps have been devised to attract mosquitoes which are thus rapidly killed.

The range of flight of a mosquito varies with the species. The males generally do not fly far but the females are able to seek their host by flying a considerable distance from their breeding places.

Natural enemies of adult mosquitoes and their larvae are numerous. The most important predacious enemies of the adult are lizards, spiders and bats, while those of larvæ are the top-feeding fish. Other natural enemies of larvæ are aquatic beetles

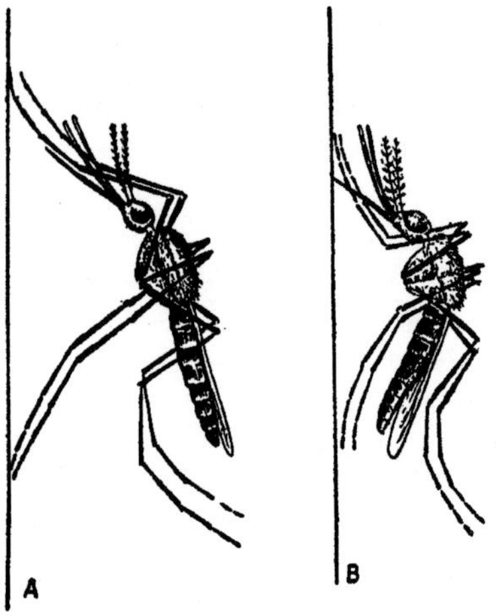

FIG. 22.
The pose of (A) *Anoph les* and (B) *Culex* mosquitoes while resting on wall.

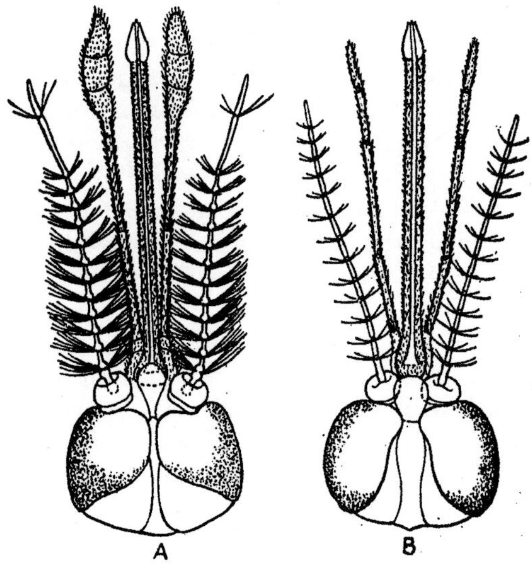

FIG. 23.
Head of male (A) and female (B) *Anopheles*.

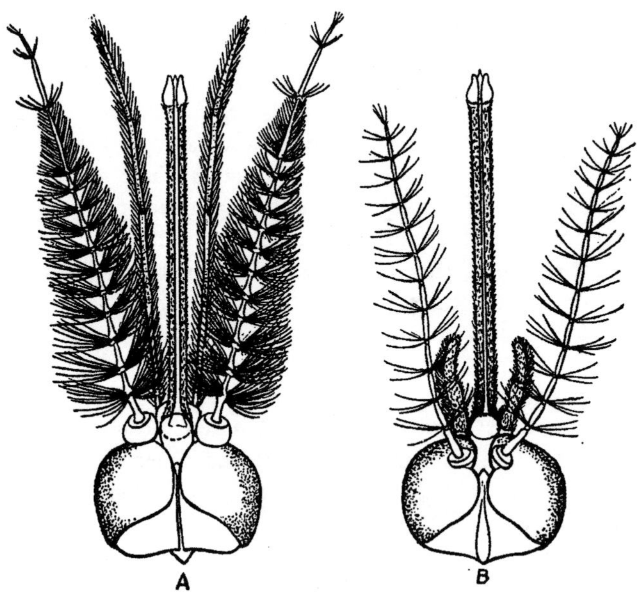

FIG. 24.
Head of male (A) and female (B) *Culex*.

and their larvæ, larvæ of dragon-flies and May flies, larvæ of *Chaoborus*, *Mochlonyx*, *Megarhinus*, *Lutzia etc.*, small crustaceas, Dytiscid beetles and their larvæ, Notonecta, tadpoles of common frog, Diptera larvæ of the family Anothomyidæ and certain molluscs

Parasites of mosquito larvæ and of adult mosquitoes are also many, *e.g.*, spirochæta, nematode, trematode, ciliata, microsporidia, bacteria, sporozoa, *etc.*, insects and mites. They have been listed by Keilin (1921).

Mosquitoes and Diseases. Mosquitoes carry a number of diseases and in this respect they act as true intermediate hosts in which the development of the parasites takes place in their bodies. The following infections are carried by mosquitoes : human, avian and simian malaria; dengue and yellow fever; equine encephalitis. *Dirofilaria immitis* and *D. repens* : these two dog filaria are also carried by mosquitoes. *D. immitis* are lodged in the heart and *D. repens* in the subcutaneous connective tissue. The filaria embryos present in the peripheral circulation develop in the malpighian tubules and make their way into the body cavity and thereafter into the labium.

The virus of equine encephalitis, principally a disease of equines and secondarily of man and of domestic animals, has been isolated from pools of trapped mosquitoes and the conviction that mosquitoes are the true carriers of the disease is growing. Ticks have also been incriminated and the virus passes to the next generation (Syverton and Berry, 1941). *Triatoma sanguisuga* and mites of the species of *Dermanyssus* and *Liponyssus* have been shown to be capable of transmitting the infection. It is probable that all types of biting flies may be concerned in the transmission of the virus.

Tribe : **Anophelini.** This tribe contains only one genus *Anopheles* though Edwards (1932) regarded the tribe as containing three genera, *Bironella*, *Chagasia* and *Anopheles.* The first is confined to New Guinea and the second to the New World. The genus *Anopheles* is found in both the Eastern and Western hemispheres of the globe. This genus has been divided by Christophers into 4 subgenera on the basis of the genital characters and pharyngeal armature. Among them only two subgenera occur in the Old World, *Anopheles* and *Myzomyia.* The former has less than four main costal spots on the wing and the pharyngeal armature is absent. In *Myzomyia* there are four or more costal spots and the pharyngeal armature is present.

The main points of distinction between the two tribes, Anophelini and Culicini, are given in a tabular form below.

	Tribe Anophelini	**Tribe Culicini**
Adult :		
Attitude	In the resting state the head, thorax and abdomen lie in a straight line.	The thorax makes an angle with the abdomen, *i.e.*, it has a hunch-backed appearance.
	A. culicifacies is an exception.	The palpi are also long and of uniform thickness throughout the whole length.
Male	In the male the palpi are long and clubbed at the end.	The terminal part is generally a little bent.
Female	The palpi in females are as long as the proboscis and are straight.	The palpi are conspicuously shorter than the proboscis.
Pupa	(*a*) The siphon tube is funnel-shaped, *i.e.* short and broad.	The siphon tube is long and narrow.
	(*b*) The accessory paddle hair lies above the origin of the paddle hair.	It is placed beside the paddle hair or may be absent.

5

Larva	(*a*) It floats parallel to the surface of water.	It hangs from the surface of water.
	(*b*) It has a characteristic swift movement.	It has much slower and snake-like movement.
	(*c*) Siphon tube absent.	Siphon tube present.
	(*d*) The respiratory openings lie on the 8th abdominal segment.	They lie at the end of the siphon tube.
	(*e*) Palmate hairs, arranged in pairs, are present on many abdominal segments and sometimes on the metathorax.	Palmate hairs are absent.
Egg	(*a*) It is boat-shaped and it is provided with lateral floats.	Elongated.
	(*b*) Laid singly.	Either laid singly (*Aedes*, *Armigeres etc.*), in rafts (*Culex*), or in a cluster (*Mansonioides*).

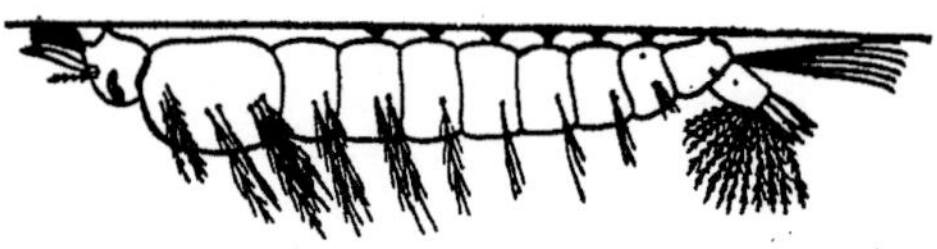

FIG. 25.
The pose of Anopheline larva.

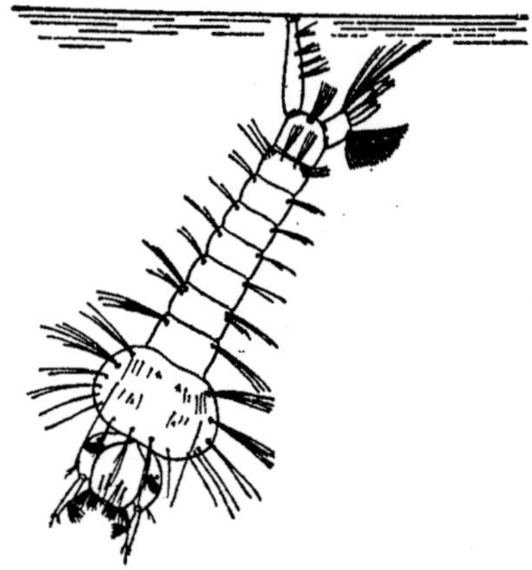

FIG. 26.
Resting attitude of Culicine larva.

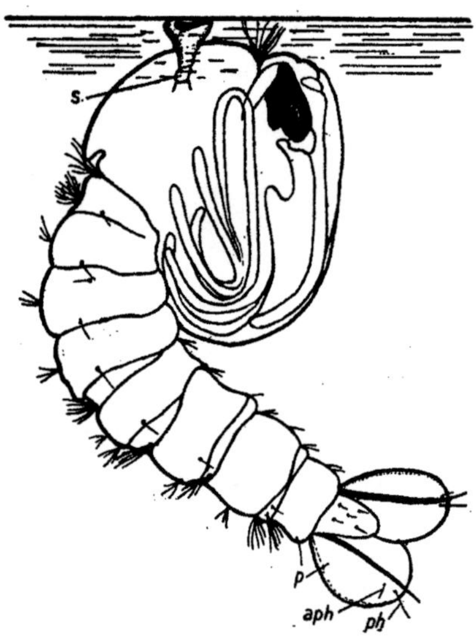

FIG. 27.
Pupa of *Anopheles* mosquito.
aph, accessory paddle hair ; p, paddle. ph, paddle hair ; s, siphon tube.

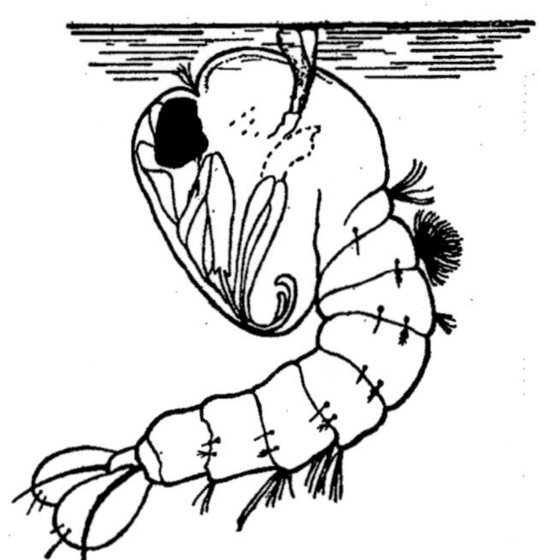

FIG. 28.
Pupa of *Culex* mosquito.

CHAPTER V

Genus : **ANOPHELES**

Certain special features in connection with the life history and biology of Anopheline mosquitoes are given below :

Egg. Anopheline eggs are always laid separately; they float on the surface of water. They are usually boat-shaped, the posterior extremity being narrower than the anterior. The upper surface or the deck is flat and corresponds to the ventral aspect of the contained larva. Surrounding the upper surface is the frill and at the sides of the egg are the floats which contain air and which help the egg to float on the surface of water. The frill is usually striated and shows a number of corrugations. The majority of the oriental species have the egg conforming to one of two main types, the whale-back and the life-boat shaped egg. Eggs should be studied under the microscope on wet cotton-wool. They may be preserved for an indefinite length of time in 4 per cent formalin solution. They may also be mounted in glycerine and the margin of the cover glass ringed.

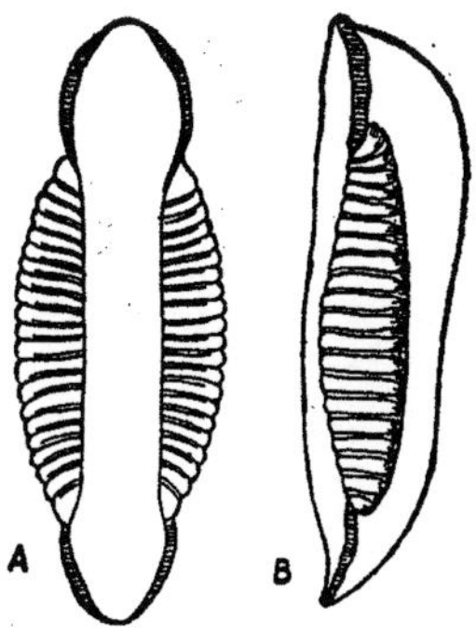

FIG. 29.

Eggs of *Anopheles philippinensis* (after Roy and Siddons).

A. Dorsal view; B. Lateral view.

So far as is known all anophelines require a blood meal for the maturation of their eggs. What portion of the different constituents of blood is directly concerned in the formation of the egg is not known.

While the intra-ovarian life of eggs in Calcutta is 4 days (Roy, 1940), Christophers (1911) found it to be 6 in the Punjab. The chronological succession of stages noticed in Calcutta in its development is given below.

Stage I : up to 24 hours—Complete filling up of the ovum with yolk granules. Stage II : up to 48 hours—Elongation of the ovum. Stage III : up to 72 hours—Appearance of floats. Stage IV : up to 96 hours—Hardening of the cuticle and maturation of the ovum.

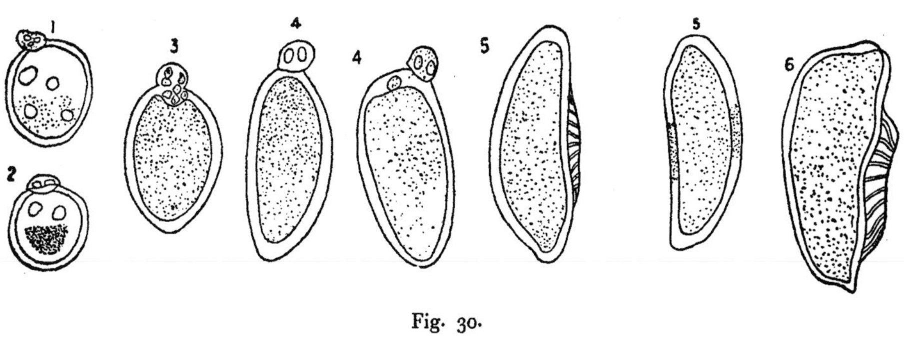

Fig. 30.

Development of eggs of *Anopheles subpictus* (after Roy).

1. Showing light peppery granules distributed irregularly indicating stimulation of the follicle. 2. Showing deposition of yolk granules. 3. Stage I up to 24 hours. 4. Stage II from 24 to 48 hours. 5. Stage III from 48 to 72 hours. 6. Stage IV from 72 to 96 hours.

According to Christophers the different stages in the maturation of eggs are as follows : Stage I. (*a*) Follicle consists of undifferentiated cells. (*b*) Follicle and central cells become differentiated : 12 hours. Stage II : (*a*) Yolk granules appear round oocyte, coarse yolk granules appear and increase in quantity. (*b*) Nearly half the follicle is filled with yolk : 36 hours.

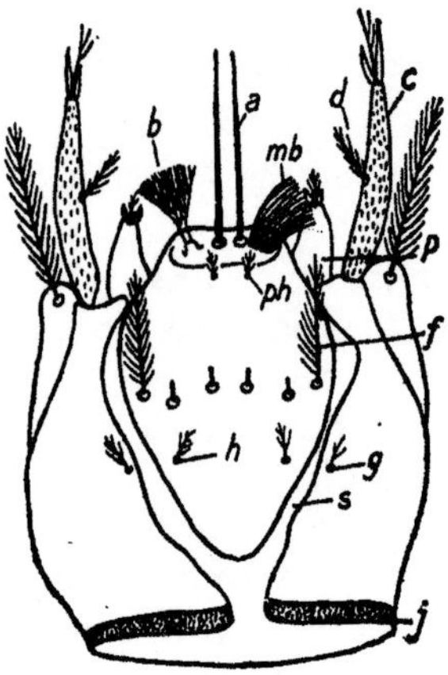

FIG. 31.

Head of *Anopheles* larva (dorsal view).

a, anterior-internal clypeal hair ; b, anterior-external clypeal hair ; c, antenna ; d, antennal hair ; f, frontal hair ; p, maxillary palp ; ph, posterior clypeal hair ; s, occipital suture ; g, external sutural hair ; h, internal sutural hair ; j, collar ; mb, mouth brush ;

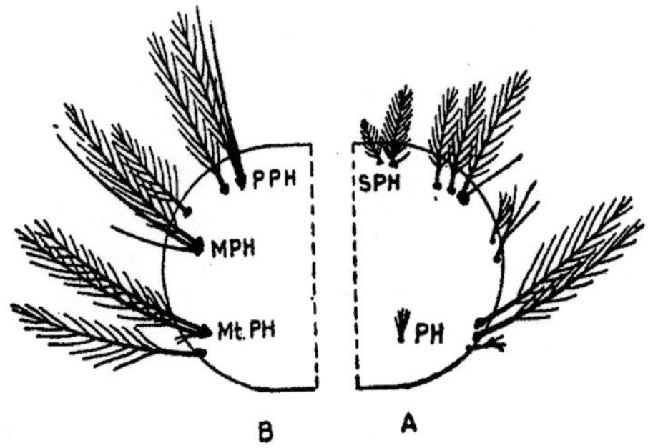

FIG. 32.

A. Dorsal view of the right half of the thorax of Anopheline larva.

B. Ventral view of the right half of the thorax of Anopheline larva.

M.P.H.—Mesothoracic pleural hairs.
Mt. PH—Metathoracic pleural hairs.
PH—Metathoracic palmate hair.
PPH—Prothoracic pleural hairs.
SPH—Submedian prothoracic hairs.

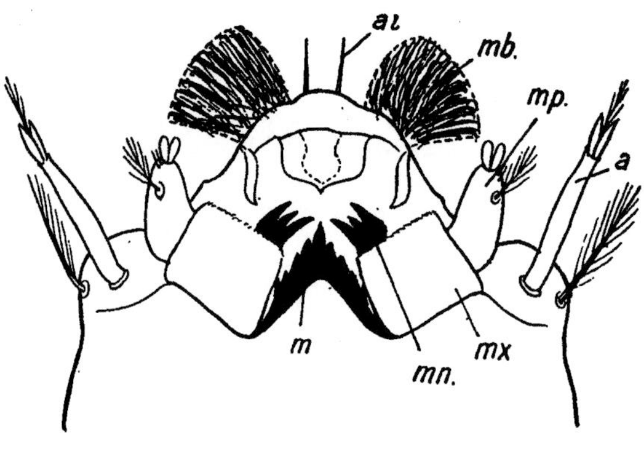

FIG. 33.

Ventral view of the anterior part of head of an anopheline larva.

a, antenna ; ai, anterior-internal clypeal hair ; m, mentum ; mb, mouth brush ; mn, mandible ; mp, maxillary pulp ; mx, maxilla.

Stage III : Gradually the yolk occupies about three-quarter of follicle : 4th day. Stage IV : Commencing elongation of follicle : 5th day. Stage V : Appearance of floats : 6th day.

Anopheline eggs like eggs of other mosquitoes, undergo some increase of size after they are laid, due to absorption of water.

In Malaya the average number of eggs in different species of *Anopheles* has been recorded by Lamborn (1922). The number varies from 29 found in *A. tessellatus* to 89 in *A. vagus*. Similar information in respect of Indian *Anopheles* is meagre. In *A. stephensi* this has been found to be on an average 80, in *A. minimus*, 118 and in *A. maculatus*, 147 in one laying.

Larva. The body is divided into head, thorax and abdomen.

Head. The median frontal plate on the head is known as the fronto-clypeus which is roughly triangular in shape. The eyes are large and reniform. The following organs constitute the appendages of the head. 1. A pair of antennæ which gradually taper from the base to the apex. 2. Mouth brushes, one on either side of the middle line. 3. A pair of toothed mandibles. 4. A pair of maxillæ which are rectangular in shape. 5. To each maxilla is attached a palp which lies internal to the antenna. The palp is much shorter and broader than the antenna. 6. A median organ, the mentum, which bears lateral teeth. The mouth lies between the mandibles laterally and the mentum posteriorly. There are a number of hairs on the head. Only those which are of taxonomic importance are given. (*a*) Clypeal hairs: anterior-internal, anterior-external and posterior hairs. (*b*) Frontal hairs: there are three pairs of frontal hairs arranged in a row and are generally branched. (*c*) Internal sutural hairs: they arise on each side of the fronto-clypeus. (*d*) Outer sutural hairs: they arise on the epicranial plates on either side of the epicranial suture.

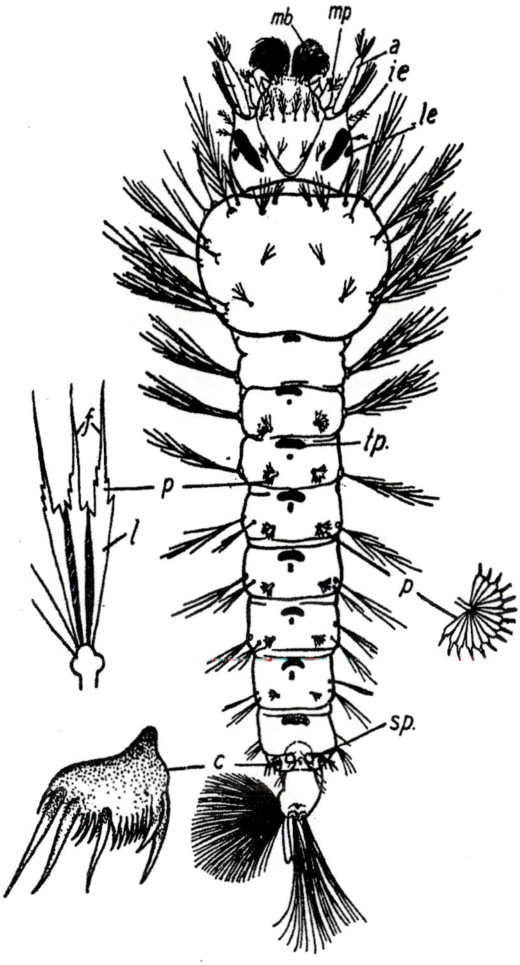

FIG. 34.
Anopheline larva (fourth stage).
a, antenna; c, comb; f, filament; ie, imaginal eye; l, leaflet; le, larval eye; mb, mouth brush; mp, maxillary palp; p, palmate hair; sp, spiracle; tp, anterior tergal plate.

Thorax. Although the thorax is fused, its segmented character is indicated by the arrangement of pleural hairs which lie on its ventral surface and are known as the prothoracic, mesothoracic and metathoracic pleural hairs. On the dorsum of the prothorax is a group of three hairs called submedian prothoracic hairs. The palmate hair may or may not be developed. It is present on the dorsal aspect of the metathorax. In the living larva a notched retractile organ will be seen occasionally protruding from the antero-lateral region of the thorax.

Abdomen. The abdomen is composed of 10 segments but 9 are visible. On the dorsal surface of each segment near the anterior border is a dark oval plate called anterior tergal plate. Posterior to the anterior tergal plate is a small rounded plate, the posterior tergal plate. At the postero-lateral region on the dorsum of certain abdominal segments lie the abdominal palmate hairs, one on each side of the middle line. They are so called on account of their resemblance to palm leaves. Each palmate hair is composed of a number of leaflets, each leaflet being composed of a blade and a filament. On the eighth and ninth abdominal segments lie the spiracular apparatus, the spiracular openings being found on the eighth segment, and on each side of this segment is a chitinous plate possessing spines of varying length and called the pecten. In addition to a large number of hairs the last or the anal segment carries four membranous organs called anal gills.

The first-stage larva is extremely small, even minute and has the egg-breaker on the posterior part of the head. In the second stage the egg-breaker is absent and the larva is larger in size. In the third stage it resembles fourth-stage larva but smaller in size and is considerably thinner. The fouth-stage larva attains a still larger size and looks quite plump.

Almost every species of *Anopheles* has some particular type of breeding place which it prefers above all others and which is essential if it is to flourish to any extent. From this the selective breeding habits of most of the common species has been determined. Some species are found in slowly running streams, some are swamp breeders and among them are those that prefer open swamps, while others prefer shade. A number of species are very fond of temporary pools, open cisterns, wells and collections of rain-water. Some like *A. subpictus* breed in extraordinary profusion in the little muddy pools. *A. sundaicus* is always found in brackish water.

Mosquito breeding in any type of water was once regarded as a physical phenomenon. The explanation of different species selecting particular types of water is now sought in the physical and chemical constituents of water. The algæ, planktons and micro-organisms all play their respective roles in the choice of the water by the female for egg-laying. The importance of food in the ecology cannot also be underestimated. In general it may be said that the presence or absence of the necessary larval food appears to be the deciding factor.

The anopheline larva is exclusively a surface feeder. At the time of feeding the head rotates through 180 degrees so as to bring the ventral surface uppermost. With the help of the feeding brushes current is set in motion whereby the food materials are drawn into the mouth. Large particles of solids are rejected.

Pupa. In the great majority of instances both pupation and emergence of the adult take place mostly during the night. The specific identification of *Anopheles* pupa can often be made from the hairs on the abdomen.

The period from oviposition to the emergence of the first offspring varies in different mosquitoes, and in different places according to the proper type of food being available to the larva and environments in which temperature plays a great part. While in India a cycle of 9 days may be taken as the average, in Egypt it is 7 days for *A. gambiae* and in Eritrea 7 to 8 days. However, the duration of the aquatic stages of *Anopheles minimus* in Assam, as observed by Ribbands (1949), is: egg—2 days, larva—7 days, and pupa—2 days, the aquatic stages lasting 11 days during the monsoon rains.

Adult. Anophelines as a rule show a decided preference for either human or animal blood. The hypothesis advanced on the food preferences of Anophelines are the following : (1) the choice of the food is determined by the maxillary index of the Anopheles; (2) the odour of the animal attracts particular species of Anopheles; and (3) feeding is induced by favourable micro-climatic conditions.

After feeding, the activity of the mosquito is greatly curtailed, when it rests in some sheltered place for the purpose of digesting the blood and for development of eggs. In practice what is usually noticed is that while a very large number of Anopheles may be caught at night, they are scarce in the day time when only a few are seen resting in dwelling houses and cowsheds.

Dispersal. The carriage of *A. gambiae* from Africa to Brazil in 1930 where it proved a great menace is an example of mosquitoes being transported in boats. Similar dispersal of *Anopheles* particularly in country boats and railway trains has been noticed in other countries. Not only *Anopheles* but mosquitoes in general are similarly carried from one place to another.

The potentialities of transportation of exotic anophelines by air plane have been discussed by Miller and his co-workers (1947).

Flight. A valuable method of studying the dispersion of larvæ, pupæ and adults has been described by Weathersbee and Hassell (1938). The larvæ are put in Giemsa stock solution diluted 1 :250 or in methylene blue, Congo red, equally diluted, for two days. Every adult after emergence will be unmistakably stained and one remained so for 39 days. The larvæ mortality with these stains is not great. The dye colours the thorax and abdomen blue.

Flight is generally determined by breeding them out in cages, and then staining them with an aqueous solution of an aniline dye. An atomiser is used to direct a fine spray upon the mosquitoes. Majid (1937) has found gold powder, commonly used in press work, extremely useful for marking mosquitoes. The powder is dusted after the insects have been lightly anæsthetised. The dusts are detected with the aid of the microscope. After a period of rest the mosquitoes are liberated to test their range of flight.

It is the commonly accepted belief that male mosquitoes do not fly far from the breeding places. Anopheline females can often cover a long distance by their own activity either by a single flight or a succession of flights. The intensity of their breeding also influences the extent of their dispersion. Much also depends on the proximity to the breeding places of suitable sources of food and shelter and also on the presence of natural barriers such as a large sheet of water or high hills.

In Malaya a distance of half a mile is considered the maximum effective range of *A. maculatus*. In Indonesia a control area of one and a half mile is considered necessary. In the latter country *A. sundaicus* is the chief danger and this mosquito is strong on the wing. In the Philippines *A. funestus* is believed to migrate a long distance from the breeding place, possibly more than 1.5 kilometers. Barber and Hayne (1925), Fisher (1923), Kligler (1924) have all pointed out that anophelines may even disperse and fly a distance greater than one mile. Kligler and Mer (1930) have published evidence that in Palestine *A. elutus* may migrate *en masse* to a distance of some 8 miles. *A. superpictus* is known to fly 4½ miles. *A. albimanus* in Panama can fly about 19 kilometer or 12 miles. *A. punctipennis* and *A. qudrimaculatus* can similarly cover a long distance, even as much as 8 miles on the second day.

Many observers have noticed in other countries that there are definite flights in the spring and in the autumn.

Indian mosquitoes are supposed to have a limited flight range. Though it is impossible to fix the range of flight, as every area must be considered separately, yet it may be safely asserted that a few individuals in India seldom stray much more than half a mile from the pool in which their larvæ are found.

The literature relating to the flight and dispersion habits of Anopheline mosquitoes has been critically reviewed by Eyles (1944).

Hibernation. In the case of Anophelines hibernation generally occurs in the adult female. It is also thought that some hibernate in the larval stage about which there is considerable doubt.

In India where the temperature is quite favourable for mosquito breeding throughout the year, hibernation is extremely uncommon, although its occurrence has been suggested by a number of workers from laboratory studies alone. It is probable that in the cold regions in the Northern Punjab and N.W.F. Province, mosquitoes pass the winter in the larval form.

Length of life. The length of life of the adult female *Anopheles* is an important factor in the epidemiology of malaria. The longer the *Anopheles* female lives, the greater chance it has to become infected and infective. While a hibernating mosquito may even live for six months or more, it is extremely doubtful if the ordinary *Anopheles* in India, at least in the plains, can live much longer than a month, the majority dying much earlier. Russell and Rao (1942) observed that the maximum longevity of female *A. culicifacies*, the most widely prevalent malaria-carrying mosquito in India, varies from 8 to 34 days and of male 8 days.

The transmission of malaria as also the length of life of the mosquito are dependent upon the co-existence of certain conditions of temperature and humidity. The normal span of life of female anophelines in nature depends on many factors.

Age. As regards the determination of age Perry (1912) divided Anophelines into four grades according to the condition of the wings. In grade I are placed specimens with the wing well marked and wing fringe practically complete; in grade II those with the wing fairly well marked, but the fringe somewhat worn; in grade III those with the wing decidedly shabby, and the fringe very much worn; in grade IV those with the wing actually thread-bare. The largest percentage of infection of the salivary glands is generally found in grades III and IV mosquitoes.

The nulliparous ovary can often be distinguished from the multiparous. Mer (1932) has found distinct changes in the common oviduct in *A. elutus* after eggs have once been discharged.

Parasites. The parasites which are known particularly to attack *Anopheles* larvae and adults are *Microsporidia*, certain *Mermithids* and fungus of the genus *Coelomomyces* of which two species have been described by Iyengar (1938). *C. indiana* and *C. anophelisica*, both of which infest *Anopheles* larvae and adults, cause considerable mortality. Three genera of *Microsporidia*, *Thelohania*, *Nosema* and *Plistophora* have also been recorded by him as attacking larvae. The life cycle of *Mermis* (Genus *Thelohania*), parasitic in *Anopheles* larvae, has been traced by him (1929). Kudo (1929) and Sen (1941) have described four species from Indian anopheline larvae. *Thelohania grassi* infects the eggs also (Missiroli, 1929). Sinton (1932) has recorded a list of trematode infection found in Indian anopheline mosquitoes.

Several genera of mites belonging to the family Hydrachnidae are found as ectoparasites of adults.

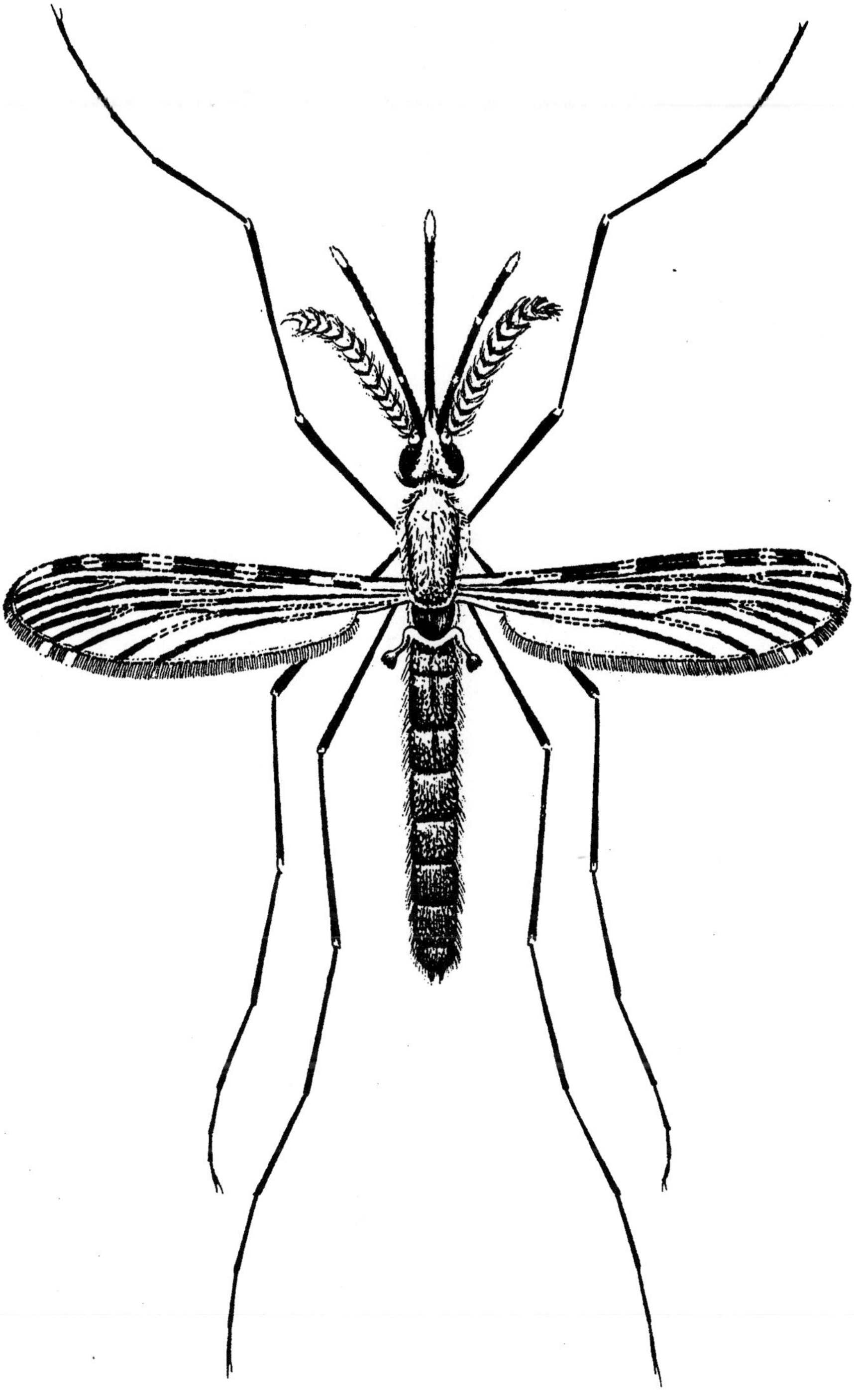

Anopheles culicifacies.

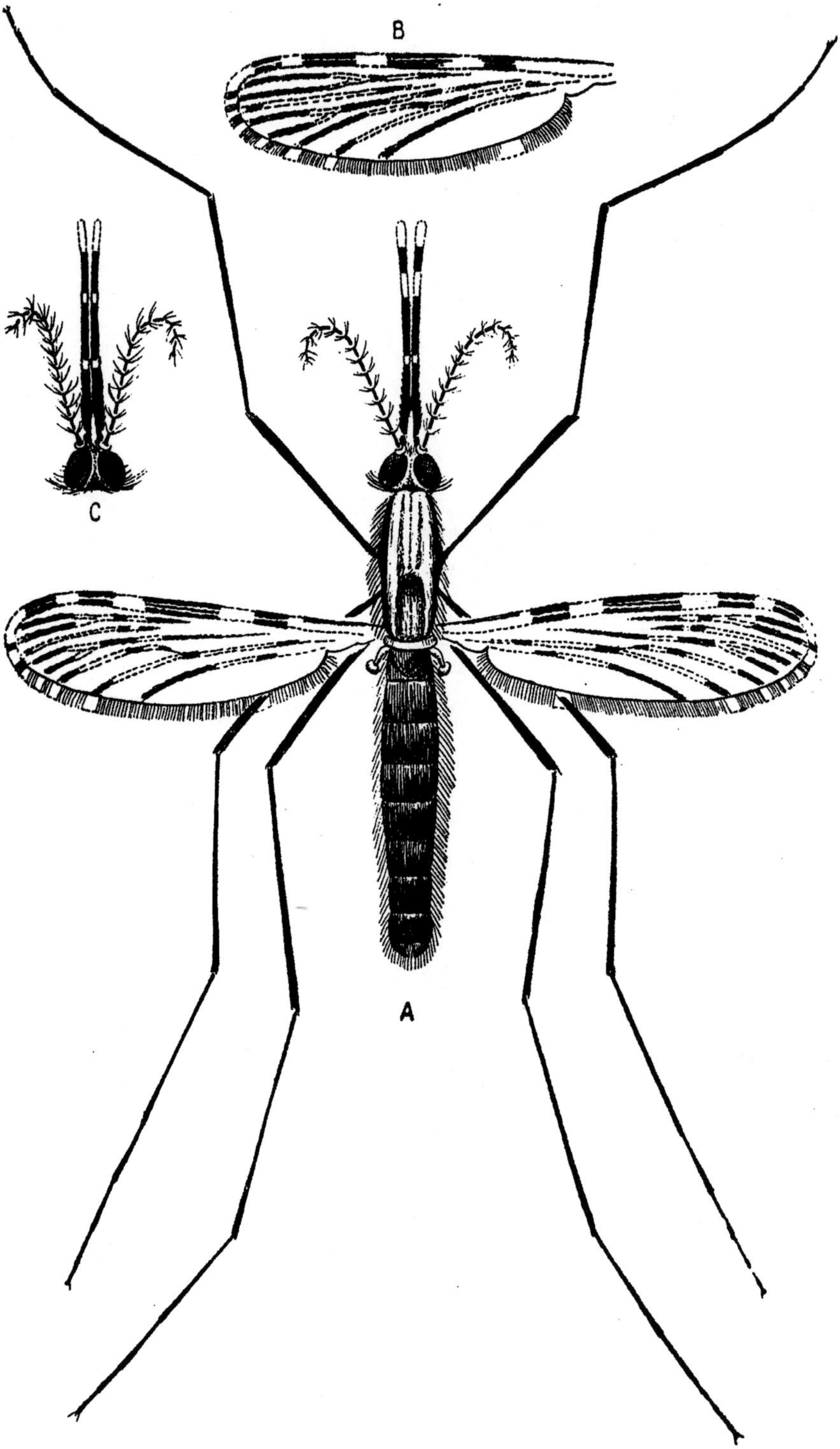

A. *Anopheles minimus.*

B. & C. Wing and head of *A. fluviatilis.*

The forms found on mosquitoes are the larval stages of these acarines; their adult stages are aquatic and free-living. Mites of the genus *Arrenurus* are encountered in India.

These mites attach themselves to the ventral surface of the adult female *Anopheles*, their larvae and pupae. They are seldom found on males but never on culicines or other mosquitoes. These parasites do not appear to interfere with the development of the larva and when the larva enters into the pupal stage, the mites transfer themselves to the latter and later from the empty case to the imago. The mouth parts penetrate in the form of a tube within the abdomen to distances varying from 0.5 to 1.0 mm. These tubes are the products of protective reaction occurring in the mosquito and are styled as stylosomes which are similar to those noticed in human beings which are produced by the bites of larval Trombidiidae.

Only one insect, *Culicoides anopheles*, is known to attack the female *Anopheles*. It has not been recorded from the male or from any Culicine. It seems that the *Culicoides* sustains itself on the blood ingested by the mosquito.

Races or Varieties of Anopheles

The presence of different races within *A. maculipennis* was first suggested by Roubaud from the fact that one of them had lost touch with man. In 1921 he proposed a method of distinguishing the races according to the number of maxillary teeth. Those that favour cattle were thought to be provided with a strong dental armature as opposed to a weak one in those which prefer to bite man. In 1924 Fulleroni drew attention to the existence of a difference in the markings of certain eggs from others and simultaneously van Thiel in Holland showed that the females of *A. maculipennis* were of two varieties, one short winged and the other long winged. In this way the existence of six different races or varieties of *A. maculipennis* was clearly recognised. They showed structural, biological and physiological differences, and among them only two, *var. labranchiae* and *var. elutus*, have been found to be always dangerous vectors of malaria; others are associated with malaria transmission only under exceptional circumstances. In the laboratory, on the other hand, no differences in their susceptibility to malarial infection have been found among them.

The conception of races in the Indian species has arisen from the diverse physiological behaviour on the part of the same mosquito in different localities and often in contiguous localities. The following may be taken as exhibiting varieties or races within the species: *A. stephensi*, *A. culicifacies*, *A. philippinensis*, *A. maculatus*, *A. aconitus*, *A. fluviatilis*.

A. stephensi. Sweet and Rao (1937) have established from measurements of wings and eggs that two distinct races or varieties are comprised in *A. stephensi*. No difference in their vector potentiality has, however, been discovered.

A. culicifacies. In consequence of the finding of salivary gland infection in *A. culicifacies* in the Punjab and in other parts of India it was assumed that *A. culicifacies* was a homogeneous species and its behaviour in regard to its infectivity potentiality would be constant wherever it was found. It is now known that this postulation does not hold good for all places.

Senior White showed that in the notoriously malarious places like Jeypore Hills, *A. culicifacies*, though the commonest species of *Anopheles* in that area, played no part in malaria transmission. It also takes no part in malaria transmission in the East Coast plains, South of Kistna, Madras, Bengal, Kutch, etc.

A. philippinensis. This is another species whose malarial transmitting potentiality is not constant everywhere. Although this species has been reported from Bengal, Assam, Southern India, Burma and Indo-China, infection has been recorded only in Bengal and Burma, the infection rate in Bengal being quite high. It is considered as the chief vector in the plains of Bengal.

A. maculatus. It is claimed that *A. maculatus* in India is different from the Malayan form where its power to transmit malaria is well known. In India, on the other hand, although it is common in all hilly areas, it appears to be a non-vector.

A. aconitus. This mosquito has a wide distribution in the Oriental Region. In India it has been recorded from the eastern and southern areas. Although it has been considered to be a dangerous species in Indonesia and in Malaya, the Indian variety has so far been regarded more or less innocuous.

A. fluviatilis. The racial peculiarities exhibited by *A. fluviatilis* (*listoni*) are as clear as noticed in *A. culicifacies*. All the earlier works had indicated that it was a vector of considerable importance. Covell

(1940) pointed out that *A. fluviatilis* in the U.P. Terai showed a poor preference to feed on man and the malaria was due entirely to *A. minimus*. Viswanathan (1950) has discovered that *A. fluviatilis* in the Bombay Presidency do not exhibit the same vectorial capacity on an uniform basis. Thus in Kanara District as well as in the hills and foot hills of Thana District, *fluviatilis* has a high anthropophilic index ranging from 65 to 80 per cent and likewise a high sporozoite index. Elsewhere in the Bombay State, this species has a very poor vectorial capacity.

Genus **Anopheles.** **Anopheles Species.**

Nearly 42 species with 10 varieties representing local forms or subspecies have been recorded from the Indian area. From their genital characteristics this genus has been further subdivided by Edwards into 4 subgenera, Stethomyia, Anopheles, Nyssorhynchus and Myzomyia: Among them only two, Anopheles and Myzomyia occur in India.

(*a*) Subgenus *Anopheles* : Wings are mainly dark with less than 4 or at the most 4 dark spots involving both the costa and the subcosta. Females without a pharyngeal armature.

Subgenus *Anopheles* Meig.: *A. aitkeni* James, *A. aitkeni var. bengalensis* Puri, *A. gigas* Swell. and Rodenw., *A. lindesayi* Giles, *A. barbirostris* v.d. Wulp, *A. hyrcanus v. sinensis* Wied., *A. umbrosus* Theo., *A. annandalei* Prashad.

(*b*) Subgenus *Myzomyia* : Wings usually with extensive pale markings; there are at least four dark spots involving both the costa and the subcosta. Females with a pharyngeal armature.

Subgenus *Myzomyia* Blanch. : *A. kochi* Don., *A. leucosphyrus* Don., *A. tessallatus* Theob. ; *A. aconitus* Don., *A. culicifacies* Giles, A. *fluviatilis* James, *A. minimus* Theo., *A. varuna* Iyengar, *A. annularis* v.d. Wulp, *A. karwari* James, *A. maculatus* Theob., *A. philippinensis* Ludlow., *A. pallidus* Theo., *A. ramsayi* Covell, *A. splendidus* Koidzumi, *A. ludlowi* Theo., *A. sundaicus* Rodenw., *A. subpictus* Grassi, *A. vagus* Don.; *A. stephensi* Liston, *A. jeyporiensis* James, *A. jeyporiensis* var. *candidiensis* Koidzumi, *A. majidi* Young and Majid ; *A. pulcherrimus* Theob.

Anopheles of India

A. culicifacies Giles. *A. culicifacies* has a very wide distribution in India, Burma and Ceylon. It is the most important malaria-carrying species in India, except in Bengal, Assam, the Jeypore Hill Tracts and a few other places. It is undoubtedly the species concerned in epidemic malaria. The atmospheric temperature and humidity have profound influence on the infectivity of *A. culicifacies* (Siddons, 1944). The characteristic feature about this species is its *Culex*-like colour and *Culex*-like attitude. *A. culicifacies* has a tendency to hide itself in well-sheltered places, such as in holes, among dung cakes, in chaff etc. This species is more zoophilic than anthropophilic. It cannot be colonised and is slow to take blood feed in captivity. The sporozoite index in this species is generally low. It is extremely susceptible to artificial infection in the laboratory.

A. culicifacies is ubiquitous in its selection of breeding places but generally prefers clear water. It breeds in open irrigation channels, river beds, pools, ponds, lakes, springs, in leaks and back water, burrow-pits, in fruit and vegetable gardens, and even in wells. The larvae are also found in seepages and other collections of still water in which grassy vegetation is abundant. It may occur in rice fields when the plants are very small, also in fallow lands. Eggs are generally deposited in the first part of the night (Pal, 1945).

It is commonly believed that *A. culicifacies* does not ordinarily disperse to distances much in excess of half a mile. But where food and shelter are scarce, this distance may be extended. On the range of flight of *A. culicifacies*, the literature has been reviewed by Afridi and Puri (1940).

A. minimus Theob. It is very widely distributed in the Oriental Region. In India it is found in Assam, Dooars, North Bengal and Southern India. It is a dangerous malaria carrier wherever it is found in large numbers. The sporozoite index in this species is generally high.

In Assam it breeds throughout the year in perennial streams close to human habitations. The larvae are also found in irrigation channels, seepages, slow-moving drains, tanks, rice-fields, burrow-pits, etc. It does not breed in low-lying areas flooded by water with silt or clay in suspension. Two essential requirements for its breeding are clear, unpolluted water in motion and grassy edges.

The adult has a tendency to rest in the day-time in dwelling houses and shows a decided preference

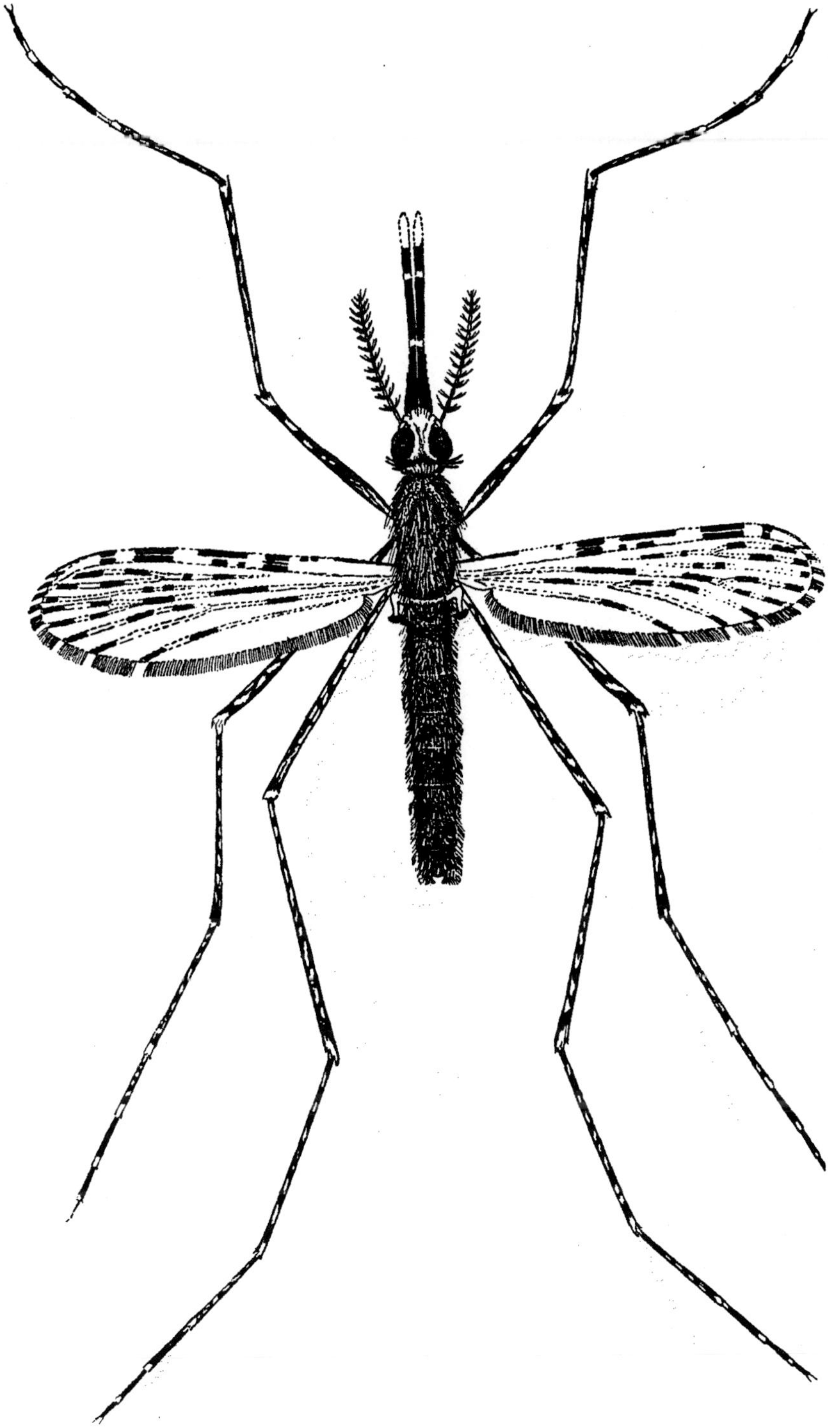

Anopheles sundaicus.

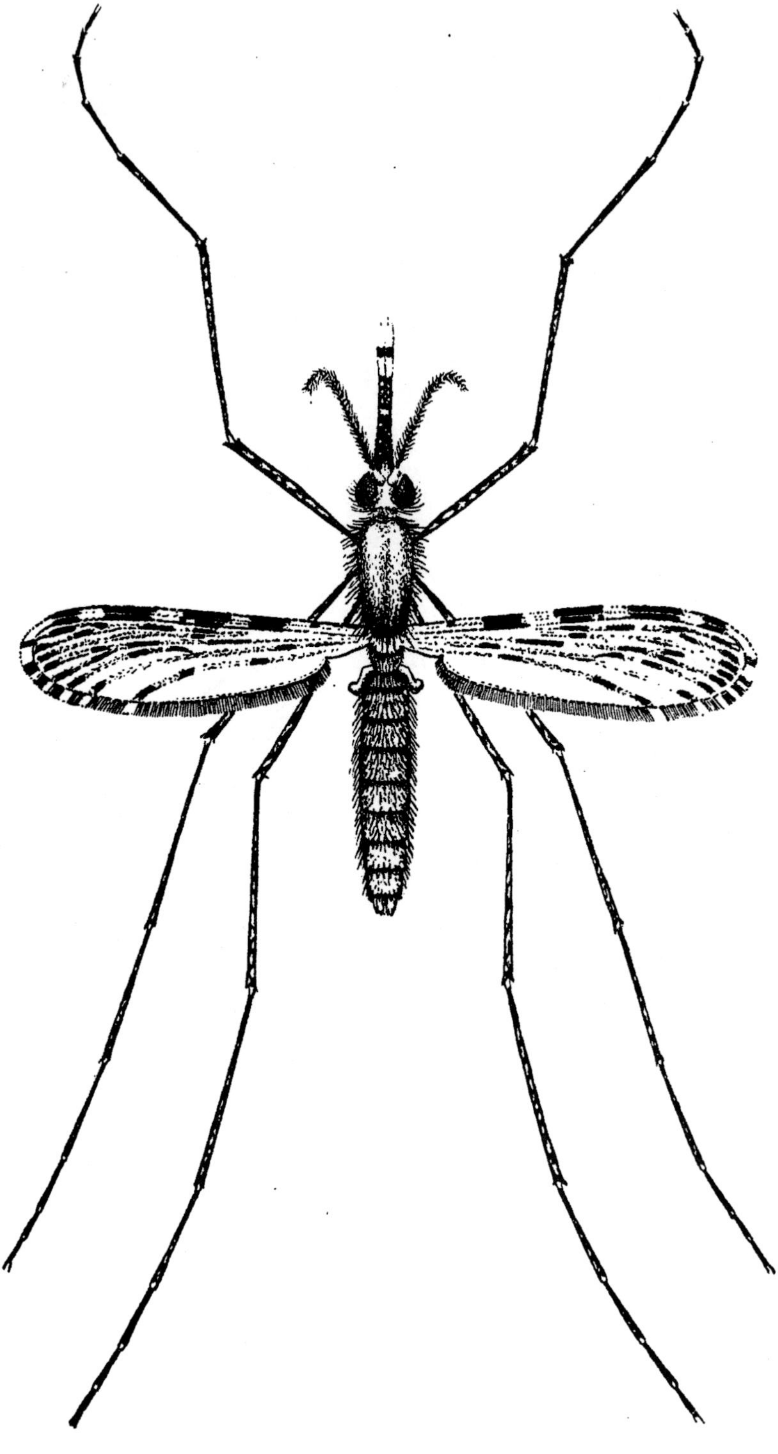

Anopheles stephensi.

for feeding on human blood. Oviposition normally takes place at night and 69 per cent of eggs are laid in the first three hours of the night (Thomson, 1941). The flight distance, according to Ramsay (1929), is one mile.

Outside India *A. minimus* is an important malaria-carrying mosquito in Burma, Indo-China, Southern China and Yunnan.

A. fluviatilis James. It is essentially an Indian species, and outside the Indian area it has been recorded only from Siam, Tonkin and Turkestan. It has a wide distribution in India from North-Western Frontier to Assam, especially in foot-hill areas and hilly and rocky tracts. It closely resembles *A. minimus* in its habits, its larvae being found in slowly running perennial streams. In Southern India and also in the South-Western parts of India this species breeds in streams and channels, rice fields, uncultivated swampy fields, sugarcane and betel ground trenches and shallow earthen field wells. The role of paddy fields as a definite habitat for *A. fluviatilis* in North Kanara was pointed out by Singh and Jacob (1944) and Rao (1945), and elsewhere by Venkat Rao (1938) and Mondol (1938).

Marked racial peculiarities are noticed in this species. Its vectorial capacity varies greatly in different places, the sporozoite rate varying from 0 to 50 per cent (Viswanathan, 1950).

A. fluviatilis breed in significant numbers up to at least half a mile from the place where they feed and this distance is presumably their usual flight range in search of food. It prefers to rest in human habitations in the day time.

Its importance as a carrier of malaria has been stressed by Senior White (1937, 1938, 1943), and Senior White and Adhikari (1940) in the Singhbhum hills, in Hazaribagh and Ranchi plateau. It is also a proved carrier in many places in southern India including Mysore State, Nilgiris District and Kanara District.

A. varuna Iyengar. The adult of this species was first described by Iyengar (1924) from specimens collected in Bengal. It has now been found to have a wider distribution outside Bengal in Peninsular India and Singhbhum Hills.

It breeds in stagnant fresh water, in ponds and ditches and is most prevalent during and soon after the monsoon when it breeds in collections of storm water by the road side. It generally prefers shade, e.g., overhanging branches of trees.

Natural infection has been recorded from lower Bengal. It is an important carrier of malaria in the Jeypore Hills, in the Singhbhum Hills and in the Eastern Satpura range (Senior White, 1937; Senior White and Das, 1938; Senior White and Adhikari, 1940). It is more attracted to human habitations than to cattle sheds, and overflows to cattle sheds only in the peak months. This species does not leave the house at dawn.

A. sundaicus Rodenw. *A. sundaicus* is distinguished from *A. ludlowi* by the phallosome. True *A. ludlowi* occurs in the Philippines and Formosa, the Indian species being *A. sundaicus*. The latter is also found in the Andamans, Malaya Peninsula, Burma and Sumatra. In fact it is widely distributed in the Oriental Region occurring along the sea coast. In India it is found on the coastal regions of Bengal, Orissa and Andamans, and is responsible for malaria outbreaks in these localities.

It breeds in brackish water generally in association with floating algae. It will also breed in water in which no algae are present, or at any rate none visible to the naked eye. This species is distinctly alkaliphile. *A. sundaicus* unlike the majority of malaria-carrying species can withstand a considerable degree of organic pollution in its breeding places (Rodenwaldt and Essed, 1925).

Opinions vary as to the optimum amount of salinity for its breeding. Rodenwaldt and Essed in the Dutch East-Indies considered that from 1.2 to 1.8 per cent was the most favourable concentration, whilst Iyengar (1931) found it 0.15 to 0.26 per cent, and Christophers (1912) 0.4 per cent. Its adaptibility to less and less saline water has been exemplified in its gradual extension into inland places in the Calcutta area far away from the sea. Similar spread of this species into North Madras along the coast and far into the inland places has been pointed out by Senior White (1947).

It has a tendency to bite man whenever it has the opportunity to do so and is found both in cattle sheds and in domestic houses.

The outbreak of malaria in relation to this mosquito takes place suddenly and spreads very rapidly. This is invariably associated with a rapid increase in its breeding.

The estuarine regions of the sea coast are relatively free from malaria, but the clearing of the virgin mangrove forests in these areas has resulted in a greater incidence of this disease, due to increased breeding of *A. sundaicus*. In the Malayan sub-region it is the *Anopheles* of the mangrove belt, and only when this forest area is opened up and when there is interference with the free flow of the tidal water, this mosquito breeds in prolific numbers in the numerous pools thus formed and which become exposed to sunlight.

A. stephensi Liston. It is essentially an Indian species and has been recorded from outside the Indian area only from Afganistan, Arabia and Mesopotamia. This species has a wide distribution throughout the whole of India. It also occurs in Burma, and Yunnan but has not been recorded from Ceylon.

It is normally an urban mosquito though occasionally also found in the rural areas. It is a highly efficient malaria carrier and is responsible wholly or partly for malaria in cities like Bombay, Madras, Bikaner, Calcutta, Quetta, Sind, Kohat, Mysore, etc.

It prefers small collections of clean water for ovipositing such as overhead cisterns, wells, small water reservoirs, pools, springs, marsh, and collections of rain water. In some localities it has been noticed to breed in large numbers even in river beds. In fact it is found in a wide variety of breeding places. The larva does not always require aquatic vegetation for its food. It can even adapt itself to saline water.

Amongst those which are known to breed in wells are *A. bifurcatus* in Palestine, *A. stephensi*, *A. culicifacies* in India and *A. gambiae* in Africa.

A. stephensi has been separated into two races on the evidence of the size of eggs, wing lengths, and the character of the offspring (Sweet *et al*, 1938; Rao, Sweet and Rao, 1938). The claim of Sweet and Rao (1937) that there is a difference in the malaria carrying potentiality between them has not been substantiated.

A. philippinensis Ludlow. Outside India it is distributed in Burma, Malaya and the Philippines and in India, Assam, Bengal, and Southern India. It breeds mainly in tanks, burrow pits and to a small degree in rice fields. It is not particularly attracted to *Pistia*. It often shows a high sprozoite index and is essentially a domestic mosquito being found more in dwelling houses than in cowsheds.

A. annularis v.d. Wulp. *A. fuliginosus* of earlier writers really comprised two species, *A. annularis* and *A. philippinensis*. The difference between them has been made clear by Christophers (1924). It is widely distributed throughout the Oriental Region. In India it occurs up to an altitude of 3,000 feet. It prefers stagnant water such as weedy tanks, rice-fields, swamps, lakes, burrow pits, etc. There is always much vegetation in the water. The adults are found particularly in cowsheds and they generally feed on cattle, its anthropophilic index being extremely low. Although infection has been recorded especially in Bengal, the infection rate is so low that it is considered to play an insignificant role in the transmission of malaria. Its importance in the Orissa coastal plains, where it is the principal carrier, has been pointed out by Sarathy (1932) and Senior White *et al.*, (1943). Although it has been found naturally infected in Yunnan it is not an important vector.

A. pallidus Theob. It is generally a rice-field breeder. It also breeds in pools, tanks, lakes etc., in the presence of vegetation. It is mostly found in Central India and also in South India. It rests in cowsheds. Though infection has been reported from Bengal, Central Provinces, and other places, it is considered to be of only local importance and plays a minor role in the epidemiology of malaria.

A. maculatus Theob. It is widely distributed in the Oriental Region and though it is a proved carrier in Malaya, direct proof of its acting as a vector in India, except in Shillong, has not been obtained. In India and elsewhere it occurs in the hilly regions such as North-West Himalayas, Assam, South India and Ceylon. It breeds in clear water in connection with hill streams exposed to sunlight. It is also found in seepages. It is very sensitive to organic pollution.

A. jeyporiensis James (var. *candidiensis* Koidzumi). In India it is distributed in the eastern, central and southern parts and the west coast and breeds in seepage pools, clear water streams and in swamps. The adults rest in the day time not only in cowsheds but also in human dwelling houses. Salivary gland infection in *A. jeyporiensis* (type form) has not yet been found in nature. The varietal form, var. *candidiensis*, has been proved to be the vector of malaria in Indo-China and in the Arakan District of Burma and also in extreme south-eastern parts of East Pakistan.

Anopheles majidi Young and Majid. It was at first thought to be a varietal form of *A. karwari* but has now been recognised to be a valid species. It has been described from Bengal and Southern India. It probably takes no part in malaria transmission.

Anopheles subpictus Grassi. It is one of the most widely distributed *Anopheles* not only in India

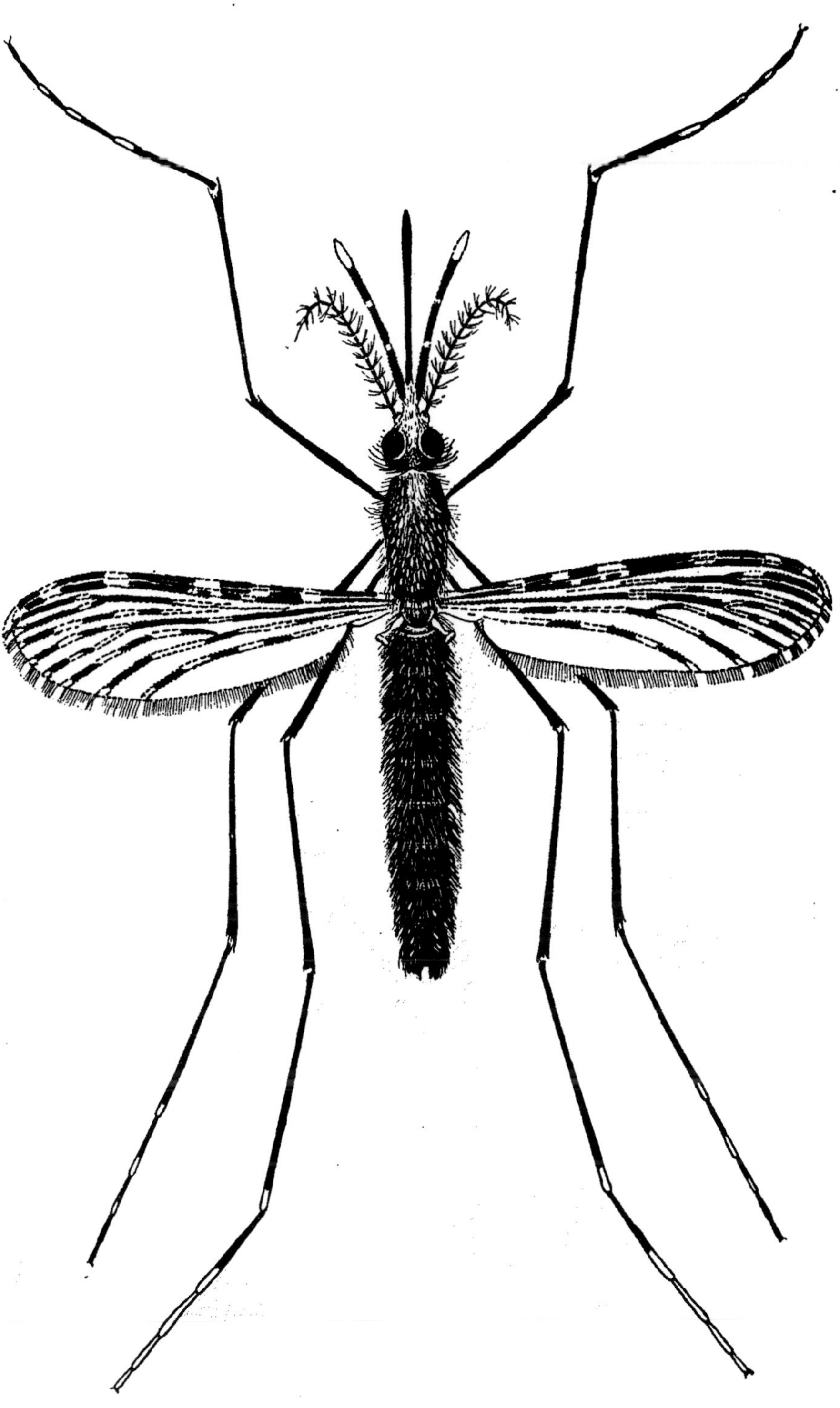

Anopheles annularis.

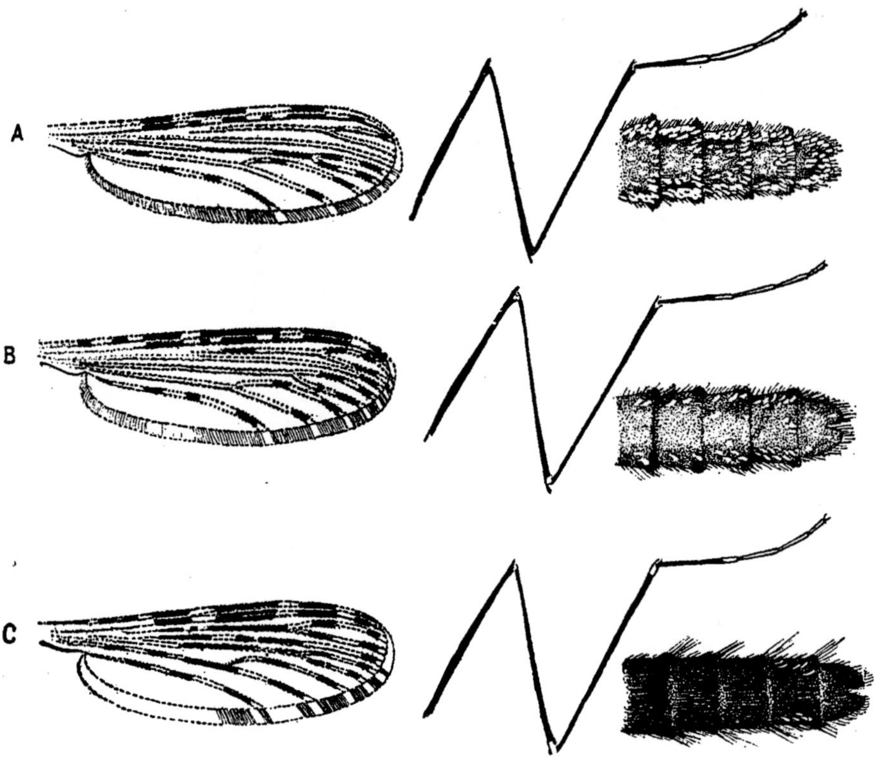

Wing, hind leg and dorsum of abdomen of (A) *A. pallidus*. (B) *A. philippinensis and* (C) *A. annularis*.

but also in the Oriental Region. It breeds in almost any type of still water, e.g., in brick-pits, muddy pools, drains, rice-fields, etc. This species, though found in the neighbourhood of human dwellings, is a marked cattle feeder, and can be caught in large numbers in cattle houses.

Anopheles vagus Don. It is found mainly in the eastern and southern parts of India. It is found in Burma also. The breeding habits are very similar to those of *A. subpictus.* It plays no part in the transmission of malaria.

Anopheles leucosphyrus Don. Though widely distributed in the Oriental Region, in India it is mainly confined to Assam, the Peninsular India and Ceylon. It is a jungle species and breeds in forest pools, and nullahs in partially shaded places. It tends to disappear with the clearance of shade.

Its importance as a malaria carrier is only local, e.g., in Digboi in Assam. In Kabaw and Kab Valleys in Burma (between Chittagong and Chindwin River, South of Imphal and North East of Akyab), *A. leucosphyrus* is a vector of major importance (Macan, 1948). Outside India natural infection has been recorded in this species by McArthur (1947) in Borneo, Toumanoff (1936) in Indo-China, and Kuitert and Hitchcock (1948) and Macan (1948) in Burma. It has now become clear that this species is a dangerous, if not the principal vector, over quite large areas. Its importance as a malaria carrier in the different Oriental countries has been discussed by McArthur (1951). In all the above publications the vector has been recorded simply as *A. leucosphyrus.* It is now known through the works of Reid (1949) and Colless (1950) that it is a composite species which includes 7 different forms or sub-species, the principal malaria vector in North Borneo being *A. leucosphyrus balabacensis* Baisas and according to Colless, the one that carries malaria in Burma also belongs to this form.

A. tessellatus Theob. In India it is distributed in Assam, Central India and South India. It is also found in Burma and Ceylon. It breeds in pools and though it has been infected in the laboratory, in nature it does not carry malaria. It is, however, the only local species and is the cause of malaria in the Maldive Islands, which during the war, was of some consequence. Infection in this species has also been recorded in Formosa by Anazawa (1949).

A. aconitus Don. In India it is mainly confined to the eastern and southern parts. Outside India it has a wide distribution in Burma, Malaya, Indo-China, Java and Sumatra. It breeds in grassy edges of irrigation canals and small streams and prefers to bite cattle rather than man in India, whereas in Indonesia this species is more anthropophilic than zoophilic. Though both gut and gland infections have been recorded in this country, it is considered, except in a few places, to be an unimportant species.

A. pulcherrimus Theob. *A. pulcherrimus* is widely distributed in Central Asia and has also been recorded from Caucasus, Iran, Turkestan and N.W.F. Province. In India it has been recorded from the Punjab, Sind and Gujrat. It is essentially a swamp breeder. Though it is a vicious biter, biting both men and animals indiscriminately, only gut infection has been recorded in this species.

A. lindesavi Giles. A large mosquito which has been recorded from altitudes of 2000-5000 ft. or over. It is principally a Himalayan species, though also found in Assam Hills and Nilgiris. It breeds in mountain streams. It is occasionally caught in cattle sheds.

A. gigas Swell. and Rodenw. This also is a large mosquito and has been recorded from high altitudes especially from the Himalaya, Khasia Hills, Pachmari Hills, Nilgiri Hills and Ceylon Hills. It breeds in small pools along hill springs and is essentially a wild species.

A. umbrosus Theob. A large and black mosquito and resembles *A. barbirostris* in general appearance. It is widely distributed in the jungle regions of Malaya where it is a carrier. In India it has been recorded only from Assam where it breeds in muddy pools containing rotten vegetation. It is a rare mosquito in this country.

A. barbirostris v.d. Wulp. It is a large and black mosquito and is widely distributed in India. For all practical purposes it is regarded as a wild species. It breeds in swamps, tanks etc., always in shady places and containing decayed vegetation. The larvae are found also in rice fields. The adults are seldom noticed in dwelling houses and are always found in cattle and buffalo sheds. While it is a wild species in Java, it has been found infected in nature in Celebes.

A. hyrcanus (var. *nigerrimus* Giles and var. *sinensis* Weid). Two varieties of *A. hyrcanus* are known. The one found in India is represented by *nigerrimus* which is a cattle feeder and breeds in water polluted with faeces and urine of cattle and buffalo. It is also found in rice fields. It has a wide distribution in India. From the malaria point of view it is totally innocuous. In Indonesia these two subspecies have not always been distinguished in reports on their ability to transmit malaria in nature, though in Sumatra they are considered to play some role in the malaria transmission. However, they have been experimentally infected with malaria in that country. Recently the subspecies *sinensis* has come to be regarded as a carrier of malaria of some importance under certain circumstances in Malaya (Hodgkin, 1933). It is the only vector in Formosa, Korea and in the plains in China. It is essentially a rice-field breeder.

A. kochi Don. This mosquito is mainly an eastern species being recorded from Assam and Burma. It is a cattle feeder. It breeds in small ground pools in the same situation as *A. subpictus*, and though it has been found naturally infected in Indonesia, no infection has been recorded in this species in India.

A. theobaldi Giles. Found mainly in west coast and Central India and breeds in shallow pools of clear water in the presence of green algae, also in rocky pools in connection with streams. Occasionally it is found breeding in wells.

A. splendidus Koidzumi. It is widely distributed in India but is generally never caught in large numbers at a time. However at Clement Town (Dehra Dun, U.P.) it can be caught in very large numbers in cattle sheds. Its malaria-carrying potentiality is probably nil. It breeds in small pools in the presence of aquatic vegetation.

A. ramsayi Covell. Previous to 1927 it was confused with *A. jamesi.* In India it is found in the eastern areas and breeds in tanks and swamps in association with *Pistia.* Though it has been found infected in nature, it is considered to play no role in the transmission of malaria. The adults are found in cowsheds.

A. turkhudi Liston. This is essentially a cattle feeder and breeds in pools and in sluggish streams in the presence of aquatic vegetation. At one time it was thought to have a limited distribution in India, being found in Quetta and sometimes in Gujrat. It is, however, a common mosquito in the Western Ghats, including Poona.

A. superpictus Grassi. This is distributed from the Mediterranean region to the N.W. of Pakistan. Though it is an important carrier of malaria in Palestine, Greece and Mesopotamia, it has not yet been proved to play any role in malaria transmission in this country. It breeds in hilly river beds and prefers flowing water. It has a very limited distribution in India.

A. jamesi Theob. Occurs in north Bengal, Assam, Southern India and in Ceylon. It breeds in ditches, ponds and tanks in the presence of aquatic vegetation. It rests principally in cowsheds and takes no part in the transmission of malaria.

A. karwari James. Reported from Bengal, Assam, Central India, the east coast of Madras and Ceylon. It breeds in clear water pools, ravines and seepage pools.

A. moghulensis Christ. It was originally described by Christophers (1924) as a varietal form of *A. jeyporiensis.* Though widely distributed in the Central Provinces, it is a rare mosquito there. It is found in numbers in the Western Ghats. It is also found in the western portion of Kashmir (Jacob, 1950). It breeds in pools and nullahs. It is a wild species and the larvae are found far away from human habitations.

A. culiciformis Cogill, **A. barianensis** James and **A. annandalei** Prashad. All the above three are rare mosquitoes and all breed in tree-holes. *A culiciformis* has been recorded only from the hills of the West Coast of Peninsular India. *A. annandalei* is known from the eastern Himalayas, Khasi Hills and Ceylon. *A. barianensis* occurs in the Punjab and Kashmir at elevations of over 6,000 feet.

A. aitkeni James. This is essentially a jungle mosquito and is found in Burma, Assam, South India and West Coast. It is common in the Nilgiris. It occurs in India where there is extensive dense tropical jungle or forest. It breeds in ground pools and takes no part in malaria transmission.

A. dthali Patton. It was previously known in this country as *A. rhodesiensis* as described by Christophers and Khazan Chand.

It is an Arabian species and occurs in this country in the hilly north western parts such as Baluchistan, Quetta and Kohat. It has recently been recorded from Kashmir. It breeds principally in pools in hill streams.

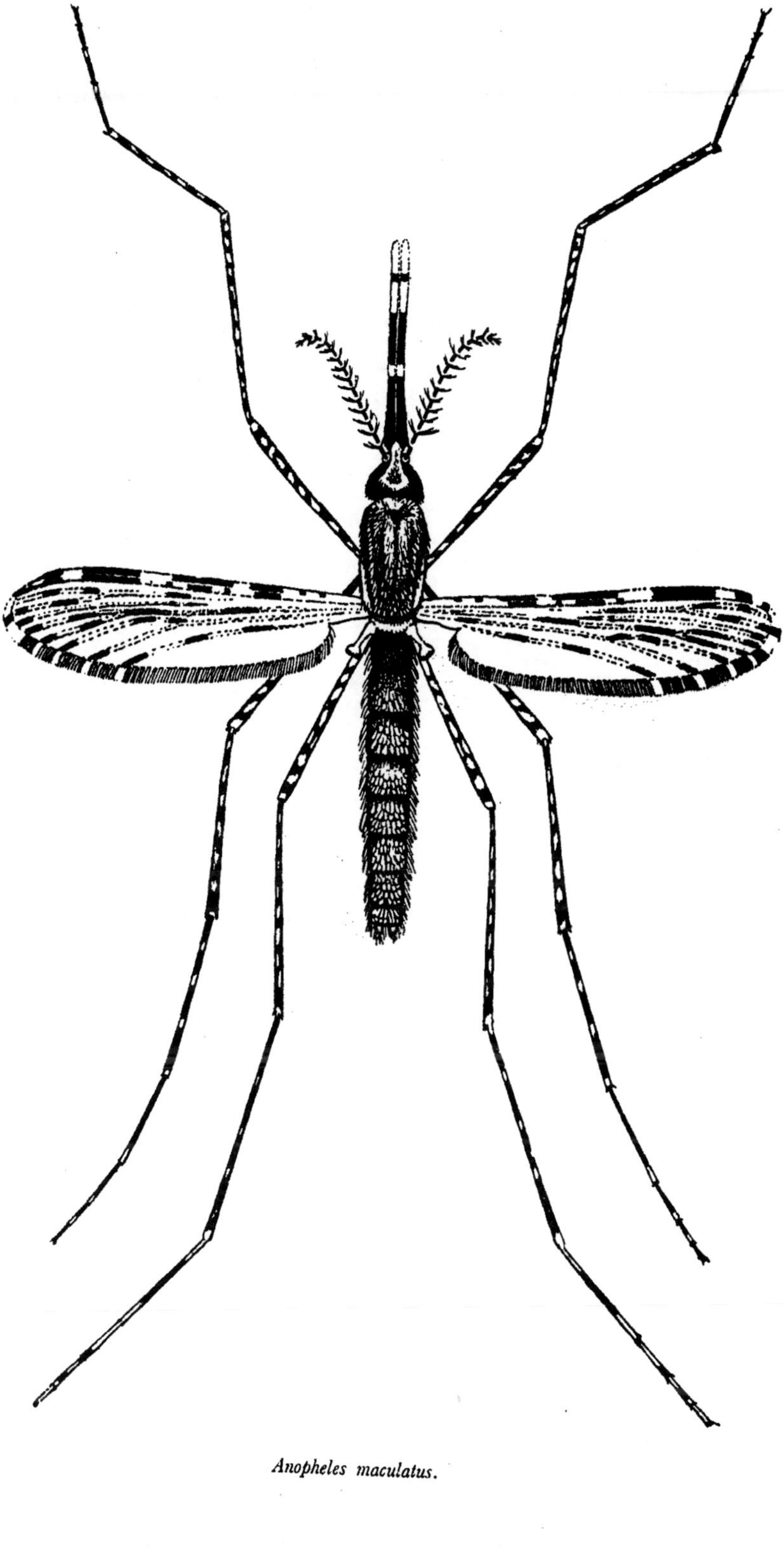

Anopheles maculatus.

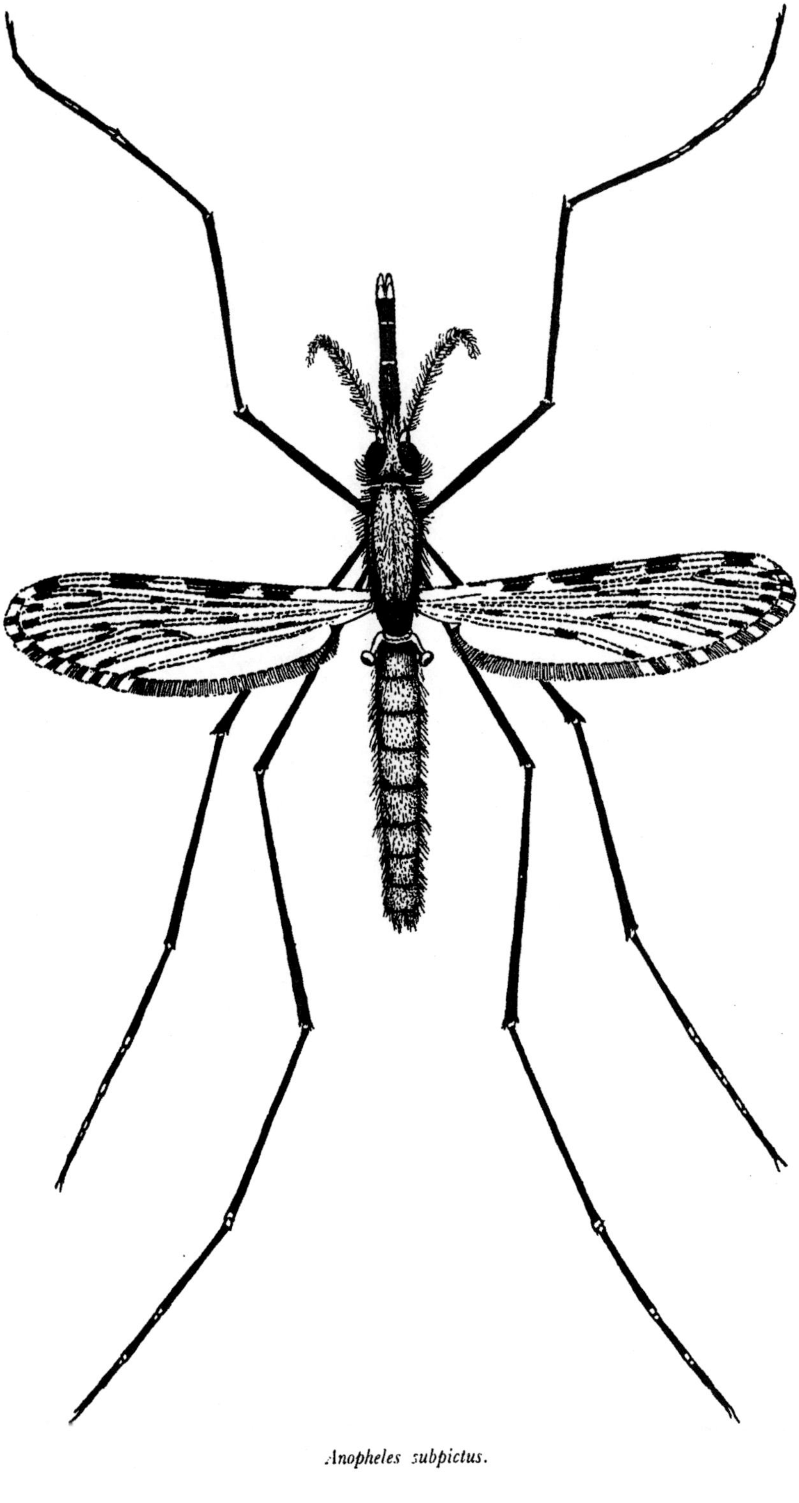

Anopheles subpictus.

MALARIA

Anopheles and Malaria. As early as 1897 it was demonstrated that the *Anopheles* mosquito was the transmitter of malaria. The parasites of malaria live and multiply within the red blood cells and produce both asexual and sexual forms. The latter are called gametocytes and they remain within the blood corpuscles. The parasites are seldom found in the blood before 9 to 10 days after the discharge of sporozoites into the circulation by the feeding mosquito. When the gametocytes reach the alimentary canal of a suitable *Anopheles* host, the sporogony cycle begins. The ookinetes bore through the wall of the insect's stomach, especially its posterior part, penetrating between the columnar epithelial cells until they lie in the space between the epithelium and the elastic membrane covering it. Here they become globular in form called oocysts, which vary from 6 to 12 microns and may even grow to a size of 30 to 80 microns in diameter according to their age. Pigment granules which are more marked in older oocysts are not noticed in mature oocysts. They grow gradually into sporoblasts and when they burst, sporozoites are liberated in the haemocoele. They penetrate the salivary glands of the mosquito and are injected into the wound at the time of biting. Some amount of maturation of sporozoites in the salivary glands takes place before they become infective. With each feed sporozoites are discharged with the saliva, and so there is a natural tendency for the glands to become gradually depleted of infection.

Certain abnormalities in the development of malarial oocysts have been observed by Basu (1947) in serial sections in *A. stephensi.* In some the oocysts, packed with sporozoites, were found developing towards the lumen of the midgut. Such oocysts would rupture into the gut, and the sporozoites liberated would have no chance of invading the salivary glands. It generally takes 4 days for the oocysts to appear on the stomach wall and the glands become infected within 8 to 14 days. Environment, e.g., temperature and humidity, exercises a marked influence on the development of the parasites in the mosquito. To date there is no proof that malaria may be transmitted mechanically by insects or any other mosquito except by the genus *Anopheles.* All the oocysts do not mature at the same time and as they mature and the cyst wall ruptures, fresh infection of the salivary glands takes place. The mosquito once infected retains the infection for a long time. This, however, depends on the frequency at which the sporozoites are liberated from the oocysts. The number of oocysts are also of importance in this respect. It has now been conclusively proved that the development of oocysts can be arrested for 2 or 3 weeks by cold and revived by appropriate warmth. Bruce Mayne (1922) showed that individual insects infected with the malaria parasite could be actually infective to man for 55 days after the original infective meal.

Not all *Anopheles* are suitable hosts for the plasmodia nor are those species which have been incriminated on either epidemiological or experimental grounds of importance in the transmission of malaria.

A mosquito should not be declared as a vector unless sporozoites have been detected in the salivary glands in the wild state as in many nonvectors the development of the parasite stops at the oocyst stage and does not proceed any further.

The Use of Infected Anopheles in Therapeutics. The simple method of inducing malaria by the inoculation of blood from a malaria patient has been criticised because of the possibility of transferring neurotropic strains of Treponema pallida or other infections from which the donor may be suffering. There is also a marked difference in the reaction following the introduction of the malarial parasite into the body of the patient between the two methods of blood injection and mosquito bite. Infected Anophelines have been employed for the transfer of malaria to paretics. For some time it has been known that the high temperatures induced by malarial fevers have a therapeutic value in the treatment of paresis and certain other nervous disorders.

In other countries mosquitoes intended to be employed for transmission of malaria, after they are infected with benign tertian parasites and which are during the course of developing infection, are kept in a cool incubator, adjusted to maintain as closely as possible a temperature varying from 20 to 22°C. (68-76°F.) with a relative humidity of 85 per cent. After they become infective, which seldom takes place before 8 days, they are stored in a Frigidaire at a temperature varying from 2 to 17°C. (25.6-62.6°F), except when they are used for transmission of infection. Such mosquitoes are allowed to feed on a rabbit only once a week. In order to give the insects an opportunity to feed, pledgets of moist cotton and split raisin are kept in each cage.

If the temperature ascends above 22°C., the mortality of the stored mosquitoes progressively increases.

In Calcutta *A. stephensi* have been employed by us for the same purpose. After the infective feed, the mosquitoes are kept in an air-conditioned room maintaining a constant temperature varying between 65-70°F., and are sustained on raisin and water. A temperature of 4°C. is fatal to *A. stephensi.*

Black Spore. During dissections of the stomach for malarial oocysts one frequently encounters characteristic black bodies called black spores, which were first described by Ross in 1898. These structures have been referred to in the literature under various names. They have been thought to have an association with the oocyst of the malarial infected mosquito, be it a culicine or an anopheline. The process of chitinisation may occur previous to the formation of sporozoites or when the cyst is full of sporozoites. Their pleomorphic forms are a marked feature, some of the bodies small and round, others spherical, flattened, banana-shaped, tangled skeins or sheaves of acicles. The shape is thought to be due to the manner in which the chitin has been deposited around oocysts or sporozoites either in a single layer or in successive layers.

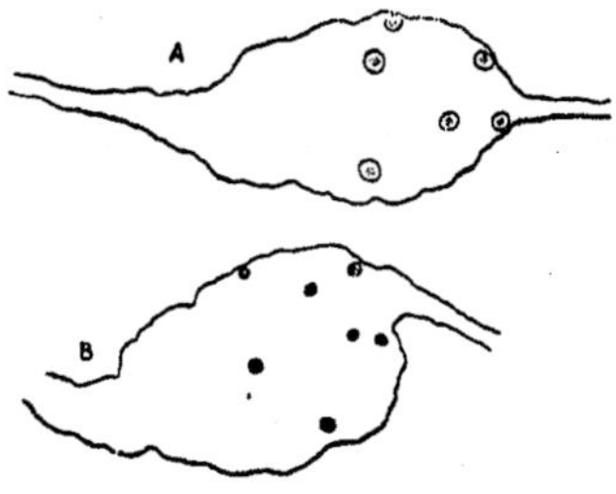

FIG. 35.

Diagramatic representations of (A) Oocysts showing extremely fine pigment granules, and (B) Black-spores in the stomach wall.

While a large majority of workers hold that the chitin bodies represent the products of death and degeneration of the malaria parasite, others believe that the black spores are nothing but the chitinous thickenings of tracheal tubes found in many insects and in mosquitoes other than *Anopheles* (Mayne, 1929). On the other hand, a more marked coincidence was noticed in the appearance of black spores in the wall of the stomach of infected mosquitoes during the rainy season than at other times (Strickland and Roy, 1933).

The following papers give useful information in regard to the formation of black spores in mosquitoes : Brug (1916); Knowles and Basu (1933); Brumpt (1938).

Geographical Distribution of Malaria-Carrying Anopheles. It is not possible to present in its proper perspective a true picture of the distribution of the malaria carriers in the World. Carriers are of different types. Some are classed as efficient vectors as their presence in any locality is always regarded as dangerous. There are others which are poor carriers and they carry under exceptional circumstances, e.g., by their numbers and non-availability of animals on which they normally feed. In some epidemics species which are innocuous may show infection in nature.

However, the following is a brief survey of the distribution of the principal malaria-carriers in Asia.

Palestine—*A. superpictus, A. sacharovi* (*A. bifurcatus,* now controlled); Arabia—*A. stephensi, A. culicifacies*; Turkey—A. *sacharovi,* A. *superpictus*; Syria—*A. sacharovi, A. claviger*; Persia—*A. stephensi, A. superpictus*; Iraq—*A. sacharovi, A. stephensi*; Mongolia—*A. m. atroparvus*; Turkestan—*A. superpictus, A. sacharovi*; Tadzhikistan—*A. superpictus, A. maculipennis*; Tashkent—*A. sacharovi*; Azerbaijan—*A. maculipennis*; PACIFIC OCEAN—Guadeloupe Island—*A. albimanus, A. tarsimaculatus*; Australia—Queensland—*A. punctulatus farauti*; Solomon Islands—*A. punctulatus*; Milney Bay—*A. punctulatus*; Wewak, Lelapipi—*A. p. farauti*; Rabaul, Rama Valley—*A. punctulatus*; Moluccas—*A. punctulatus*; Indonesia—*A. aconitus* (Java, Bali), *A. sundaicus* (Java, coast), *A. maculatus* (Java, hill), *A. hyrcanus* var. *sinensis* (southern Sumatra), *A. leucosphyrus* (forest, Sumatra); Philippines—*A. minimus* var. *flavirostris, A. sundaicus* (coast); Corregidor—*A. minimus, A. sundaicus*; Borneo and Celebes—*A. umbrosus* (west), *A. leucosphyrus, A. sundaicus*; Southern Celebes —*A. hyrcanus* v. *X.*; China—north—*A. pattoni* (hills : presumptive carrier); central—*A. hyrcanus* v. *sinensis*; south—*A. minimus, A. jeyporiensis* var. *candidiensis*; Manchuria—north—*A.m. atroparvus*; west—*A. elutus*; Formosa—*A. minimus, A. hyrcanus* v. *sinensis*; Japan—*A. h. sinensis*; Korea—*A. h. sinensis*; Indo-China—*A. minimus, A. jeyporiensis* v. *candidiensis, A. aconitus, A. maculatus, A. sundaicus*; Malaya—*A. maculatus, A. umbrosus, A. sundaicus*; Thailand—*A. minimus, A. sundaicus, A. aconitus, A. j. candidiensis*; Burma—*A. sundaicus, A. minimus, A. philippinensis*; Burma-China Border—*A. minimus, A. j. candidiensis, A. maculatus, A. h. sinensis*; Andamans—*A. sundaicus*; Arakan—*A. j. candidiensis*; Ceylon—*A. culicifacies*; India—*A. culicifacies, A. philippinensis, A. minimus, A. fluviatilis, A. varuna,* A. *sundaicus, A. stephensi*; Afganistan—*A. superpictus, A. culicifacies*; Kashmir—*A. culicifacies, A. fluviatilis.*

In determining the vector species in any locality the following points have to be taken into consideration. (1) A sufficiently large number of dissections of all species have to be performed. (2) The mere presence of oocysts is without much significance unless infection of the salivary glands is also found. The accuracy of the oocyst rate as a measure to test the suitability of *A. maculipennis* to transmit malaria holds good as far as conditions in Europe are concerned (Swellengrebel and De Buck, 1931). In India where mosquitoes are known to behave in a diverse manner, often in contiguous localities, the oocyst rate alone cannot be taken as an indication of the transmitting capacity of a mosquito unless the association of the particular species with malaria has already been proved. Experimental infections are not now used so much as formerly and less emphasis is placed on the results of such infectivity tests. They merely show the susceptibility of a mosquito to the different species of *Plasmodia* or to any particular strain of parasite. Experimental infections should not be taken as a guide showing the natural capabilities of a species to transmit malaria. (3) In order to act as a vector of major importance the infected species must be present in sufficient numbers during the time of malaria outbreaks and must have a span of life of at least 12 days. (4) A large number of mosquitoes of that particular species should as a rule show infection in the salivary glands. (5) Its high anthropophilic index is a pointer to its inclination to feed more on human than on any other blood.

Regarding the susceptibility of different anophelines in this country it may be pointed out that : (*a*) *A. stephensi*, *A. minimus*, *A. fluviatilis*, *A. varuna* and *A. sundaicus* are particularly susceptible to infection and if they exist in sufficiently large numbers, they are extremely likely to prove harmful. (*b*) *A. culicifacies* is a vector of major importance except in certain regions, *e.g.*, Assam, Bengal and in limited areas in the south of the Peninsula. (*c*) *A. aconitus*, *A. maculatus*, *A. annularis*, *A. subpictus* and *A. jeyporiensis* var. *candidiensis*, although found infected in nature, have only local importance and some, e.g., *A. subpictus* may be neglected altogether.

Malaria and Malaria-Mosquito Survey

A considerable amount of knowledge has accumulated regarding the habits and bionomics of mosquitoes. It has now been fully realised that the dangerous ones are those that bite man and among them many are able to carry diseases. It is therefore logical to hold that attention should be concentrated on the eradication or control of these mosquitoes alone instead of an indiscriminate attack against all. In the beginning Ross advocated anti-mosquito measures against all mosquitoes prevailing in the locality with the idea that not only malaria but all mosquito-borne diseases would thereby be controlled. Such assault against mosquitoes in general cannot, however, be carried out in practice owing to the fact that it would not be financially feasible. It has been demonstrated by Watson in Malaya that it is not only possible but is also economical to eradicate or greatly reduce the incidence of malaria in a locality by adopting measures against the proved carriers only.

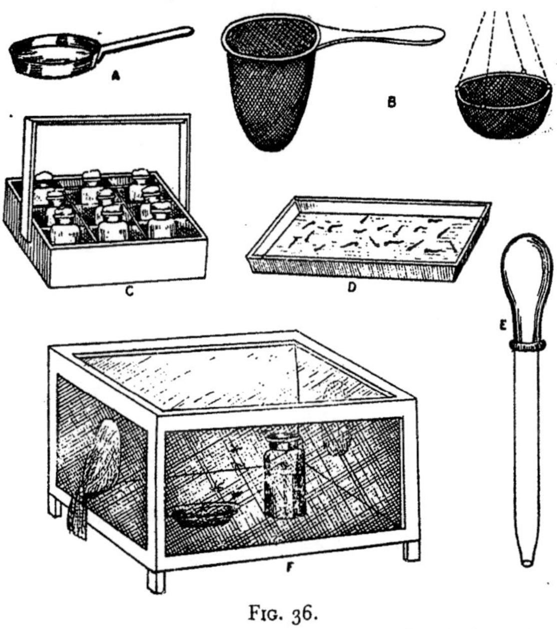

Fig. 36.

Laboratory equipment for mosquito and malaria work. A. Ordinary fry pan. B. Net for catching larvae. C. Convenient way of carrying larvae in bottles. D. Tray. E. Pipette for picking out larvae from tray. F. A glass-topped cage: used for (a) keeping adult mosquitoes alive on raisin and water, (b) for breeding adults from mature larvae or pupae, and also for (c) feeding mosquitoes on the blood of volunteers.

In undertaking species control it is necessary to be provided with adequate knowledge of (*a*) the species that are dangerous, (*b*) their distribution and prevalence, (*c*) their life history and habits, and (*d*) of other factors pertaining to malaria in that particular locality. These constitute a malaria survey which should not only precede but also accompany anti-mosquito measures. In order to judge the progress of anti-malaria operations, quantitative estimation of the incidence of the disease is also necessary. The various factors involved in malaria survey are dealt with below.

(*a*) Collection of anopheline larvae. (*b*) Collection of adult anophelines. (*c*) The study of the habits of the mosquitoes. (*d*) Dissection of mosquitoes. (*e*) Detection of infection in mosquitoes. (*f*) Spleen-index of children. (*g*) Parasite index. (*h*) Infantile malaria index. (*i*) Transmission season. (*j*) Laboratory infectivity experiments. (*k*) Infective density of anophelines. (*l*) Accessory factors, e.g., meteorological conditions, general configuration of the country, economic conditions of the local inhabitants, birth rate, etc. (*m*) Detection of the origin of blood meals in mosquitoes.

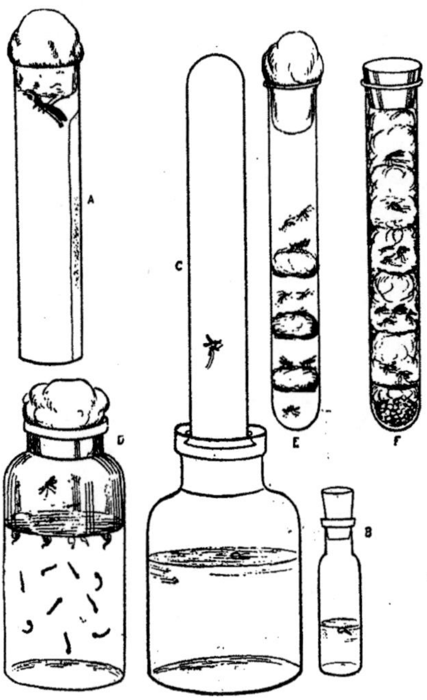

FIG. 37.

A, A glass tube lined with filter paper and well soaked in water is used for inducing mosquitoes to lay their eggs in the laboratory. As far as possible blood-fed wild females should be used for this purpose.

B and C, for collecting larval skin ; mature larvae must be reared separately. The larval skin is preserved in spirit (B), and the pupa is bred out into an adult (C).

D, A wide-mouthed bottle is generally used for breeding a large number of adult mosquitoes from larvae.

E, More than one living adult mosquito can be captured in a single tube.

F, Mosquitoes may be preserved if necessary in corked tubes. At the bottom a few bits of naphthalene are placed ; these will prevent other insects, which habitually prey on such preserved specimens, from entering the tube.

Before the actual work of the survey is undertaken it is necessary to prepare a sketch map of the locality up to an area of at least half-a-mile from the outer perimeter of the area to be controlled and the above data as far as possible should be plotted on the map. The malaria survey should also be extended to that area which is supposed to cover the distance of flight of anopheline mosquitoes in general.

(*a*) **Collection of Anopheline Larvae.** Bearing in mind that *Anopheles* mosquitoes may breed in all types of water such as in pools, tanks, ditches, burrow-pits, cisterns, drains, wells, cattle hoof-marks, grassy-edged streams, irrigation channels, river beds, rain-water puddles, and in puddles in hill streams etc., it is essential that all water collections, large and small, clean or dirty, should be properly searched. The larvae are collected from the edges by means of an ordinary enamel fry pan, against the white back-ground of which the mosquito larvae show up very well. They are transferred by a wide-mouthed pipette to a collecting bottle containing some water and a few aquatic plants from the same source as the larvae. Any handy wide-mouthed bottle will serve as a collecting bottle. Quinine bottles are very commonly used for this purpose. Each bottle should be properly numbered and all records in regard to the breeding habits of different batches of larvae are to be entered separately in the book.

Larvae from wells may be collected with a properly devised net. When larvae are suspected of breeding some distance from the edge, it may be necessary to affix the fry pan to the end of a long handle.

When the larvae have been brought to the laboratory they should be sorted out. Only a few mature larvae are to be bred out separately in wide-mouthed bottles, the open end being closed with cotton wool. Examination must be made daily. The dead larvae and pupae together with larval skins are to be preserved in 80 per cent spirit. Immature larvae, whether dead or living, must also be preserved in the same way. A point to remember is that when larvae are crowded in a small place and when adequate food is lacking, they become cannibalistic. It may be necessary to change the water from time to time and as far as possible natural water should be used.

When the adult emerges it is collected in a test-tube, the open end of which is placed against the open end of the bottle. It is advisable to let the adult die a natural death in the test tube rather than kill it with chloroform. The mosquito then dies with its wings and legs fully stretched.

(*b*) **Collection of Adult Anophelines.** (*i*) For collecting adult mosquitoes, dwelling houses, thatched huts, cattle sheds, pig styes, cellars, etc. are to be searched. The mosquitoes generally take shelter especially in dark places. They often sit on thatch or on strings and soots hanging from the roof. Search should be made, if possible, twice, once in the early morning and again in the evening, otherwise in the morning only. An electric torch will be of invaluable assistance. The usual practice is to place the open mouth of a test-tube over the resting mosquito. In this way mosquitoes may be separately enclosed in different compartments of the same tube by pushing a cotton wool plug as far down as possible after a mosquito has been captured. (*ii*) When the mosquitoes are out of reach, they may be caught with a small suction tube. This is an ordinary tube, the open end being so designed as to allow the mosquito to run in but it cannot escape; to the other end is attached a rubber tube for suction. (*iii*) The upside down umbrella in collecting mosquitoes in thatched ceilings is a modification by Barbar and Rice (1938) of the well known procedure by which mosquitoes killed or stupefied in a room by direct hit with pyrethrum sprays are collected on a sheet spread over the floor. The mosquitoes fall on an umbrella held with the bottom side up under the thatch and in close contact with it. The umbrella should be specially made with white cloth and with a considerably smaller number of gadgets than usual. (*iv*) By using a mosquito trap. Those which give satisfactory results are: (*a*) a good mosquito net inside a torn net with a human bait inside; and (*b*) house screen having a few holes to allow the mosquitoes to enter.

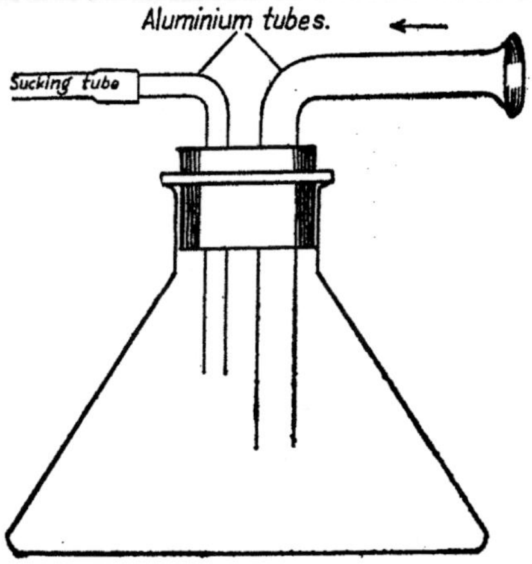

FIG. 38.
Suction apparatus (diagramatic).

(*c*) **Habits of Mosquitoes.** Considerable information of the habits of different mosquitoes particularly in regard to their preference for human or cattle blood, the time of their feed, their daytime resting places, the time of their migration into dwelling houses at night, their breeding habits etc. may always be gathered from the behaviour of the adults. Such information will be of invaluable help not only in evolving the best means of control measures but also in judging the progress of anti-mosquito operations. Mosquito catching stations are widely used as a guide to the mosquito breeding conditions.

(*d*) **Dissection of Mosquitoes. Salivary Glands.** The object of dissection of an *Anopheles* mosquito is to study the sites of development of malarial parasites in its body. These are the salivary glands and the stomach.

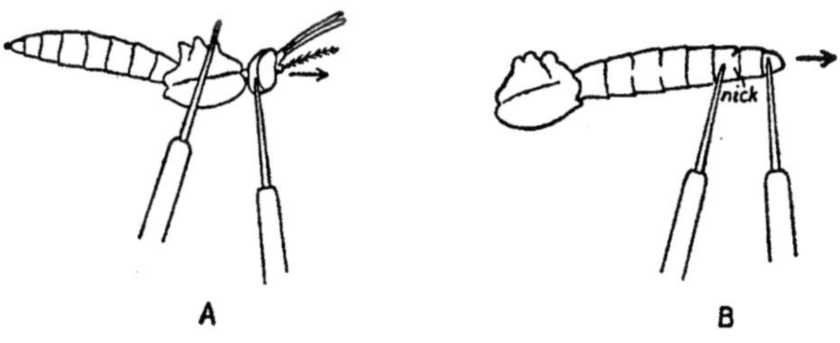

FIG. 39.
Dissection of (A) salivary glands and (B) alimentary canal of a mosquito.

Dead and dried mosquitoes are quite unsuitable for dissection. Fresh ones should be killed with chloroform.

As far as possible sufficient time must be allowed to pass for the digestion of the blood meal otherwise the detection of oocysts in the stomach containing fresh blood becomes a little difficult. Such mosquitoes may be sustained on raisins and water for two or three days. Dissection for salivary glands is better practised in females than in males as the organs are better developed in the former and it is easier to expose the glands. The mosquito should first be identified with the help of a synoptic table; the wings and legs are next removed. In order to eliminate the surface tension it is necessary to immerse the insect in rectified spirit for 10 seconds in a tube and to shake it. Mosquitoes when kept in spirit for a long time are liable to become hard. It is next immersed in normal salt solution in a watch-glass. Dissection is best done with needles under a dissecting microscope against a dark background. The insect is placed on a glass slide in a drop of salt solution with either the head pointing towards the right and its back towards the dissector or the head pointed towards the dissector and abdomen away from him. A blunt needle held by the left hand is placed across the thorax in order to steady the insect and with another needle held in the right hand and placed on the back of the head, gentle traction is exerted on the head away from the body along the axis of the abdomen; simultaneously gentle pressure on the thorax with the other needle will cause the glands to bulge forward. When the head is carefully separated from the body in the way stated above, the glands will be found projecting out of the tissues at the cut end of the neck. The glands are easily recognised under the low power of the microscope by their glandular structure, and their transparent glistening appearance. The glands are now separated from the neck tissue using very clean needles. They may now be covered with a cover-glass, the edges of which are ringed with vaseline to prevent desiccation.

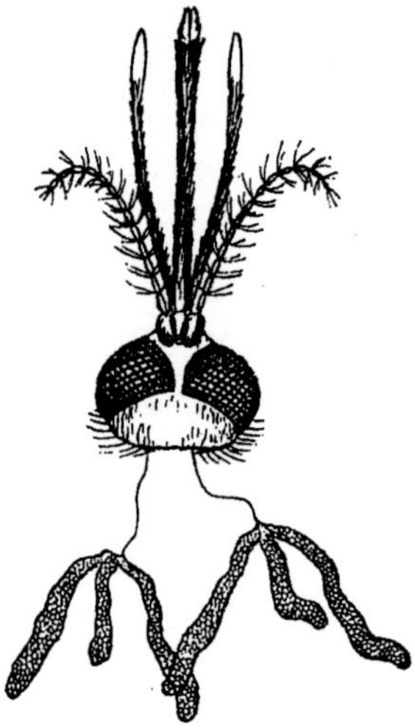

Fig. 40.
Salivary glands of a mosquito attached to the head.

The glands of either side consist of three acini, the ducts of which join soon after leaving the acini to form a single common duct. Oesophageal diverticulae are liable to be confused with salivary glands. They always contain bubbles of gas whereas the latter are glistening, highly refractile bodies and have a uniform glandular appearance.

Oesophageal Diverticulae. The dissection of the diverticula is difficult. The best way to isolate them is by the anterior method, because by the posterior route the sacs nearly always break. The first step is the separation of the dorsal shield, i.e., the chitin and the muscles of the dorsum of the thorax, care being taken not to injure the underlying tissues. There is considerable pressure inside the sacs, which cannot suffer much greater distension than that already existing and even a little pressure with the needle is likely to press out some of the gas bubbles. Success therefore greatly depends on a good and complete separation of the scutum of the thorax. The sternite and the pleura are likewise carefully removed. The chitin of each segment of the abdomen is then split open. The abdomen is cut transversely at the lower level of the ventral sac. The tissues surrounding the gut and the diverticula are then gently separated.

Alimentary Canal. After the salivary glands have been dissected, the same mosquito may be utilised for dissection of the alimentary canal. The mosquito is placed in a drop of salt solution on the slide and two lateral cuts involving the cuticle only are made with a needle between the 7th and 8th abdominal segments taking care not to injure the alimentary canal. The soft tissues may be pressed by a blunt needle to the opposite side before the cuticle is cut. The insect is to be kept steady by placing one needle across the thorax and with the other needle the last segment is drawn away from the body by making steady traction. As this segment is pulled away, the alimentary tract and the reproductive system, which are attached to the hind end of the body, will gradually come into view. In this way the entire alimentary system can be exposed.

(*e*) **Detection of Infection in the Mosquito.** (*i*) **Oocysts in the Wall of the Stomach.** These generally appear on the 4th day after the infective blood feed. They are oval, cannot be detatched from the gut wall by rolling or by flooding, and are placed within the wall of the stomach. They moreover contain distinct pigment granules. They do not take ordinary stains, e.g., eosin, but are easily stained with methylene blue. When mature, they contain a very large number of sporozoites; these escape when the oocyst is ruptured.

The oocyst stage of the different plasmodial parasites can be identified without much difficulty by noting the character and arrangement of the pigment. In *P. vivax* grains of pigment are numerous (more than 30 or 40), lighter brown than in any other species, their rods and specks scattered throughout the parasite into definite pattern. The pigment grains in *P. malariae* are numerous, but larger, thicker and much darker than *P. vivax.* In shape they are more like cocci than rods or specks and tend to be bunched together in a clump round the contour of the cyst. The contour of the oocyst is remarkably well defined in *P. falciparum.* Grains of pigment are few, about 10 or 12, large in size and black. They are characteristically arranged as a streak across the oocyst or as a half circle. In *P. ovale* the grains are more like those of *P. malariae.* They are always arranged in one or other of several definite patterns of which the most common is two rows of dots crossing each other at right angles.

Preservation and Staining of Oocyst. (1) Pass Bouin's fixative between the slide and cover-glass. After 15 minutes carefully lift the cover-glass, pass the stomach through 70, 50 and 30% alcohol allowing 5 minutes for each treatment and thereafter leave the object in distilled water for at least one hour, preferably two. It is then stained in Mayer's acid haemalum for at least 2 hours, preferably the object should be left in the stain overnight. Then wash in water, pass through ascending grades of alcohol 30, 50, 70, 80, 90, absolute alcohol, alcohol and xylol, clove oil, and mount in canada balsam. (2) Place the dissected stomach in a mixture of half formalin and half glycerine for 24 hours; then in glycerine for 24 hours; change and mount in pure glycerine. The margin of the cover-glass may be ringed with coloured wax and thereafter with 95 per cent solution of wax in alcohol. (3) Pour a few drops of saturated solution for 15 minutes when the stomach is likely to adhere to the slide. Leave in distilled water overnight; stain with Mayer's haemalum, then wash in running water. For dehydration, xylol and carbolic acid may be used. Mount in Canada balsam. (4) An infected stomach may be preserved in pure formalin indefinitely. Before mounting, it should be washed in water, stained for about six minutes with haemalum, washed in a current of water and mounted on a slide in the usual way. The stomach after washing in water may be passed through ascending grades of alcohol from 30 p.c. to 80 p.c. and then mounted permanently in Berles' fluid. (5) A fresh stomach preparation should be flattened out by pressing with a cover-glass; the latter is now removed and the salt solution should be allowed to dry up. As soon as the drying is nearly complete, pour two drops of Bless's fluid or Bouin's fixative and allow the fixative to act for 5 minutes. The slide is then reflooded with the fixing fluid and left inside a Petri dish for 1 hour. The preparation is washed in water. Next stain with Delafield's haematoxylin overnight. The specimen will look dark. It should be carefully washed in tap water and examined under the microscope to find out the depth of the staining. Excessive staining is removed by acid alcohol used well diluted and the process watched under the microscope. As the necessary depth of staining is reached, i.e., when the oocysts stand out prominently and the granules inside them are distinctly visible, wash off the alcohol with water; pass through ascending grades of alcohol beginning with 30 p.c. and ending with absolute alcohol; clear in xylol and mount in Canada balsam.

(*ii*) **Demonstration of Ookinetes in the Mosquito Stomach.** The mosquitoes after being fed on a patient bearing a large number of crescents in the peripheral blood, are placed in an incubator, the temperature being 25°C. and 80 per cent humidity. At intervals, between 10 and 20 hours later, one or two mosquitoes are taken out. The entire stomach is dissected out in Locke's fluid. The blood clot is lifted on the point of a dissecting needle and transferred to a drop of Locke's fluid on another slide. The clot is now teased with a needle so that as the blood escapes, it becomes quickly mixed with the Locke's fluid. If the blood is confined within a small area, it will prevent any part of the blood from drying. When the blood and the fluid are thoroughly mixed to form a small drop, several thin films from it on clean slides are made. The films are dried in the air and stained with Leishman's or Giemsa's stain. Search for ookinete should be made with the oil-immersion lens.

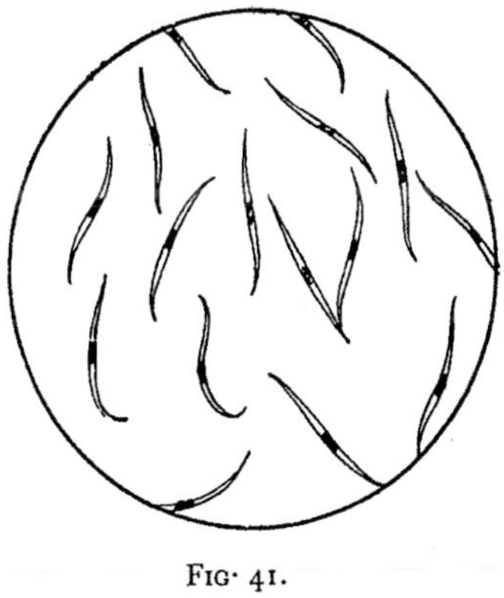

FIG. 41.
Sporozoites stained with Giemsa's stain as seen with oil immersion lens.

(*iii*) **Detection of Sporozoites in the Glands.** The detection of sporozoites in the salivary glands is easy when they are present in large numbers. A heavy infection of the glands is a very common feature unless the mosquito has become old and the infection is dying off due to continued loss of sporozoites with each feed. As soon as the glands are ruptured by pressure the sporozoites escape in streams from inside the gland. When examined under 1/6 inch lens they appear as thin, curved rods and have a characteristic slow movement, on account of which they do not always maintain a uniform shape. Under unfavourable conditions the movements of sporozoites may become sluggish. In size they are twice the diameter of a red blood corpuscle.

Dissection of salivary glands may be done in a solution of 0.5 per cent brilliant cresyl blue in physiological salt solution; this will stain the living sporozoites which will aid the diagnosis of infection in the mosquito.

Sergents' Method of Detection of Infection in the Mosquito. The sporozoites after their liberation from the oocysts into the haemocoele or the "body-cavity" of the mosquito invade all the organs of the body except the ovaries, particularly the musculature of the palps and scutellum which become crammed with parasites, at least for some time (Mayer, 1921; Muhlens, 1921). This has been practically utilized in evolving an easier and much quicker method for the detection of sporozite infection in the body of the mosquito. This is known as Sergents' method (1910) and has been reported upon by Soesilo (1928) and also by Strickland and Roy (1931) as quite reliable for adoption in field malariology. By this method 361 individuals per diem were dealt with by Strickland and Chowdhury (1933). After the avulsion of the legs and wings, the mosquito is placed on a glass slide and the head decapitated by means of a sharpedged needle. A blunt needle is then placed across the thorax, and gentle pressure is applied, so that some "body fluid" mixed with the juice from the crushed salivary glands exudes through the cut end of the neck into a minute drop of normal saline, previously placed upon the slide; the smaller the drop of saline into which the haemocoele fluid escapes, the more quickly one finds the sporozoites, for a smaller area has to be searched. A cover-glass is then applied and the preparation examined under 1/6th inch objective with No. 8 ocular. The sporozoites in fresh specimens are either straight, a little curved, sickle or S-shaped, and are feebly motile.

Staining of Sporozoites and Salivary Glands. (*a*) The slide containing sporozoites in salt solution should be dried in air and then fixed in Merck's methyl alcohol for 10 minutes. When nearly dry, the sporozoites are stained with Giemsa's; the cytoplasm stains blue and the nucleus red. The stain fades away within a short time. (*b*) The salivary glands should be fixed with Bless's fluid for 24 hours and left in Delafield's haematoxylin overnight. Differentiation is done with weak, acid-alcohol. It is then dehydrated in alcohol and mounted in Canada balsam.

(*f*) **Measurement of Enlarged Spleen.** Different methods have been used. Examination for splenic measurements must be made in the erect position. The spleen rate, unless otherwise stated, is meant to indicate the percentage of children aged from 2 to 10 years with enlarged spleen.

(*a*) By *finger-breadth* : This notation is the one most commonly used and for all practical purposes it gives us the necessary information. A finger breadth may be taken as 2 cms. (*b*) *Average enlarged spleen* and *average spleen.* Mark the apex of the spleen below the costal margin and then measure (*i*) the distance in centimeters between this mark and the umbilicus, (*ii*) the distance between the mark and the mid-line, i.e., umbilicus and xyphoid notch line, (*iii*) the distance between the mark and the costal margin. If the apex is below a horizontal line passing through the umbilicus, the measurement is to be recorded as "minus". For the purposes of field work the *average enlarged* spleen can be sufficiently indicated by taking the mean value for (*i*), *i.e.*, all the positive figures for the apex-umbilicus measurement are to be added from which the sum of the minus figures is to be subtracted. The result is to be divided by the numbers of children whose splenic measurements are recorded. Children without palpable spleens are not counted in this process. The *average spleen* is obtained by dividing the sum of (*iii*) all measurements by the total number of children examined, instead of only by the number with recorded splenic enlargement. In the case of (*i*) and (*ii*) the smaller the measurement, the larger the spleen.

The degree of the endemicity of malaria in any locality is generally determined by the spleen index. Thus hyperendemic conditions are defined as those in which the spleen rate in children is permanently over 50 per cent. In moderate or high endemicity it does not exceed 40 per cent and low endemicity is represented by the spleen rate between 0 and 10 (Christophers, 1924).

(*g*) **Parasitic Index.** The parasite index is determined from examination of blood film, either thin or thick, for malarial parasites. The index represents the number of positive cases out of 100 persons examined. The parasite index is as a rule inversely proportional to the spleen index.

Preparation of blood films. Slides for preparing blood films should be first boiled in soap and water, rinsed in water, dried and kept in rectified spirit. For use dry them, do not touch the surface with the fingers; they should be held by their edges while wiping. When ready for use place them on a sheet of paper on a table. Take thick film at one end and thin film on the rest of the same slide. Thin films are stained by Leishman's or Giemsa's method. The rapid method of Singh and Bhattacharji and of Field also gives very satisfactory results.

(*h*) **Infantile Malaria Index.** The study of the examination of blood for malarial parasites in newly born infants during the first attack of fever after their birth is useful from several points of view, the most important being the determination of the actual season of transmission of malaria, also whether different seasons are concerned in the transmission of the different species of parasites.

(*i*) **Transmission Season.** The transmission season is determined from (1) the malaria index in newly-born babies, (2) from sporozoite indices in wild mosquitoes correlated with different seasons, and (3) from artificial infectivity results carried out in room conditions.

(*j*) **Artificial Infection of Anopheles.** Mosquitoes may either be bred out from eggs laid in the laboratory or they may be reared from larvae caught in the field. In India where larvae are so easily obtained, the latter method is more convenient. The adult mosquitoes may be sustained on split raisins and cotton wool soaked in water till they are required. Mosquitoes live much longer if kept in a cold place than in the laboratory room, preferably at a temperature of 65-70°F. in the presence of a moderate amount of humidity.

Whether a person harbouring parasites is suitable for infecting mosquitoes will depend principally on the presence of sufficient numbers of mature gametocytes. At one time stress was laid more on the number than the quality. It is now held that the presence of the flagellating forms of the parasites will alone determine their development in the stomach of the mosquito provided other conditions are favourable. It is also noticed that gametes may exflagellate *in vitro* and yet fail to infect.

[For enumeration of gametocytes by Sinton's method, see *Ind. J. Med. Res.*, **15**, 341, 1924.]

It is necessary to point out that if other conditions are favourable, one blood meal will be sufficient to infect mosquitoes. For artificial infectivity experiments we seldom allow the mosquitoes more than one feed on a volunteer.

From the fourth day onward the dead mosquitoes are examined for oocysts but the living ones should not be touched before the 9th day. If the glands do not show signs of infection till the 15th day, no further dissections should be performed.

The time required for the complete development of the parasites in anophelines is usually given as 9 or 10 days, but it is subject to variations depending on the temperature to which the mosquitoes are exposed. It has been shown by Jansco (1904) and others that cessation of development of oocysts takes place when the atmospheric temperature remains below 60°F. King (1929) found in his experiments that the shortest period from the time of the blood meal to the invasion of the salivary glands with sporozoites was 12 days with malignant tertian parasites at a mean room temperature of 79°F., and 14 days with benign tertian parasites at a mean temperature of 74°F. James (1931) observed that *A. maculipennis* became infected with benign tertian parasites in 15 to 23 days when the mosquitoes were kept in a room in England in July and August. Control batches in the incubator at 76 to 80°F., dry bulb, and 74 to 78°F., wet bulb, showed sporozoites after 9 days.

An idea of the time for the malarial infection to appear in the salivary glands of *A. stephensi* under the normal room conditions in Calcutta can be obtained from the work of Strickland, Roy and Chaudhuri (1933). Knowles and Basu (1936) made elaborate studies on the infection time also in *A. stephensi* but the mosquitoes after the infective feed were subjected to different controlled conditions of temperature and humidity.

The following factors are known to influence the infectivity of a mosquito: (1) The unknown intrinsic factors which favour the complete development of the parasites in certain mosquitoes, partial development in some and absence of development in others. (2) Density of gametocytes. Different observers have noted different numbers of gametocytes as the minimum necessary to infect mosquitoes with different species of *Plasmodia*. Thus Strickland, Roy and Chaudhuri (1933) and Knowles and Basu (1943) give 40 per cmm. in the case of *A. stephensi*. Green (1929) succeeded in infecting *A. maculatus* in Malaya with only 10. (3) The simultaneous presence of male and female gametocytes in equal numbers in the peripheral blood has been stressed by Boyd (1930). (4) The flagellating forms of gametocytes may be taken as an indication of their maturity and the presence of mature gametes is perhaps more important than immature forms. (5) The co-existence of suitable temperature and humidity is regarded as the principal dominating factor influencing the infection in the mosquito. The findings recorded by Strickland, Roy and Chaudhuri (1933) clearly indicate that mosquitoes cannot be infected during April, May and June in Calcutta as the temperature is too high. Although Bentley (1911) has laid stress on humidity more than on temperature as a factor in the infection of mosquitoes, later observers like Gill (1921) and Wenyon (1926) regarded temperature as a much more important factor than humidity, provided the latter is not too low to endanger the life of the mosquito seriously. (6) In order that the parasites may develop to the sporozoite stage the mosquito must live for at least 10 to 20 days.

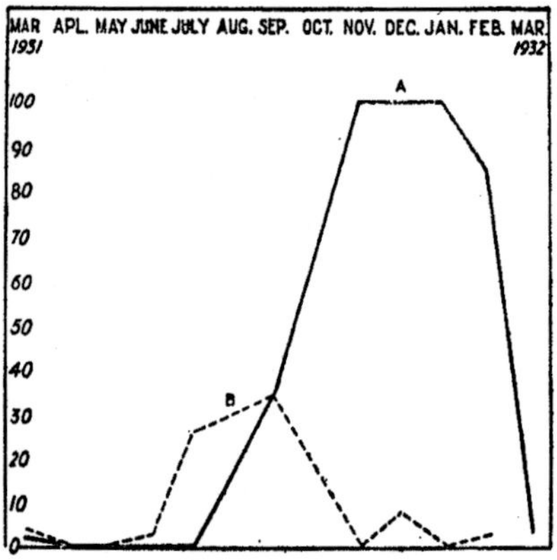

FIG. 42.

Malarial infection of the gut of *Anopheles stephensi* in Calcutta. (After Strickland and Roy).

A. Sporozoite rate; B. Zygotes showing chitinous bodies, rate%.

(*k*) **Infective Density of Anophelines.** Davey and Gordon (1933) have described a method for use in interpreting the anopheline density and sporozoite infection rate of infective anophelines. Infective density is the average number of infective anophelines per room per night.

(*l*) **Source of Stomach Blood of Mosquitoes.** The precipitin test is the only means of obtaining exact information of the feeding habits of blood-sucking insects. The principle of this test is that when clear serum extracted from dried blood from the stomach of insects is brought into contact with a series of sera containing precipitin for blood of the possible animal sources, a precipitation or flocculation occurs in the specimen in which the blood and antiserum are homologous. At present the ring test of Fornet and Müller is universally applied for the detection of the precipitation. The precipitating sera are produced in either rabbits or fowls. [For details of the preparation of the serum, consult Roy and Ganguli (1943). *Ind. Med. Gaz.*, **78**, 81.]

Malaria Control

This is classified under the following headings : (I) Personal prophylaxis. (II) Destruction of larvæ. (III) Destruction of adult mosquitoes. (IV) Drainage. (V) Drug prophylaxis. (VI) Cattle prophylaxis. (VII) Naturalistic method.

(*I*) **Personal prophylaxis.** (*a*) Use of mosquito nets : Nets should be neither too small nor too large. Nothing coarser than a mesh of 16 per inch is effective for protection against *Anopheles*. (*b*) Use of fans, trousers, boots, etc. (*c*) Repellents or culicifuges : (*i*) Bamber oil—Oil citronella 1 part, liquid paraffin 1 part, cocoanut oil 2 parts, carbolic acid 1 per cent. (*ii*) Citronella oil, lemon grass oil, pyrethrum extract diluted in kerosene and mixed with cocoanut oil. (*iii*) Pure dimethyl phthalate to which a little Java citronella is added is the best repellent for mosquitoes and midges. It should be remembered that the protective power of dimethyl phthalate is diminished when mixed with other substances like diethyl phthalate, dibutyl phthalate, pine oil, bentonite and pyrethrum solution. (*d*) Screening of houses : In the tropics with much heat and high atmospheric humidity and particularly near the coast a mixture of copper and tin for making the wires is the best. The aperture of the screen should not be over 0.046 inch sq.

(*II*) **Destruction of larvae by larvicides.** (*a*) Paris greening : Paris green, at one time the most widely used larvicide, is a double compound of the arsenite and acetate of copper. It is insoluble in water. In the dosage used, it is harmless to man and animals. Paris green floats on the surface of water and is readily taken by anopheline larvæ but not by culicines which are bottom feeders. With the introduction of DDT, Paris green is seldom used now. (*b*) Oiling: Previous to the discovery of pyrethrum and DDT, anti-larval measures were the principal sheet anchor of anti-malaria measures. Taking other things into consideration anti-larval measures were then the least expensive. Kerosene and its substitutes were the most important substances used against mosquito larvæ. In India "Malariol" was once extensively used.

(*III*) **Destruction of adult mosquitoes by the spraying method.** (*a*) Pyrethrum : The active principles, pyrethrins I and II, are contained mostly in the flowers of *Pyrethrum cynerariafolium.* Fully mature flowers are harvested, dried and powdered and pyrethrin is extracted in petroleum ether. The extract is diluted with kerosene to 0.1 per cent solution before use. It is essentially a space spray and hence used in a closed room. It is deadly poisonous to insects in general. It has a marked and an immediate effect but no residual effects. (*b*) Freon-Pyrethrum Aerosol or Freon Bomb : This is a mixture of liquid freon 12 with 0.8 per cent wax-free pyrethrins and 6.0 per cent sesame oil and is contained in an iron drum with a suitable nozzle. The aerosol cloud from one bomb is sufficient to kill all flies and mosquitoes in 1,000,000 c.ft. (*c*) DDT : Amongst a large number of synthetic insecticides that have been prepared, only two, DDT (Dichlordiphenyl-trichloroethane) and B.H.C. (Benzene hexachloride), are being extensively used for controlling a large variety of insect pests. DDT is regarded as an insecticide of outstanding value for the control of insect-borne diseases, especially malaria and plague. It presents the particular advantage over others for its continued action lasting for days, weeks and months after a single application. The term residual insecticide has been applied to it on account of this property.

DDT acts slowly and an insect, though exposed to the lethal dose, yet takes some hours to die. The mosquito acquires the minimum lethal dose when it rests for a sufficiently long time on a surface on which DDT has been sprayed. The length of contact necessary to produce the lethal effect varies greatly according to the period which has elapsed since the time of application.

DDT in addition to being a contact repellent, also acts as a direct repellent. This is manifest from the mosquito's disinclination to settle on places which have been treated with DDT solution.

It is used in 5 per cent *solution* prepared in kerosene. It is also used as an *emulsion* which is prepared by dissolving DDT in xylene and emulsifying the solution with triton x 100. For *suspension* concentrate, gum acacia, even as high as 10 per cent. may be used.

The dosage for prolonged residual action lasting several months is near 200 mg. per sq. ft. The residual action of DDT varies widely according to the nature of the treated surface, exposure to dust and light, moisture and above all to the resistance of individual insects to the action of the insecticide. However, it is an accepted fact that anophelines compared to culicines are most susceptible and Aedes the most resistant not only to DDT but to other insecticides in general.

In recent years considerable light has been thrown on the behaviour of the different Anophelines particularly those that frequent domestic houses against DDT and

Gammexane. Some species are able to enter the treated rooms and bite but may not rest on the wall afterwards and thus avoid imbibing the lethal dose.

The control of mosquitoes by using DDT and other chemicals has been exhaustively dealt with in Part II.

Estimation of DDT in Wall Scraping. The scraping from 4 sq. in. of the surface and measuring 200 to 500 mg. are taken in a test tube and extracted with 2 c.c. of benzene. It is filtered and the residue re-extracted twice with 2 c.c. of benzene and again filtered. The combined filtrate is evaporated to dryness on a water bath and mixed with 5 c.c. of N/2 alcoholic KOH and refluxed on a water bath for 30 minutes. The mixture is then cooled to room temperature and made acidic with 5 c.c. of 2 N. HNO_3., 5 c.c. of N/100 Ag NO_3 and 2 c.c. of pure nitrobenzene are added and the mixture titrated with N/100 NH_4CNS using 2 c.c. of 5 per cent solution of ferric alum as an indicator. A control is run at the same time with the same quantities of the reagents used as above.

Let c.c. of N/100 NH_4CNS required for the blank .. b.
c.c. of N/100 NH_4CNS required for the sample .. a.
factor of normality of NH_4CNS .. f.

Then mg. DDT per sq. ft. of the wall

$$\frac{(b-a) f \times 0.355 \times 10 \times 36 \times 1000}{1000} = (b-a)f \times 127.8.$$

Estimation of p-p″ DDT in the Technical Powder. 2.0 gms. of DDT technical are dissolved by boiling under reflux in 150 c.c. of a saturated solution of pure p-p″ DDT in 75% ethanol. When the solution is complete, the liquid is allowed to cool down and kept for 4 hours in a thermostat bath at 30 ± 0.1° C. with occasional shaking. The mixture is filtered through tared Gooch crucible keeping the crystals always under the liquid surface. When the filtration is over, the residue is washed with 20 c.c. of the saturated solution and then dried in the air oven to constant weight at 80-90° C. The melting point of the crystals is determined and for every degree lowering of the melting point below 109°C., 2·5% is subtracted from the percentage found. The percentage of pure p-p″ DDT in a good sample of DDT technical varies from 60-90 per cent. (O-p'-DDT has 25 per cent of the insecticidal activity of p-p″DDT).

DDT technical, which is sold in the market as DDT, is a mixture of the above two compounds with a small quantity of DDD (Dichlorodiphenyl dichloroethane).

Alessandrini (1950) has described a rapid and a fairly sensitive colorimetric method which may be easily applied, even in the field, with limited facilities and which makes it possible to detect and determine small amounts of DDT on various surfaces treated with this product in the forms of solutions in kerosene, aqueous emulsions, wettable powders, aerosols or in any other form. A colour-scale of great practical value, particularly in the field, makes it possible to read off directly amounts of DDT between 0.0001 and 0.005 g. corresponding to the colourations obtained by the above method. This has been put to extensive field trials in various parts of Italy and has been reported as a fairly dependable method.

DDT as larvicide. 5 per cent. DDT solution in oil may be used for destroying mosquito larvæ. Emulsions of DDT with turpentine as a solvent or suspension of DDT are more effective than oil solutions as larvicides. When used in dust form it is about 25 times as toxic as Paris green. Control of anopheline larvæ can be obtained with 0.1 pound or less per acre. When used against culicines a much higher dose is required.

DDT in powder form or in suspension is the ideal larvicide for use against anopheline larvæ in filter beds. In the dosage used for this purpose it is totally harmless to man and to aquatic insects.

Gammexane. The discovery of this very useful insecticide was made in the laboratories of the Imperial Chemical Industries in 1942. It is now used all over the world as an insecticide which has its applications in medicine and public health, in veterinary practice, in the preservation of food grains and in plant protection against insects. Though it has been used against adult anophelines and declared to be superior to DDT as regards its immediate effects, the residual properties are decidedly inferior. It has also been used against mosquito larvæ but does not seem to possess any greater advantage over DDT. [For details see in Part II] On account of its peculiar smell, it is not very popular as a household insecticide.

(*IV*) **Drainage.** Drainage for the control of malaria is a permanent measure and it is the most economical way by which mosquito breeding can be totally eliminated. If ill-planned and ill-executed, it may prove a source of great nuisance and it itself may lead to breeding sometimes of a dangerous type of mosquito. Antimosquito drainage may be accomplished in a great variety of ways : namely by the use of surface drainage, underground drains, vertical drains, pumps, sluices, tide-gates and other devices.

(*V*) **Mass drug prophylaxis.** Prophylaxis by drugs is only possible if it has effect on the sporozoite itself or on the stage or stages of the parasite before its manifestation in the peripheral circulation. There is yet no drug known which can render absolute protection to an individual against malarial infection. The anti-malarial drugs have therefore mostly therapeutic values. As some of them are endowed with the property of destruction of gametocytes, they can be used for gametocyte prophylaxis. Suppression of malaria by drugs has certain advantages. It is the best method of dealing in sporadic or endemic localities. In places where anti-malaria operations cannot be instituted or have not been instituted, it can be tried during the transmission season only. A material reduction of fever rate, spleen and parasite rates and death rate, always follows institution of such measures in a community. With the cessation of drug administration, however, there is sharp rise of these indices.

(*a*) **Quinine, Plasmochin and Atebrin.** The gametocidal action of plasmoquine once offered the only possibility of grappling with the problem of anti-malarial sanitation from a new angle, that is, by drug treatment of the human being, whereas previously the only methods available were protection from mosquitoes and extirpation of the *Anopheles,* since quinine has proved useless in this connection. It was found that with drug treatment, e.g., plasmoquine, mepacrine, quinine, etc. it was impossible to eradicate malaria parasites or to reduce them to the point at which transmission was lessened. No doubt with mepacrine alone it had been possible to maintain the armed forces fighting against the Japanese in highly malarious areas in the Pacific and East Asian regions in a state of fitness. The dosage employed was 0.1 gm. daily. (*b*) **Paludrine**: The recommendations of the Medical Research Council on Paludrine published in the *British Medical Journal* of April 2, 1949, emphasize the value of a single weekly dosage of 0.3 gm. of paludrine for prophylaxis and suppression or for clinical cure of malaria in village or rural populations. For children between 5 and 12, the dosage should be 0.15 gm., and 0.075 gm. for children between 2 and 5. years (*c*) **Chloroquine** (SN 7618) : Though its properties are somewhat identical with those of mepacrine, it, however, possesses about three times the activity of the latter. Field trials have been conducted in dosage of 1 tablet containing 0.48 gm. which is adequate in suppressing malarial attacks. The drug has, however, a toxic action. (*d*) **Cam-Aqui or Camaquin** : is another quinoline derivative which in avian malaria has proved several times more effective than quinacrine, while the toxicity in laboratory animals is only half. A single dose of 0.5 gm. of CAM-AQI may be employed as a suitable remedy for mass treatment of malaria in rural areas. (*e*) Primaquine, has also marked curative and preventive properties.

The paper by Chaudhuri, Chakravarty and Chaudhuri published in the *Brit. Med. Jl.*, 1952, **1**, p. 568 gives a review of drug prophylaxis in malaria.

(*VI*) **Cattle prophylaxis.** The chief factor in the epidemiology of malaria is the proportion of mosquitoes that succeed in biting human beings, and this factor must certainly be influenced by the presence of cattle. The close association with cattle does not always prevent a human epidemic of malaria but it is also true that a ring of cattle sheds on the outskirts of a village with dwelling houses in the centre would reduce the chances of malaria.

It has been clearly demonstrated by Walch (1932) that the anthropophilic index of a species of anopheline varies greatly in localities where cattle are present and where they are scarce.

(*VII*) **Naturalistic measures of anopheline control.** Such measures of malaria control have been reported from various parts of the world. The measures are chemical, physical and biological. The most significant chemical measure is the alteration of the pH of the water by organic substances. Naturalistic methods of malaria control include : (*a*) *Fouling the water* with sullage, sewage and industrial wastes of vegetable matters etc., in order to make it unfit for anopheline breeding. It is not, however, a practicable proposition. This always leads to a great increase of non-carriers and culicine mosquitoes. (*b*) *Alteration of salt content of water.* The alteration of the salt content of the water in some cases has also been found useful. (*c*) *Physical measures* : Physical measures are many, such as removal of shelter of larvæ by eliminating aquatic vegetation, deepening the edges of tanks and ponds, converting marsh into numerous small and tidy tanks stocked with fish and confining water into definite channels in which it can flow freely. (*d*) *Shading* : Shading is a measure extensively practised in Malaya in dealing with seepages which are a prolific source of *A. maculatus* in the hills. On the other hand, the breeding of *umbrosus* which takes place in peaty marshes in virgin jungle, can be eliminated by clearing the forest. In Assam shading the edges of perennial streams by growing jungle as a measure of controlling the breeding of *A. minimus*, has proved successful. This has also been successfully practised against *A. fluviatilis* in Southern India. (*e*) *Herbage packing*: This method of malaria control aims at the suppression of the breeding of the malaria carriers by altering the quality of the water. It is well known that the malaria carriers, with the possible exception of *A. sundaicus*, prefer clean water, and some of them exceptionally clean water. Williamson was the first to experiment with herbage packing in hill streams in Malaya against the breeding of *A. maculatus*. Herbage packing has only a theoretical value. (*f*) *Extension of agricultural activities* will lead to betterment of the economic conditions of the people. (*g*) *Biological control by fish* : The employment of fish in the control of mosquitoes is nothing new. During the aquatic life of a mosquito, it is vulnerable to attacks by its enemies, among which fish may be considered as the most important. It has its limitations as the movement of the fish is greatly restricted in the presence of aquatic vegetation. They, also, do not particularly attack mosquito larvæ in the presence of other types of food. *Lebistes reticulatus*, or "millions," have been successfully employed in Barbados and *Gambusia affinis* in America and other places. In Java, both *A. sundaicus* near the coast, and *A. aconitus* in rice fields have been controlled cheaply and effectively by this method.

Fish is particularly useful for controlling mosquito breeding in small confined spaces such as wells, filter beds, pits, cisterns, etc. *Anabas scandens* and *Saccobranchus fossilis* are ideal for employment in wells. *Gambusia affinis*, which is an imported variety, may be used in filter beds, cisterns etc. but if food is scarce, they will have to be fed.

IDENTIFICATION OF FEMALE ANOPHELES SPECIES OF INDIA

General Characters of a Mosquito. (*a*) It is a delicately built midge-like insect. (*b*) It has long filamentous many-jointed antennae. (*c*) Proboscis is long and adapted for piercing the skin and for sucking blood. (*d*) Body and wings are clothed in scales. (*e*) Wing venation is characteristic, e.g., the second, fourth and fifth longitudinal veins are branched. (*f*) Presence of a row of scales (wing fringe) along the whole length of the posterior border of the wing.

Identification of Female Anopheline Mosquito. Sexes are differentiated from antennae which are plumose in males and pilose in females. The female *Anopheles* has long, rod-like palpi whereas in the male *Anopheles* the palpi are clubbed at the end.

Instruction for Examination of Anopheline Mosquito for Identification. (*a*) It is very convenient to examine a mosquito when it has been mounted on a pin. (*b*) When there are no facilities for pinning, it can be examined on some teased cotton wool. The legs and wings are very important for identification and therefore they must be fully stretched taking care that this does not cause any damage to either the leg or the wing. It must be remembered that scales on the wings when rubbed off will render identification extremely difficult. The specimens are examined with a hand lens x 15 unless a dissecting microscope is handy. Examination of minute structures like hairs and scales is always made under $\frac{2}{3}$ objective of a compound microscope. (*c*) It is essential that mosquitoes should always be examined in subdued light. When examined in ordinary light, banding and speckling may be missed. It may even be necessary to interpose the hand or a paste board in order to cut off light. (*d*) For the identification of an unknown *Anopheles* proceed first with the key and for confirmation use the table. (*e*) Mosquitoes have been placed under six different groups according to certain distinctive characters common to every member of that group.

Certain terms must be properly understood before any attempt is made in the examination of a mosquito. (*i*) Banding on palpi. There are generally three pale bands on each palp and these are arranged as follows. When examined from the tip of the palp there are two white bands placed near to each other and close to the apex; the third band is placed near the base. The first band is known as the apical pale band and the second band, the subapical pale band (Fig 33). In a few species there are four pale bands on the palp. (*ii*) Speckling of legs and palpi. Speckling is due to numerous minute white spots on a dark background. (*iii*) Base of wing is that part of the wing which is in contact with the thorax. The basal part of the costa means that part of the anterior border of the wing which lies in close proximity to the thorax. (*iv*) White-footed mosquitoes, *i.e.*, those which have at least the last tarsal segment of the hind legs white. (*v*) Unless otherwise mentioned abdominal scales mean scales on the dorsum of the abdomen. (*vi*) "Subcostal pale area" is the pale area situated almost at the middle of the anterior border of the wing. This is of some importance in the identification of *A. annularis*. (*vii*) Hair. It has a round circumferance and is of nearly uniform thickness throughout. Scale. A scale is a modified hair and is flat.

N.B.—All figures relating to the identification of *Anopheles* adults refer to those on pp. 65-67.

TABLE FOR THE IDENTIFICATION OF ADULT ANOPHELES OF THE FEMALE SEX ONLY OF INDIA

Mosquitoes have been placed under six different groups according to certain distinctive characters common to every member of that group.

GROUP I

Chief characters. **Costa uniformly dark**, i.e., **not interrupted by any pale spot ; also no white spot on the wing field.**

A. Conspicuous white scaling on vertex of head; milk-white frontal tuft present; white scales on the anterior third of the mesonotum; distinct white banding at the distal end of hind femur.

A. barianensis. (Fig. 1).

B. No white scaling on vertex of head; white frontal tuft absent; mesonotum of uniform colour and scaleless; banding at the distal end of hind femur absent.

(*a*) Head scales very long, narrow, erect and notched at the free end; anterior forked cell much larger than the posterior; long filamentous palpi. *A. aitkeni.* (Fig. 2).

(*b*) Head scales normal and broad; anterior forked cell slightly larger than the posterior. *A. culiciformis.*

GROUP II

Chief characters. **Less than 4 dark spots involving the costa, subcosta and the first longitudinal vein. Femora and tibia not speckled.**

A. Palpi with distinct pale bandings.

(*a*) Inner quarter of costa with pale interruptions; a large pale spot on the wing fringe between veins 5.2 and 6. *A. gigas.* (Fig. 3).

(*b*) Inner quarter of costa dark without any pale interruptions; wing fringe between veins 5.2 and 6 dark; a broad golden spot at the tip (apex) of the wing; the proximal half of vein 6 pale. *A. hyrcanus* var. *nigerrimus.* (Figs. 4 & 35).

(In *A. hyrcanus* var. *sinensis* a fringe spot at 5.2 present).

(*c*) Presence of prominent tuft of scales, black above and white below, about the femoro-tibial joint of hind legs. *A. annandalei.* (Fig. 25).

B. Palpi unbanded.

(*a*) Hind femur with a conspicuous broad white band. *A. lindesayi.* (Figs. 5 & 26).

(*b*) Hind femur without any band.

(*i*) Palpi shaggy; fringe spot at 5.2; scattered black scales on the proximal half of vein 6; presence of both dark and white scales on wing veins; 7th abdominal segment with a conspicuous tuft of black scales on the ventral surface. *A. barbirostris.* (Figs. 6, 19 & 36).

(*ii*) Palpi thinner; fringe spot at 5.2 absent; proximal half of vein 6 pale; only dark scales on the wing field; 7th abdominal segment without tuft of scales. *A. umbrosus.*

GROUP III

Chief characters. **At least 4 dark spots involving the costa, subcosta and the first longitudinal vein. Femora and tibia not speckled. All are dark-footed, i.e. at least the tips of the hind legs are dark.**

A. Apex of palpi dark (not definitely white).

(*a*) Only 2 indefinite dark spots on vein 6, the distal one being very long; extreme base of costa dark; complete absence of dark scales on the mesonotum. *A. turkhudi.* (Figs. 8 & 37).

(*b*) Three definite dark spots on vein 6; extreme base of costa pale; scattered scales on mesonotum. *A. multicolor.* (Fig. 7).

(*c*) Palpi inconspicuously banded; mesonotum scaleless; no pale scales on wing veins except on costa and vein 1 and no fringe spot present. Head scales narrow and erect. Winter specimens having 3 dark spots involving the costa and the first longitudinal vein may be found. *A. dthali.*

B. Apex of palpi white.

I. Two broad pale bands of equal length at the apex, the intervening dark area being narrow.

(*a*) Costa with one pale interruption at the base; no fringe spot at vein 6; proboscis uniformly dark. *A. minimus.* (Figs. 9 & 38).

(*b*) The basal area of the costa usually with an interruption, fringe spot at vein 6 present; proboscis golden at its distal half. *A. aconitus.* (Figs. 11 & 40).

(*c*) The basal area of the costa without any interruption; no fringe spot at vein 6; proboscis may have a light golden appearance at the distal part. *A. varuna.* (Figs. 10 & 39).

II. The pale apical band nearly of the same size or a little broader than the subapical pale band, the intervening dark area being much broader than either of the two bands.

(*a*) Nearly the whole of the third vein dark.

(*i*) Only two fringe spots 4.2 and 5.1 present; basal dark area on the costa interrupted by a white spot opposite which the first longitudinal vein is dark; mesonotum scaleless.
A. culicifacies. (Figs. 12 & 41).

(*ii*) Fringe spots at all veins except the 6th; basal dark area on the costa uninterruptedly dark opposite which there is a large pale area on the first longitudinal vein; palpi comparatively long, about $1\frac{1}{2}$ times the length of the thorax. *A. sergenti.*

(*b*) Nearly the whole of 3rd vein pale.

(*i*) Inner third of costa uninterruptedly dark. No fringe spot opposite vein 6.

(*α*) 2 dark spots on vein 6, the distal one being fairly long; tibio-tarsal joints dark; scales confined to the anterior third of the mesonotum.

Winter variations from the type are known. Such specimens may show 3 dark bands involving the costa and the first longitudinal vein. In some specimens the 6th vein may be all dark. Palpi in winter specimens of *A. fluviatilis* may occasionally show some abnormalities in the banding.
A. fluviatilis. (Figs. 13 & 42).

(*β*) 3 dark spots on vein 6; tarsal joints of front legs narrowly banded; white scales distributed over the whole mesonotum along the middle line. A. *moghulensis.* (Fig. 43).

(*ii*) Inner third of costa interrupted by pale spots.

(*α*) Tarsal joints of front and midlegs narrowly banded white; white scales in the form of a tuft in the middle of the mesonotum; fringe spot at vein 6; 3 dark spots on vein 6.

A. jeyporiensis var. candidiensis **is distinguished from the type form by markings on the palp (Fig. 45).**
A. jeyporiensis. (Fig. 14, 32 & 44).

(*β*) Tarsal banding and mesonotal tuft absent; no fringe spot at vein 6; 2 dark spots on vein 6, the distal one being fairly long; palpi long and thin; mesonotum heavily scaled.
A. superpictus. (Fig. 47).

III. The pale apical band very much broader than the subapical pale band, the intervening dark area being either of the same size as the apical pale band or smaller than the latter; front tarsal joints broadly banded. A prominent T-shaped dark spot involving the costa, subcosta, and the first vein about the middle of the anterior border of the wing.

(*a*) The dark intervening area of the same length as the pale apical area.
A. subpictus. (Figs. 15, 27 & 46).

(*b*) The dark intervening area extremely narrow, even less than half the apical pale area.
A. vagus. (Fig. 48).

GROUP IV

Chief characters. **At least 4 dark spots on the costa, etc., as in Group III. Femorae and tibiae speckled. Dark-footed series (tip of hind legs definitely dark).**

A. Three pale bands on the palp; 3 or less than 3 dark spots on vein 6.

(*a*) Two equally broad white bands at the apex, the intervening dark area being very small; palps speckled (easily seen in fresh specimens); thoracic scales very broad. *A. stephensi.* (Fig. 49).

(*b*) The apical pale area is much broader than the subapical pale band which is very narrow, and the former is of almost the same length as the intervening dark area. Thorax is mostly hairy, no broad scales being present. *A. sundaicus.* (Fig. 50).

B. Four pale bands on the palp. More than 3 dark spots on vein 6.

(*a*) Broad tibio-tarsal white band on the hind leg. One broad and three narrow white bands on the palp as seen from the tip to the base. *A. leucosphyrus.* (Figs. 24 & 51).

(*b*) Three broad apical and one narrow band; the three apical bands being placed close to each other and separated by two narrow black bands; apical half of proboscis golden yellow.
A. tessellatus. (Fig. 52).

GROUP V

Chief characters. **At least 4 dark spots on the costa etc. as in groups III and IV. Femora and tibia not speckled. White-footed series (tips of hind legs at least white).**

A. At least the third, fourth and fifth tarsal segments of hind legs completely white.

(*a*) 5th vein mainly dark with a dark spot at its bifurcation, the dark area extending over the anterior branch for a short distance; the subcostal pale area bridged by a dark area on vein 1; a conspicuous white spot at the end of segment 1 of hind tarsus. *A. annularis.* (Figs. 16 & 29).

(*b*) 5th vein extensively pale without the dark spot on the stem at its bifurcation; the subcostal pale area not bridged by a dark area on vein 1.

(*i*) Distal end of segment 1 of hind tarsus inconspicuously marked white; no white scales on ventral aspect of abdomen except on last 2 or 3 segments only: scaling on the dorsum of the abdomen confined to the last two or three segments. *A. philippinensis.* (Figs. 17 & 30).

(*ii*) No white spot at the end of segment 1 of hind tarsus; broad white scales scattered over ventral aspect of most of the abdominal segments; scaling on the dorsum of abdomen extends over the last 5 or 6 segments. *A. pallidus.* (Figs. 18 & 31).

(*c*) Conspicuous white scales on the dorsum of the thorax and abdomen, the scale tufts projecting laterally from each abdominal segment. *A. pulcherrimus.* (Figs. 34 & 55).

B. Only tarsus 5 and one-third of tarsus 4 of hind legs completely white.

(*a*) Three pale bands on the palp. The two apical bands are broad and equal. *A. majidi.* (Fig. 53).

(*b*) Four pale bands on the palp. The three distal bands are broad and the proximal one is very narrow. *A. karwari.* (Figs. 28 & 54).

GROUP VI

Chief characters. **At least 4 dark spots on the costa etc., as in group III, IV and V. Femora and tibia speckled. White-footed series.**

A. At least the third, fourth and fifth tarsal segments of hind legs completely white.

(*a*) The apical pale band broad, the subapical pale area being narrow, palps not speckled.

(*i*) At least the last two segments of the abdomen covered with golden scales; wings pale; a yellowish mosquito. *A. jamesi.* (Fig. 56).

(*ii*) No golden scales on the abdomen; wings much darker; a small dark mosquito. *A. ramsayi.*

(*b*) Two equally broad pale apical bands; palpi speckled. *A. splendidus.* (Fig. 57).

(*c*) Four pale bands on the palpi; conspicuous white scales on the dorsum of the abdomen and the thorax; lateral scale tufts projecting from each abdominal segment.

(*A. pulcherrimus* may have speckled front legs).

B. The whole of fifth and fourth tarsal segments of hind legs completely white; no dark band on fourth tarsal segment. *A. theobaldi.* (Fig. 22).

C. Fifth and a third of fourth tarsal segments of hind legs completely white; a dark band on the fourth tarsal segment present.

(*a*) Abdominal scales only on the last 2 or 3 segments. *A. maculatus.* (Figs. 23 & 58).

D. Only half of the last tarsal segment of hind legs white; four pale bands on the palp; prominent tuft of black scales on the ventral surface of each abdominal segment; outer half of proboscis golden. *A. kochi.* (Figs. 20 & 21).

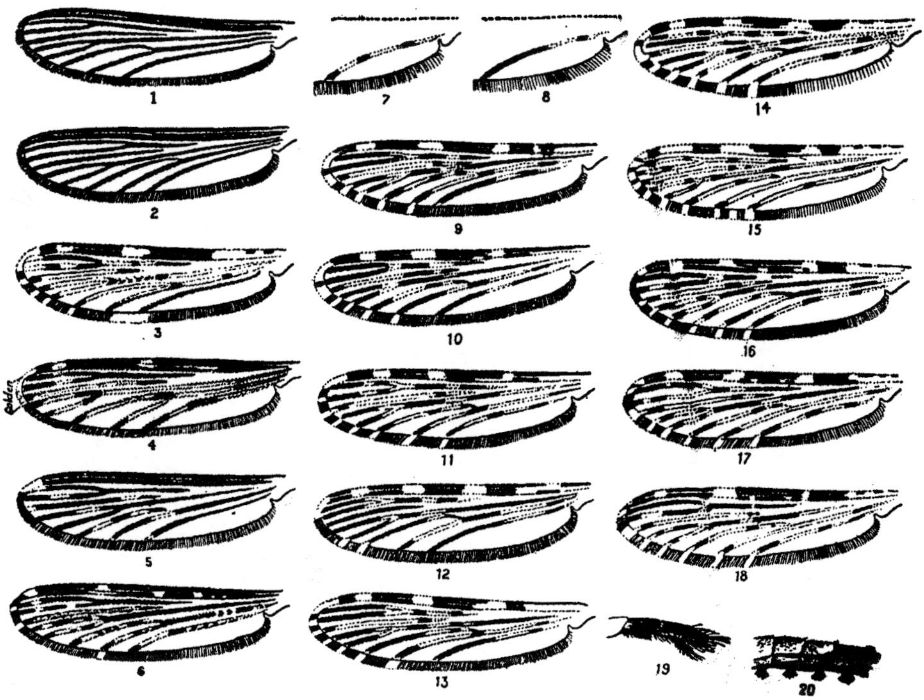

FIG. 43.

1. Wing of *A. barianensis.*
2. ,, ,, *A. aitkeni.*
3. ,, ,, *A. gigas.*
4. ,, ,, *A. hyrcanus* var. *nigerrimus.*
5. ,, ,, *lindesayi.*
6. ,, ,, *A. barbirostris.*
7. ,, ,, *A. multicolor.*
8. ,, ,, *A. turkhudi.*
9. ,, ,, *A. minimus.*
10. ,, ,, *A. varuna.*
11. Wing of *A. aconitus.*
12. ,, ,, *A. culicifacis.*
13. ,, ,, *A. fluviatilis.*
14. ,, ,, *A. jeyporiensis.*
15. ,, ,, *A. subpictus.*
15. ,, ,, *A. annularis.*
17. ,, ,, *A. philippinensis.*
18. ,, ,, *A. pallidus.*
19. Abdomen of *A. barbirostries.*
20. ,, ,, *A. kochi.*

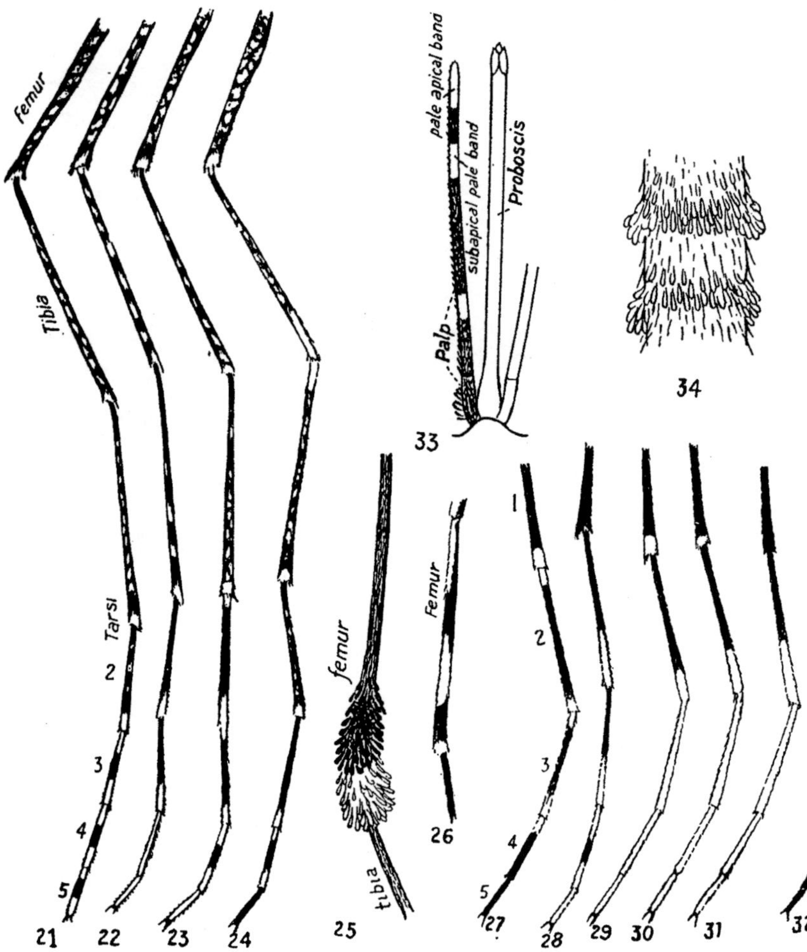

FIG. 44.

21. Hind leg of *A. kochi*.
22. „ „ „ *A. theobaldi*.
23. „ „ „ *A. maculatus*.
24. „ „ „ *A. leucosphyrus*.
25. „ „ „ *A. annandalei*.
26. Femur of *A. lindesayi*.
27. Frontal tarsal joints of *A. subpictus*.
28. Hind foot of *A. karwari*.
29. „ „ „ *A. annularis*.
30. „ „ „ *A. philippinensis*.
31. „ „ „ *A. pallidus*.
32. Foreleg of *A. jeyporiensis*.
33. Proboscis and palp of mosquito.
34. Abdomen of *A. pulcherrimus*.

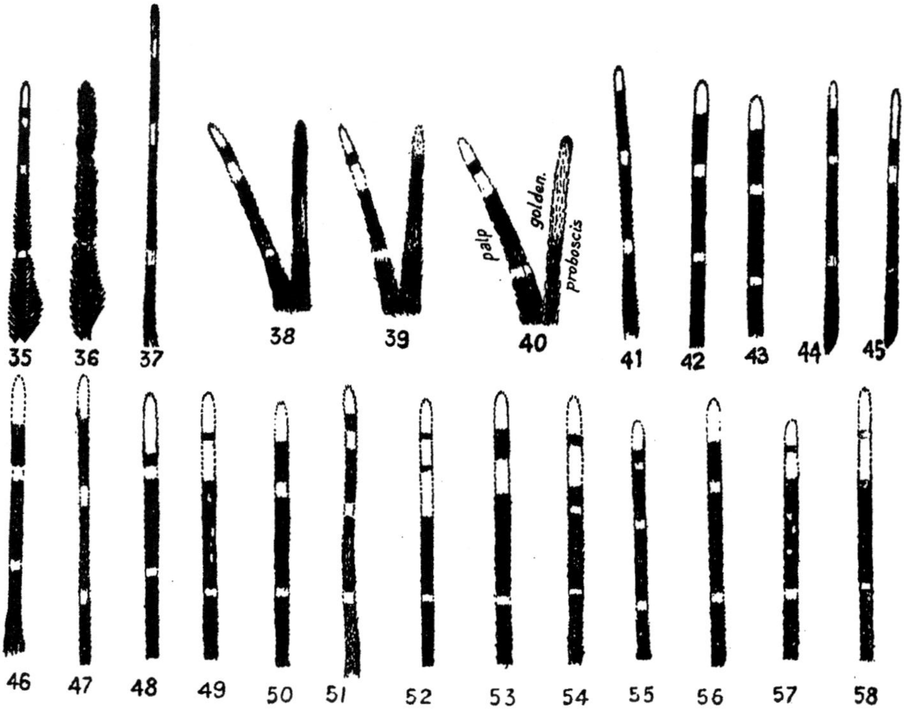

FIG. 45.

35. Palp of *A. hyrcanus* var. *nigerrimus*.
36. „ „ *A. barbirostris*.
37. „ „ *A. turkhudi*.
38. Palp and proboscis of *A. minimus*.
39. „ „ „ „ *A. varuna*.
40. „ „ „ „ *A. aconitus*.
41. Palp of *A. culicifacies*.
42. „ „ *A. fluviatilis*.
43. „ „ *A. moghulensis*.
44. „ „ *A. jeyporiensis*, type.
45. „ „ *A. jeyporiensis* var. *candidiensis*.
46. „ „ *A. subpictus*.
47. Palp of *A. superpictus*.
48. „ „ *A. vagus*.
49. „ „ *A. stephensi*.
50. „ „ *A. sundaicus*.
51. „ „ *A. leucosphyrus*.
52. „ „ *A. tessellatus*.
53. „ „ *A. majidi*.
54. „ „ *A. karwari*.
55. „ „ *A. pulcherrimus*.
56. „ „ *A. jamesi*.
57. „ „ *A. splendidus*.
58. „ „ *A. maculatus*.

KEY TO THE IDENTIFICATION OF ANOPHELES (FEMALES ONLY) OF INDIA

Group I

Costa uniformly dark, *i.e.*, not interrupted by any pale spot; also no white spot on the wing field.

- Anterior forked cell much larger than the posterior.
  A. aitkeni.
- Anterior forked cell of nearly the same size as the posterior.
  - Distinct white banding at the distal end of hind femur. Frontal white scale tuft present.
    A. barianensis.
  - Banding absent. Scale tuft absent.
    A. culiciformis.

Group II

Less than four dark spots involving the costa, subcosta and vein 1.

- Palpi with distinct pale bandings.
  - A prominent tuft of scales, black above and white below, about the femoro-tibial joint of hind legs.
    A. annandalei.
  - No such tuft of scales present.
    - Inner quarter of costa with marked pale interruptions. Wing fringe between 5.2 and 6 white. A brightly coloured mosquito.
      A. gigas.
    - Inner quarter of costa dark. Tip of wing golden. A dark mosquito.
      A. hyrcanus var. *nigerrimus.*
- Palpi unbanded.
  - Hind femur with a conspicuous white band.
    A. lindesayi.
  - Hind femur without any white band.
    - Palpi shaggy. Fringe spot at 5.2 present. Presence of both dark and white scales on wing veins. A common species.
      A barbirostris.
    - Palpi thinner. Fringe spot at 5.2 absent. Only dark scales on the wing field. A rare species in India.
      A. umbrosus.

Groups III—VI have at least 4 dark spots involving the costa, subcosta and vein I.

Group III

Dark-footed series (of hind legs only). Femorae and tibiae not speckled

- Tips of palpi dark.
  - Wing veins except on costa and V. 1 contain only dark scales.
    A. dthali.
  - All wing veins contain both dark and white scales.
    - 2 indefinite dark spots on V6., the distal one being very long.
      A. turkhudi.
    - 3 definite dark spots on V6.
      A. multicolor
- Tips of palpi pale.
  - V3 mostly dark.
    - Only two fringe spots 4.2 and 5.1 present. (Palpi of normal length).
      A. culicifacies.
    - Fringe spots at all veins except V6. (Palpi very long).
      A. sergenti.
  - V3 mostly pale.

V3 mostly pale

- The two apical pale bands are of equal or nearly equal lengths and the intervening dark area is small.
  - No fringe spot at V. 6.
    - Basal area of the costa dark without any pale interruption. *A. varuna.*
    - Basal area of the costa with one pale interruption. *A. minimus.*
  - Fringe spot at V6. *A. aconitus.*
- The two pale apical bands are definitely unequal.
  - The intervening dark area on the palp is much larger than either of them.
    - Inner third of costa uninterruptedly dark.
      - 2 dark spots on V6. Tibio-tarsal joints dark. *A. fluviatilis.*
      - 3 dark spots on V6. Tibio-tarsal and tarsal joints of front legs narrowly banded. *A. moghulensis.*
    - Inner third of costa interrupted.
      - Tarsal joints of leg 1 banded. Fringe spot at V6. *A. jeyporiensis.*
      - Tarsal banding of leg 1 absent. No fringe spot at V6. *A. superpictus.*
  - The intervening dark area is either of the same size as the apical pale band or very much smaller. Broad white bands at the tarsal joints of the front legs.
    - The dark area of the same size as the apical area. *A. subpictus.*
    - The dark area is very much smaller. *A. vagus.*

Group IV

Dark-footed series (hind legs only). Femorae and tibiae speckled.

- 3 pale bands on the palp.
  - The apical and the subapical bands are equally broad. *A. stephensi.*
  - The apical and the subapical bands are unequal, the former being broader than the latter. *A. sundaicus.*
- 4 pale bands on the palp.
  - Broad tibio-tarsal white band on the hind leg. *A. leucosphyrus.*
  - The three distal bands are much broader than the proximal band which is narrow. *A. tessellatus.*

Groups V

White-footed series (hind legs only). Femorae and tibiae not speckled.

- Only tarsus 5 and 1 3rd. of tarsus 4 of hind legs completely white.
  - 3 pale bands on the palp. *A. majidi.*
  - 4 pale bands on the palp. *A. karwari.*
- At least tarsal segments 3, 4 and 5 of hind legs completely white.
  - Vein 5 mainly dark with a dark spot at its bifurcation. *A. annularis.*
  - Vein 5 extensively pale and no dark spot at the bifurcation.
    - Distal end of tarsus 1 of hind legs inconspicuously marked white. *A. philippinensis.*
    - Distal end of tarsus 1 of hind legs dark. *A. pallidus.*
  - Conspicuous white scales on the dorsum of the abdomen and thorax. *A. pulcherrimus.*

Group VI

White-footed series (hind legs only). Femorae and tibiae speckled.

- ½ of tarsus 5 of hind legs white. Prominent scale tufts on the ventral surface of each abdominal segment. *A. kochi.*
- Whole of tarsus 5 and one-third of 4 of hind legs white with a dark band on tarsus 4. *A. maculatus.* *A. maculatus* var. *willmori.*
- Whole of tarsi 5 and 4 of hind legs white. *A. theobaldi.*
- At least tarsi 3, 4 and 5 of hind legs completely white.
  - 2 equally broad pale apical bands, palpi speckled. *A. splendidus.*
  - The two pale apical bands unequal.
    - At least the last two segments of the dorsum of the abdomen covered with golden scales. *A. jamesi.*
    - Dark scales on the abdomen. *A. ramsayi.*
    - Conspicuous white scales on the dorsum of the thorax and abdomen. *A. pulcherrimus.*

IDENTIFICATION OF ANOPHELINE LARVAE OF INDIA

Methods of Examination of Anopheline Larvae: 1. They may be examined in the living state in a drop of clean water and covered with a cover-glass. 2. Dead specimens are mounted in a drop of carbolic acid and covered with a cover-glass. In this way a large number of specimens can be examined within a short time. 3. The specimens may be permanently mounted on a glass slide. It is absolutely useless for the purpose of identification if the mosquito larva has not been mounted with the dorsal surface pointing upwards. Clypeal hairs, palmate hairs, etc., lie on the dorsal surface and are always examined with 1/6 objective. For clearing and permanently mounting use Berles's fluid.—(Chloral hydrate—160 g., Gum arabic—15 g., Glucose—10 g., Glacial acetic acid—5 c.c., water—20 c.c.).

Characters of the Dorsal and Ventral Surfaces of Anopheline Larva: Examination is made in a good drop of either carbolic acid or lactophenol under 2/3rd. objective of a compound microscope. The specimen may or may not be covered with a coverglass. On the dorsal surface lie the clypeal hairs, frontal hairs, antennae, palmate hairs and spiracles, whereas on the ventral surface are the mandibles, mentum and the maxillary palps. Therefore in order to determine whether the dorsal or the ventral surface lies uppermost, the object is placed under the microscope and focussed and if the mouth organs and the maxillary palps are seen first, the larva has to be turned with a needle in order that the dorsal surface may point upwards. (Lactophenol is composed as follows: Carbolic acid—20 c.c., Lactic acid—20 c.c., Glycerine—40 c.c., Water—20 c.c.).

N.B.—All figures relating to the identification of Anopheline larvae refer to those on pp. 78 and 79.

How to Proceed with the Identification.: (1) Find out under which group the mosquito larva can be placed. To facilitate identification anopheline larvae have been placed under seven separate groups on the basis of certain distinctive features. (2) Then consult the key. (3) The Table should be referred to whenever necessary or when confirmation of the identification already made with the help of the key is required. (4) As a preliminary to routine identification as a part of malaria survey, an idea of the local anopheline fauna should be obtained by breeding out adult mosquitoes from larvae collected from scattered places within the area to be surveyed. (5) In case of doubt one should study the adults bred out from larvae and the larval skins under such conditions should always be preserved for verification of the identification. (6) Before the final identification is declared the regional distribution of the mosquitoes concerned and the characters of the breeding places should always be kept in mind.

Table for the Identification of fourth stage or mature Anopheline Larvae of India

(A few have been left out as they are not only uncommon but are also unimportant).

Group I	Group II	Group III	Group IV	Group V	Group VI	Group VII
(*a*) Internal clypeal hairs placed close together. (*b*) Antennal hair branched and placed dorso-internally. (*c*) Frontal hairs long and thickly branched. (Figs. 1 & 45).	(*a*) Internal hairs placed close to each other. (*b*) Antennal hair simple and placed dorso-externally. (*c*) Most of the frontal hairs are short and simple or sparingly branched. (Figs. 18 & 46).	Anterior abdominal tergal plates exceptionally well developed (Figs. 7 & 8).	(*a*) Internal clypeal hairs so finely and sparingly frayed that they appear simple. (*b*) The external hair is nearly one quarter the length of the internal. (*c*) The posterior clypeal hair placed far posteriorly. (*d*) Shape of mid-abdominal tergal plates characteristic. (*e*) The filamentous part of the mid-abdominal palmate hairs is hazy and has a ground glass-like appearance. (Figs. 25, 26 & 30).	Internal clypeal hairs finely but conspicuously frayed.	Internal clypeal hairs thickly frayed.	Internal clypeal hairs simple.
A. aitkeni type. *A. aitkeni* var. *bengalensis*. *A. gigas*. *A. lindesayi*. *A. barbirostris*. *A. hyrcanus* var. *nigerrimus*. *A. umbrosus*.	*A. annandalei*. *A. barianensis*. *A. culiciformis*. *A. sintoni*.	*A. aconitus*. *A. jeyporiensis*. *A. minimus*. *A. fluviatilis*. *A. varuna*.	*A. kochi*. *A. tessallatus*. *A. leucosphyrus*.	*A. maculatus*. *A. karwari*. *A. stephensi*. *A. moghulensis*. *A. superpictus*. *A. splendidus*. *A. theobaldi*.	*A. annularis*. *A. philippinensis*. *A. pallidus*. *A. jamesi*. *A. ramsayi*. *A. pulcherrimus*. *A. splendidus*.	*A. subpictus*. *A. culicifacies*. *A. vagus*. *A. majidi*. *A. sundaicus*. *A. stephensi*. *A. multicolor*. *A. turkhudi*. *A. dthali*. *A. sergenti*.

Group I

Chief characters :	(*a*) Internal clypeal hairs placed close together. (*b*) Antennal hair branched and placed dorso-internally. (*c*) Frontal hairs long and thickly branched. (Figs. 1 & 45).
1. *A. aitkeni.* A jungle species and found mainly in Assam and Nilgiris.	The two common forms are (1) *A. aitkeni* type and (2) *A. aitkeni* var. *bengalensis.* *A. aitkeni* type. (*a*) Internal clypeal hair bifid. (*b*) External hair short and branched. (*c*) Posterior hair a short tuft, the branches arising close to the root. *A. aitkeni var. bengalensis.* (*a*) Internal clypeal hair has more than two branches. (*b*) External hair longer than in the type form and has also more than two branches. (Fig. 1).
2. *A. gigas.* A hill species found at high altitudes.	(*a*) Internal clypeal hair simple. (*b*) External hair simple but occasionally may have one or two fine branches. (*c*) Posterior hair very small, either simple or bifid. (*d*) Abdominal palmate hair developed on segments 3 to 7, the filament being poorly differentiated. (*e*) Metathoracic hair not differentiated. (Fig. 2).
3. *A. lindesayi.* Recorded at places from high altitudes.	(*a*) Internal and external clypeal hairs long and simple. (*b*) Posterior hair short and simple or may be bifid. (*c*) Abdominal palmate hairs developed on segments 2 to 7, the filament being long and sharply pointed. (*d*) Metathoracic palmate hair well developed. (Fig. 3).
4. *A. barbirostris.* Widely distributed in India.	(*a*) Internal clypeal hair long and simple. (*b*) External hair characteristically branched. (*c*) Posterior hair small and branched. (*d*) Inner submedian prothoracic hair branched from the root. (Figs. 4 & 10).
5. *A. hyrcanus* var. *nigerrimus.* Widely distributed in India.	(*a*) Clypeal hairs same as in *barbirostris.* (*b*) Inner submedian prothoracic hair simple; occasionally the tip may be bifid or trifid. (Figs. 4 & 11).
6. *A. umbrosus.* A very rare species recorded from dense forests in Assam.	(*a*) Internal clypeal hair long and simple. (*b*) External hair branched more or less dichotomously with about 7 to 10 long branches. (*c*) Posterior hair a short tuft with generally three branches arising from the base. (*d*) Abdominal palmate hairs undeveloped on any segment. (Fig. 5).

Group II

Chief characters.	(*a*) Internal clypeal hairs placed close to each other. (*b*) Antennal hair simple and placed dorso-externally. (Fig. 46). (*c*) Most of the frontal hairs are short and simple or sparingly branched. (Fig. 18).
1. *A. annandalei.* Recorded from Darjeeling forest and Nilgiris (Gudalur).	(*a*) Internal clypeal hairs long and simple. (*b*) External hairs branched. (*c*) Inner frontal hairs simple and very long. (*d*) Innumerable minute spines on the thorax and abdomen even visible under the low power.
2. *A. barianensis.* An Himalayan species occurring at high altitudes.	(*a*) All clypeal hairs simple. (*b*) Frontal hairs longer with one or two branches. (Fig. 18).
3. *A. culiciformis.* Confined to Southern India.	(*a*) Clypeal hairs like those of *barianensis.* (*b*) Frontal hairs larger with one or two branches.
4. *A. sintoni.* Southern India	(*a*) Clypeal hairs like those of *culiciformis.* (*b*) Frontal hairs have 5 to 7 branches.

N.B.—All are tree-hole breeders.

Group III

Chief Characters.

Anterior abdominal tergal plates exceptionally well developed. (Figs. 7, 8 & 13).

1. *A. aconitus.*
Confined to the eastern and southern areas of India

(*a*) Both internal and external clypeal hairs stout snd branched,
(*b*) the branches giving the appearance of thorns.
Posterior hair with generally three branches arising from the root. (Figs. 6 & 8).

2. *A. jeyporiensis.*
Distributed in the eastern and southern areas of the Peninsula.

(*a*) Both internal and external clypeal hairs thickly branched.
(*b*) Posterior hair has two or three branches a short distance from the root.
(*c*) Abdominal tergal plate has a different appearance from that of *aconitus*, *minimus* or *varuna*. (Its posterior border is concave and it does not include the small posterior tergal plate). (Figs. 7 & 9).

3. *A. minimus.*
Extremely common in Eastern India, *e.g.*, Assam and Dooars.

(*a*) Internal and external clypeal hairs stout, simple and long.
(*b*) Posterior hair long and simple.
N.B.—See below under *A. varuna* for points of distinction between *minimus* and *varuna*. (Figs. 8 and 12).

4. *A. fluviatilis.*
A foot-hill species and widely distributed in India.

Difficult to distinguish from *A. minimus.*

5. *A. varuna.*
Widely distributed in the eastern and southern areas of India.

Resembles *A. minimus* even in minute details but the presence of a minute branched hair on the tergal plate, one on each side of the middle line, is characteristic of this species.
In *minimus* the hair is placed near the postero-lateral angle of the tergal plate.
(The internal clypeal hair is so finely frayed in this species that it often appears simple). (Fig. 13).

Group IV

6. *Chief characters.*

(*a*) Internal clypeal hairs so finely and sparingly sprayed that they appear simple.
(*b*) The external hair is nearly one-quarter the length of the internal.
(*c*) The posterior clypeal hair placed far posteriorly.
(*d*) Shape of mid-abdominal tergal plates characteristic.
(*e*) The filamentous part of the mid-abdominal palmate hairs is hazy and has a ground glass-like appearance. (Figs. 25, 26 & 30).

1. *A. kochi.*
A jungle species found in Assam.

Inner submedian prothoracic hair without any prominent root and with 5 to 10 branches. (Figs. 25 & 26).

2. *A. tessellatus.*
Recorded from eastern and southern India.

Inner submedian prothoracic hair with less than 4 branches, the root being quite conspicuous. (Figs. 27 & 28).

3. *A. leucosphyrus.*
A wild mosquito recorded mainly from Assam and South India, also from Ceylon.

Inner submedian prothoracic hair with a prominent dark root and more than ten branches. (Fig. 29).

Group V

Chief characters.

Internal clypeal hairs finely but conspicuously frayed. (Fig. 31).

1. *A. maculatus.*
Recorded from the Himalayan areas, hilly parts in South India, also from Ceylon and breeds in hill streams.

(*a*) Both internal and external hairs finely frayed.
(*b*) Posterior long and simple and is often slightly bent towards one side which is very characteristic.
(*c*) Metathoracic palmate hair undeveloped (may be slightly differentiated.
(*d*) Abdominal palmate hairs developed on segments 3-7, the filament being long but not very long as in *subpictus* or *moghulensis*.
(*e*) In about 50 per cent one of the two inner sutural hairs may at times show bifurcation at the distal end. (Figs. 31 & 32).

2. *A. karwari.*
Assam and Southern India.

Closely resembles *A. maculatus* but the filaments are practically absent and blunt at the tip. (Figs. 33 & 34).

3. *A. stephensi.*

See under group VII.
(It may occasionally show minute frayings).

4. *A. moghulensis.*
Found in hill streams in North Western and Central parts of India.

(*a*) Internal clypeal hairs stout and very minutely frayed.
(*b*) Both external and posterior hairs long and simple.
(*c*) Metathoracic palmate hair well developed.
(*d*) Abdominal palmate hairs well developed on segments 2 to 7, the filament being long with pointed ends. (Fig. 24).

5. *A. superpictus.*
Baluchistan and N.W.F. Province.

(*a*) Internal clypeal hair slender and minutely frayed.
(*b*) External and posterior hairs long and simple.
(*c*) Metathoracic palmate hair well developed.
(*d*) Abdominal palmate hairs as in *moghulensis.*

6. *A. splendidus.*
Has a wide distribution in India.

(*a*) Both internal and external clypeal hairs prominently frayed.
(*b*) Posterior hair long and simple.
(*c*) Internal sutural hair long and with about three branches.
(*d*) Palmate hairs on abdominal segment I undeveloped but well developed on segments 3 to 7, and the filaments are short and not sharp pointed.
(*A. splendidus* bears many points of resemblance to *A. ramsayi*). (Figs. 22 & 23).

7. *A. theobaldi**
Scattered especially over Southern India.

Similar to *A. maculatus.* In about 25 per cent cases one inner sutural hair may be bifid.

8. *A. maculatus* var. *willmori.*

Also similar to *A. maculatus.*

* *A. maculatus,* and *A. theobaldi* may be difficult to separate when found in the same area.

Group VI

Chief characters.

Internal clypeal hairs thickly frayed. (Fig. 14).

1. *A. annularis.*
Widely distributed in India.

(*a*) Internal clypeal hair stout, long and thickly frayed.
(*b*) External clypeal hair long and characteristically branched.
(*c*) Posterior hair with two or three branches at the end of a short stalk.
(*d*) Internal sutural hair long and simple or bifid at the tip.
(*e*) Palmate hair on the first abdominal segment well differentiated. (Figs. 14, 15, 16, & 17).

2. *A. philippinensis.*
Recorded from Bengal, Assam, North Eastern parts of Bihar, and west coast of India.

(*a*) Internal and external clypeal hairs as in *annularis.*
(*b*) Posterior clypeal hair has 7 to 9 branches arising from the root without the stalk.
(*c*) Internal sutural hair may have 2 to 4 branches.
(*d*) Abdominal palmate hair on segment I well developed. (Figs. 48 & 49).

3. *A. pallidus.*
Found mostly in Central Provinces and Orissa.

(*a*) Internal and external clypeal hairs as in *annularis.*
(*b*) Posterior clypeal hair with 3 to 5 branches at the end of a short stalk.
(*c*) Internal sutural hair has dichotomous branching consisting of about 5 branches.
(*d*) Abdominal palmate hair on segment I well developed. (Figs. 50 &51).

4. *A. jamesi*
Found mainly in Bengal and Assam.

Closely resembles *A. annularis* from which it can be easily separated by palmate hair on abdominal segment I, which is well developed in *annularis* but is not so in *jamesi.* (Figs. 17 & 19).

5. *A. ramsayi.*
Recorded from Bengal, Assam, Orissa and Ceylon.

(*a*) Both internal and external clypeal hairs have conspicuous spine-like side hairs.
(*b*) Internal clypeal hair is very long and the terminal part vanishes away during focussing and cannot be followed.
(*c*) Posterior hair long and simple.
(*d*) Internal sutural hair long and simple.
(*e*) Palmate hairs developed on abdominal segments 3 to 7.
(*f*) The filament of abdominal palmate hairs poorly developed and the end pointed. (Figs. 20 & 21).

6. *A. pulcherrimus.*
North-Western Pakistan.

(*a*) Internal clypeal hairs long and have fine branches on distal two thirds. The branches are simple and long.
(*b*) External hair has long fine branches approaching the branching in *A. annularis.*
(*c*) Posterior hair short and has 2-3 branches from a short stalk.
(*d*) Metathoracic hair well develped.

Group VII

Chief characters.
Internal clypeal hairs simple. (Fig. 35).

1. *A. subpictus.*
Widely distributed in India.

(*a*) Internal, external, and posterior clypeal hairs long and simple.
(*b*) External sutural hair has more than three branches arising from the end of the stalk. (Sometimes there are three or less branches on the outer sutural hair on one side or rarely on both sides).
(*c*) Metathoracic palmate hair undeveloped.
(*d*) Abdominal palmate hairs have long filaments. (Figs. 35, 36 & 41).

2. *A. culicifacies.*
Distributed all over India and Ceylon and found also at high altitudes in Kashmir.

Closely resemble *A. subpictus* except that the metathoracic palmate hairs are well developed. Inner submedian prothoracic hair has a dark root. (Figs. 40 & 47).

3. *A. vagus.*
Confined to Eastern and Southern India.

(*a*) Internal clypeal hair long and simple.
(*b*) External hair less than one-third the length of the internal hair and simple.
(*c*) Posterior hair short like anterior external hair and placed between the internal clypeal hairs.
(*d*) Metathoracic palmate hair not developed.
(*e*) Abdominal palmate hairs have long filaments. (Fig. 37).

4. *A. majidi.*
Distribution limited to Darjeeling and Jalpaiguri Districts in Bengal, and Mysore and Malabar districts in Southern India.

(*a*) Clypeal hairs like those of *A. subpictus* except that the internal clypeal hairs are stouter and are exceptionally long.
(*b*) Metathoracic palmate hair well developed.
(*c*) Filament of abdominal palmate hairs short.

5. *A. sundaicus.*
A brackish water species and found near sea coast.

Closely resemble *A. subpictus* except that hair No. 5 on mesothorax, has usually three branches arising from the end of a a short stalk.
(*A. subpictus* may have the same number of branches but they always arise from the end of a comparatively long stalk). (Figs. 42. 43 & 44).

6. *A. stephensi.*
Found generally in large cities in India.

Closely resemble *A. subpictus* except that the external sutural hair has 2 to 3 branches at the end of a long stem.
The second hair dorsal to the lateral hair on abdominal segment I with 3 to 5 branches.
(Internal clypeal hairs may occasionally show minute frayings). (Figs. 52, 53 & 54).

7. *A. multicolor.*
A desert species and found in Baluchistan.

Clypeal hairs resemble *A. subpictus.* The metathoracic palmate hair is also undifferentiated. Its distribution, habitat and the presence of characteristic head spots must be taken into consideration in differentiating this species from *A. subpictus.*

8. *A. turkhudi.*
Recorded from the North-Western and Central parts of India.

(*a*) All clypeal hairs are simple but the posterior hair is abnormally long.
(*b*) Abdominal palmate hairs undeveloped on segments 1 to 3.
(*c*) The filaments of abdominal palmate hairs short. (Figs. 38 & 39).

9. *A. dthali.*
Baluchistan, North-Western India.

Difficult to distinguish from *A. culicifacies.*

10. *A. sergenti.*
North-Western India.

Similar to *A. dthali.*

Key to the Identification of Mature Anopheline Larvae

Group I

- Internal clypeal hairs long and simple.
  - External hair heavily and characteristically branched.
    - Inner submedian prothoracic hair branched. *A. barbirostris.*
    - Inner submedian prothoracic hair simple *A. hyrcanus* var. *nigerrimus.*
  - External hair dichotomously branched (8-10 branches). *A. umbrous.*
  - External hair simple.
    - Abdominal palmate hairs developed on abdominal segments 2-7. *A. lindesayi.*
    - Abdominal palmate hairs developed on abdominal segments 3-7. *A. gigas.*
- Internal hairs branched into 2 or more branches. *A. aitkeni* and varieties.

Group II

Tree-hole breeders.

- Inner clypeal hair simple; outer hair branched. *A. annadalei.*
- Both inner and outer hairs simple.
  - All frontal hairs very small and simple. *A. barianensis.*
  - Frontal hairs larger with 1 or 2 branches. *A. culiciformis.*
  - Frontal hairs with 5-7 branches. *A. sintoni.*

Group III

Anterior tergal plates well developed.

- Both inner and outer clypeal hairs simple.
  - Posterior hair small and consists of three branches. *A. aconitus.*
  - Posterior hair long and simple.
    - Presence of a minute bifid hair on the tergal plate. *A. varuna.*
    - The bifid hair lies outside the tergal plate at the post. lat. region of the abdml. segment. *A. minimus.* *A. fluviatilis.*
- Both inner and outer clypeal hairs thickly branched. *A. jeyporiensis.*

Group IV

Inner submedian prothoracic hair

- with a prominent root
  - and with less than 4 branches. *A. tessallatus.*
  - and with more than 10 branches. *A. leucosphyrus.*
- without a prominent root, and with 3 to 10 branches. *A. kochi.*

Group V

Internal clypeal hair finely frayed. Posterior clypeal hair long and simple.

- External clypeal hair also finely frayed. Metathoracic palmate hair not differentiated.
  - Filament of abdominal palmate hairs long. *A. maculatus.*
  - Filaments of abdominal palm. hair absent and the tip is blunt. *A. karwari*
- External hair has side branches like thorns. Metathoracic palmate hair not differentiated. Filaments of abdominal palm. hairs short. *A. splendidus.*
- External clypeal hair simple. Metathoracic palmate hair differentiated. *A. moghulensis* *A. superpictus.*

N.B.—*A. superpictus* and *A. moghulensis* can be differentiated by the spots on the head or fronto-clypeus. This can be learnt from experience.

Group VI

Both internal and external clypeal hairs prominently and even thickly branched.

- Branching of external hair is very heavy and is of a characteristic type. Posterior clypeal hairs short and branched.
  - Internal clypeal hairs have fine lateral branches on distal two thirds. Posterior hair has some 2-3 twigs from a short stalk. Metathoracic palm. hair well developed. *A. pulcherrimus.*
  - Palmate hair on abdominal segment I undeveloped. *A. jamesi.*
  - Palmate hair on abdominal segment I developed.
    - Posterior clypeal hair has over 7 branches arising from a root without a stalk. *A. philippinensis.*
    - Posterior hair has less than 5 branches.
      - Internal sutural hair long and simple. *A. annularis.*
      - Internal sutural hair has a dichotomous branching consisting of about 5 branches. *A. pallidus.*
- External hairs have side branches resembling thorns. Posterior clypeal hairs long and simple.
  - Internal sutural hair long and with about 3 branches. Filaments of abdominal palmate hairs short and not sharp-pointed. Metathoracic palmate hair not differentiated. *A. splendidus.*
  - Internal sutural hair long and simple. Filaments of abdominal palmate hairs short but sharp pointed. *A. ramsayi.*

Group VII

Internal and external clypeal hairs simple.

- Metathoracic palmate hair well developed.
  - Filament of abdominal palmate hair long. *A. culicifacies.*
  - Filament of abdominal palmate hair short. *A. majidi.*
- Metathoracic palmate hair undeveloped.
  - Posterior clypeal hair exceptionally long, at least as long as the internal hair. Frontal hairs have 4-5 branches and never reach the roots of the internal clypeal hairs. *A. turkhudi.*
  - Posterior clypeal hair not longer than the external clypeal hair.
    - Posterior clypeal hair very short and placed behind and between the anterior internals. *A. vagus.*
    - Posterior clypeal hair long, and placed between anterior internal and anterior external hairs.
      - External sutural hair has 2-3 branches. *A. stephensi.*
      - External sutural hair has more than 3 branches.
        - Mesothoracic hair No. 5 has usually 3 branches arising close to the root of the hair. *A. sundaicus.*
        - Hair No. 5 may have 3 branches but they always arise from the stem some distance from the root. *A. subpictus.*

N.B.—(*a*) It is not easy to distinguish *A. multicolor* from *A. subpictus*. However, the former is a desert species and possesses certain spots on the head by which the two can be separated.

(*b*) It is also difficult to distinguish *A. dthali* and *A. sergenti* from *A. culicifacies*.

(*c*) *A. maculatus* cannot be separated from *A. theobaldi* and *A. willmori* unless their distribution and habitat are known.

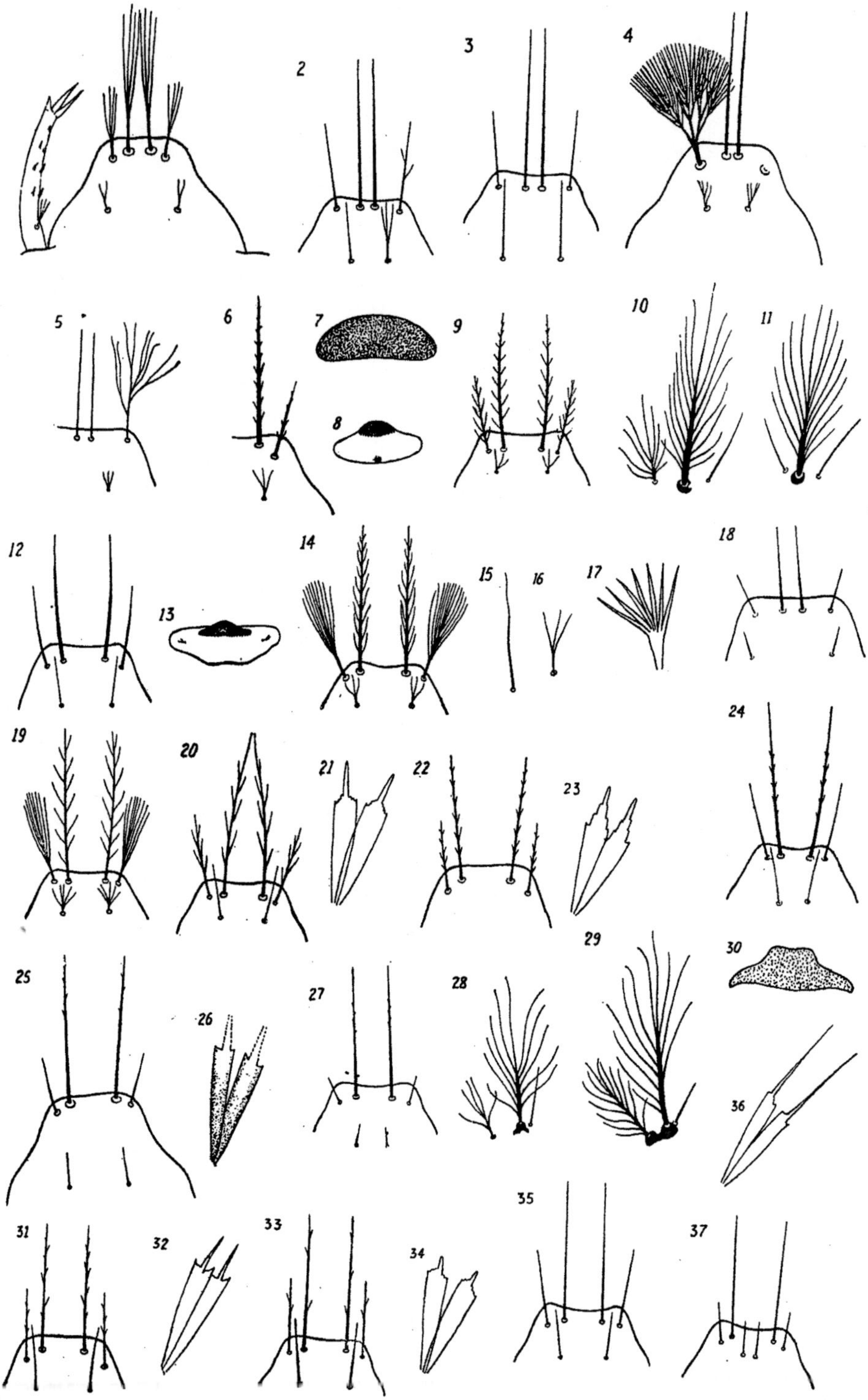
2
3
4
5
6
7
8
9
10
11
12
13
14
15
16
17
18
19
20
21
22
23
24
25
26
27
28
29
30
31
32
33
34
35
36
37

Fig. 46.

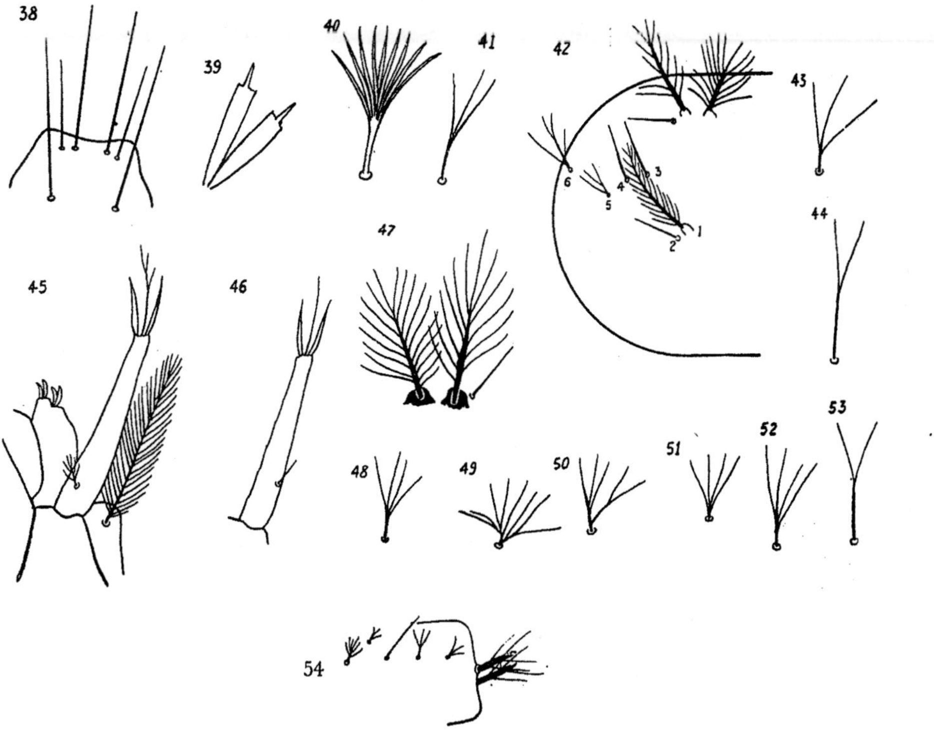

FIG. 47.

1. Antennal hair and clypeal hairs of *A. aitkeni* var. *bengalensis*. 2. Clypeal hairs of *A. gigas*. 3. Clypeal hairs of *A. lindesayi*. 4. Clypeal hairs of *A. hyrcanus*. 5. Clypeal hairs of *A. umbrosus*. 6. Clypeal hairs of *A. aconitus*. 7. Anterior tergal plate of *A. jeyporiensis*. 8. Anterior tergal plate of *A. aconitus* and *A. minimus*. 9. Clypeal hairs of *A. jeyporiensis*. 10. Submedian prothoracic hairs of *A. barbirostris*. 11. Submedian prothoracic hairs of *A. hyrcanus* var. *nigerrimus*. 12. Clypeal hairs of *A. minimus*. 13. Anterior tergal plate of *A. varuna*. 14. Clypeal hairs of *A. annularis*. 15. Internal sutural hair of *A. annularis*. 16. Posterior clypeal hair of *A. annularis*. 17. Palmate hair on first abdominal segment of A. *annularis*. 18. Clypeal hairs of *A. barianensis*. 19. Clypeal hairs of *A. jamesi*. 20. Clypeal hairs of *A. ramsayi*. 21. Mid-abdominal palmate hair of *A. ramsayi*. 22. Clypeal hairs of *A. splendidus*. 23. Mid-abdominal palmate hair of *A. splendidus*. 24. Clypeal hairs of *A. moghulensis*. 25. Clypeal hairs of *A. kochi*. 26. Mid-abdominal palmate hair of *A. kochi*. 27. Clypeal hairs of *A. tessellatus*. 28. Submedian prothoracic hairs of *A. tessellatus*. 29. Submedian prothoracic hairs of *A. leucosphyrus*. 30. Mid-abdominal anterior tergal plate of *A. leucosphyrus*. 31. Clypeal hairs of *A. maculatus*. 32. Mid-abdominal palmate hair of *A. maculatus*. 33. Clypeal hairs of *A. karwari*. 34. Mid-abdominal palmate hair of *A. karwari*. 45. Clypeal hairs of *A. subpictus*. 36. Mid-abdominal palmate hair of *A. subpictus*. 37. Clypeal hairs of *A. vagus*. 38. Clypeal hairs of *A. turkhudi*. 39. Mid-abdominal palmate hair of *A. turkhudi*. 40. Metathoracic palmate hair of *A. culicifacies*. 41. Metathoracic palmate hair of *A. subpictus*. 42. Showing the position of mesothoracic hair no 5. 43. Mesothoracic hair no. 5 of A. *sundaicus*. 44. Mesothoracic hair no 5 of *A. subpictus*. 45. Antennal hair of *A. aitkeni*. 46. Antennal hair of *A. barianensis*. 47. Submedian prothoracic hairs of *A. culicifacies*. 48. Internal sutural hair of *A. philippinensis*. 49. Posterior clypeal hair of *A. philippinensis*. 50. Internal sutural hair of *A. pallidus*. 51. Posterior clypeal hair of *A. pallidus*. 52. External sutural hair of *A. subpictus*. 53. External sutural hair of *A. stephensi*. 54. Dorsal view of part of 1st abd. segment of *A. stephensi* showing the lateral hairs with 3 to 5 branches.

CHAPTER VI

Tribe: **CULICINI**

The tribe Culicini can be broadly divided into two groups, one representing *Culex* and the other *Aedes* mosquitoes. In the *Culex* group eggs are laid in rafts and the siphon tube of the larva is long; in the *Aedes* group eggs are laid singly and the siphon tube is short and broad. The *Culex* adult is as a rule not ornamented whereas in *Aedes* there are snow white scales on the body and legs.

This tribe is represented in India by fifteen genera, the important ones being *Culex* and *Aedes*; these two comprise nearly half the numbers of mosquitoes belonging to this tribe.

Genus **Culex:** The genus *Culex* comprises about 6 subgenera. Among them *Culex* and *Lutzia* are most widely distributed. The adults are distinguished from all other genera by the presence of well developed pulvilli (these are best seen under a binocular microscope against a dark background when they appear pale).

Subgenus **Culex:** Mosquitoes are of moderate or small size; scales on vertex narrow and upright and no border of flat scales round eye margins in front. Palpi of male always longer than proboscis with last two segments upcurved and hairy. The subgenus is widely distributed over the world, and comprises one of the two largest Culicine fauna in India.

Barraud has classified this subgenus into two groups according to the presence or absence of the lower mesepimeral bristles. Those in which this bristle is lacking have pale rings on proboscis and tarsi. In the second group where the lower mesepimeral bristle is present, the proboscis and tarsi are usually dark. Eggs are laid in rafts and the larva has a long and narrow siphon tube. All the species of this subgenus found in India breed in ground and rocky pools.

Amongst the *Culex* species of mosquitoes which have no pale ring on the proboscis, the most important mosquito is *Culex fatigans*.

FIG. 48.
Culex head.

Genus **Cuxlex L.** These mosquitoes are important for the reason that a large number of them frequent dwelling houses and are terrible pests of man. At least one species, *C. fatigans*, has a wide distribution and has been proved to be the principal intermediate host of *Wuchereria bancrofti*.

Characters :—They are usually moderate-sized mosquitoes; male palpi are longer than the proboscis and curved upwards at the extremities; female palpi are quite short. The genus *Culex* is distinguished from all other mosquitoes by the possession of distinct pulvilli.

Eggs are laid in rafts and the larva has a long and narrow siphon tube.

Synopsis for the identification of the Indian species of *Culex* L. with a pale ring on the proboscis in both sexes (from Barraud, 1923).

1.	Wings speckled with light and dark scales, or with patches of pale scales	2
	Wings dark scaled	4
2.	Wings with patches or spots of white scales as in many species of *Anopheles*	3
	Wings speckled all over with light and dark scales	*bitaeniorhynchus.*
	Wings with a yellowish patch near the tip over the upper fork cell, rather large yellowish mosquito, with pale band on the proboscis occupying nearly two-thirds of the length	*epidesmus.*
3.	Third longitudinal vein with a pale area in the middle ..	*mimeticus.*
	Third longitudinal vein entirely dark	*mimulus.*
4.	Anterior two-thirds of mesonotum conspicuously whitish or pale greyish ochreous, posterior one-third dark	5
	Anterior two-thirds of mesonotum not conspicuously paler than the posterior one-third	7
5.	Abdomen with basal bands, anterior two-thirds of mesonotum with white scales, small species	6
	Abdomen with apical bands, anterior two-thirds of mesonotum with greyish or ochreous scales	*sinensis.*
6.	Tarsi with basal rings only, pale area of mesonotum produced posteriorly in four lines to scutellum, wing scales broad	*whitmorei.*
	Tarsi with basal and apical rings, pale area of mesonotum not produced to scutellum, wing scales narrow	*gelidus.*
7.	Abdominal tergites with apical bands	*cornutus.*
	Abdominal tergites with basal bands	8
8.	Mid and hind femora with a pale stripe anteriorly	*edwardsi.*
	Mid and hind femora not striped	9
9.	Femora dark, speckled with pale scales	*sitiens.*
	Femora dark, without a speckling of pale scales	10
10.	Mesonotum clothed uniformly with dark brown scales, tibiae without any pale stripe on the outside	*tritaeniorhynchus.*
	Mesonotum with brown and some lighter scales, mid and hind tibiae with a more or less distinct pale stripe on the outside	*vishnui, barraudi, whitei* (separated by male hypopygium).

C. fatigans : The most important species, *C. fatigans*, has no pale ring on the proboscis which is of uniform colour throughout its whole length.

Breeding habits.

C. bitaeniorhynchus : a common mosquito; breeds in weedy pools, ponds and ditches.

C. sinensis : though widely distributed, it is never found in large numbers in any place.

C. sitiens : extremely common; breeds in ground pools, rice fields, marshes etc.

C. tritaeniorhynchus : as widely spread as *vishnui* and has the same breeding habits.

C. whitmorei : breeds in ground pools.

C. fuscocephalus : widely distributed; breeds in ground pools, rice fields etc.

C. mimeticus : a hill species.

C. barraudi : recorded from the Punjab and Western Himalaya.

The culicines found commonly in cowsheds or in dwelling rooms are *Culex fatigans*, *C. vishnui*, *C. bitaeniorhynchus*, *C. gelidus*, and *C. sitiens*; of these *fatigans*, *vishnui*, and *sitiens* are known to bite man.

Culex fatigans Wied. They are the commonest among the culicines found at least in the urban areas in India. They are of importance for the reason that a large number of them frequent dwelling houses and are terrible pests of man. *C. fatigans* has

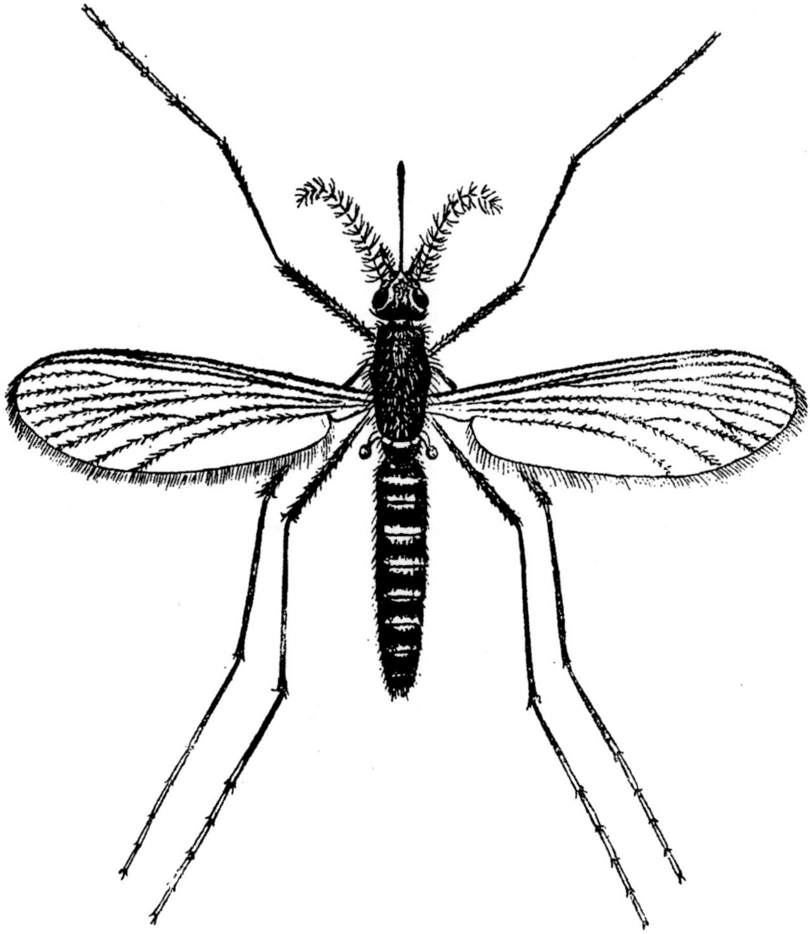

FIG. 49.
Culex fatigans female.

a wide distribution in the plains and has been proved to be the principal intermediate hosts of avian *Plasmodia* and of *Wuchereria* (*Filaria*) *bancrofti*.

Characters. They are usually moderate-sized mosquitoes; male palpi longer than the proboscis and curved upwards at the extremities; female palpi are quite short; the adult mosquito is characterised by the absence of a distinct pale band or ring in the middle of the proboscis. The abdomen is provided with ochreous or yellowish basal bands; pleuræ pale brown and the venter of the abdomen pale. There are very small knee spots especially on hind femora.

Life history. This species is one of the commonest of Indian mosquitoes and is found everywhere in this country up to 6000 ft. or more. It alone is responsible for the winter and spring plague of mosquitoes experienced almost all over India. It is essentially a domestic species and the larvæ are found in localities inhabited by man.

The egg raft of *C. fatigans* is boat-shaped and may contain upwards of 200 eggs. The egg is ovoid and dark in colour; it floats with the narrow end pointed upwards and the broad end, which corresponds to the head of the larva, lies on the surface of water. To the broad end is attached a knob-like micropylar process through which the sperm travels at the time of fertilisation of the ovum. Oviposition always occurs at night. The maturation of the ovum inside the ovary takes at least 72 hours following a blood meal and the earliest time eggs are deposited is at the end of four days after emergence. The larva has a long and narrow siphon tube; it is nearly four times as long as it is broad near its base. The antenna has a tuft of hairs at the junction of the proximal two-thirds and the distal third.

The peak of *Culex* breeding takes place in spring. It breeds in almost any kind of water but where it breeds in prolific numbers, it chooses water contaminated with sewage, e.g., in cesspools, septic tanks, sewage farms, etc. They also breed in ground pools, cement pits, drains, etc. The larvæ subsist on bacteria and other organic materials. The period from the egg to the adult stage takes from 8 to 10 days during January in Calcutta. The biting reflex is at its maximum at 31 per cent relative humidity. This species is unable to stand wide variations of temperature and humidity. It avoids both high and low temperatures. A relative humidity of 80 per cent is not only favourable to the life of *C. fatigans* but the development of *Proteosoma* up to the sporozoite stage may be completed in 6 days though infection of the salivary glands is seldom seen until the ninth day.

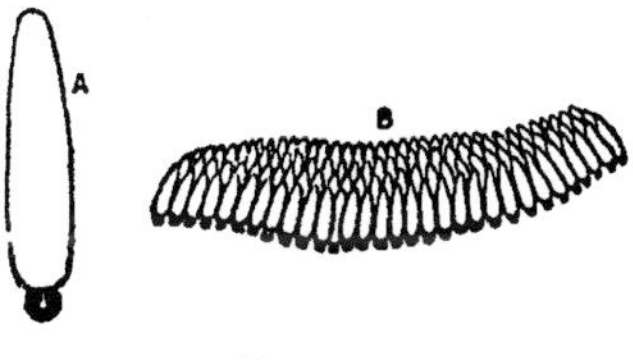

FIG. 50.
A, An individual egg of *Culex fatigans* with micropylar process.
B, Culex egg raft.

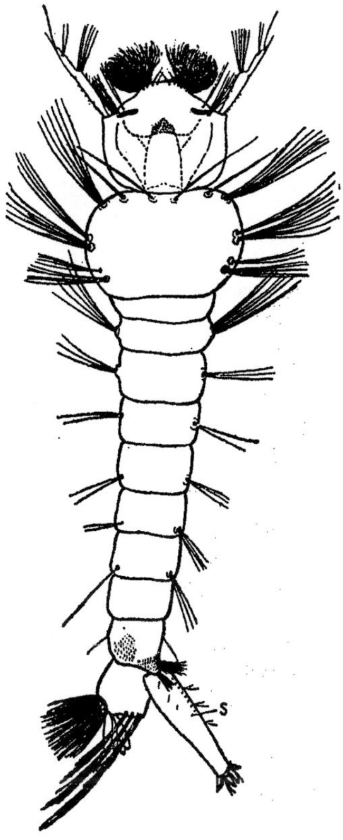

FIG. 51.
Larva of *Culex fatigans*.
s, Siphon tube.

This mosquito is capable of dispersing over a wide area and it is able to cause nuisance even at a distance of 3 miles from the breeding area. The adult is generally reluctant to feed in the cage till 3 or 4 days have elapsed after its emergence. Mating takes place in captivity. It is most active towards dusk and during the first part of the night.

Culex fatigans breeds in stagnant water containing a high concentration of decaying matter. The association with urban areas of *C. fatigans* and consequently of infection with *W. bancrofti* appears to be due to the fact that the density of the population and therefore the quantity of the sullage is greater than in rural areas. An insufficient system of open concrete drains in a town usually favours breeding. The utilisation of sullage water for irrigation sometimes produces intensive breeding, if no precautions are taken to prevent the stagnation of the water. Cesspits and open drains connected with service privies unless adequate precautions are taken, will continue to produce *C. fatigans* even when other measures, however thorough they may be, are taken.

No improvement of the *Culex* position in a city is possible unless a properly planned drainage system is installed. An efficient system of subsoil sewers with treatment of sullage at the outfall is indeed the most effective. No sewage farming should be allowed within a distance of at least three miles from the city. In other situations such as ponds, ditches or other stagnant collections of water, oil should be used. It may be pointed out that the ordinary mixture of kerosene and crude oil does not act so well when the water is grossly contaminated with organic matters. When DDT or Gammexane is used as a larvicide, pyrethrum extract and rosin should be added to the solution. Against the adult mosquito DDT and pyrethrum should be used for indoor spraying.

Defective septic tanks or effluents from septic tanks should be periodically inspected, for they may be the source of prolific breeding of *C. fatigans*.

Culex pipiens Lin. It resembles *C. fatigans* very closely from which it can be separated with certainty by examination of the male genitalia. Both are house mosquitoes. *C. pipiens* is a mosquito of the temperate regions, while *C. fatigans* is a tropical and subtropical species. Although they have been regarded as different species, Weyer (1936) obtained fertile offspring when he crossed the autogenous race of the two species using males of either. Roubaud (1929) obtained adult offspring from crosses between these species only when the autogenous variety of *C. pipiens* was used. Both *C. fatigans* and *C. pipiens* are proved vectors of *W. bancrofti* in China.

Genus **Mansonia** (*Taeniorhynchus*). The mosquitoes of this genus are readily recognised by their well built body, speckled wings and legs. The speckling of the wings is due to the presence of broad scales of both dark and light colour; many of these scales are asymmetrical; the palpi of the male are longer than the proboscis and the apical segment minute; tergite VIII of female has a row of short tooth-like spines; claws of female simple.

The females readily attack man and are at times very troublesome pests. All are swamp or tank breeders especially those thickly over-grown with vegetation.

This genus is composed of four subgenera. Amongst them *Mansonia* and *Mansonioides* are found in India.

Subgenus. **Mansonia. M. crassipes.** There are purplish scales on first four or five abdominal tergites; a patch of silvery flat scales on lower part of sternopleura. It is common in the Punjab.

M. ochracea. Abdomen yellow scaled; wing scales all yellow. This species has been recorded in Assam. It also occurs further eastwards.

Subgenus **Mansonioides.** These are moderate-sized mosquitoes, of robust build and have an yellowish brown appearance. The wings are clothed in broad asymmetrical scales which are mostly light coloured though some are dark. This gives a speckled appearance to the wings. The legs have pale markings and white rings on tarsi. The arrangement of hooks on tergite VIII in females is of specific importance.

FIG. 53.
Pupa of *Mansonioides*.

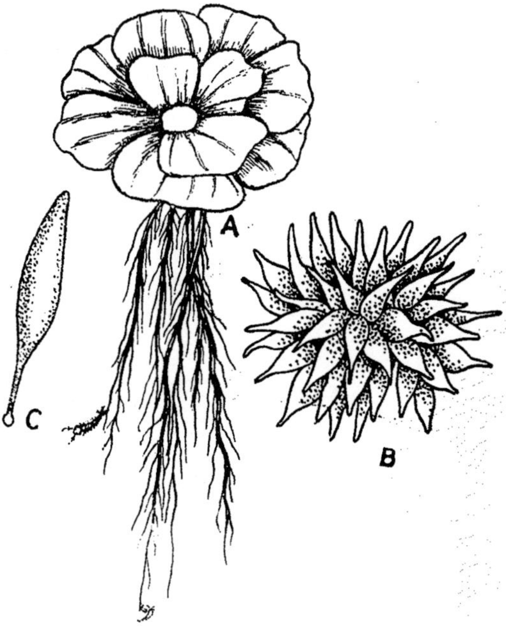

FIG. 52.

A, *Pistia stratiotes* showing how larvae and pupae of *Mansonioides* mosquitoes remain attached to the roots.

B, Egg cluster of *Mansonioides*.

C, Individual egg.

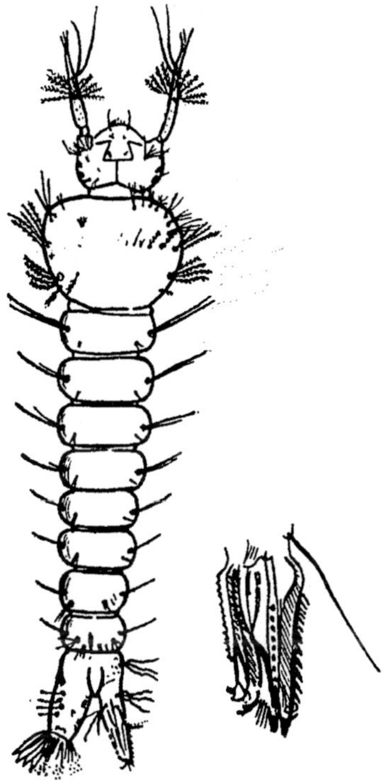

FIG. 54.
Larva of *Mansonioides* showing the saw-like apparatus on the siphon tube.

Key to adult Mansonioides of India
(after Barraud)

1. Mesonotum marked with distinct round spots of white scales. 2
   Mesonotum not so (though there may be small indefinite white spots or patches). 3
2. Yellowish-brown; mesonotum marked with 4 (or more) distinct round white spots; rather broad white scales on mid-lobe of scutellum. *annulifera.*
   Brownish-black; mesonotum marked with 2 (or 3) round white spots; scales on mid-lobe of scutellum narrow. *longipalpis.*
3. Mesonotum marked with a pair of sublateral greenish stripes on a brown ground. *uniformis.*
   Mesonotum dark brown, not marked with greenish stripes; some white scales, tending to form indistinct spots or patches, in some specimens. *indiana,*

M. annulifera Theo. It is widely distributed in India and occurs in Bihar and eastwards; also in Orissa and along the coasts of Peninsular India including Ceylon. Outside India it is found in Borneo, Malaya, Siam, Philippines and Dutch East Indies.

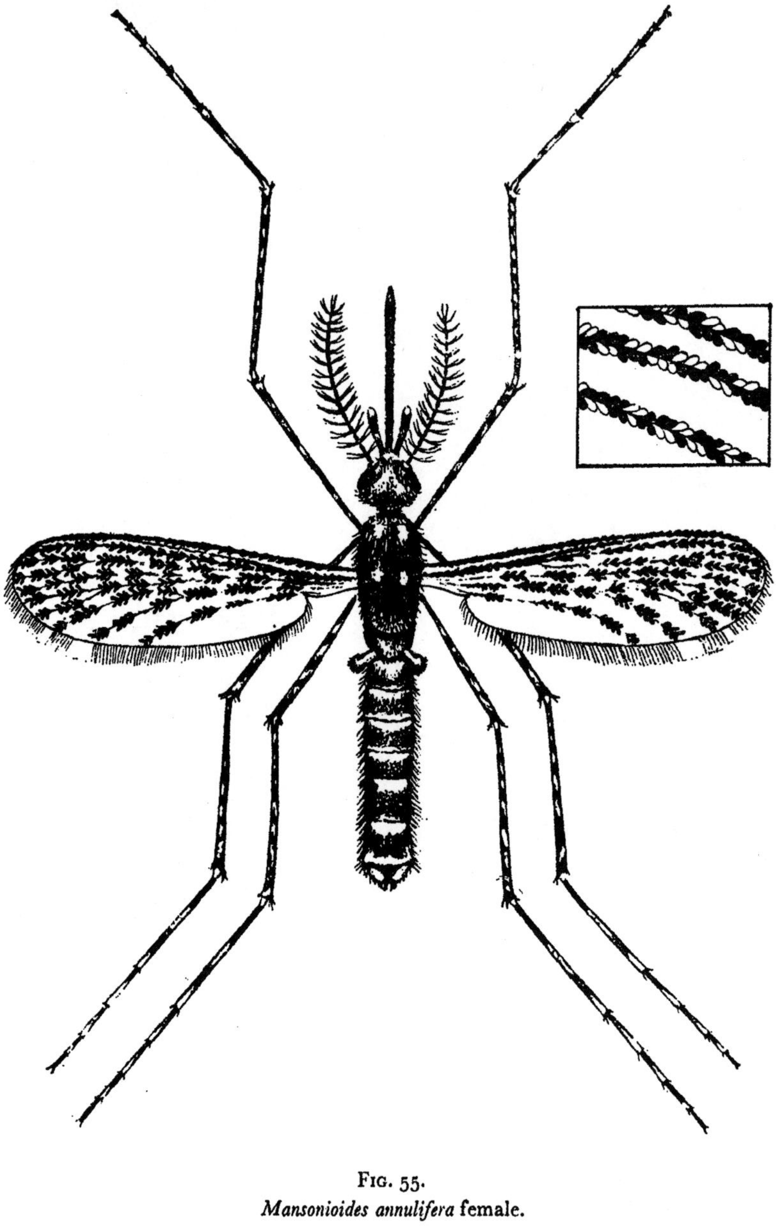

FIG. 55.
Mansonioides annulifera female.

M. longipalpis v.d. Wulp. It is larger in size than *M. annulifera*. It has been recorded in Assam and further east.

M. uniformis Theo. It is the most widely distributed and the commonest mosquito among the

Mansonioides found in India. This species occurs in the comparatively wet parts, such as eastern and southern India.

M. indiana Edw. It is less common than *M. uniformis* and occurs mostly in Assam, Bengal and in parts of South India. The characteristic feature about the breeding particularly of *M. uniformis* and *M. indiana* is that they depend on certain aquatic plants especially *Pistia stratiotes* for the supply of oxygen. Larvae and pupae do not come to the surface until shortly before the emergence of the mosquito. The larva is provided with a pair of long curved spines or hooks on the siphon tube and a kind of saw apparatus at the extreme tip. The larva remains anchored to the plant by means of the hooks and the saw apparatus is used for boring. The same mechanism also exists in the pupa. The eggs are always laid in clusters on the undersurface of the leaf.

Control. The existence of *Mansonioides* larva without the associated aquatic plant on which it depends for the supply of oxygen, is utterly impossible. In the case of *M. uniformis* and *M. annulifera*, Iyengar (1937) and Sweet and Pillai (1937) have shown that clearance of pistia will at once stop their breeding.

Genus **Armigeres.** Though they are not known to convey any disease, they are fearless and intolerable biters and migrate in numbers in dwelling houses. They are distinguished from *Aedes* by their large size, absence of silvery white patches or spots on the mesonotum, and by the form of the proboscis, which is curved downwards at the tip, and distinctly flattened laterally; the female claws are toothed. They are fairly large mosquitoes with flat scales on vertex and scutellum.

A. obturbans Walk. This species is widely distributed in India. It is a large, dark mosquito with a distinct pale border to the mesonotum, and with apical pale bands on the 3rd to the 6th abdominal sternites; these bands are usually wide and are all about the same width; palpi of the female about one-third the length of the proboscis; abdomen shows patches of white scales laterally; scutellar scales are entirely dark.

Adults are intolerable biters. They migrate towards dusk and bite even in the day time. They breed in enormous numbers in small pools and earthen drains grossly contaminated with urine. The earthen drains along which the blood from slaughter houses flows often contain millions of larvae. Another prolific breeding place is the vat used for soaking hides in the manufacture of leather. In these situations they breed throughout the year. In China and Indo-China enormous numbers develop in latrines; larvae may occasionally be found in bamboo stems, tree-holes, etc.

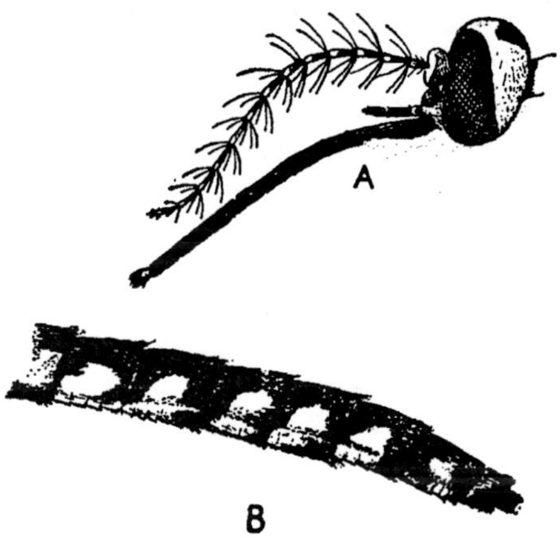

Fig. 56.
A. Armigeres head, and B. abdomen.

The larvae are predators and prey on other mosquito larvae. The tracheal gills are particularly well developed. The antennae are of uniform width and are almost bare. The siphon-tube is short and broad. The chitinous hooks on the 8th segment are arranged in a row as in *Aed. aegypti* but they have a different shape.

Both larvae and adults are particularly resistant to ordinary chemical larvicides but succumb quickly to DDT-pyrethrum-kerosene mixture. Gammexane may be used in place of DDT.

A. obturbans has been successfully infected with *Plas. knowlesi* in India.

Genus **Leicesteria.** They are similar to *Armigeres* in external appearance except that there are no postspiracular bristles and instead, this area is covered with flat black scales anteriorly and white scales posteriorly.

They all breed in bamboo stumps and do not attack man.

Genus **Lutzia.** The adult mosquitoes resemble the large species of *Culex* in general appearance and structure and scaling. The adults are, however, distinguished from *Culex* by the presence of more numerous lower mesepimeral and pro-epimeral bristles. The female palpi 1/4 length of proboscis; fore and mid femora and tibiae mottled with spots of pale scales.

Larvæ are predacious and live in pools and domestic collections of water.

L. fuscana Wied. Abdominal tergites VI-VIII, entirely yellow scaled or with broad apical bands; tergites II-IV entirely dark or with very narrow apical bands. It is widely distributed chiefly in the plains. The adult has been recorded as biting man.

The larvae are predators and also cannibalistic in habits, preying upon the larvae and pupae of other mosquitoes breeding together. They are found in domestic collections of water, ground pools etc.

The mouth brushes of the larva are specialised and adapted for holding prey, each brush consisting of a number of strong curved rods with minute hairs along one side of the apical half; the clypeus is armed with a number of processes.

L. raptor Edw. Very similar to *L. fuscanus* except that the abdominal tergites have whitish or ochreous apical bands. The larvae are found in the same situation as those of *L. fuscanus*. This species occurs mostly on the western side of India.

Mosquitoes and filarial diseases of man. (**Culex fatigans and Wuchereria (Filaria) bancrofti.** Though the disease had existed for centuries, nothing regarding its cause was known until 1863 when the embryos were discovered in the hydrocele fluid by Demarquay and in the chylous urine by Wucherer. In 1872 Lewis discovered the embryos in the blood. In 1876 Joseph Bancroft confirmed the earlier work of Lewis who had first noticed the presence of microfilaria in the peripheral blood of man. The same year Bancroft demonstrated the presence of the adult worm *Wuchereria bancrofti* (*Filaria bancrofti*) in a lymphatic abscess. Later Patrik Manson formulated his hypothesis of *Culex* transmission of the disease when he found larval filaria in the stomach and noticed development of the embryo taking place in the body of the mosquito. According to Manson, the larvæ escaped from the body of the mosquito into water in which the mosquitoes died and thus entered man's body through his alimentary canal. The exact mode of transmission was later described by T. L. Bancroft, the son of Joseph Bancroft, who noticed that the microfilaria required about 16 to 17 days in the mosquito before it could reach the infective stage. The filarial embryo left the body of the mosquito through its proboscis and human infection occurred at the time of biting.

Filariasis is distributed throughout the world from about 41°N. to about 28°S. latitude in the Eastern Hemisphere and from about 30°N. to about 30°S. latitude in the Western Hemisphere.

Culex fatigans is a proved carrier of the microfilaria of *W. bancrofti*. This mosquito is, however, refractory to that of *Wuchereria malayi* which develops in *Mansonioides* species. It is worth noting that both *Microfilaria bancrofti* and *Microfilaria malayi* may occur in the same individual.

The commonly accepted view regarding the mode of entry of larvæ of *W. bancrofti* into the body cavity of the mosquito is that microfilaria from human blood enter the stomach of the mosquito and after casting their sheath penetrate through the wall of the stomach, enter the haemocœle and then find their way into the thoracic muscles. According to Iyengar (1936), the worms enter the hæmocœle in a remarkably short time frequently in less than half an hour after the infective meal. He also demonstrated that the microfilariæ ingested with the blood into the stomach of the mosquito tend to stay in the cardiac portion of the midgut immediately behind the proventriculus. It thus travels forwards until it reaches the proventriculus itself and escapes into the perivisceral cavity of the thorax. Those which have moved to the pyloric region do not appear to have much chance of entering the hæmocœle. On escaping into the perivisceral cavity from the proventriculus, the larva enters the muscle bundles, either the vertical or the longitudinal ones. It moves about in the spaces between the muscle bundles for some time before finally penetrating them. Inside the muscle it stretches out parallel to the muscle strands and then becomes dormant. They are found in different stages of development in the thoracic muscles and during their development they undergo extraordinary transformation in their outward appearance. As the worm gets older its activity increases. During the last stages of its metamorphosis the filaria, which up to this time has been at rest in the thoracic muscles, migrates and finds its way either to the connective tissues of either the thorax, legs, head or the proboscis. During the course of development of larvæ of *W. bancrofti* in the mosquito, chitinous encapsulation of filarial larvæ may take place. The encapsulation may be complete or partial. When partial, the filarial larvæ are able to break loose from its chitinous sheath (Hu, 1938).

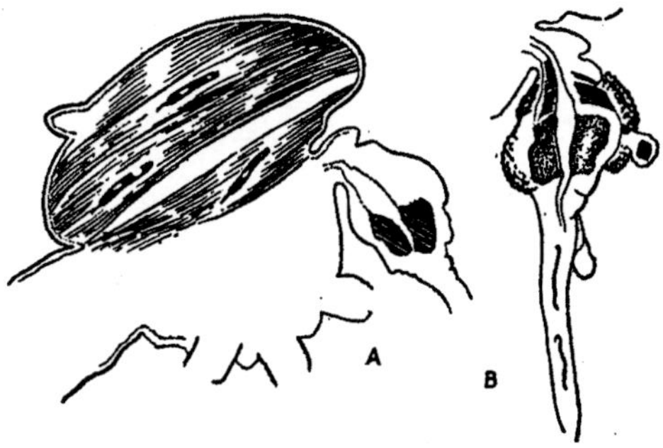

FIG. 57.
Development of *Wuchereria bancrofti* in *Culex fatigans*.
A. After 3 days ; B. After 10 days.

It was once thought that the larvæ escape by rupturing a membrane at the apex of the labium on its dorsal side at the junction of the labium and the labella; this membrane has been called Dutton's membrane. The conjectures of Yamada and Komori (1926) that the filaria escapes through the extreme tip of the labella have been confirmed by Rao and Iyengar (1932). Fully developed larvæ of filaria are visible to the naked eye and when set free from the proboscis of the mosquito, they show very active movements. Menon and Ramamurti (1941) found that mature larvæ of *W. bancrofti* were unable to penetrate through unbroken skin. They suggest that the mechanism of transmission has two parts, the provision of a stimulus for the larvæ to penetrate the proboscis and the direct inoculation of the larvæ into the tissues. The stimulus is the act of sucking of warm blood by the mosquito. At the time of biting some larvæ are deposited on the dry margins of the puncture and these apparently die with the drying of the parts.

According to Basu and Rao (1939), both temperature and relative humidity play a very important part in the transmission of *W. bancrofti* in man by *C. fatigans*. The optimum conditions for transmission found by them are 80°F. and 90 per cent relative humidity.

The minimum microfilarial density sufficient for infection of mosquitoes is important in order to determine the efficiency of certain drugs in rendering persons harbouring microfilariæ non-infective to the mosquito. The minimum number varies according to the finding of different workers. Thus Bahr thought three to be the required number per c.mm. for this purpose. A m.f. count of one or less gives unsuccessful results. According to Hewitt *et al* (1940), it should be at least one or more per 2 c.mm. of blood. Opinion is, however, unanimous that the presence of large numbers of microfilariæ, even 10 per c.mm., will invariably kill a considerable number of the infected mosquitoes (Cruickshank and Wright, 1914). It is, however, a well recognised fact that on an average, mosquitoes take up a relatively greater concentration of microfilariæ, about 40 to 50 times more than is to be found in finger prick blood (Ashburn and Craig, 1907).

Examination of Blood for Microfilaria. (1) Collect blood between 10 p.m. and 2 a.m. The patient should be in darkness for about at least an hour before collection of the sample. (2) Four or five drops of blood are taken by finger prick and made into a thick film about one inch square and allowed to dry. The film is dehaemoglobinised the next day by distilled water and examined wet under low power of the microscope for presence of microfilariae. If positive, the film can be stained by Leishman stain or by haematoxylin and eosin or Giemsa, after fixation in a mixture of equal parts of alcohol and ether. (3) 5 c.c. of venous blood is collected in 10 cc. of normal saline containing 1.5 per cent sodium citrate. This is thoroughly mixed and left overnight. This is haemolysed the next day by addition of 1 per cent saponin in normal saline drop by drop till haemolysis is complete. It is next centrifugalised and the deposit examined under the low power of the microscope without staining. If necessary, a smear may be made with the deposit, just dried and stained with Leishman stain.

Although the population of *C. fatigans* is at its lowest during the monsoon, this is the most favourable season for the transmission of filaria by this species, the development of the worm during this time taking place within 10 to 12 days while during the winter the embryos take 18 to 20 days to develop (Rao and Iyengar, 1932). Rao (1929), on the other hand, had previously found the highest natural infection during the winter months. These observations were made both in the laboratory and in the field. The infective stage of *Litomosoides carinii*, the cotton-rat filaria, on the other hand, is reached after the 14th day.

Culex fatigans is not the only mosquito which acts as the intermediate host of *W. bancrofti*, though it plays the dominant role in its transmission. In China *C. pipiens* as well as *C. fatigans* act as efficient vectors. A large number of *Anopheles* species are able to carry the parasite and infect man. This is surmised from the ability of the filarial embryos to undergo the fullest development in these mosquitoes and from the discovery of filaria-like parasites in the body of wild *Anopheles*. Almost in every country where filariasis is endemic certain *Anopheles* species are labelled as efficient though subsidiary transmitters of the parasite in nature. In Africa *A. gambiae* and in South America *A. albimanus* and *A. darlingi* are regarded as filaria carriers. Newton, Wrighty and Pratt (1945) have investigated the potential mosquito vectors of *W. bancrofti* in the Continental United States. In addition to *Anopheles albimanus* and *Aed. aegypti*, the parasite has been noticed by them to undergo the fullest development in a large number of culicine mosquitoes. Rao and Iyengar (1932), on the other hand, found *Aed. aegypti* an inhospitable host for *W. bancrofti*.

In the Dutch East Indies *A. barbirostris* has been shown to have an artificial infective index of 83 per cent and a natural one of 8.9 per cent. In certain localities in China and especially in the rural areas *A. hyrcanus var. sinensis* is regarded as a more efficient vector in nature than *C. fatigans*. However, it is noticed in India that *W. bancrofti*, when discovered in *Anopheles* mosquitoes, are generally present in very small numbers.

Rao and Iyengar (1932) have tested the susceptibility of the common Indian species in this respect under experimental conditions and have found complete development taking place in *Anopheles philippinensis*, *A. pallidus*, *A. annularis*, *A. stephensi* and *A sundaicus*. According to them *Armigeres obturbans* and *Aed. vittatus* are totally refractory. Partial development was noticed to proceed in a number of other mosquitoes.

In Fiji and Polynesia *Aed. scutellaris* acts as an efficient transmitter of *W. bancrofti* which is non-periodic there. *Aed. scutellaris* is a day-biting mosquito.

Hetrazan. In testing filaricides in the laboratory the usual laboratory animals employed are the dog and the cotton rat and the filarial parasites on which the tests are made are *Dirofilaria immitis* and *Litomosoides carinii* respectively.

It has been universally observed that Hetrazan when administered per mouth has a marked effect in freeing the peripheral blood of microfilaria within 24 to 48 hours. The total amount to be given is 1 gm. for an adult and the duration of administration should be 5 days given twice a day. Following its use in some cases, more marked in those infected with *W. malayi*, there are untoward reactions of varying severity. Within 24 to 36 hours after commencement of the treatment, there may be fever, nausea, vomiting, headache, body pains; inflammation of the lymph nodes. These as a rule disappear within 3 to 4 days of the onset. This presents some drawback in the use of the drug for mass treatment.

The so-called anti-microfilarial effect of the drug is, however, short lasting. Within about 8 months after cessation of the treatment, the microfilariae reappear in the peripheral blood and this is generally accompanied by recurrence of the symptoms of the disease.

The literature on the action of Hetrazan in *bancrofti* filariasis has been reviewed by Raman, Ramamurthy and Pinakapani which has been published in the *Jour. Ind. Med. Assoc.*, Vol. 19, 1950, p. 163.

Control of filariasis. (1) Anti-Culex measures by spraying with a mixture of DDT, pyrethrum and citronella oil: this should be introduced on a wide scale. It should be remembered that the residual effects of DDT against *C. fatigans* are not so pronounced as against anophelines. The spraying should therefore be frequently repeated. Brown and Williams (1949) and later Kohler, Brown and Williams (1951) noticed that by DDT residual house spraying the population of *Culex fatigans* (*C. quinquefasciatus*), the vector of filariasis in St. Croix (Virgin Island), was reduced approximately 50 per cent in the houses and 50 per cent reduction of infected mosquitoes. The infection rate in school children also dropped. In India the output of *Culex fatigans* is so enormous during late winter and early spring, particularly in cities and towns, that it would appear that DDT spraying has hardly any marked effect on this mosquito. However, spraying with DDT alone is ineffective in coping with this nuisance. It is therefore always necessary to mix an appreciable quantity of pyrethrum with DDT and to which some citronella oil should be added. The mixture when sprayed will appreciably reduce the number of mosquitoes entering the house but, as has been pointed out previously, its action does not last longer than 6-8 hours. Unlike *Anopheles* adults, the control of *C. fatigans*, when the attack is against the adult mosquito alone, is therefore a very tedious and expensive operation unless it is accompanied simultaneously with anti-larval control measures. (2) Anti-larval measures by using a 2.5 per cent DDT in Malariol at the rate of 1.75 gallons per acre and applied once a week. There is a significant reduction in the density of adult *Culex fatigans* and at the same time in the filarial infection rates both in the mosquito and in man. (3) General anti-mosquito measures such as the use of mosquito-nets and application of culicifuge on the exposed parts of the body. (4) Permanent anti-*Culex fatigans* measures such as the institution of a well planned drainage system.

Mansonioides sp. and **W. malayi.** *W. malayi* has a patchy distribution in India. Outside India it occurs in Malaya, Java, Sumatra, Borneo, Celebes and South China. It does not seem to spread further eastward than the Moluccas and the lesser Sunda Islands.

Mosquitoes of the genus *Mansonioides* are normally the true intermediate hosts of *W. malayi* wherever the latter occurs. It does not undergo any development in *Culex fatigans* which is the intermediate host of *W. bancrofti*. In India *M. uniformis* is the most important carrier.

In Java the four common carriers of *W. malayi* are *M. annulifera*, *M. indiana*, *M. uniformis* and *A. hyrcanus* var. *sinensis*. The last one is extremely susceptible to infection by *W. malayi* having an index of 86 to 90 per cent in the laboratory feeding experiments. In South China the same *Anopheles* is regarded as a carrier in nature of both *W. bancrofti* as well as *W. malayi*, especially in the rural areas.

Areas from where *W. malayi* infection has been reported in India are: Orissa, Madhya Pradesh, Madras, Travancore, Cochin and Assam.

Control. *Mansonioides* mosquitoes cannot propagate without associating themselves in the larval stage with certain types of aquatic plants. Removal of the plants will at once stop their breeding.

Genus **Aedes** (Stegomyia). Mosquitoes of the genus *Aedes* are responsible for the transmission of the causative agents of two important diseases of man, dengue and yellow fever. A less important part is played when they carry *Dirofilaria immitis* and *D. repens* to dogs. Only one species of *Aedes* is definitely known to carry the microfilariæ of *Wuchereria bancrofti* in Fiji. The natural host of *Plasmodium gallinaceum* is the jungle fowl and the parasite is transmitted by various species of *Aedes* including *Aed. vittatus* and *Armigeres obturbans*. It, however, does not develop in *Anopheles* and *Culex* species of mosquitoes.

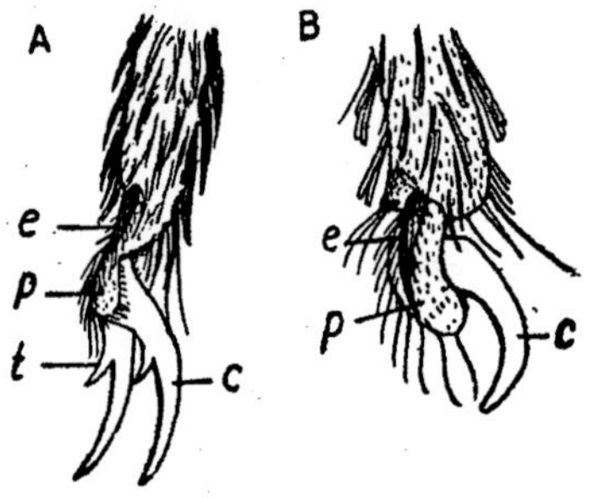

Fig. 58.
Front leg of (A) *Aedes aegypti* and (B) *Culex fatigans*.
c, claw; e, empodium; p, pulvillus, t, extra tooth.

They lay their eggs singly, and the larvæ have a short and broad siphon tube. They are all bottom feeders and breed in small collections of water especially in artificial containers. The adult mosquito is ornamented with silvery white scales all over its body.

Characters. Colour black or blackish brown; body ornamented with snowy white or silvery spots, patches and markings; spiracular bristles absent; female palpi short ; the palpi in the male are normally longer than the proboscis; the last two joints slender, upturned with very few hairs; pulvilli absent; tarsi of hind legs with basal white rings only, the apices of the segment being dark except the last segments which are wholly white; front and middle claws of the female have an extra tooth; 8th abdominal segment of female retractile and in some cases hidden within the 7th.

Important species of this genus are, *Aedes aegypti*, *Aed. vittatus*, *Aed. albopictus* and *Aed. variegatus* (*pseudoscutellaris*).

Key to the Common Species of Aedes.

Mesonotum with 4 to 6 small round white spots; femora with complete pre-apical white rings. *vittatus* (*sugens*).

Mesonotum with a white median stripe; mesonotal stripe narrow; all joints of hind tarsi white ringed; scutellar scales all white; mid-femora without a pre-apical white spot in front; white lateral markings of abdomen broadest in middle of segments; white scales of pleurae arranged in lines. *variegatus* (*pseudoscutellaris*).

White lateral markings of abdomen broadest at bases of segment; white scales of pleurae arranged in patches. A small patch of flat, silvery-white scales just in front of each wing base; no definite pale line on margin of mesonotum; median thoracic stripe narrow and more silvery white. *albopictus* (*scutellaris*).

Tibia not ringed with white; last segment of hind tarsi completely white; midfemora without a pre-apical white spot in front; mesonotum with two median yellowish lines and a pair of sublateral curved silvery white lines. *aegypti* (*argenteus*).

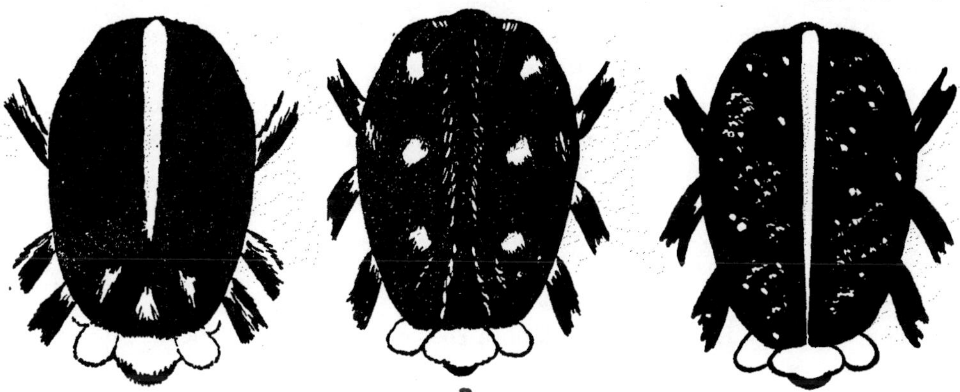

A B C

FIG. 59.
A, *Aedes albopictus*.
B, *Aedes vittatus*.
C, *Aedes variegatus*.

Aedes aegypti L. It is a domestic mosquito. It can be easily distinguished by the characteristic lyre-shaped markings on the thorax, while the larva has a pair of lateral hooks on each side of the thorax.

Aedes vittatus Bigot. (**Stegomyia sugens.**) It is not a domestic mosquito neither is it so common as *Aed. aegypti*. Mesonotum has four or six small round white spots. It is widely distributed in the Ethiopian and Oriental regions occurring as far south as Ceylon.

Aedes albopictus Skuse. (**Stegomyia scutellaris**) : It is also an Oriental species and possesses a single median silvery white stripe on the mesonotum and separate patches of white scales on the pleura of the thorax. Barraud (1934) gives tree-holes, bamboos, leaf-axils and only rarely artificial receptacles or rock-pools as their breeding sites. Sen (1935) found it breeding in broken glass jars and metal drums.

Though it lives away from man, it bites freely whenever it gets the chance to do so. In Java and Malaya it has been reported breeding in sewage disposal installations.

In Manila *Aed. albopictus* is, next to *Aedes aegypti*, the most common mosquito and has similar breeding habits to *Aed. aegypti*. It has been proved to be capable of producing dengue in volunteers (Simons, St. John and Reynolds, 1932). It is the carrier of dengue in Japan, Formosa, Sumatra and Philippines.

The progeny of female *Aed. albopictus* when mated with *Aed aegypti* gives rise to hybrid adults resembling *Aed. albopictus* (Toumanoff, 1936).

Aedes variegatus Dol. It is a widely spread species in the Australasian region, and under its synonym of *Stegomyia pseudoscutellaris* is well known as the only carrier of the non-periodic filarial parasites in the Pacific Islands such as Fiji, South New Guinea, Ellis Islands, South Gilbert Island and whole of Polynesia. On the other hand, nocturnal periodic *Wuchereria bancrofti* prevails in Melanesia and Micronesia, the Bismarck Archipelago, New Caledonia, Solomon Island and New Hebrides. This is carried by *Culex fatigans*. It is a widely spread species not only in the Australasian region but also in India and has been recorded from the Andaman Islands. In the Hawaiian Island it is the only representative of this group.

The usual breeding places of *Aed. variegatus* are in cocoanut husks, old tins, tree holes and any other small sheltered collection preferably of rain water.

Aed. variegatus is a day-biting mosquito and bites most avidly in the early morning. It has the same breeding habits as *Aed. aegypti*. It flourishes in the bush as well as in and around houses. Like *Arm. obturbans* it is an unsuitable host for *W. bancrofti*. King, Pandit and Menon (1929) have reported the finding of filaria without any developmental forms in this species. It is, however, capable of transmitting dengue (Simons, St. John and Reynolds, 1932). It differs from *Aed. albopictus* in having three parallel white stripes on the pleura of the mesothorax and the white abdominal cross bands incomplete.

Aedes aegypti L. is the common tiger or yellow-fever mosquito and is widely distributed both in the tropical and sub-tropical regions. In colder countries it appears in the summer but with the advent of cold weather it disappears.

Adult. *Head* : dark with distinct double white median line; palps black, white at the tip, proboscis black. *Thorax* : brown with characteristic lyre-shaped marking on the dorsum; scutellum completely covered with silvery white scales; pleura with several patches of brilliant white scales; the first and second pairs of legs with white bands on the tarsi, the hind pair with five white bands, the last joint being wholly white. *Abdomen* : dark with white bands on the bases of the segments and also laterally.

Egg. The egg is elongate, blackish in colour and the shell is highly chitinised. The outside is sparsely studded with minute hemispherical bodies of whitish secretionary matter.

Larva : The larval head and thorax are broad; antenna is almost bare except a single short hair on the shaft near the middle and a few delicate hairs at the apex. There is a pair of chitinous hooks on each side of the thorax. On each side of the 8th segment of the abdomen are the lateral combs arranged in a line, and composed of 8 to 10 serrated spines. The tracheal gills are long.

Pupa : The respiratory trumpets are shorter and broader than in *Anopheles*. In mature pupa the adult is visible through the pupal skin.

Life history. The eggs are conspicuously dark objects and are always laid separately. The production of eggs has a direct bearing on the amount of blood ingested by the female. Buxton (1927) recorded the number of egg per laying as maximum 37.3, mean 20.5 and minimum 8.3. The maximum number of eggs recorded by Roy is 83 (1936).

The eggs are deposited not only on the surface of the water but also along the sides of the containers above the water level and they adhere closely thereto. Eggs

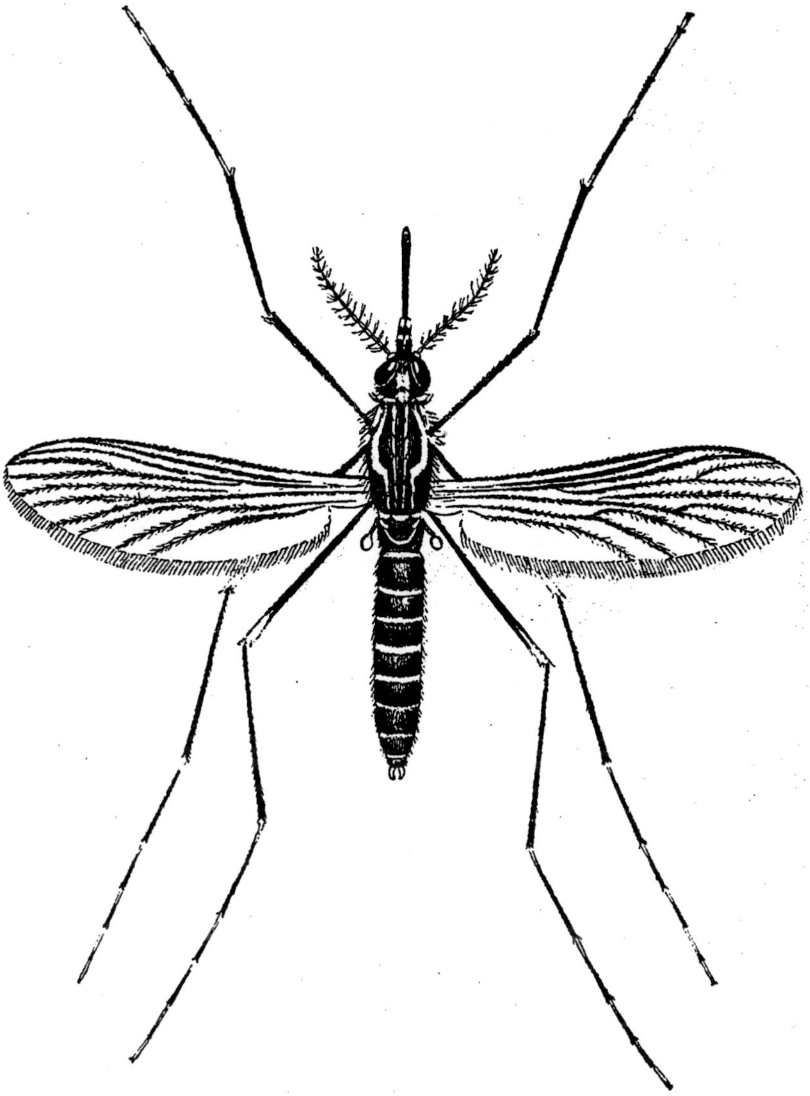

Fig. 60.
Aedes aegypti female.

generally float on the surface but sometimes they sink to the bottom though this does not interfere with subsequent hatching. The eggs are very resistant to unfavourable environment and will hatch after storage for several months in a dry place. It is not, however, a partly developed egg but a fully developed resisting larva enclosed in serosa and amnion. The larvæ are ready to emerge under suitable conditions. They can thus be safely transported to distant places. For this purpose eggs should be picked up from water with a camel hair brush 40 hours after they have been deposited, and placed on dry filter paper.

FIG. 61.
Egg of *Aedes aegypti* (highly magnified).

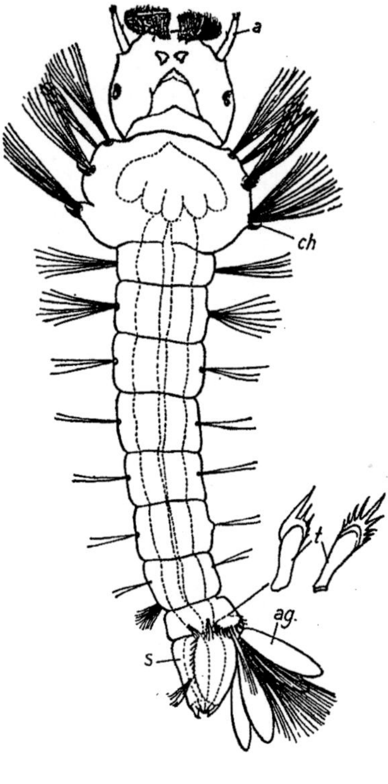

FIG. 62.
Larva of *Aedes aegypti*:
ch, chitinous hook ;
ag, anal gill ;
s, siphon tube ;
a, antenna ;

As *Aedes aegypti* is a highly domesticated mosquito, its breeding places are never far from the abode of man. It seldom breeds in any natural collection of water but for reasons unknown it prefers small artificial collections. Its larvæ are found in discarded tins, broken bottles, cocoanut shells, flower pots, flower vases, broken crockery, earthen pots, cut ends of bamboo, tree holes, rock holes, roof gutters of houses, choked drains, ant-guards, water cisterns, fire buckets, drinking water pitchers, barrels, leaf axils of pineapple and *Dressina* etc. Sometimes enormous numbers of larvæ are found in rain-water collecting in country boats. A faintly saline water or a moderately foul water is not always rejected by the female for oviposition though this species is regarded as a clean water breeder showing a decided preference for rain water. It should therefore be borne in mind that a very small quantity of water from a tea-spoonful upwards may suffice as a breeding place for Aedes. It is particularly attracted to hay infusion and Buxton and Hopkins (1925) thought that the larvæ might be destroyed by poisoning the infusion with arsenious oxide or copper sulphate and thereby check its breeding. *Aed. aegypti* breed continuously throughout the year provided the temperature does not remain below 68° F. In the tropics the eggs generally hatch within two days after deposition. When covered by frost they retain their vitality for a long time. Atkin and Bacot (1917) demonstrated that the presence of bacteria, yeasts and less definitely moulds, exerts a stimulus which causes eggs to hatch. This stimulus is less powerful if killed cultures or sterile filtration and extracts of bacteria and yeasts are used though it is possible to rear larvæ under sterile conditions.

The larva has characteristic worm-like movement and is a bottom feeder. What it actually feeds upon is not known but it is apparent that it can sustain itself on solids dissolved in the water though in the laboratory it greedily ingests solid particles. The retardation of the natural development of larvæ of *Aed. aegypti* in the absence of micro-organisms has been noticed by Hinman (1933). They will grow, however, when autoclaved liver extract and yeast are added (Trager, 1935). Similar observations have been made by Rozeboom (1934). The larvæ grow very rapidly during the rainy season especially when they are kept agitated by rain drops. The larva is resistant to immersion and can be resuscitated after two hours' submersion. The larval life under the most favourable conditions lasts four days.

The pupæ of *Aed. aegypti* are not able to descend unharmed to such great depths as the larvæ. They usually descend only an inch or two whereas the *Aedes* larva can go down even to 5 feet (Macfie, 1923).

Sunlight is not at all essential for the development of *Aed. aegypti* (Jobling, 1937).

The adult female is a vicious biter and will persistently haunt man. They are apparently fearless. In presence of man it will never feed on any other animal. It is characteristically a day-biter taking blood between day light and dusk but when no blood is available in the day, it will bite at night. It can easily bite through thin socks and shirts.

In attractancy tests Brown (1951) found that moist clothing attracted 2 to 4 times as many as dry clothing, and sweat, CO_2 increased attractancy. White clothing was much less attractive than red or blue.

Whereas several blood meals are required before oviposition when *Aed. aegypti* are fed on rabbits, guineapigs or monkeys, it is exceptional for more than one necessary for the purpose when fed on man and in the latter the average number of eggs laid is always higher (Mathis, 1934).

Egg formation has not been known to occur without a suitable amount of blood feed. Fertilised ova may sometimes lie latent in the body for some time. The female lays her eggs on the fourth night after the blood feed and only one blood meal is required for laying. Oviposition will not take place in a solution of common salt in 2 per cent strength and if eggs are laid at all, the larvæ are readily killed (Macfie, 1921).

The adult mosquito definitely prefers dark and black material on which to rest. It is a strictly domestic mosquito and is incapable of flying any great distance at one time. Its range of flight is probably restricted to 50 yards. Both males and females can be sustained on honey, raisin, sugar, banana etc, for a long time extending to several months. The female lives longer than the male. The length of life of male *Aed. aegypti* in captivity, as noticed by Beeuwkes, Kerr and their coworkers (1933), was up to a maximum of 135 days and the mean length of life varying from 40 to 61 days. Female mosquitoes survived for a maximum period of 225 days with mean length of life of from 70 to 116 days. In captivity pairing takes place about 12 hours after emergence from the pupa and is followed by a meal of blood. Under certain laboratory conditions the female feeds readily on frog and turtle and produces viable eggs. Blood from different species of vertebrates may differ in its effect on the egg production (Woke, 1937). Hibernation does not take place in this species. They can be easily transported in aircraft to distant places without producing any ill effect to them (Hicks and Dewanchand, 1936). Similar observations have been made in America and in the Belgium Congo. This species especially the blood-fed female will survive for 3 to 4 hours when it is subjected to a pressure of 27 mm. of Hg. It is now fully realised that aerial transport involves new risks of introducing diseases into areas previously exempt from them, and of these yellow fever is the most feared. It is carried from one place to another and even from one country to another in steamers, aeroplanes, railway trains and country boats. It is the most common mosquito found in ships and freely breeds in country boats and steamers.

Aed. aegypti is the intermediate host of *Plasmodium gallinaceum*, and a detailed account of the development of the parasite from sporozoite to erythrocytic trophozoite has been given by Huff and Coulston (1944).

For the biological testing of antimalarial synthetic compounds *Aedes aegypti* are employed for infecting 10 days old chicks with *P. gallinaceum*. The drug under test is administered orally to the chick by means of a catheter tube inserted into the gizard one or two days after the appearance of the parasites in the peripheral blood which generally takes place from about 9 to 10 days after the mosquito bite. The parasites seldom appear in the salivary glands before the 7th day.

Chick embryos can similarly be inoculated with sporozoites of *P. gallinaceum* by inducing infectious mosquitoes to feed through shell membrane and the foetal membranes immediately beneath. Chick embryos which have been incubated at 99-100°F. for 10 to 15 days are used and the drug is introduced intravascularly.

Besides *P. gallinaceum* other parasites that are also employed for the same purpose and the mosquitoes used for transmitting the parasites are the following : *P. relictum* (*P. praecox*) found in canaries and sparrows—*Culex fatigans*; *P. lophurae* in chicks and ducks—*C. pipiens*, *C. fatigans*; *P. berghei* in white rats and golden hamsters—*A. stephensi*; *P. knowlesi* in *Rhesus* monkeys—*A. annularis*, *A. stephensi*; P. *cynomalgi* in *Rhesus* monkeys—*A. m. atroparvus*, *A. annularis*, *A. subpictus*, *A. quadrimaculatus*; *P. inui* in *Rhesus* monkeys—*A. m. atroparvus*, *A. stephensi*; *P. vivax*, *P. falciparum* and *P. malariae* in man—*A. stephensi*, *A. quadrimaculatus*, *A. maculipennis*, *A. punctulatus*, *A. gambiae*.

Breeding. A method of rearing large stocks of Stegomyia mosquitoes in a closed cage for continuous breeding has been described by Leeson (1932). There is a compartment for accomodating a guinea pig from which the female mosquitoes obtain their blood meals. The tray at the bottom of the cage permits it to be taken out for cleaning without letting out the mosquitoes. In another compartment there is a shelf for water tank. There are arm holes for passing the arm for collection of mosquitoes. It has been noticed by us that larvae of *Aedes* thrive well on a diet of a mixture of dried cowdung cakes well powdered and yeast powder.

The larvæ of *Aedes aegypti* are particularly liable to be attacked by the gregarine parasite, *Lankesteria culicis* Ross.

Disease relationship. It is the most important and perhaps the only mosquito concerned in the transmission of dengue and yellow fever. The microfilariæ of *Wuchereria bancrofti* do not seem to develop in this species (Rao and Iyengar, 1932).

Dengue. It had been recognised that the outbreaks of dengue were widely distributed throughout the tropics. Graham in 1905 was the first to bring forward evidence of the mosquito transmission of dengue. Later Bancroft carried out transmission experiments and clearly showed that if the transmitting agent was a mosquito, it was a day-biting species.

Aedes aegypti as the carrier of dengue has been proved by experimental transmission in Australia by Bancroft (1906) and by Cleland, Bradley and MacDonald (1916), in Texas by Chandler and Rice (1923), and in the Philippines by Siler, Hall and Hitchens (1926). Experimental transmission has been attempted with other mosquitoes and though success has been claimed with *Culex fatigans*, such findings have not been confirmed. In Japan positive results have been obtained with *Aedes albopictus* and *Armigeres obturbans*.

The blood of a dengue patient is usually infective to the mosquito only during the first three days of the disease. The incubation period for the development of the virus in *Aed. aegypti* is from 11 to 14 days, and once infected, the infection lasts during the remainder of its life. There is a no hereditary transmission of the virus. Schule (1928) later observed that *Aed. aegypti* may transmit dengue as early as the 8th day after the ingestion of the virus, and suggested that the rapidity of maturation of the virus in the mosquito may be influenced by temperature.

Though dengue and yellow fever are carried by the same mosquito, they bear no relationship with each other. Anti-yellow fever serum is without action on the virus of dengue and the serum of dengue convalescents is without action on the yellow fever virus.

The widespread distribution of dengue is attributed to the carriage of infected mosquitoes on ships. The epidemic generally starts in a port city from where it spreads. The infected mosquitoes are carried mainly in country crafts, steamers and railway trains. A plan for international agreement regarding dengue was approved at the session of the Permanent Committee of the International Office of Public Hygiene held in May, 1929. These included : (1) Prompt notification to the other participating countries, also the International Office of Public Hygiene; (2) Prompt notification to the Port authorities when an epidemic would break out in a boat and isolation of the patients in a safe area of the boat where they could not be bitten by mosquitoes; and (3) Inspection of ships for ascertaining that no mosquito was breeding and taking prompt measures when mosquito breeding was detected.

Yellow fever. The history of yellow fever is indeed tragic as there are few diseases which have cost the lives of eminent men engaged in their investigations as yellow fever has done. As early as 1884 Nott suggested that yellow fever was possibly transmitted by mosquitoes and in 1853 Beaupathuy announced that the disease was spread by mosquitoes. In 1881 Carlos Finlay carried out experiments to prove mosquito transmission. Credit is due to Reed, Carrol, Argamonte and Lazear for showing to the world its exact method of spread and as a result of the knowledge thus acquired it has been possible to adopt preventive measures against this disease. The first organised campaign for the mosquito concerned was initiated by Gorgas in the Panama Canal Zone, at one time a hot bed of yellow fever which soon afterwards was made more or less free from this deadly disease.

The original cradle of yellow fever is South America. It has been recorded by Augustin in 1909 that yellow fever attacked the followers of Columbus in San Domingo in 1493, and the next year Canary Island became infected. Probably not until 1759 that the disease appeared in Africa. It spread from St. Louis at the mouth of the Senegal River. From there it soon spread to Gambia and then to Sierra Leon. Yellow fever soon became endemic in West Coast of Africa. In Africa yellow fever smoulders, and it is widely scattered. While African Negro suffers in a mild way the foreigner usually pays a heavy penalty. This was also true in the case of the Brazilians in South America arriving at Rio de Janeiro and infalliably becoming victims of the disease.

The chief mosquito vector of yellow fever is definitely *Aed. aegypti* which was known until recently to be the only vector of this disease; so too was man believed to be the only animal host of the virus and not until 1927 was it demonstrated that the disease could be given to monkeys, *Silenus rhesus* and *S. sinicus*, with fatal effects (Stokes *et al*, 1928).

The distribution of yellow fever covers wide areas in West Africa. Successful mouse protection tests have shown that yellow fever is more prevalent in West Africa and in the Anglo-Egyptian Sudan than it was thought to be. The yellow fever area runs from the west coast at latitude 15°N. along the Southern border of the Sahara as far as the Blue Nile. The eastern limit is from the Sudan-Abyssinian border to the western part of Uganda and the southern border is along the river Congo. It has recently spread to Belgian Congo. The increase of mechanical transport in Africa has added to the danger of its spread to new areas where the disease did not hitherto exist. It may also spread to the sea coast and then may extend to Asia as the sea route is short and trade is largely carried on by Arab dhows and coasting vessels which are difficult to control. Of course the danger to India is more threatening with the development of air traffic.

The only mosquito found infected in nature in Africa is *Aed. aegypti*, but several other African mosquitoes are capable of transmitting the disease by bite in the laboratory. These include *Aed. simpsoni*, *Aed. vittatus*, *Taeniorhynchus* (*Mansonioides*) *africanus*, *Aed. africanus*, *Aed. stokesi*, *Aed. taeniorhynchus*, *Aed. fluviatilis*, *Aed. luteocephalus*, *Aedes albopictus*, *Aed. scapularis*, *Culex fatigans*, *Culex thalassius*, *Eretmopodites chrysogaster*. However, differentiation should be made between those which are potentially dangerous and those that are actually dangerous. Only in the southern Sudan is there epidemiological evidence to suggest that *T.* (*Mansonioides*) *africanus* may act as a vector.

Two different types of yellow fever have now been clearly recognised. These are the urban type existing in Africa and the rural or jungle type found in South America and also in Africa. In the urban type man is the reservoir while in the rural type jungle animals particularly monkeys serve as reservoirs of the disease.

In South America the jungle yellow fever involves vast areas of the continent where yellow fever was never recognised as well as in regions apparently free of yellow fever for many years. It is widely distributed in Brazil, Venezuela, Ecuador, British Guinea, Dutch Guiana and Panama. It has been demonstrated in South America that yellow fever can exist and be transmitted in epidemic form in the absence of *Aed. aegypti* by two species of jungle mosquitoes, *Aedes leucoceloenus* and *Haemogogus capricorni*, as both these mosquitoes have been found infected in nature. Jungle strains can be transmitted in the laboratory by *Aed. aegypti* just as the urban strains can be transmitted by various species of jungle mosquitoes found in Africa and others found in South America.

The infective material of yellow fever circulates in the blood of the patient for the first three days and during this time mosquitoes feeding on them become infective. It takes about 12 days before the mosquito is able to propagate the disease to non-immunes. The incubation period of yellow fever in mosquitoes depends to a considerable extent on the temperature. If kept continuously at 37°C, *Aed. aegypti* may be infected in 4 days after its infecting blood meal. If kept at tropical room temperature averaging 23.4° C. it is infective after 11 days, and at 28°C., 9 days. At low temperatures the incubation period may be definitely prolonged (Hindle, 1930).

The possibility of the introduction of yellow fever into the United States or in India has now been increased on account of the tremendous advance in rapid transport by aeroplanes. Either infected mosquitoes or yellow fever patients may be carried this way and may start an epidemic in the above countries. Infected human beings are, however, more important than mosquitoes.

If ever the disease by chance enters this country the danger will be all the greater on account of :(1) the wide distribution of *Aed. aegypti* (Barraud, 1928, 1934); the Indian strain of this mosquito is as infective to yellow fever as the African strain (Hindle,

1929); (2) both the two common monkeys of India, *Silenus rhesus* and *S. sinicus*, are susceptible to the virus of yellow fever; these animals are not only widely distributed in this country but in many places they are domestic animals; (3) the circulation of the virus in the peripheral blood of the monkey throughout the course of the disease (Hindle 1932); monkeys infected with yellow fever virus generally die within 4 days; (4) the discovery that in addition to *Aed. aegypti*, other species of *Aedes*, *Aedes vittatus*, *Aed. albopictus*, *Aed. scutellaris*, *Mansonioides uniformis*, *M. africanus* and other Culicines have been proved to be capable of transmitting the disease experimentally. *M. uniformis* has a wide distribution in India and bites man freely ; (5) suitable climatic conditions; (6) the presence of both susceptible human and insect populations.

Control of Aedes aegypti. An *Aedes* is deemed essential to the proper appreciation of the potential danger to which a locality may be exposed to the actual danger. The *Aedes* index is defined as the percentage of houses or quarters inspected in which stegomyia mosquitoes are found breeding. The control measures include the following : (1) Eliminate breeding places; all receptacles in which rain water may collect in the house or vacant land must be either destroyed or removed to at least a safe distance. Defective gutters, blocked pipes, faulty drains should be removed. (2) All accumulations of stored water should be either screened or covered with cloth; an 18-mesh screen cloth made of wire of 30 or 28 Imperial Standard Wire Gauze should be used. (3) Institution of 24 hours supply of filtered water will overcome the necessity of storing water in houses. (4) All holes in trees within 50 yds. of human habitation be properly sealed up. (5) Sufficient common salt may be left in flower vases, ant-guards etc. to bring the concentration of the solution up to 2 per cent NaCl. (6) A better way of dealing with ant-guards is to add some phenyle, cresol or crude oil to the water. Once treated, it will remain unfit for *Aedes* breeding for about a fortnight. (7) The establishment of a mosquito controlling section of the Health Department and organised mosquito control campaigns should be inaugurated. Inspection of every house should be made at intervals of not more than 7 days and the *Aedes* index charted. A careful watch should also be maintained on their breeding in steamers and country boats. (8) Insect repellents and insecticides do not work so well in the case of adult *Aedes*. However, after trying the classical control measures against *Aedes aegypti* in George town, British Guiana, which did not produce very satisfactory results, De Caires (1951) used 5 per cent. DDT in kerosene at 100 mg. per sq. ft. and this reduced the *Aedes* index considerably.

The different ways by which yellow fever may spread from one village to another or from one country to another are as follows : (*a*) Slow village to village spread along trade routes. (*b*) More rapid spread along motor routes. (*c*) Spread by railways. (*d*) Spread by aircraft. (*e*) Spread by boats and steamers along river channels. (*f*) Spread by sea. Methods (*a*) and (*b*) would chiefly be by means of the carriage of infected cases in the incubation period; method (*c*), (*d*) and (*e*) might cause spread both by carrying infected cases in the incubation period and by carrying infected mosquitoes.

Methods of control of yellow fever in endemic areas. (1) Direct mosquito control. (2) Indirect mosquito control by broad general sanitation and piped water supplies. (3) Vaccination. (4) Segregation of infected cases. (5) Usual sanitary precautions to be adopted as in other infective diseases.

Sanitary Aerodrome. Proper sanitary control over aircraft is essential. Those working on international lines should depart from and arrive at certain authorised aerodromes where the requisite sanitary provisions will be maintained. A sanitary aerodrome should be provided with adequate personnel and apparatus necessary to undertake disinsectization. The aerodrome must be situated at an adequate distance from the nearest inhabited centre. The hospital and other buildings situated in the aerodrome area must be screened. Such sanitary aerodromes are called "anti-amaryl" aerodromes.

For the destruction of mosquitoes in aircrafts pyrethrum extract or DDT and pyrethrum mixture is commonly used.

Yellow Fever Control in Brazil. That stegomyia-control and yellow fever-control are possible has been amply demonstrated by the International Health Board of the Rockefeller Foundation. The anti-stegomyia service has been so successful in Brazil that in Rio de Janeiro and in many other towns which formerly had the worst reputation for yellow fever, it is now difficult to discover the presence of a single stegomyia mosquito. Not long ago certain countries refused to allow their nationals to emigrate to certain parts of Brazil or to land at certain ports on the coast owing to the risk of contracting yellow fever. To-day there is no reason for fear in landing in any port in Brazil. The present programme of yellow fever service aims at complete eradication of *Aed. aegypti* from the country. Vaccination has proceeded during this period and the duration of immunity appears to be very long. Where *Aed. aegypti* has been completely eliminated, vaccination is no longer needed. A special service has been organised for performing visceretomy in all jungle region countries where yellow fever has been endemic for ascertaining where vaccination is required. Crews and passengers of aircrafts passing through endemic areas are compulsarily vaccinated. Aircrafts coming from yellow fever areas are inspected on their arrivals and disinsectized.

Steps taken by the Government of India against the introduction of yellow fever into this country. Twenty-two different places have been declared infected with yellow fever. These are (1) Angola; (2) Anglo-Egyptian Sudan south of Khartoum; (3) Belgian Congo; (4) Cameroons; (5) Dahomey; (6) French Equitorial Africa; (7) French Guinea; (8) French West Africa; (9) Gambia; (10) Gold Colony; (11) Ivory Coast; (12) Liberia; (13) Nigeria; (14) Senegal; (15) Sierra Leone; (16) Togoland; (17) Uganda; (18) Upper Delta Territory; (19) Abyssinia; (20) Spanish Guinea; (21) Eritrea, and (22) Kenya Colony.

(*a*) Entry of individuals from yellow fever areas : No one who has been to a declared yellow fever area may enter India unless such individual (*i*) Had had the disease or has been inoculated against yellow fever not less than 10 days and not more than two years before entering a yellow fever area; or (*ii*) Has been inoculated against yellow fever not less than 19 days before arrival in India; or (*iii*) Nine days have elapsed between the date of leaving a yellow fever area and arrival in India. Failing this he will be subject to quarantine for the balance of the required period. (*b*) Entry of mosquitoes and animals from endemic areas : An aircraft which has started from or landed in a yellow fever area, unless disinsectized at an intermediate yellow fever aircraft station is disinsectized on arrival in India. The entries of mosquitoes and other animals from endemic areas is prohibited. (*c*) Immunization : This is carried out by the yellow fever vaccine which contains living attenuated strains of yellow fever virus. The virus is desiccated and kept in vacuum in ampoules at 0°C. During transport special precautions are necessary to ensure the same low temperature. The vaccine is diluted in cold sterile salt solution to give it a strength of 1 in 10. The syringe and the needle must be cool before the vaccine is taken up. A single dose of 0.5 ml. of the diluted vaccine is used. The rest of the vaccine not used in half-an-hour is thrown away. As a rule there is little or no reaction except in a few who may complain of headache, malaise some 5 to 8 days after injection. Immunity develops in 10 to 14 days and lasts for at least 2 years. In India stocks of yellow fever vaccine are maintained at many centres some of them being All-India Institute of Hygiene and Public Health, Calcutta; Central Research Institute, Kasauli; Provincial Public Health Laboratory, New Delhi; King Institute, Guindy; Haffkine Institute, Bombay; Air Port Health Office, Santa Cruz Air Port and also at Porabander and Navanagar.

Similar restrictions are imposed in other countries on the entry of persons coming from or passing through a yellow fever area.

CHAPTER VII

OTHER NEMATOCERA

Family : **Psychodidae**

(Sand-Flies)

The insects of this family are widely distributed. In size they are small or minute; the body and wings are thickly covered with hairs. The antennæ are long and filamentous and carry whorls of hairs at the joints. Metamorphosis is complete. The larva is a hairy maggot which lives on solids. Some larvæ are aquatic and others live in damp, dark places. The adult is a fluid feeder. All are characterised by the second longitudinal vein branching twice.

The Psychodidæ are grouped into two families, namely, (1) Psychodinæ, and (2) Phlebotominæ. They can be easily distinguished by the following characters : (1) Pose : While at rest the wings in Psychodinæ are held roof-like over the abdomen, whereas in Phlebotominæ they are held-up. (2) Wings : The wings in the former are much broader and have the second longitudinal vein branching for the first time

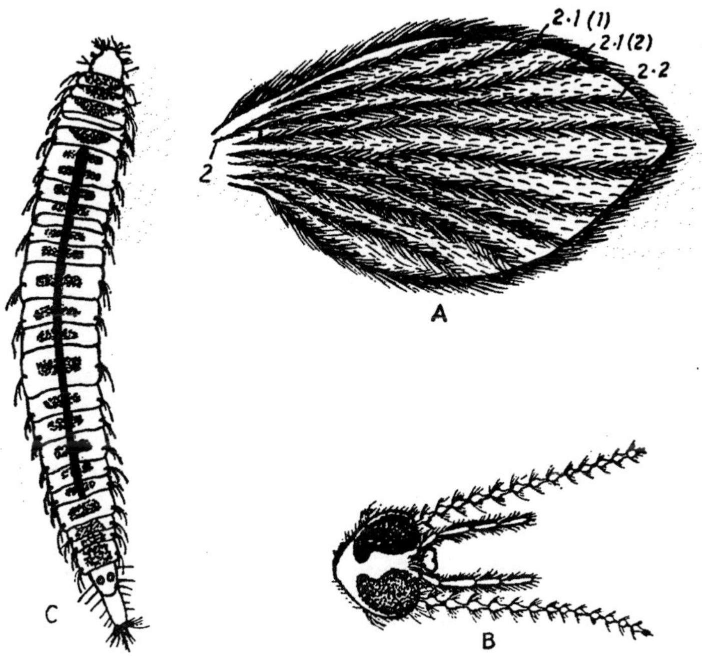

Fig. 63.
Psychoda.
A. Wing ; B. Head; C. Larva.

very close to the root of the wing, whereas in the latter the wings are lance-shaped, i.e., narrow and pointed and the first branching of the second longitudinal vein takes place in the middle of the wing. (3) The external genitalia in Psychodinæ consist of two pairs of claspers while three pairs are present in Phlebotominæ.

Subfamily **Psychodinae.** A large number of species exist but none has yet been reported with certainty to suck blood. These midges are popularly known as "moth flies" because of the peculiar roof-like position of the wings when at rest which gives them the appearance of tiny moths. This group of insects, also known as "owl midges," includes the commonest species of flies that breed in the bacterial films of filter beds used for the purification of sewages. They breed also in drains filled with decaying vegetable and organic matter. As a rule they are harmless.

Its breeding can be checked by means of bleaching powder. The adults may be controlled by spraying their resting places with 5 per cent DDT emulsion or suspension at the rate of 1 quart. to 250 sq. ft. area.

Subfamily **Phlebotominae.** This contains only one genus *Phlebotomus*. They

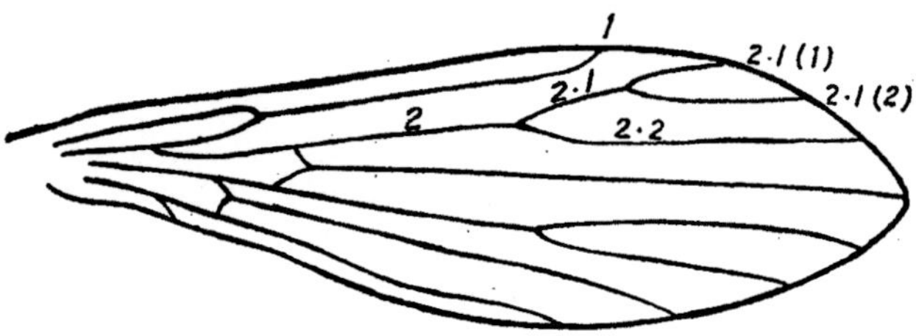

FIG. 64.
Wing venation of *Phlebotomus*,

are important as the females are blood-suckers and some of them are proved carriers of certain disease organisms.

Genus **Phlebotomus.** They are small and are of a greyish-brown colour. They are extremely delicate. They have a wide distribution in the moderately warmer regions of the world. All sand-flies are incapable of flying any long distance unlike mosquitoes; they move by hopping from one place to another and by this character alone they can be readily recognised in nature.

The females only have piercing mouth parts and are blood-suckers. They feed on a variety of animals and except the lizard, these are all warm blooded. They are active only at night and seek protection by day in shelters. The antennæ on the head are long and slender and are composed of 16 segments. They carry whorls of hair at the joints and are similar in the two sexes. The palpi are long and flexible, and consist of 5 segments which are clothed with scales and hairs. The proboscis is composed of (1) the labium which consists of true cutting and piercing organs, (2) a pair of mandibles which are broad and serrated at the edge, (3) a pair of maxillæ which are much narrower than the mandibles and bear a number of teeth on one edge, (4) the labrum-epipharynx, and the hypopharynx, the tip of the latter being pointed and the edges near the tip serrated.

The thorax is mainly composed of mesothorax. The prothorax is distinctly small. The scutellum and post-scutellum are well developed. The legs are much longer in proportion to the size of the insect and slender.

The wing is densely hairy. The second longitudinal vein branches twice, the first time in the middle of the wing. Scales are entirely absent.

Abdomen: The abdomen which is clothed with hairs consists of 10 segments, the last two being modified for sexual purposes. The sexes are differentiated by the

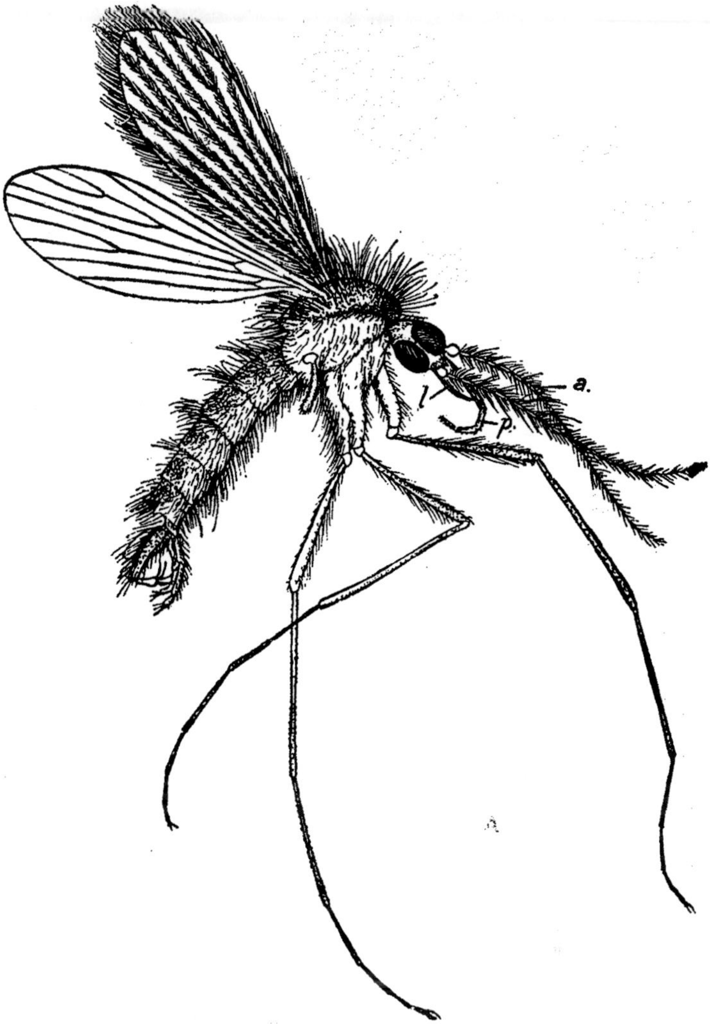

FIG. 65.
Phlebotomus argentipes, male.
a, antenna ; p, maxillary palp ; l, labium.

abdomen which is narrow and is often slightly upturned posteriorly in males which also possess claspers. These are attached to the last abdominal segment. In females the abdomen is rounded and claspers are absent. The male terminalia are conspicuously large and are complex structures; they are of great importance in the diagnosis of the species. The appendages are arranged in five pairs as follows:

(*a*) Superior claspers: One pair, and placed dorsally. Its distal segment is provided with spines arranged differently in different species. Their characters are helpful in the identification of the male fly. (*b*) Inferior claspers : One pair and situated ventrally. They are much shorter. (*c*) Submedian lamellæ : These lie

Fig. 66.
Phlebotomus argentipes, female.

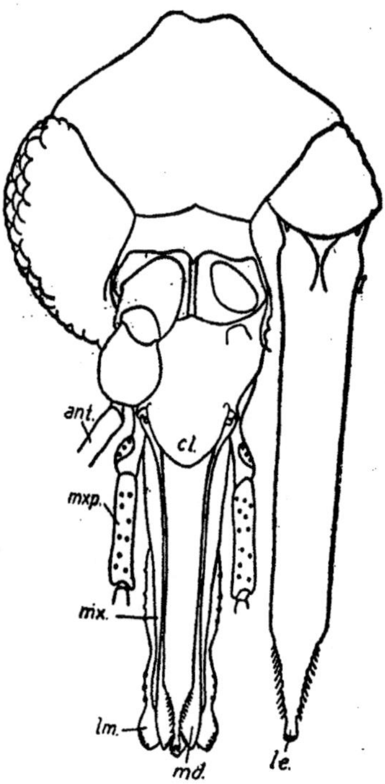

Fig. 67.
Mouth parts of *Phlebotomus*.
ant, antenna; cl, clypeus; le, labrum-epipharynx ; l, labium ; mx, maxilla ; mxp, maxillary palp ; (After Christophers, Shortt and Barraud).

between the inferior claspers. They are usually short, thin, leaf-like structures. (*d*) Intermediate appendages : These occupy a median position; usually one pair, may be even three. (*e*) Intromittent organ : One pair, this has been described as the penis. They lie between the intermediate appendages.

Alimentary canal. (1) Buccal cavity which lies at the base of the clypeus; (2) Pharynx; (3) Oesophagus with an elongated diverticulum; (4) Midgut; (5) Hind-gut; (6) Two pairs of Malpighian tubules; and (8) Rectum. The salivary glands consist of two spherical organs. The ducts unite near the mid-region of the head forming a common duct which enters the buccal cavity. The buccal cavity in sand-flies consists of a ventral plate, and two dorsal ones which unite posteriorly to form a single thick pigmented plate. The ventral plate bears at its posterior margin two systems of teeth which project horizontally into the lumen. The armature of the pharynx consists of ridges and blunt teeth in the posterior part. In the male the armature of the pharynx and the buccal cavity is much less developed. The midgut consists of two parts which differ histologically. The upper part or cardia lies in the thorax. The œsophageal valve lies at the anterior end of the cardia.

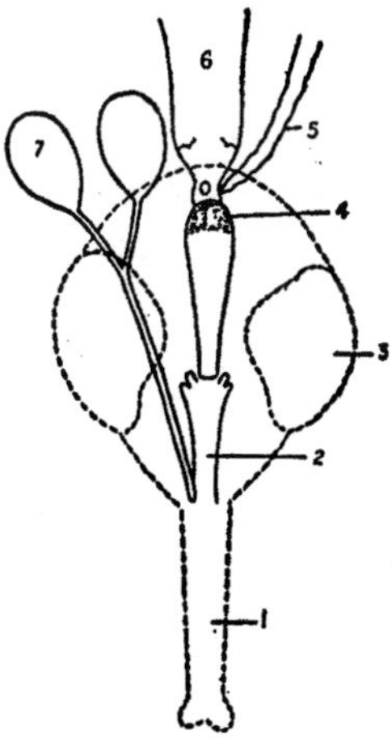

FIG. 68.

Anterior part of the alimentary canal of *Phlebotomus argentipes*.
1. proboscis; 2. buccal cavity; 3. eye; 4. posterior part of pharynx; 5. duct of diverticulum; 6. midgut; 7. salivary glands.
(After Shortt, Barraud and Craighead.

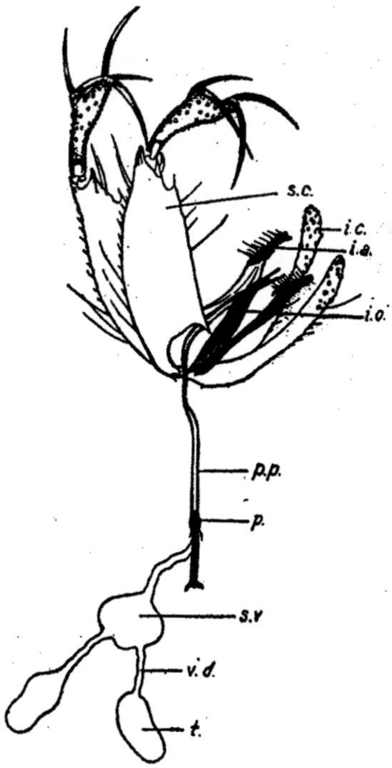

FIG. 69.

Generative organs and hypopygium of male *Phlebotomus antennatus*.
i.a., intermediate appendage; i.c., inferior clasper; i.o., intromittent organ; p., pompetta; p.p., paired penis; s.c., superior clasper; s.v., seminal vesicle; v.d., vas deferens; t., testes.

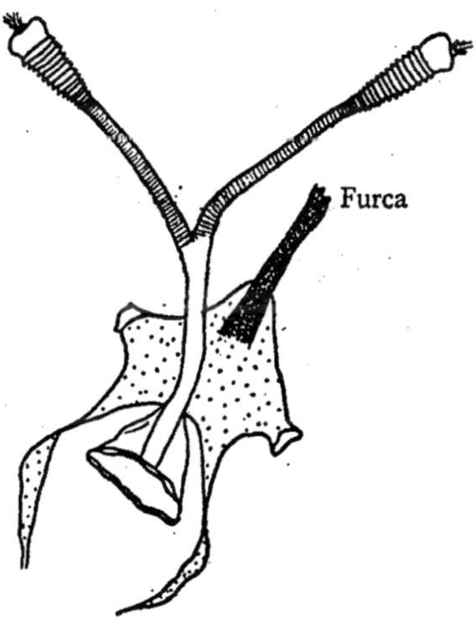

FIG. 70.

Receptaculum seminis and furca of *P. argentipes*.

Reproductive system. *Male*: (1) Testis—elongate; one pair. (2) Seminal vesicle—a large pyriform sac; (3) Ejaculatory duct—two in number—connected with a pump (pompetta) which regulates the flow of spermatozoa; (4) The ejaculatory duct is then continued with the penis or the intromittent organs which lie between the intermediate appendages. *Female*: (1) Ovary; (2) Spermathecæ which differ widely; their structure is of specific importance. The spermathecæ lie in the median line in various species in the region of the last two abdominal segments. Each consists of a single thin-walled sac and is relatively very large. At their junction with the oviduct they are strongly chitinised and are transversely ridged on their inner surfaces. The spermathecal ducts may remain separate along their entire lengths as in *Phlebotomus papatasii*, *P. perniciosus*, *P. chinensis*, *P. sergenti*, *P. avellari*, *P. travassosi* (the last two occurring in Brazil) or may unite to form a common duct as seen in the majority of sand-flies.

Life history and bionomics. The egg is comparatively large and torpedo shaped. It possesses some longitudinal wavy lines on the outside. Any damp moist place, where there is sufficient moisture, absence of light and where organic matters especially excrement of animals are present in sufficient amount, is selected by the female for oviposition. The egg, on account of its small size, is never noticed in nature but sites of oviposition can be guessed by finding larvæ. In the laboratory the egg stage lasts in the case of *P. argentipes* from 3 to 4 days during the rainy season and is extended to 7 days in winter.

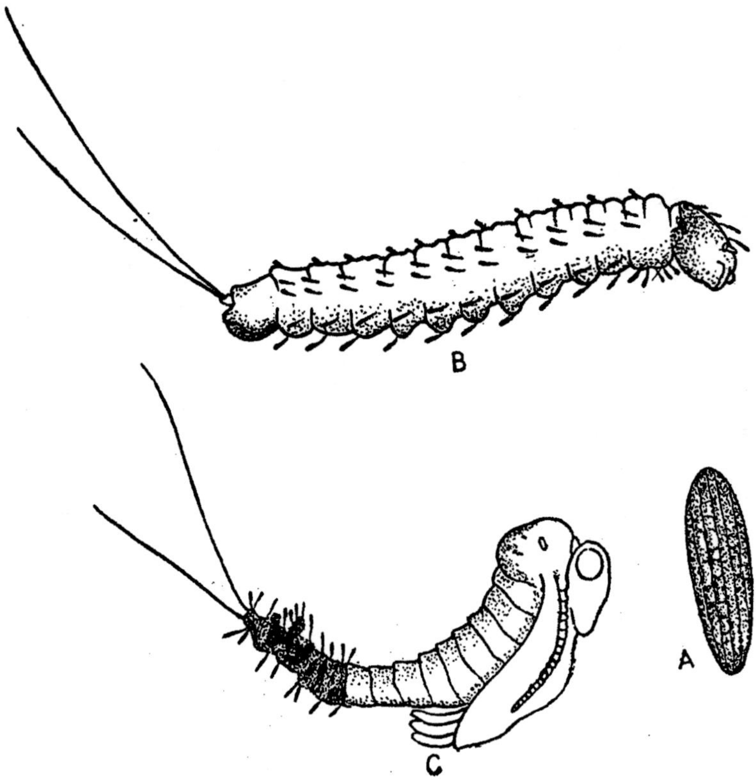

FIG. 71.
Showing complete metamorphosis of a sand-fly.
A, egg ; B, larva ; C, pupa.

The larva is an apodous hairy maggot and has four moults. It has a distinct head, 3 thoracic and 9 abdominal segments. The last segment carries two pairs of long bristles which are kept erect; among them one pair is remarkably long. The mouth is provided with mandibles; eyes are absent. The body is covered with spines in more or less regular transverse rows. This stage lasts from 2 weeks to a longer period depending on the temperature. The pupa has always the discarded last larval skin enveloping the two distal segments of the pupal body. The abdomen of the pupa is curved upwards. The integument is clothed with minute spines. The minimum life of the larval and pupal stages combined is about 30 days in Calcutta during the rainy season.

The blood feeding habit of the adult is restricted to the female alone; the males live entirely on vegetable sap. An average female *P. papatasii* weighs 0.3 mg., and after a full blood feed, weighs 0.4 mg. The weight of successive blood meals by sand-flies has been shown in a table by Theodor (1936). The length of life of sand-flies depends on many factors such as environments, mating and successive blood feed. Inspite of the best efforts it would hardly be possible to keep an appreciable number alive in the laboratory for more than 15 days though Wenyon (1913) succeded in keeping a captured sand-fly alive for 40 days by feeding it on water alone.

The presence of the sand-fly is scarcely observed. It is the bite which is first felt. The bite is irritating and the sensation is unlike that experienced in the case of mosquitoes. Sand-flies are particularly annoying to man at night when there is little or no breeze and the atmospheric temperature is high. They are nocturnal in habits and only a few species are attracted to light. They usually attack the lower extremity rather than the upper and the back in preference to the front part. The face is occasionally attacked. They can bite through thin clothing. They cannot fly far and remain confined to within about 50 yds. of their breeding places. All sand-flies are exceptionally small and it is only the finest net that will prevent them from entering through the meshes. Anong the common natural enemies of sand-flies are lizards and spiders. Sand-flies can be forcibly fed and the technique has been described by Napier (1930).

Phlebotomus argentipes. This midge is so named on account of the silvery white feet which is due to the presence of white scales on the tarsi. So far as the writer knows it is the only species which has white feet. By this character alone this species can be readily recognised even in the living state. This species is found only in India and Ceylon. Its distribution in India which was once thought to be confined to the east of an imaginary line drawn from Bombay to Simla is not strictly so.

Larvae are found in soil collected within a radius of 20 yds. of a dwelling house or cattle shed. In cattle sheds a favourite breeding site is under the earthenware feeding troughs. The larvae penetrate to a depth of 3 or 4 inches in loose and sodden soil. Rat burrows seldom yield many larvae. Cattle sheds are their favourite breeding sites.

Under the most favourable conditions the whole life cycle takes about 30 days. In winter it takes two to three months.

Sand-flies do not feed again till they have completely digested the previous blood meal and oviposited; this process takes approximately 3 days in mated flies and much longer in unfertilized ones. About 30 per cent. of *P. argentipes* will survive until the third feed about 6 to 7 days after emergence. Smith, Halder and Ahmed (1940) found that about half of the sand-flies that take a first blood meal and then feed on raisins are available at the end of 10 days for feeding. *P. argentipes* is known to feed in nature on man, cattle and goats but it shows a decided preference for cattle. This sand-fly is found more in cattle sheds and chicken houses than in human sleeping quarters. It is rarely attracted to artificial light.

For marking sand-flies use finely powdered methylene blue, basic fuchsin or gentian violet which should be sprinkled over the bottom and walls of a specimen tube. When the insects jump about, the stain becomes entangled in their hairy covering and is retained there even after 10 days after liberation of the insects. The identification of the captured sand-flies should be made by placing the killed insect on a piece of filter paper and moistening it with alcohol. The alcohol dissolves the dye and the filter paper shows a blot of the appropriate colour. They can also be marked by puncturing the wings.

Phlebotomus papatasii. While *P. argentipes* are more numerous in cattle sheds, *P. papatasii* are caught in large numbers in human dwellings, its breeding being found in places inside houses such as in cracks in the floor and in walls, even in rat holes. It is also a notorious biter of man. It is far more active in its habits than *P. argentipes*. On water alone *papatasii* can be kept alive longer than *P. argentipes.* Whittingham and Rook (1922) have worked out its life cycle.

Phlebotomus minutus. This is so named on account of its minute size. Though this is the most widely distributed sand-fly in India, it can never be captured in large numbers. It is found even at a height up to 6000 to 7000 ft.

This group is a very confused one and in India it shows great variations in its characters in different areas. At least three forms have been described. *P. minutus* var. *montanus* is now ranked as a distinct species named *P. montanus*.

This species does not readily feed on man but feeds chiefly on cold-blooded animals especially lizards though it may be found in human sleeping quarters.

P. sergenti. Evidence has been collected by Sergent brothers etc. (1921) and later by other observers suggesting that this species bites man in nature.

Disease relationship. (1) **Kala-azar.** The flagellate stage of *Leishmania donovani* has been demonstrated in the anterior part of the midgut of *P. argentipes* when they are dissected from the third day after feeding on a case of kala-azar (Knowles, Napier and Smith, 1924). Free forms are also seen as far forward as the middle of the pharynx. Later Swaminath, Shortt and Anderson (1942) succeeded in transmitting the disease to volunteers by the bites of *P. argentipes*. These sand-flies after feeding on the blood of K.A. patients were maintained at 28°C. on raisins until they were fed on the volunteers. The distribution of kala-azar also coincides with the distribution of *P. argentipes*. In China, Patton and Hindle (1927) found *L. donovani* developing in the foregut also of *P. chinensis* and consequently this species has been regarded as a vector of kala-azar in China. Natural infection has been recorded in *P. argentipes* in India and *P. chinensis* in China. The mechanism of infection was once thought to be by regurgitation, as blocking of the œsophagus and the buccal cavity with the flagellates of *Leishmania* was observed by Shortt, Barraud and Craighead (1926) in connection with the development of the parasites in the sand-fly. The actual method of the transfer of the parasites is, however, simple.

The parasites after development in the pharynx make their way towards the buccal cavity and proboscis about the 7th day, and pass into the circulation of the healthy host at the time of the second blood feed about this time. During this interval between the first or the infective blood meal and the second blood meal the insects must depend for their nourishment on vegetable diets only and not on blood, as in the latter case all the flagellates will be killed.

The following papers also give useful information in this subject : Christophers : Report of the Kala-azar Commission, 1926; Patton and Hindle, 1927; Shortt, Smith and Swaminath, 1930; Smith, Haldar and Ahmed, 1940, 1941.

(2) **Oriental sore.** The epidemiological in addition to experimental evidence points to the fact that *P. papatasii* and *P. sergenti* can both act as carriers of oriental sore. Acton (1919) showed that the distribution of oriental sore on the body corresponds very closely with the area more liable to bites by sand-flies. Sergents, Parrot, Donatien and Beguet (1921) produced oriental sore in man by the inoculation of macerated sand-flies. Sinton (1924, 1925, 1927) in a careful study of the distribution of oriental sore in India was able to show a close correlation between

it and that of *P. sergenti*, and later Shortt, Sinton and Swaminath (1936) discovered a definite host parasite relationship between *P. sergenti* and the Punjab strain of *L. tropica* and which readily develops in this species. They produced oriental sore in a monkey with flagellates from *P. sergenti* five days after it was infected. Therefore all indications point to the transmission in nature taking place by the bite of the sand-fly, the mechanism being similar to that observed in the case of the parasites of visceral leishmaniasis developing in *P. argentipes*.

The only reservoirs of cutaneous leishmaniasis were known to be man and dogs till Latyshev and Kriukova (1941) discovered rodents especially gerbeles, to be the reservoir of *L. tropica* in Turkmenistan. Phlebotomus have been found breeding in the burrows of these rodents and have been found infected in nature. Man becomes infected during his sojourn through these uninhabited regions.

(3) **Sand-fly fever.** Both Taussig (1905) and McCarrison (1906) drew attention to a special type of short fever which they had suspected of being carried by sand-flies, hence the name sand-fly fever was given to it. Through the researches of Doerr (1908) and Doerr and Russ (1909) it became known that the blood is infective during the first two days of the illness and *P. papatasii* is the carrier of this disease, infection taking place directly by the bite. Whittingham and Rook (1922, 1924) observed that the recently bred adult was capable of infecting human beings and that the midge was more than a direct vector. Infected sand-flies brought to England from Malta oviposited before death and the resultant adults were found capable of infecting human beings. This suggestion of Whittingham that the virus is transmitted to the next generation has, however, been disputed by Moshkovski (1937) and Sabin *et al.* (1944). Whether such natural conveyance of infection from the adult fly to its breed, or the alternative, that the larva acquires the virus by feeding on the infected faeces passed by the parent or by consuming the dead body of the parent, is the true explanation has not yet been determined. It should be taken into consideration that in India breeding is more or less continuous and there are several broods in a year. There is no evidence yet that any other insect or other species of sand-fly can transmit the disease.

In the Punjab and North Western Frontier Provinces two seasonal outbreaks of this disease occur during the year, one in the early summer and one in the autumn. The insects are most numerous during the hot and relatively moist atmosphere and are found in small numbers during the very hot months when the humidity is very low. In Northern India the area of distribution of *P. papatasii* fairly corresponds with that of the disease which holds true all over the globe. It is doubtful if sand-fly fever occurs in places where *P. argentipes* is the most prevailing species.

(4) **Leishmania infantum and Canine leishmaniasis** (due to *L. canis*). Both infantile leishmaniasis and canine leishmaniasis are common in the Mediterranean littoral. It is now known that *P. perniciosus* is the vector of this type of *Leishmania* found in man in Algeria, Malta and Marseiles, *P. major* in Greece and *P. sergenti* and varieties of *P. major* in other places. In certain parts of North Africa the vector is probably *P. papatasii* as the local strain of *Leishmania* has been found to multiply better in this species than in *P. sergenti.* Adler and Theodor (1930) found *P. perniciosus* infected with *L. infantum* in nature.

Canine leishmaniasis though common in the Mediterranean has not yet been reported from India. The *Leishmania* parasites in dogs are distributed throughout the body in the skin where they are lodged in the histiocytes and fibroblasts though the parasites are not numerous in the smears from the skin. Chodukin and his collaborators (1931) have demonstrated the transfer of the *Leishmania* parasite from dog to dog through sand-flies. In China where canine leishmaniasis is common *P. chinensis* and *P. sergenti* var. *mongolensis* are the principal carriers of this disease. It has been demonstrated that the parasites of canine leishmaniasis are able to develop into the flagellated stage also in the intestine of the dog-louse *Linognathus setosus.*

(5) **Oroya fever and Verruga** (Carrion's disease). Sand-flies incriminated with Carrion's disease in Peru are: *P. verrucarum, P. noguchii* and *P. peruensis.* According to Townsend (1913) *P. verrucarum* is the transmitter of both Oroya fever and verruga. Noguchi (1926) has shown that *Bartonella bacilliformis* can be transmitted from infected to healthy monkeys by bites of *Dermacentor andersoni.* There is, however, some indication that Oroya fever is transmitted to man through the agency of some blood-sucking arthropods. Later researches by Hertig (1942) showed that when *P. verrucarum* and *P. noguchii* were fed on suitable patients suffering from the disease, there usually developed a marked infection of the mid-intestine, pharynx and buccal cavity. This theory of the transmission of the disease by sand-flies, however, cannot have any logical support as wild sand-flies of both sexes exhibit a condition of massive infections of the tip of the proboscis with organisms which on culture though closely resemble *B. bacilliformis*, the causative organisms of this disease.

Lately Ray and Idnani (1940) have drawn attention to bartonellosis occurring in dogs in India but the carrier has not been mentioned which is likely to be a tick.

(6) **Espundia.** The transmitting agent of this disease which is peculiar to certain parts of South America and which is caused by *L. brasiliensis* (*L. tropica*) is not known though natural infection has been recorded by Pessoa and Coutinho (1941) in many species of sand-flies. Aragao (1922) thought *P. intermedius* was the vector.

Breeding of P. papatasii. Three conditions need to be satisfied for breeding *P. papatasii* : (1) A mould-free diet; (2) Optimal temperature conditions; and (3) A minimum of humidity, particularly at the time the larvae pupate.

Food. Ashner (1927) obtained good results when rabbit faeces was mixed with soil as the growth of moulds is suppressed. One part of sifted garden soil is mixed with two parts of crushed faeces added to a thick paste and the mixture dried in the air with repeated stirring. The larvae are allowed only a few particles in the beginning and the older ones larger amounts according to their number and size, until the bottom of the vessel is covered with a layer of 1 to 2 cm. (2) *Optimal temperature* : A temperature of 30 to 32°C. is much more favourable to the development of the larva than 23 to 25°C. (3) *Humidity*: The larvae should be allowed to pupate in the minimum amount of moisture. Normal conditions of moisture for the larvae is indicated if the grown-up larvae do not emigrate but pupate in the food substrate.

McCombie Young and his collaborators (1926) have described the following method for breeding *P. papatasii* which gives better results than that described by Ashner. A small earthenware pot is half filled with dry mule litter and a thick layer of mud over the litter. A crucial incision is made in the mud and the pot placed out in the sun, so that the mud hardens and further cracks form in it. When the mud is thoroughly sun baked, it is stood in a petridish of water and the dish and pot placed in a muslin cage with some fed females of *P. papatasii.* The flies enter through the cracks and oviposit in the deeper layers of the litter. The larvae hatch out in *situ* and develop uninterruptedly. The full grown larvae usually come to the surface of the mud and exposed sides of the pot to pupate.

For the breeding of *P. papatasii*, the papers written by Unsworth and Gordon (1946) and published

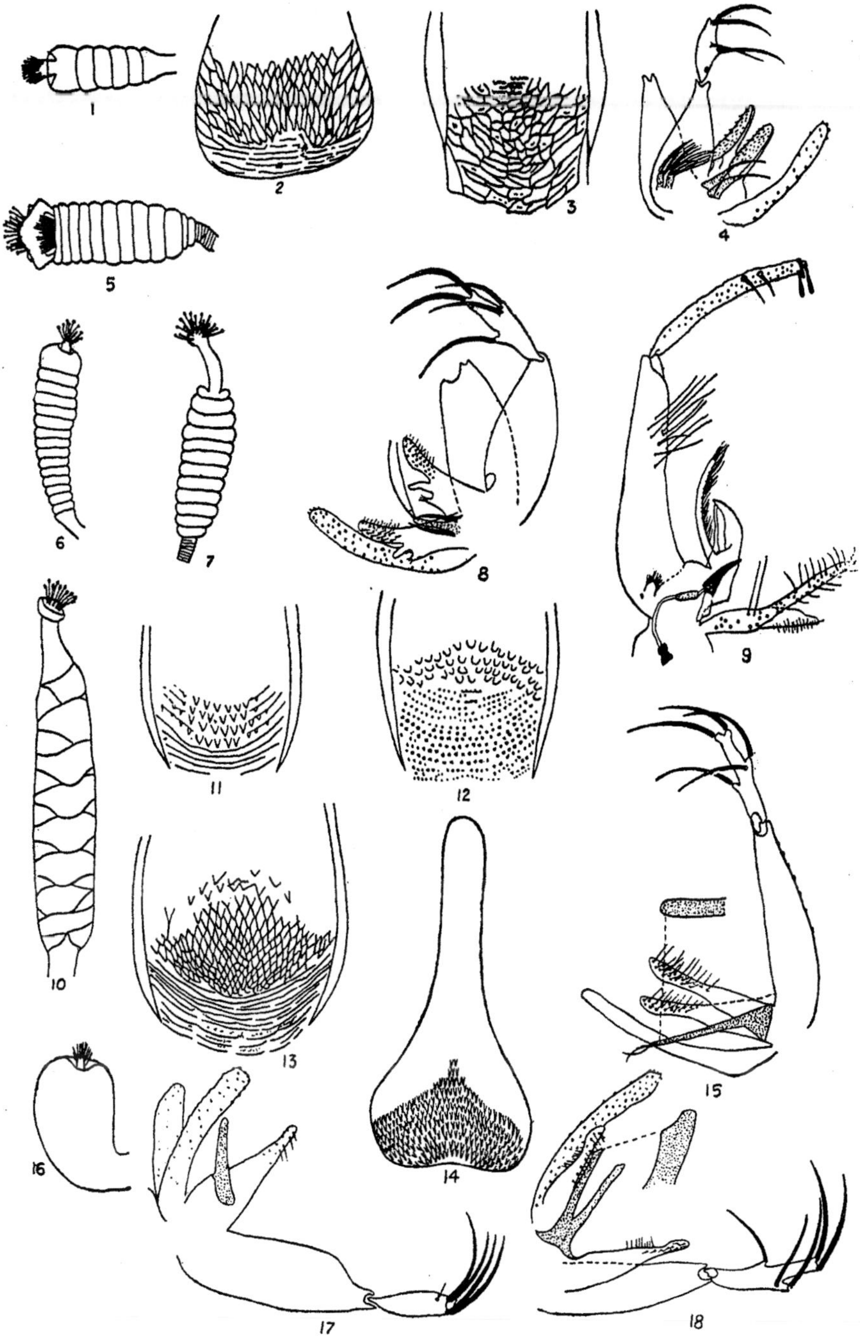

FIG. 72.

Spermatheca, pharyngeal armature and male hypopygium of some important disease-carrying *Phlebotomus*. (After Sinton.)

1. Spermatheca of *P. sergenti*. 2. Pharynx of female *P. sergenti*. 3. Pharynx of female *P. papatasii*. 4. Hypopygium of *P. sergenti*. 5. Spermatheca of *P. papatasii*. 6. Spermatheca of *P. argentipes*. 7. Spermatheca of *P. major*. 8. Hypopygium of *P. argentipes*. 9. Hypopygium of *P. papatasii*. 10. Spermatheca of *P. chinensis*. 11. Pharynx of female *P. argentipes*. 12. Pharynx of female *P. major*. 13. Pharynx of female *P. chinensis*. 14. Pharynx of female *P. minutus*. 15. Hypopygium of *P. major*. 16. Spermatheca of *P. minutus*. 17. Hypopygium of *P. minutus*. 18. Hypopygium of *P. chinensis*.

in *Annals of Tropical Medicine and Parasitology*, Vol. 40, p. 219; by Najera, L. E. (1950) published in *Rev. Sanidad Hig. Publica*, 24, p. 28; by Hertig (1940) in *Science*, 92, p. 91; and Ashner (1927) in *Trans. Roy. Soc. Trop. Med. Hyg.*, 20, 452, may also be consulted.

Breeding of Phlebotomus argentipes. Ordinary earthenware pots, as these are porous, are the most suitable receptacles for breeding sand-flies in the laboratory. For oviposition a fair degree of moisture must be provided. The empty pot is placed in an enamel dish containing water or preferably in wet sand and as soon as the flies have deposited eggs, a lesser degree of moisture should be allowed. Blood-fed wild flies are kept in a lamp chimney over the pots and the insects are not disturbed for 4 to 5 days. The moisture can be regulated by varying the depth of the water in which the pot has been placed or by reducing the moisture of the sand. The pot is kept covered with a piece of cloth fixed to the neck by means of a rubber band. As soon as larvae have emerged, they are provided with a small quantity of food. The young larva requires much less moisture than the egg and as the larva grows, the amount of moisture necessary decreases while during pupation practically no moisture is required. Pots containing pupae are best placed in a separate breeding-out cage from which the flies may be caught and used as required. The most convenient food for the larva is the crushed faeces of rabbits to which some powdered blood may be added but blood is not an essential requirement. The pots are examined from day to day and the food and the moisture are regulated. *P. argentipes* breed best at a temperature of 28°C. In the tropics, except during the winter, it will not be necessary to use an incubator for this purpose.

Recovery of Sand-fly Larvae and Other Insect Larvae from Soil. It is possible to recover insects, insect larvae and other arthropods contained in soil after the suspected sample is washed in water and the washing is passed through a series of sieves with meshes of decreasing size (Morris, 1922). Mc-Combie Young, Richmond and Brendish (1926), by applying the floatation method, reported their success in finding a large number of larvae of sand-flies from the soil or their breeding places. The routine procedure described by Shortt, Smith and Swaminath (1930) should be adopted for this purpose.

Identification of Sand-flies. The genus *Phlebotomus* has been divided into three groups by Sinton (1931, 1933) as follows :

A. Erect-haired group. Numerous erect hairs on dorsum of abdominal segments II-VII; dorsal recumbent hairs very scanty or absent. *Spermatheca and pharynx in the female and hypopygium in the male have diagnostic morphology.*

This group includes : *P. papatasii* Scop; *P. argentipes* Ann. and Brun.; *P. sergenti* Parrot; *P. chinensis* Newst; *P. major* Ann.; *P. colabaensis* Young and Chalam; *P. maynei* Sinton; *P. elanorae* Sinton; *P. newsteadi* Sinton.

B. Intermediate group. Only a few dorsal abdominal hairs are erect, these being mostly recumbent. *In the female the buccal armature and the pigmented area are well developed and these have specific morphology.* This group includes : *P. squamipleuris* Newst; *P. christophersi* Sinton; *P. clydei* Sinton; *P. hospitii* Sinton; *P. eadithiae* Sinton; *P. hodgsoni* Sinton; *P. iyengari* Sinton.

C. Recumbent-haired group. On the dorsum of segments 2 to 7 the hairs are mostly recumbent. *In the female the spermathecae are much less chitinised. The pharynx and the buccal cavity which is armed with teeth and which has a pigmented area have diagnostic importance.*

This group includes : *P. minutus* Rond.; *P. babu* Ann.; P. *africanus* Newst.; *P. montanus* Sinton; *P. shortti* Adler and Theodor; *P. barraudi* Sinton; *P. baghdadis* Adler and Theodor; *P. bailyi* Sinton; *P. zeylanicus* Ann.; *P. himalayensis* Ann.; *P. malabaricus* Ann., *P. sylvestris* Sinton; *P. purii* Sinton; and *P. arboris* Sinton.

In the case of the male of groups A, B and C, the hypopygium has diagnostic characters.

From the public health point of view only five species of sand-flies are important and their identification can be easily made with the help of the illustrations. However, for other species the diagnostic tables constructed by Sinton (1932, 1933) for both males and females should be consulted.

For identification especially of female sand-flies, the spermathecae, the pharyngeal and the buccal armatures have to be exposed.

Dissection of Spermathecae.

(A). Freshly chloroformed specimen after dipping in rectified spirit is transferred to a clean slide and placed in a drop of Langeron's fluid.* Dissection is performed under a dissecting

* Langeron's Fluid has the following composition. Chloral hydrate (crystal) 2 parts. Acid carbolic (crystal) 1 part. Lactic acid, 1 part.

finest microscope with the help of fine dissecting needles. Such dissecting needles are made of the entomological pins No. 20 after the head has been chopped off and the pin inserted into a wooden handle. The spermathecae lie in the 8th segment. Two nicks involving the cuticle only are made between the 7th and 8th segments and the 8th segment is drawn away from the rest of the abdomen by giving tearing jerks with the help of a needle, the other needle being used for steadying the insect. If the 8th segment is now separated from the 9th after making similar nicks on both sides of the abdomen between these two segments, the spermathecae remaining attached to the spermathecal ducts will now be seen lying free, projecting from the severed end of the 9th segment. The preparation is covered with a cover glass and examined under the high power or 1/6th objective.

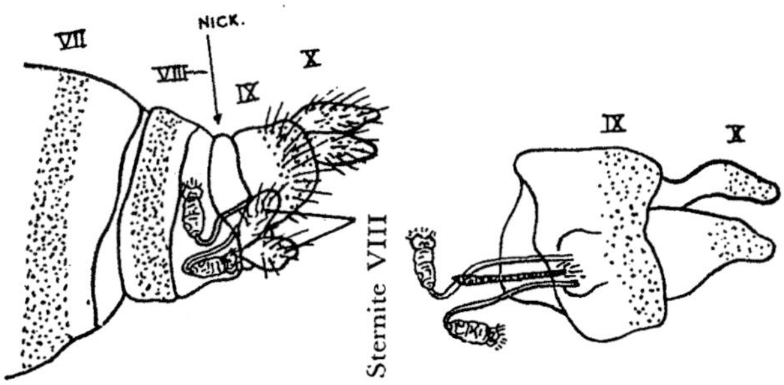

FIG. 73.
Dissection of Receptaculum seminis (*P. papatasii*).

The advantage of making the dissection in two stages, *i.e.*, separation of the posterior part of the abdomen from the 7th segment and thereafter the 9th from the 8th segment, lies in the fact that under this condition the chances of loosing the spermathecae are small. If by accident one or both the spermathecae become detached, they can easily by traced in the 8th segment.

The accessory glands not shown in the illustration, as they can only be seen in fresh and unmounted specimens, are two conspicuous organs proportionately much larger than the size of the insect and they lie on either side of the spermathecal ducts.

Those who are experienced in the technique of fine dissections will find it convenient to expose the spermathecae in a single operation by merely separating the 9th from the 8th segment.

(*B*). Specimens preserved in a dry state may be placed directly in Langeron's fluid in a watch glass and they will be ready for dissection in 4 to 6 hours. Langeron's fluid not only softens dried objects quickly, but also clears them and in these respects it is superior to lactophenol.

Dissection of Pharynx and Buccal Cavity. (1) After the spermathecae have been dissected out as stated above, the head is separated from the body and placed in a drop of Langeron's fluid. The head is turned in order that it lies on its dorsal surface and the ventral surface looks upwards. A cover glass is applied which is gently pressed with a needle. The object is now examined with the high power of the microscope. (2) The head lying on its dorsal surface is placed in a drop of Langeron's solution on a slide. With the help of fine needles the eyes on both sides are removed. This will leave the pharynx exposed. The antennae and the palps in good preparations will remain intact. It is covered with a cover glass and examined under the high power.

Staining and Mounting. Either for identification or for permanent mounting it is not necessary that the object should be stained. However, if necessary, the method described by Sinton (1932) can be followed. It is useful for staining both pharynx, spermathecae and wing and also for mounting them permanently.

Dissection of Alimentary Canal. The method of dissection of the alimentary system is the same as in mosquitoes. After the skin on both sides between the last and the penultimate segments has been cut, the abdomen is pulled and thereafter the head is severed from the neck. By pulling the last segment, the whole alimentary canal may be exposed.

Collection of Sand-flies. A clear idea of the sites where sand-flies should be looked for and the appearance of this fly in nature are important as a preliminary to their collection. In cattle sheds which are their favourite haunts, the dark places in the corner and on the walls near the floor should be searched. Sand-flies often take shelter under the earthenware trough from which cattle take their feed. Next to cattle sheds they may be found in sleeping rooms occupied by man. These flies often hide in cracks and holes in walls of both pucca and mud houses usually within three feet from the floor. Other places where they can be obtained are goat and sheep enclosures and fowl houses. Rat-holes and tree-holes may at times provide very useful collections.

Their presence is readily detected as soon as a puff of cigarette smoke or air is gently blown on suspected places or in cracks; this will cause them to become agitated at once: Their tiny size, erectly-held wings, and hopping flight are very characteristic features by which they can be easily recognised.

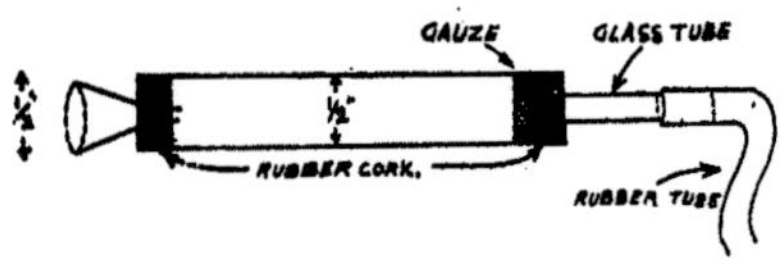

FIG. 74.
Aspirator for catching midges (the funnel is detachable).

The most important equipment necessary for catching sand-flies are : (1) a torch ; (2) an aspirator; and (3) test-tubes.

(i) They can be collected in the morning or evening in dark and damp rooms with the help of a torch. They are captured in a test-tube by inverting it over the insect. (ii) Buxton (1928) recommends the use of an aspirator which consists of a glass barrel 3 inches by 1 inch, a nozzle of glass tubing and an outgoing piece of tubing, protected by gauze and connected by rubber tubing to a mouth-piece. It can be used for catching any small insect. The insects are sucked into the barrel. (iii) Traps: traps have often been used to obtain sand-flies from the burrows of animals. These consist of strips or flags of paper smeared with castor oil and held in cleft sticks by means of which they can be fixed across the mouth of the burrow. Sand-flies entering or leaving the burrow and accidentally touching the paper are immediately caught on the paper.

Preservation. Pinning always injures flies to such an extent that identifications are rendered impossible. When preserved in spirit, the hairs are all lost; these are particularly important for identification.

The insects should be placed on a web-like layer of teased cotton-wool contained in a shallow pill box or a corked tube but they must not be covered with wool as the appendages are thereby likely to be damaged. When specimens are to be preserved for a long time, Kahn's tubes are very suitable. Each tube should contain only one sand-fly and the open end is plugged with cotton-wool. These tubes are placed in a cigarette tin on cotton-wool on which a drop of creosote is put. After the cover of the tin has been placed back, it is sealed with a mixture of paraffin and naphthalin.

Preventive measure and control against sand-flies and kala-azar: (1) All breeding places within the house and within 50 yds. should be eliminated. Particular attention should be paid to cattle sheds and poultry houses which are the common source of sand-flies found in the house. If the breeding places are directly attacked, this will greatly reduce the number of sand-flies coming into the building. Any place that provides moisture, darkness and organic matter, the three conditions that favour breeding, should be dealt with. (2) The earthen floor of cattle sheds should be stamped down so as to make conditions difficult for the larvæ to burrow. (3) All fissures in the walls and floors must be attended to. Cracks and holes in walls up to 3 ft. from the ground can be effectively sealed up by applying a mixture of cow-dung and earth. (4) Sleeping under a sand-fly net with a mesh of 45 to the square inch is worth the discomfort. However, a DDT impregnated mosquito net is more convenient to use. (5) Regarding the two sand-fly repellents Dibutyl phthalate and Dimethyl phthalate, and the effects of DDT residual spray on the breeding of sand-flies, see in Part II.

Geographical Distribution of Sand-flies in India. *P. papatasii* is widely distributed in the plains particularly in places where the climate is hot and dry. It has been found as far east as Calcutta and as far south as Madras. *P. argentipes* is found where the climate is warm and moist. It is confined to the area East and South of a line joining Bombay and Simla. It has been recorded at altitudes as high as 4,000 ft. (Simla Hills). It is the principal sand-fly in Ceylon. *P. major* is a hill (Himalayan) species at heights of 5000 to 7000 ft. It occurs all along the Himalayan foot hills. *P. chinensis* has a distribution similar to that of *P. major*. *P. sergenti* is confined to the plains on an area north and west of a line joining Bombay and Simla. It prefers dry and hot climate. *P. newsteadi* is like *P. major* a hill species. *P. colabaensis* has been recorded from Bombay City and Vizagapatam Agency of Madras, *P. maynei* from Saharanpur (U.P.), *P. elanorae* from Karnal (Punjab), and *P. hospitii* from Western foot-hills of the

Himalayas. *P. christophersi* is confined to the Punjab. *P. clydei* has been recorded widely from the plains of India. *P. minutus* var. *antennatus* is widely distributed over the plains of India and Pakistan, and *P. minutus* in the North Western places of India. *P. africanus* is confined to Sind. *P. babu* is widespread over the plains and in Himalayan foot-hills. *P. baghdadis* is found in North Western parts of India and *P. shortti*, North Eastern parts of India and Burma. *P. barraudi* is found only in Assam and Burma. *P. bailyi* has a wide distribution over the plains and at heights up to 6000 ft. above sea level. *P. montanus* is a Western Himalayan species. *P. zeylanicus* is a wild species chiefly associated with jungle and with moist warm climate. *P. malabaricus*, type only has been collected from Travancore. *P. sylvestris*, *P. himalayensis*, *P. purii* and *P. arboris* are rare flies which have been recorded from the Darjeeling areas. *P. smithi* and *P. thapari* have been collected in Poona regions. *P. squamipleuris* has a wide distribution all over India up to 6,000 ft. and is essentially a garden species.

Qutubuddin (1952) has recently drawn attention to a new form of *P. antennatus* Newstead, found in Deccan, and named by him as *decanensis* exhibiting close affinities with the African form *antennatus* var. *dubius*. He has also described a few other sand-flies found in N. W. F. Province (1951).

Regional Distribution. Jammu Province: *P. minutus* var. *antennatus*, *P. papatasii*, *P. major*, *P. sergenti*, *P. baghdadis*, *P. chinensis*, *P. squamipleuris*. Kashmir: *P. babu*, *P. papatasii*, *P. sergenti*, *P. chinensis*, *P. montanus*. Ladakh: *P. chinensis*, *P. sergenti*. **Poona:** *P. papatasii*, *P. argentipes* (differs from the type in having dark feet), *P. squamipleuris*, *P. thapari*, *P. smithi*. **Calcutta:** *P. argentipes* (type), *P. papatasii*, *P. squamipleuris*, *P. minutus*. **Rajputana:** (Ajmer City)—*P. sergenti*, *P. papatasii*, *P. colabaensis*, *P. clydei*, *P. babu*, *P. baghdadis*, *P. minutus* var. *antennatus*, *P. bailyi* var. *campester*, *P. argentipes*. (Mount Abu)—*P. argentipes*. (Sambar town and Salt Lake)—*P. argentipes*, *P. baghdadis*, *P. colabaensis*, *P. sergenti*, *P. babu*.

Family **Chironomidae.** The insects belonging to this family are delicate midges which resemble mosquitoes very closely but can be easily distinguished by (*i*) the thorax, (2) wing venation, and (3) proboscis. The anterior part of the thorax extends over the head. The wing is devoid of scales and the venation is quite different from that of mosquitoes. The proboscis is much shorter.

The eggs are laid in a transparent jelly-like mucus in stagnant water. The larvæ are aquatic. The larvæ of one species on account of the presence of a red pigment in the hæmolymph are popularly called blood worms. They are found in pools, ponds, ditches, drains, etc. They are also found in settling-tanks of waterworks and may at times by their numbers partly block some pipes. They generally live in tubes formed of mud.

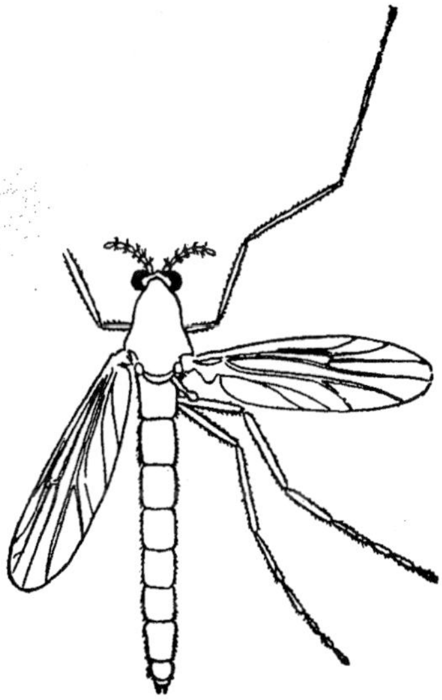

FIG. 75. Chironomus.

The adults are harmless and all live on vegetable sap.

Family **Ceratopogonidae.** The members of this family are of minute size and some of them are blood sucking. They lay their eggs in a single row on vegetation in overhanging swamps, in decaying vegetation near moist situations, in tree holes, or in manure heaps, so that the larvæ may be aquatic, semiaquatic or terrestrial. This family was formerly considered a subfamily of Chironomidæ but has been raised to the status of family by Malloch in 1917. The difference between the two is as follows. In Ceratopogonidæ the thorax does not usually project over the head. In Chironomidæ the mandibles are absent and the anterior part of the thorax arches forwards over the head. Out of numerous genera of this family only four are blood suckers; these are: (*a*) *Culicoides*, (*b*) *Lashiohelea*, (*c*) *Leptoconops*, and (*d*) *Acanthoconops*. The females alone bite, males being vegetable feeders. They have a world-wide distribution.

Genus **Culicoides** Lat. The wings are blotchy or dappled; macrotrichia of wings distinct but very scanty; empodium very short. It is almost world-wide in distribution and they are found in damp places and about decaying vegetation. The larvae are generally aquatic or else are found in rotting vegetation.

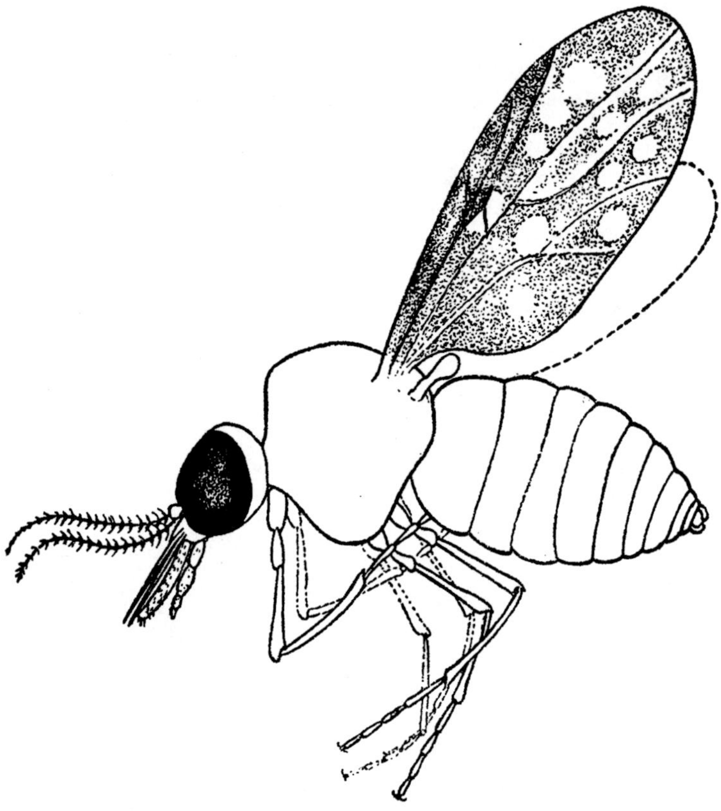

Fig. 76.
Culicoides.

Genus **Lashiohelea** Kieff. Wings are not dappled; they have dense macrotrichia all over; costa extends to about the middle of the wing; 2nd radial cell very long; there are bare lines adjoining the veins; empodium very long. The larvae are terrestrial and breed in moist decomposing vegetable matter.

Genus **Leptoconops** Skuse. Costal vein does not extend beyond the middle of the wing; the first and the third veins are fused; presence of an extra vein (really a fold of the wing) between the third and the fourth longitudinal veins. The antennae are composed of 14 segments. In genus *Holoconops* the antennae of the female consist of 13 segments. The females of several species are known to bite man. They sometimes attack domestic animals in swarms. The bites are painful and the subsequent local reaction is irritating and persistent. The majority are found in the tropics.

Genus **Acanthoconops** Carter. This is distinguished from *Leptoconops* in having the front clothed with bristles or spines.

Genus **Culicoides.** They are minute insects and the females only are able to inflict painful bites and at times in certain localised areas they constitute a greater nuisance than mosquitoes. They sometimes attack in swarms. They generally occur out of doors and are seldom found inside the house. They bite mostly after sunset. In addition to cattle and other domestic animals, some bite man. Some even bite caterpillars, snails, earthworms, lizards, etc. The adult is a small, delicate midge. The sexes are differentiated by the antenna which is plumose in males and pilose in females. The eyes in the male are more widely separated than in the female. The labium is fleshy. The labrum is armed with teeth on both its lateral edges near the tip. The mandibles and maxillae are each armed with several strong teeth. The hypopharynx is finely toothed at the tip. The wing is clothed with fine hairs. There is a well-marked fringe. There are numerous oval spots on the wing itself and large dark spots on the costal border. The larvae are found in such places as pools, tree holes, rotting of banana plants, leaf moulds, etc. They are generally aquatic. The mature larva consists of the head, three thoracic and ten abdominal segments. The last segment has 4 pairs of appendages which correspond to tracheal gills of other aquatic larvae. The larva of *Culicoides* has no spiracles. The body is slender and is sparsely hairy. Its presence is readily detected by its rapid vibratile movements which closely simulate those of a spirochaete. The pupa has well developed spines at the sides. The duration of the different stages of *C. oxystoma* in confinement is as follows : egg stage lasts 3 to 11 days, larval stage 2 to 10 weeks and the pupal stage 3 to 7 days (Patel, 1922).

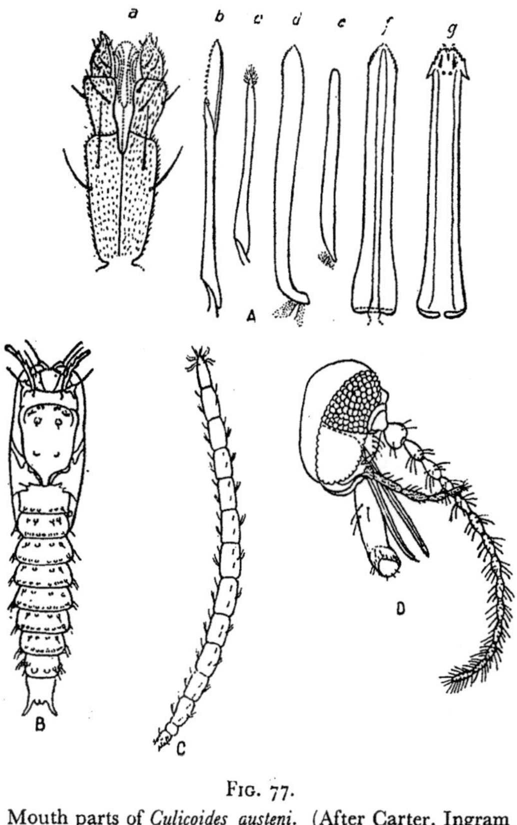

FIG. 77.

Mouth parts of *Culicoides austeni*. (After Carter, Ingram and Macfie).

A. a,—labium, ventral view (female) ; b,—maxilla female) ; c,—maxilla (male) ; d,—mandible (female); e,—mandible (male); f,—hypopharynx (female); g.—labrum (female);

Culicoides occur throughout India particularly in Assam and Bengal. They are identified mainly by the wing pattern and the male hypopygium.

Among others the following species are common in this country. *C. oxystoma* Kieff: feeds on mammals and seems to prefer horses, cattle, buffaloes, and goats. *C. fulvithorax* Aust. : feeds on large demestic animals, also bites man. *C. anopheles* Edw. : normally it bites cattle but occasionally it will be found attached to the abdomen of female *Anopheles* mosquitoes and sand-flies. For reference of *Culicoides* as ectoparasites of *Anopheles*, see Sinton and Harwood Little (1925) : *Trans. Roy. Army Med. Corps.*, 45, p. 47.

Several species of *Culicoides*, especially *C. austeni* and *C. grahami* (Sharp, 1928), have been reported to serve as intermediate hosts of *Acanthocheilonema perstans* (*Filaria perstans*), a non-periodic filaria. *F. perstans* is widely prevalent in tropical Africa, Algeria, Tunis, New Guinea, British Guiana but its principal area of infection is in and around the Congo basin. The method of invasion of the body of the insect by the filarial nematode is the same as is noticed in the case of *Wuchereria bancrofti* in *Culex fatigans.*

FIG. 78.
Wing of *Culicoides oxystoma.*

FIG. 79.
Wing of *Culicoides anopheles.*

Generally within 6 hours the microfilariae have disappeared from the lumen of the stomach and after penetrating the stomach wall they lie in the thoracic muscles of the fly. The time taken for the microfilaria to undergo complete development is about 7 to 10 days.

Onchocerca cervicalis Railliet and Henry, is widely spread and occurs in the ligamentum nuchae of the horse and the mule. It is transmitted by species of *Culicoides.* It takes 24 days for the parasite to attain the full development in *C. nubeculosus* (Steward, 1933). Causey (1938) found a specimen of *C. peregrenus,* collected from a bed room in Siam, heavily infested with filaria larvae.

For capturing these and other small midges, a sucking tube is very useful.

Family **Simuliidae.** Flies of this family though ethnologically grouped under midges are called blackflies on account of their resemblance to flies. They are dark in colour and are of robust build. They are readily recognised by their general appearance, wing venation and by the annulated antennæ which project out like horns of an animal. The antennæ are similar in both sexes and each consists of 11 joints. The sexes are distinguished by the eyes which are close together in males and widely separated in females. The females are vicious biters and suck blood, mostly of cattle. Some species also attack man. The mouth parts in *Simulium* are like those of *Culicoides* but shorter. The wings are large and hyaline.

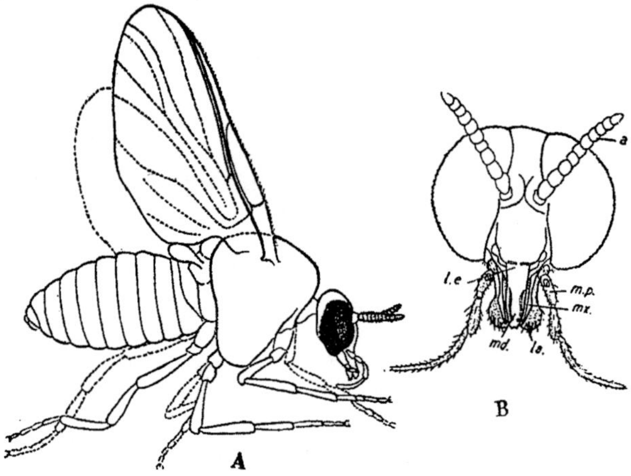

FIG. 80.
A. *Simulium indicum.*
B. Mouth parts of Simulium (after Gibbins).
a., antenna ; la., labium ; le., labrum-epipharynx ; md., mandible ; m.p., maxillary palp ; m.x., maxilla.

The anterior border is very prominent on account of the presence of well developed costal and first longitudinal veins. Other veins are poorly developed. There are no scales or hairs on the surface of the wing. Species is identified by colouration, tarsi and claw.

The metamorphosis is complete. Eggs are laid on aquatic plants submerged in water. The female fly sits on a twig and oviposits by extending its ovipositor below the surface of water and glues the eggs to some submerged twigs or some aquatic plants. Sometimes the female dives and lays eggs on the object which may be a stone and which lies well below the surface of the water. The larvae occur in swiftly running streams usually where there is a strong current and the water falls over an impediment but the water is not deep. The larvæ fix themselves to stones or plants sometimes at a depth of 8 to 10 ft. but generally at a depth of 1 ft. The larvæ require much oxygen without which they cannot live. In India they breed in hill streams. The larva is cylindrical in shape and possesses a sucker at the posterior end. It has very strong mouth organs for chewing solids, e.g. crustaceans and other small animals. It breathes by means of gills. It moults six times. Before pupation the larva spins a cocoon inside which the pupa rests. It breathes by means of thoracic gills. The fly emerges while the pupa lies immersed in water and quickly appears on the surface. The larval and pupal stages occupy nearly a month.

The adults are very active on bright sunny days. They are vicious biters. Swarms of them attack domestic animals particularly horses. They have been reported as causing loss of live stock from many parts of the world. The dead cattle exhibit reddish spots which are very numerous near the genitalia and on the inner side of the legs. The death is due to toxin injected while the flies are biting.

Disease relationship. Simulium has been proved to carry onchocerciasis to man and in some localities it also carries the same disease to cattle and horse.

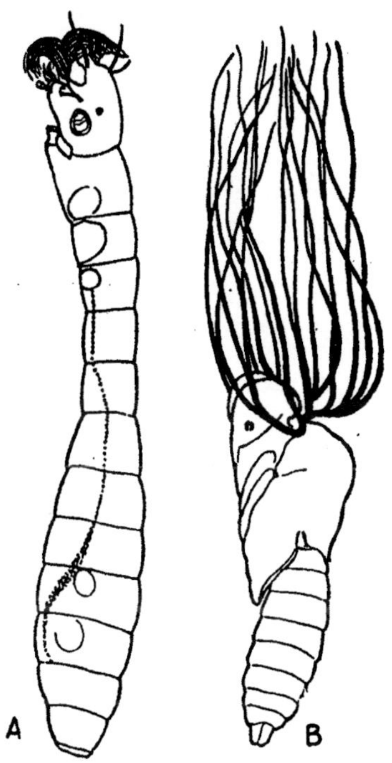

FIG. 81.
Larva and pupa of Simulium. (After Castellani and Chalmers).

Human Onchocerciasis. The adult male of *O. volvulus* (*O. caecutiens*) is about one inch in length and the female is much longer. These worms are found in the subcutaneous tissues of man in West Africa and in Tropical America and produce painless tumours. Embryos are found in the peripheral blood. In 1926 Blacklock discovered the development of *O. volvulus* in *Simulium damnosum* in West Africa. In Kenya, *S. neavei* Roub. is the carrier. It takes 6 to 7 days to complete the development.

O. caecutiens is endemic in Guatemala and Mexico in tropical America. It produces tumours mostly on the head and the embryos also invade the skin. In Mexico the carrier is *S. mooseri* (*S. callidum*, Dyer and Shannon) (Hoffman, 1930). The Central American species, *S. metallicum* Ball (*avidum*), *S. ochraceum* and *S. callidum* are the vectors in Guatemala.

Animal Onchocerciasis. Onchocerciasis in animals is much more widespread. It occurs in India, Australia, Ceylon and in many countries in the east and in the west. *O. gibsoni* (*O. indica*) is prevalent in Australia, India, Ceylon, Malaya, Java, Siam and other countries in the east. It produces tumours in the hind legs of cattle and buffaloes. In Southern Rhodesia the infection is very mild. Its vector is probably a blood-sucking arthropod (Cleland, 1927). It has been proved in Malaya to be carried by *Culicoides pungens* de Meij., (Buckley, 1938). *O. gutterosa* (*O. bovis*) is found in cattle in France, England, Egypt, Sudan and in the Argentine Republic. In England it has been shown to be transmitted by *Simulium ornatum* (Steward, 1937). *O. armillata* is extremely common in Egypt and Sudan. Its vector is not known. *Culicoides nubeculosus* has been proved to be the carrier of *O. cervicalis* of the horse in England (Steward, 1933).

The insects concerned in the transmission of animal onchocerciasis are not yet exactly known. It has been clearly established that the disease is prevalent in places where *Simulium* is absent. Although a large variety of biting insects have been employed to determine the vector, all have met with negative results in Australia.

For review on Onchocerciasis in Africa and Central America, see Supplement to the *Amer. J. Trop. Med.*, 1938, Vol. 18, Number 1 : Also Mohammed, A. S. (1931). *Ann. Trop. Med.* and *Parasit.*, 25, p. 509 : and for development of *O. volvulus* in *S. damnosum* Theob., Wanson, Henrard and Peel (1945). *Rev. Appl. Ent.*, 36 : 148.

Simulium damnosum is widely distributed in equitorial Africa. It can be easily recognised by the greatly flattened front tarsi and silvery pubescence of the front tibiae. It is widely distributed and is found on the banks of streams and rivers. *Simulium indicum* is the common Indian species. Short yellow hairs are present on the abdomen for at least the last two segments.

CHAPTER VIII

ORTHORRAPHA—BRACHYCERA

Flies belonging to Orthorrapha-Brachycera comprise a very large group of insects which include also a large number of families. They are of stout build and only a few are blood suckers. The maxillæ are well developed and they all possess rigid maxillary palps. The wing venation is complex. The metamorphosis is complete.

Families of Orthorrapha-Brachycera. (1) Family **Leptidae.** This family contains a large number of insects which bear a close resemblance to robber flies (Asilidae) but are much smaller. Only four species have been reported to suck blood. The wing is often blotched and the venation is more or less Tabanus-like. (2) Family **Asilidae.** They are large flies with an elongated abdomen. The proboscis is a rigid chitinous dagger. The body is bristly and hairy. The legs are strong and very bristly. They are purely carnivorous and feed on other insects having soft cuticle. (3) Family **Bombyliidae.** The flies are of stout built, the body pubescent and some resemble bees. The wings are large and have dark spots or markings. The adults live on nectar whereas the larvae are parasitic on Hymenoptera such as larvae of bees and wasps. At least two species are known to attack pupae of Glossina. (4) Family **Dolichopodidae.** These are of large size. They are so named on account of their long and slender legs. They are usually brightly metallic and greenish in colour. The wing venation is of a simple type. The arista is terminal and very long. They are predacious and are said to attack mosquito larvae.

Family **Tabanidae.** This family is widely spread throughout the world and contains a very large number of species. Although this family is of little importance in medicine, it is chiefly concerned in the transmission of surra, a fatal trypanosome disease of cattle and equines. They are of very large size and the females are very active and voracious blood suckers. The flight is rapid. They attack horses and jungle animals, rarely man.

The family Tabanidæ includes two subfamilies, Tabaninæ and Pangoninæ. In the former tibial spurs and ocelli are absent, whereas in the latter these are present. The subfamily Tabaninæ includes only two blood-sucking genera, *Tabanus* and *Haematopota* while Pangoninæ contains *Chrysops* and *Sylvius* which are blood-suckers. It is extremely doubtful if the members of the genus *Pangonia* (Pangoninæ) are also able to suck blood.

In *Tabanus* the 3rd joint of the antenna is annulated and provided with a tooth which forms an angular projection. In *Chrysops* the three segments of the antennæ are of equal length. In *Haematopota* the wings have characteristic dark spottings and markings. The second antennal segment is very short whilst the first and the third are relatively long.

Genus **Tabanus.** These flies are popularly known as gad-flies. They are large in size and though they are blood-suckers, they do not carry any human diseases. Their chief characters are the following : Eyes are large and brilliantly coloured in life. In the male the eyes meet in the middle line while in the female there is an intervening space between them on which lies a white band called the callus. Antennæ are three-jointed; the third joint is segmented into a number of smaller joints. The third segment is also furnished with a distinct tooth. The mouth parts lie enclosed in the labium which is comparatively short and is a fleshy structure. The cutting organs are the mandibles and maxillæ. The mandibles are well developed and pointed at the tip. The maxillæ have serrated edges. Other mouth organs consist of the

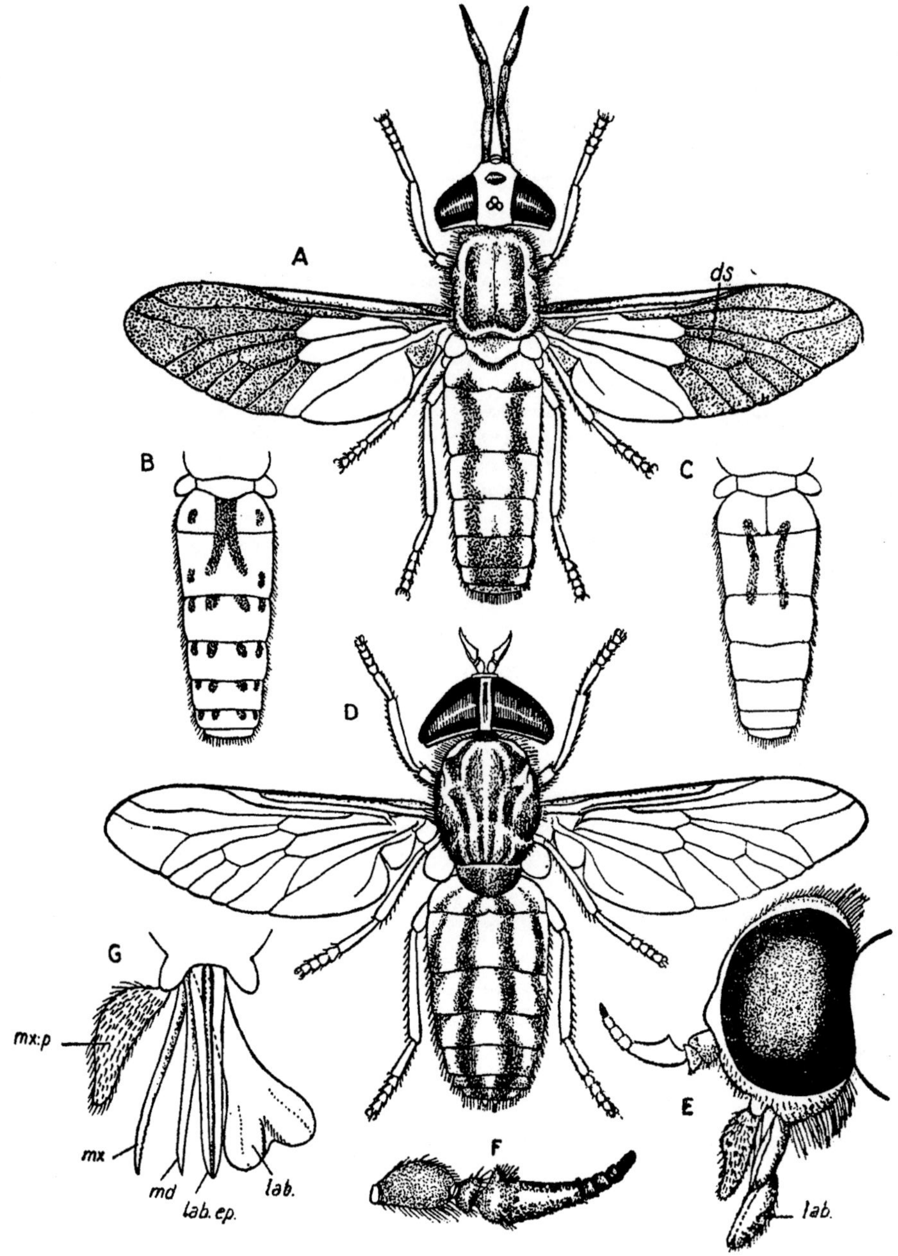

FIG. 82.

A. *Chrysops dimidiata*, female. d.s., discal cell. B. *C. discalis*. C· *C. silacea*. D. *Tabanus striatus*. E. Head of *Tabanus*. F. Antenna of *Tabanus*. G. Mouth parts of *Tabanus* of one side only. lab., labium; lab. ep., labrum-epipharynx; md., mandible; mx., maxilla; mx. p. maxillary palp. F. Antenna of *Tabanus*.

epipharynx and the hypopharynx. Wings may be clear or coloured. The venation has a complex arrangement. The abdomen is composed of 7 visible segments.

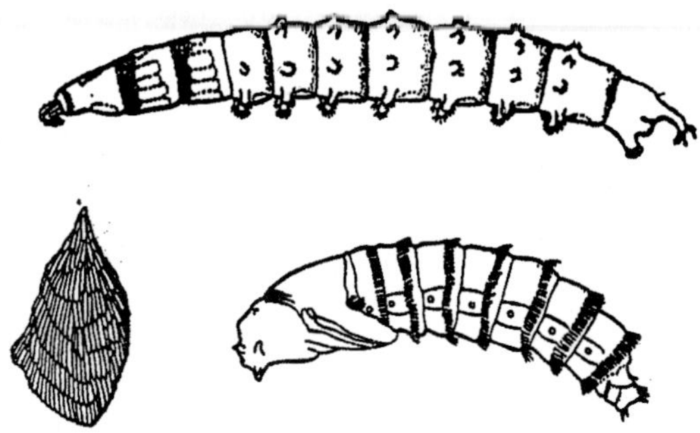

FIG. 83.
Egg, larva and pupa of *Tabanus kingi.* (After Austen).

Life history. Metamorphosis is complete. Eggs are laid in glued masses on leaves and stems of plants overhanging water surfaces or on mud borders of streams and ponds. The larvæ on escaping from the egg-shell drop into the water or burrow into wet sand or mud. The larval body is segmented and is very narrow at both ends. In addition to the head, and three thoracic segments, there are 10 abdominal segments. The head carries a pair of antennæ, a pair of mandibles and a pair of maxillæ with segmented palps. The mandibles are extremely powerful. The breathing tube is attached to the last abdominal segment. A number of body segments bear pseudopods. They thrive on mosquito and fly larvæ, annelids, molluscs, crustacea and other aquatic insects. They sometimes prey on each other. The pupa is stationary and resembles that of Lepidoptera. The spiracles of the pupa are placed on the thorax. Before pupation the larva crawls out of the water and lies buried in the loose earth. According to Mitzmain (1913), the whole life cycle in *T. striatus* Fab., takes about 50 days which includes 4 days for the egg stage, 39 days for the larva which moults three times, and 7 days for the pupa. In the case of the Indian *Tabanus* the larval life lasts 4 to 6 weeks (Rao and Mudaliar, 1935). The males in nature derive their nourishment from vegetable juice. A diet of blood is essential for the development of eggs in females. The adults generally rest on trees. Horses are more particularly attacked than cattle. Man is occasionally bitten. As the fly settles on an animal, it immediately thrusts its proboscis but with the slightest movement of the animal, its feed is interrupted and it flies away to return to the same or another animal after sometime. It seldom gets the chance to feed to repletion on one animal and in one sitting. It is difficult to induce the female to suck blood in captivity.

Disease. Surra, a disease transmitted especially by a tabanid fly, is widely distributed being found in Central Asia, India, Burma, China, Indonesia and the Philippines. It also exists in Persia and has often been imported into Africa, the United States and Australia.

The causal organism is *Trypanosoma evansi* which closely resembles *T. brucei.* Surra affects equines, cattle, buffaloes, camels, elephants and occasionally dogs. The disease generally breaks out in epizootic form in the rainy season in India and is commonest in riverine and marshy tracts or in the region of forest lands.

Evans (1880) first drew attention in Dehra Ismail Khan to surra being due to a trypanosome but the precise way in which the infection is commonly transmitted from animal was not known till Mitzmain (1913) pointed out that the common disseminating agent of surra among animals of the equine species in the Philippine Islands is *Tabanus striatus* and the mode of transmission is merely mechanical. This can take place only when the feeding is interrupted. There is no life cycle of the trypanosome in Tabanids. In India buffaloes and cattle seldom manifest any symptom of trypanosomiasis and they act as the "reservoirs" of this disease. In India Cross and Patel (1922) found *T. nemocallosus* to be extremely efficient carriers. It has been reported that the trypanosome remains viable on the proboscis for 48 to 72 hours. In actual practice, however, successful transmission can only occur if the transference of the parasites takes place within a few minutes after bite and the chances are great if it occurs almost instantly. Considering the mechanical way in which the trypanosome of surra is carried by *Tabanus*, other blood-sucking flies, e.g., *Musca crassirostris* Stein, *Lyperosia exigua* de Meij., *Stomoxys calcitrans* L., and *Haematopota* sp., may at times also be concerned in the transmission of the disease. It has also been suggested that *Tabanus* may carry anthrax in the same way. Kapur (1941) has stated that *T. orientis* Wlk., can act as a vector of rinderpest. These flies take many blood meals and the feeding is always interrupted.

There is no doubt that surra is spread through the agency of any genus of biting fly among which the many species of *Tabanus* are the principal transmitters.

Some Common Indian Species of Tabanus. T. striatus F. Callus on the forehead has a spindle shaped lineal extension; abdomen blackish or reddish brown with grey median and lateral stripes almost all the same length. **T. hilaris** Walk. Resembles *T. striatus* except that the abdomen with grey median stripe begins only on the third segment and with lateral stripes ending on the 3rd or 4th segment. **T. albimedius** Walk Forehead has no callus. Abdomen and thorax obscurely reddish brown; femora reddish brown. The median abdominal stripe composed of almost equal sized spots not very large; the side spots are large and distinct. **T. rubidius** Wied. Resembles *albimedius* except that the abdomen and thorax blackish brown and the femora blackish. **T. speciosus** Ric. Differs from *albimedius* in that the median abdominal stripe is composed of spots of unequal size, those on the 3rd and 4th segments being very large and conspicuous. The abdomen and the thorax are reddish brown and the femora blackish. **T. brunnipes** Sehsak. A brown species; the median abdominal stripe is broad with two round spots on the 2nd segment; femora reddish; wings brown on anterior border. **T. ditaeniatus** Macq. It is a particularly small species. Forehead has 2 separate callus; abdomen yellowish with median and lateral blackish and brownish stripes. **T. rufiventris** F. (**sanguineus** Wlk.). This is a brown or a blackish brown species. Abdomen is reddish brown with broad white bands and large triangular median spots. **T. orientis** Wlk. Abdomen reddish yellow, darker at apex with narrower yellowish white segmentations and median spots; legs black; tibiae yellowish.

Genus **Chrysops.** They are found all over the world and are smaller and more delicately built than *Tabanus*. There are three ocelli on the vertex. The distal part of the wing is variously coloured in different species. The antennæ are long and thin.

The life history is the same as in *Tabanus*. The larvæ are found in the mud of swamps. They mostly bite cattle. Some African species, e.g., *C. silacea* Aust., and *C. dimidiata* v.d. Wulp, bite man freely in the day time. They make no sound even when they are near. They are persistent biters and the bite is painless being seldom noticed at the time. There is considerable irritation and even swelling of the affected parts afterwards. A fly generally withdraws 0.03 g. of blood.

The suggestion that *Chrysops dimidiata* might prove to be the carrier of Loa loa was first made by Manson in 1895 on geographical grounds. Loa loa is a filarial disease common in equitorial Africa and known as Calabar swellings or fugitive swellings. This was supported later by Leiper in Nigeria and Kleine in Cameroons. Connal (1921) brought forward positive evidence in favour of *C. silacea* Aust., and *C. dimidiata* v.d. Wulp, being the vectors of this disease. The developmental forms occur in the muscles of the fly. The forms and size of the worms when they escape from the labella of the fly are similar to those of *W. bancrofti* escaping from the proboscis of the mosquito The microfilariæ are found in the peripheral circulation during the day time. The infective stage is reached in 10-12 days when the filariæ begin to travel towards the head and accumulate at the root of the proboscis and in the labium. They gain exit by the labella (Connal and Connal, 1922).

The different species of Chrysops are distinguished mainly by the markings on the abdomen, also by the character of the colouration of the wing.

Both *C. dimidiata* and *C. silacea* possess two broad parallel longitudinal dark brown lines on the dorsum of the abdomen. In *C. dimidiata* the abdomen and legs are yellow. In *C. silacea* the abdomen is reddish and the tip is brown; legs are also reddish. Both are widely distributed West African species.

It has been claimed by Francis and Mayne (1921) and by Francis (1937) that *C. discalis* Will. is one of the transmitters of tularaemia. *C. discalis* is the North American species of horse-fly.

Some Oriental Species of Chrysops. **C. dispar** F. Wings with a dark transverse band near the middle and anteriorly extending to the apex; a hyaline patch in the lower part of the transverse band; abdomen with a short black bifid stripe on the second and third segments. It is very widely distributed in the Oriental Region and is found in jungles. **C. indiana and C. stimulans.** They have no hyaline patch in the lower part of the transverse band. In *stimulans*, wings have a clear spot in the discal cell, and the abdomen is blackish. In *indiana*, wings have no clear spot in the discal cell, the abdomen is yellow with a bifid stripe on the second segment.

Genus **Haematopota.** Members of the genus *Haematopota* are readily recognised by the characteristic speckling of the wings together with the presence of ocelli

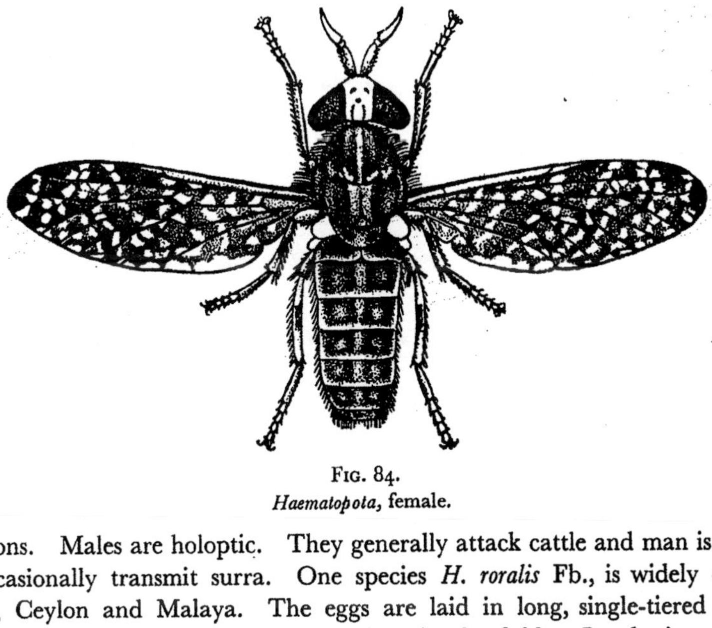

FIG. 84.
Haematopota, female.

on the frons. Males are holoptic. They generally attack cattle and man is neglected. They occasionally transmit surra. One species *H. roralis* Fb., is widely distributed in India, Ceylon and Malaya. The eggs are laid in long, single-tiered clusters of 40-100 on the leaves of rice and other foliage in rice-fields. Incubation period is 4 days. The larvæ are aquatic and feed on any soft-bodied animal. They do not attack each other. The larval stage lasts 5-7 weeks. Pupation takes place in rather drier soil but close to the larval habitat. The pupal period lasts about 7-9 days. The flies are most abundant in July-August (Isaac, 1932).

Genus **Sylvius.** They are widely distributed in the tropics. They are medium-sized flies and some of them may resemble *Tabanus*. The wings are hyaline. The third segment of the antenna has a basal tooth, ocelli and tibial spurs are absent. The proboscis is, however, considerably shorter. Only the females suck blood.

Genus **Pangonia.** The marked feature in these flies is the presence of an exceptionally long proboscis which projects horizontally. The long proboscis is used to extract the nectar from flowers. *P. longirostris* Hard., are found chiefly in forest areas in this country. The fly is very active in its habits and cannot be easily caught. They are found in jungle not far from water.

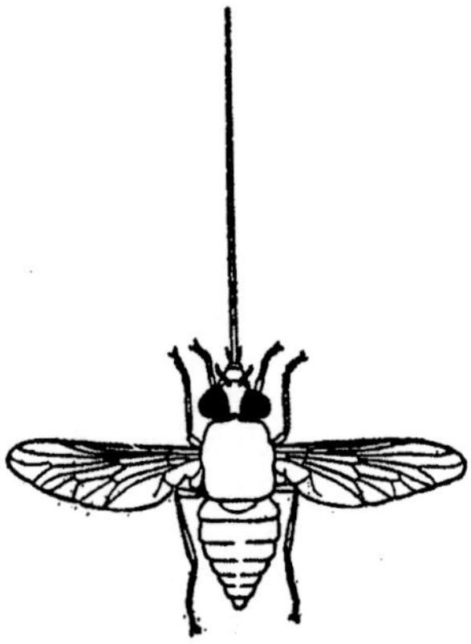

FIG. 85.
Pangonia longirostris, male.

Diptera Aschiza. These insects do not possess frontal suture and in this respect they differ from true flies. The only importance which can be assigned to this group of insects is that the larvæ of *Tubifera* are suspected of causing certain pathological conditions in man.

Family **Syrphidae.** These are robust, medium-sized flies, of dark colour and exhibiting bands and spots on the body. The presence of a spurious vein between r 4+5 and m 1+2 and traversing the cross vein r-m is characteristic in this family. The larva may be carnivorous or phytophagous.

Genus **Syrphus.** It contains the familiar hover flies. They resemble wasps in external appearance and have spots or bands on wings. The larvæ are generally carnivorous and live mostly on aphids.

Genus **Tubifera.** This genus is also known as *Eristalis*. The adult fly bears a close resemblance to the drone bee in external appearance. The eyes are pilose.

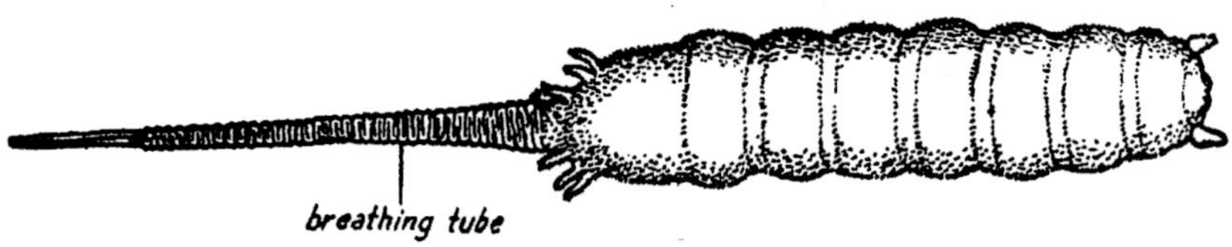

FIG. 86.
Rat-tailed larva of *Eristalis* (*Tubifera*).

The third antennal segment carries a dorsal arista. The thorax is hairy. The larva has a long tail and is familiarly known as the rat-tailed larva. The tail constitutes the respiratory tube. The larvæ are aquatic and are found in extremely foul water. They are known to find their way into human intestines possibly with drinking water and have been reported to produce unpleasant symptoms such as abdominal pain, vomiting etc., at times simulating appendicitis or intestinal obstruction. However, no case of intestinal myiasis due to the larva of *Tubifera* has yet come to our notice in this country. Nevertheless 22 cases of human intestinal myiasis by larvæ of *E. tenax* have been reported by Swartzwelder and Kali (1942).

REFERENCES (Nematocera and Brachycera)

Acton, H. W. (1919). *Ind. J. Med. Res.*, 6, 262. **Adler, S.** and **Theodor, O.** (1925). *Ann. Trop. Med. & Parasit.*, 19, 309. **Adler, S.** (1930). *Nature*, 126, 437. **Afridi, M. K.** and **Majid, S. A.** (1938). *J. Mal. Inst. India*, 1, 155. **Afridi, M. K.** and **Puri, I. M.** (1940). *Ibid*, 3, 1. **Alessandrini, M. E.** (1950). *Bull. World Hlth. Org.*, 2, 629. **Anazawa, K.** (1949). *J. Med. Assoc. Formosa*, No. 319, 1027. **Aragas, De B. H.** (1922). *Brazil-Medico*, Rio de Janeiro, 36, 129. **Ashburn, P. M.** and **Craig, C. P.** (1907). *Philip J. Sc.*, 2, 1. **Ashner, M.** (1927). *Trans. Roy. Soc. Trop. Med. & Hyg.*, 20, 452. **Atkin, E. E.** and **Bacot, A.** (1917). *Parasitology*, 9, 482. **Bancroft, T. L.** (1906). *Australian Med. Gaz.*, 25, 17. **Barber, M. A.** and **Hayne, T. B.** (1921). *Pub. Hlth. Rep.*, 36, 3027. **Barber, M. A.** and **Rice J. B.** (1938). *J. Mal. Inst. India*, 1, 37. **Barraud, P. J.** (1928). *Ind. J. Med. Res.*, 16, 377. **Barraud, P. J.** (1934). Family Culicidae. Tribe Megharhinini and Culicini.F auna Brit. India, Dipt. 5 London. Taylor and Francis. **Barraud, P. J.** and **Covell, G.** (1928). *Ind. J. Med. Res.*, 15, 671. **Basu, B. C.** (1947). *Ind. J. Malariology.*, 1, 129. **Basu, B. C.** and **Rao, S.** (1939). *Ind. J. Med. Res.*, 27, 233. **Beeuwkes, H., Kerr, Weathersbee** and **Taylor** (1933). *Trans. Roy. Soc. Trop. Med. & Hyg.*, 26, 425. **Bentley, C. A.** (1911). *Paludism*, 2, 43. **Blacklock, D. B.** (1926). *Ann. Trop. Med. & Parasit.*, 20, 1. **Boyd, M. F.** (1930). An Introduction to Malariology. Harvard University Press. **Brown, A. W. A.** (1951). *Pub. Hlth. Rep.*, 64. 863. **Brug. S. L.** (1916). *Arch. f. Protistenk.*, 36, 188. **Brumpt, E.** (1938). *Ann. Parasit. Humaine et Comparre*, 16, 220. **Buckley, J. J. C.** (1938). *J. Helminth.* 16, 121. **Buxton, P. A.** (1927). Researches in Polynesia and Melanesia. Part I-IV. London School of Hygiene and Tropical Medicine. **Buxton, P. A.** (1928). *Trans. Roy. Soc. Trop. Med. & Hyg.*, 22. 179. **Buxton, P. A.** (1935). *Parasitology*, 27, 263. **Buxton, P. A.** and **Hopkins, H. G. E.** (1925), *Bull. Ent. Res.*, 16, 151. **Causey, O. R.** (1938). *Amer. J. Hyg.*, 27, 399. **Chandler, A. C.** and **Rice, L.** (1923). *Amer. J. Trop. Med.*, 3, 233. **Christophers, S. R.** (1911). *Paludism*, 2, 73. **Chodukin, N. J., Sofieff, Schevtschenko** and **Radsivilovski** (1931). *Arch. Schiffs-u. Tropen Hyg.*, 35, 424. **Christopher, S. R.** (1911). *Sci. Mems. n. s. Govt. India*, No. 46. **Christopher, S. R.** (1912). *Ibid*, No. 56. **Christopher, S. R.** (1924). *Ind. J. Med. Res.*, 12, 273. **Christopher, S. R.** (1924). *Ibid.*, 12, 295. **Christopher, S. R.** *et al.* (1926). *Ind. Med. Res. Mems.* Memoir No. 4. **Cleland, J. B.** (1927). *Rep, Direct. Pub. Hlth.*, N.S.W., 1925, 193. **Cleland, J. B., Bradley, McDonald** (1916). *Med. J. Australia.* 2, 179. **Colless, D. H.** (1950). *Ind. J. Malariology*, 4, 377. **Connal, A.** and **Connal, S. L. M.** (1922), *Trans. Roy. Soc. Trop. Med. & Hyg.*, 16, 64. **Cornwall, J. W.** and **Patton, W. S.** (1914). *Ind. J. Med. Res.*, 2, 569. **Covell, G.** (1940). *Annual Rep. Mal. Inst. Ind.*, 1939. **Cruickshank, J. A.** and **Wright, R. E.** (1914). *Ind. J. Med. Res.*, 1, 741. **Cross, H. E.** and **Patel, P. G.** (1921). *Punjab Agric. Dept. Lahore, Vet. Bull.* No. 6. **Davey, T. H.** and **Gordon, R. M.** (1933). *Ann. Trop. Med. & Parasit.*, 27, 27. **De Caires, P. F.** (1951). *Abstract in Rev. Appl. Ent.*, B, 39, 53. **Doerr, R.** (1908). *Berl. Klin. Wochensch.*, 45, 1847. **Doerr, R.** and **Russ.** (1909). *Arch. f. Schiffs. u. Trop. Hyg.*, 13, 693. **Edwards, F. W.** (1932). *Genera Insectivorum. Diptera.* Family Culicidae. **Evans** (1880). *Rep. on Surra*, published by Military Dept., Punjab Govt. **Eyles, D. E.** (1944). *Pub. Hlth. Bull.* No. 287, Washington. **Fisher, H. C.** (1923). *Rep. of the Health Dept.* of the Panama Canal for 1922; Panama Canal Press. **Francis, E.** (1937). *U. S. Pub. Hlth. Rep.*, 52, 103. **Francis, E.** and **Mayne, B.** (1921). *Ibid.*, 36, 1738. **Gahan, J. B.** and **Lindquist, A. W.** (1945). *J. Econ. Ent.*, 38, 223. **Gill, C. A.** (1921). *Ind. J. Med. Res.*, 8, 633. **Green, R.** (1929). *Bull. Inst. Med. Res.*, F.M.S., No. 5. **Hajra, B. N.** (1948). *Ind. J. Malariology*, 2, 307. **Hertig, M.** (1942). *Amer. J. Trop. Med.*, 22, Suppl. 1. **Hewitt, R., Kenney, Chan** and **Mohammed** (1950). *Ibid.* 30, 217. **Hicks, E. P.** and **Diwan Chand** (1936). *Rec. Mal. Surv. Ind.*, 6, 73. **Hindle, E.** (1929). *Trans. Roy. Soc. Trop. Med. & Hyg.*, 22, 405. **Hindle, E.** (1930). *Lancet*, ii, 835. **Hindle, E.** (1932). *Br. J. Exp. Path.*, 13, 135. **Hinman, E. H.** (1932). *Quart. Rev. Biol.*, 7, 210. **Hinman, E. H.** (1933). *Ann. Ent. Soc. Amer.*, 26. 45. **Hinman, E. H.** (1933). *Amer J. Hyg.*, 18, 224. **Hodgkin, E. P.** (1933). *Bull. Inst. Med. Res.*, F.M.S., No. 1 of 1932. **Hoffman, C. C.** (1930). *Ann. Inst. Biol. Univ. Mexico*, 1, 59. **Howland, L. J.** (1930). *Jl. Ecology*, 18, 81. **Hu, S. M. K.** (1938). *Peking Nat. Hist. Bull.*, 13, 39. **Haff, C. G.** and **Coulston, F.** (1944). *J. Infect. Dis.*, 75, 231. **Isaac,**

P. V. (1932). *Ind. J. Vet. Sci.*, 2, 278. **Iyengar, M. O. T.** (1924). *Ind. J. Red. Res.* 12, 23. **Iyengar, M. O. T.** (1939). *Trans. 7th Cong.* F.E.A.T.M., 3, 136. **Iyengar, M. O. T.** (1931). *Ind. J. Med. Res.*, 19, 499. **Iyengar, M. O. T.** (1936). *Parasitology*, 28, 190. **Iyengar, M. O. T.** (1937). *Ind. Med. Gaz.*, 72, 300. **Iyengar, M. O. T.** (1938). *Proc. Nat. Inst. Sci. Ind.*, 4, 237. **Jacob, V. P.** (1950). *Ind. J. Malariology*, 4, 251. *James*, **S. P.** (1931). *Trans Roy. Soc. Trop. Med.* & *Hyg.*, 24, 477. **Jansco, N.** (1904). *Cent. f. Bakt.*, 36, 624. **Jobling, B.** (1937). *Trans. Roy. Soc. Trop. Med.* & *Hyg.*, 30, 467. **Kapur, H. R.** (1941). *Ind. J. Vet. Sci.*, 11, 84. **Keilin, D.** (1921). *Parasitology*, 13, 225. **Keilin, D. Tate** and **Vincent** (1935). *Parasitology*, 27, 257. **Kennedy, J. S.** (1947). *Bull. Ent Res.*, 37, 593.. **King, W. V.** (1929). *Amer. J. Hyg.*, 10, 560. **King, H. H., Pandit, Menon** and **Iyer** (1929). *Ind. J. Med. Res.*, 17, 406. **Kligler, I. J.** (1924). *Bull. Ent. Res.*, 14, 403. **Kligler, J.** and **Mer, G.** (1930). *Riv. di Malariol.*, 9, 363. **Knowles, R.** and **Basu, B. C.** (1936). *Proc. 23rd Ind. Sci. Congr.*, 492. **Knowles, R.** and **Basu, B. C.** (1943). *J. Mal. Inst. Ind.*, 5, 1. **Kowles, R.** and **Basu, B. C.** (1933). *Ind. J. Med. Res.*, 20, 757. **Knowles, R., Napier** and **Smith** (1924). *Ind. Med. Gaz.*, 59, 593. **Kohler. C. E., Brown, H. W.** and **Williams, R. W.** (1951). *Rev. Appl. Ent.*, B, 39, 161. **Kudo, R.** (1929). *Arch. Protistenk.*, 67, 1. **Kuitert, L. C.** ahd **Hitchock, J. D.** (1948). *Proc. Ent. Soc. Wash.*, 50, 77. **Lamborn, W. A.** (1922). *Bull. Ent. Res.*, 13, 129. **Latyshev, N. I.** and **Kriukova, A. P.** (1941). *Abs. in Trop. Dis. Bull.*, 40, 24. **Leeson, H. S.** (1932). *Bull. Ent. Res.*, 23, 25. **Leiper, R. T.** (1912). *Br. Med. Jl.*, *i*, 39. **Macan, T. T.** (1948). *Bull. Ent. Res.*, 39, 237. **Macfie, J. W. S.** (1921). *Ann. Trop. Trop. Med.* & *Parasit.*, 15, 377. **Macfie, J. W. S.** (1923). *Ibid*, 17, 5. **Majid, S. A.** (1937). *Rec. Mal. Surv. Ind.*, 7, 105. **Mathis, M.** (1934). *C. R. Soc. Biol.*, 115, 1624. **Mayer, M.** (1921). *Arch. f. Schiffs-u. Tropen. Hyg.*, 25, 64. **Mayne, B.** (1922). *Pub. Hlth. Rep.*, 37, 1060. **Mayne, B.** (1929). *Ibid.* 17, 109. **McCarrison, R.** (1906). *Ind. Med. Gaz.*, 41, 7. **McCombie Young, T. C., Richmond** and **Brendish** (1926). *Ind. J. Med. Res.*, 13, 961. **Menon, T. B.** and **Ramamurti, B.** (1941). *Ibid*, 29, 393. **Miller, A., Burgess** and **Carpenter** (1947). *J. Nat. Mal. Soc.*, 6, 227. **Missiroli, A.** (1929). *Riv. Malariol.*, 8, 393. **Mitzmain, M. B.** (1913). *Philip. J. Sc.*, B, 8, 197, 223. **Mondal, R. S.** (1938). *Rep. B. N. Ry. Malaria Inspectors Cönference*, 54. **Morris, H. M.** (1922). *Bull. Ent. Res.*, 13, 197. **Moshkovski, S. D.** *et al.* (1937). *Abs. in Rev. Appl. Ent.*, B, 1932, 26, 241. **Muhlens, P.** (1912). *Arch. f. Schiffs-u. Tropen-Hyg.*, 16, 66. **Muhlens, P.** (1921). *Ibid*, 25, 58. **Napier, L. E.** (1930). *Ind. J. Med. Res.*, 18, 699. **Newton, W. L., Wrighty** and **Pratt** (1945). *Amer. J. Trop. Med.*, 25, 253. **Newstead, R.** and **Carter, H. F.** (1927). *Ann. Trop. Med.* & *Parasit.*, 21, 419. **Noguchi, H.** (1926). *J. Exp. Med.*, 44, 729. **Pal, R.** (1945). *J. Mal. Inst. Ind.*, 6, 53, 217. **Patel, P. G.** (1922). *Rep. Proc. 4th Ent. Meeting*, Pusa, 1921, 272. **Patton, W. S.** and **Hindle, E.** (1927). *Proc. Roy. Soc.*, Sec. B., 101, 369. **Perry, E. L.** (1912). *Paludism*, 5, 32. **Pessoa, S. B.** and **Coutinho, J. O.** (1941). *O Hospital*, 20, 25. **Philip, C. B., Paul. J. R.** and **Sabin, A. B.** (1944). *War Medicine*, Chicago, 6, 27. **Qutubuddin, M.** (1952). *Proc. Roy. Ent. Soc.* Lond., B. 21, 79. **Qutubuddin, M.** (1951). *Pakistan J. Hlth.*, 1, 34. **Reid, J. A.** (1949). *Proc. Roy. Ent. Soc.* Lond., 18, 42. **Ribbands, C. R.** (1945). *Bull. Ent.* Res., 36, 316. **Ribbands, C. R.** (1949). *Bull. Ent. Res.*, 40, 227. **Robinson, G. G.** (1939). *Parasitology*, 31, 212. **Rodenwaldt, E.** and **Essed, W.F.R.** (1925). *Meded. Dienst. Volk. Ned.-Ind.*, 2, 89. **Roubaud, E.** (1921). *Bull. Soc. Path. Exot.* 14, 577. **Roubaud, E.** (1929). *C. R. Acad. Sci.*, Paris, 188, 735. **Roy, D. N.** (1927). *Ind. J. Med. Res.*, 14, 995. **Roy, D. N.** (1931). *Ibid*, 19, 635. **Roy, D. N.** (1936). *Bull. Ent. Res.*, 27, 423. **Roy, D. N.** (1940). *J. Mal. Inst. Ind.*, 3, 499. **Rozeboom, L. E.** (1934). *Amer. J. Hyg.*, 20, 496. **Russell, P. F., Knipe, Rao** and **Patnam** (1945). *J. Exp. Zool. Philadelphia*, Pa., 97, 135. **Russell, P. E.** and **Rao, H. R.** (1940). *Amer. J. Trop. Med.*, 22, 1517. **Sabin, A. B., Philip** and **Paul** (1944). *J. Amer. Med. Assoc.*, 125, 603. **Schule. P. A.** (1928). *Amer. J. Trop. Med.* 8, 203. **Sen, P.** (1935). *Rec. Mal. Surv. Ind.* 5, 97, 211. **Sen, P.** (1941). *J. Mal. Inst. Ind.*, 4, 257. **Senior White, R.** (1937). *Rec. Mal. Surv. Ind.*, 7, 47. **Senior White, R.** and **Adhikary, A. K.** (1940). *J. Mal. Inst. Ind.*, 3, 383. **Senior White, R.** and **Das, B. K.** (1938). *Ibid.*, 1, 169. **Senoir White, R.** and **Rao, V. V.** (1946). *Ibid*, 6, 297. **Sergent, Ed.** and **Et.** (1905). *Compt. R. S. Biol.*, 8, 56. **Sergent, Ed.** and **Et.** (1910). *Ann. Inst. Past.*, 24, 55. **Sergent, Ed.** and **Et., Parrot, Donatien,** and **Beguet** (1921). *C. R. Acad. Sci.*, 173, 1030. **Sharp, N. A. D.** (1928). *Trans. Roy. Soc. Trop. Med.* & *Hyg.*, 21, 371. **Shortt, H. E., Barraud, P. J.** and **Craighead, A. C.** (1926). *Ind. J. Med. Res.*, 13, 923. **Shortt, H. E., Sinton** and **Swaminath** (1936). *Ibid.*, 23, 271. **Shortt, H. E., Smith** and **Swaminath** (1930). *Bull. Ent. Res.*, 21, 269. **Siddons, L. B.** (1944). *J. Mal. Inst. Ind.*, 5, 375. **Siler, J. F., Hall** and **Hitchens** (1926). *Philip J. Sc.*, 29, 1. **Simmons, J. S., St. John, Holt** and **Reynolds** (1931). *Amer. J. Trop. Med.* 11, 199. **Sinton, J. A.** (1923). *Ind. J. Med. Res.*, 10, 742. **Sinton, J. A.** (1924).

Ibid, 11, 1035; 12, 341. **Sinton, J. A.** (1925). *Ibid*, 12, 701. **Sinton, J. A.** (1927). *Ibid*, 14, 941. **Sinton, J. A.** (1931). *Ibid*, 19, 107. **Sinton, J. A.** (1932). *Ibid*, 20, 55. **Sinton, J. A.** (1932). *Rec. Mal. Surv. Ind.*, 3, 347. **Sinton, J. A.** (1933). *Ind. J. Med. Res.*, 21, 221, 225. **Sinton, J. A.** and **Covell, G.** (1927). *Ibid*, 15, 301. **Simmons, J. S., St. John,** and **Reynolds** (1931). *Amer. J. Trop. Med.*, 11, 199. **Smith, R. O. A., Halder,** and **Ahmed** (1940). *Ibid.* 28, 575. **Smith, R. O. A., Halder,** and **Ahmed** (1941). *Ibid*, 29, 783. **Soesilo, R.** (1928). *Med. van der Dienst. der Volks Ned.-Ind.*, 17, 509. **Stammers, F. M. G.** and **Whitfield, F. G. S.** (1947). *Bull. Ent. Res.*, 38, 1. **Steward, J. S.** (1933). *Rep. Univ. Camb. Inst. Animal Path.*, 3, 272. **Steward, J. S.** (1937). *Vet. Rec.* (N.S.), 15, 1563. **Stokes, A., Bauer,** and **Hudson** (1928). *Amer. J. Trop. Med.*, 8, 103. **Strickland, C.** and **Chowdhury, K. L.** (1927). *Ind. J. Med. Res.*, 21, 67. **Strickland, C.** and **Roy, D. N.** (1931). *Ind. Med. Gaz.*, 66, 388. **Strickland, C.** and **Roy, D. N.** (1933). *Riv. di. Mal.*, 12, 57. **Strickland, C., Roy, D. N.** and **Chowdhury** (1933). *Ind. J. Med. Res.*, 20, 819. **Swaminath, C. S., Shortt** and **Anderson** (1942). *Ibid*, 30, 473. **Swartz Welder, J. C.** and **Kali, S. J.** (1942). *Amer. J. Trop. Med.*, 22, 159, **Sweet, W. C.** and **Pillai, V. M.** (1937). *Ind. Med. Gaz.*, 72, 730. **Sweet** and **Rao, B. A.** (1937). *Ibid*, 72, 665. **Sweet, Rao** and **Rao, A.M.S.** (1938). *J. Mal. Inst. Ind.*, 1, 149. **Swellengrebel, N. H.** and **De Buck, A.** (1931). *Proc. K. Akad. Wetensch.*, Amsterdam, 34, 183. **Syverton, J. T.** and **Berry, G. P.** (1941). *J. Exp. Med.*, 73, 507. **Taussig** (1905). *Wien. Klin. Wochenschr.*, 18, 129. **Theodor, G.** (1936). *Bull. Ent. Res.*, 27, 653. **Thomson, R. C. M.** (1941). *J. Mal. Inst. Ind.*, 4, 217. **Toumanoff, C.** (1935). *Bull. Soc. Path. Exot.*, 28, 948. **Toumanoff, C.** (1936). L' anophelisme en Extreme—Orient. Contribution faunistique et biologie. Paris. **Masson et Cie. Townsend, C. H. T.** (1913). Insecutor Inscitiae Menstruus, 1, 107. **Trager, W.** (1935). *Amer. J. Hyg.*, 22, 475. **Viswanathan, D. K.**(1950). Malaria and its Control in Bombay State, published by D. K. Viswanathan, Cannaught House, Poona 1. **Viswanathan, D. K.** and **Rao, T. R.** (1948). *Ind. J. Malariology*, 2, 157. **Vogel, R.** (1920). *Zool. J.*, 42, 259. **Wansod, M., Henrad and Peel** (1945). *Rev. Appl. Ent.*, B, 36, 148. **Wenyon, C. M.** (1913). *J. Lond. Sch. Trop. Med.*, 2, 170. **Wenyon, C. M.** (1926). Protozoology, London. **Weathersbee, A. A.** and **Hasell, P. G.** (1938). *Amer. J. Trop. Med.*, 18, 531. **Weyer, F.** (1936). *Arb. Physiol. Angew. Ent.*, Berlin, 3, 20. **Whittingham, H. E.** and **Rook, A. F.** (1922). *Trans. Roy. Soc. Trop. Med. & Hyg.*, 16, 262. **Whit- tingham, H. E.** and **Rook, A. F.** (1924). *Ibid*, 17, 290. **Woke, P. A.** (1937). *Amer. J. Trop. Med.*, 17, 729. **Yamada, S.** and **Komori, K.** (1926). *Sc. Rep. Govt. Inst. Infect. Dis.*, Tokyo, 647. **Yarke, W.** and **Macfie, J. W. S.** (1924). *Ann. Trop. Med. & Parasit.*, 18, 103, 125.

CHAPTER IX

DIPTERA—CYCLORRAPHA

Superfamily **Muscoidea.** This includes the true flies and some of them are of considerable economic importance to man. The house-fly is taken as a typical example of an insect belonging to the Order Diptera. It has a retractile proboscis, it sucks liquid food and has a complete metamorphosis. The different stages in its life are egg, larva, pupa and adult. The larva is maggot-like and the pupa develops inside a hard chitinous shell from which the adult fly escapes. A short description of the structure of the fly and its different stages together with its habits are given below.

The body of an adult fly consists of head, thorax and abdomen. The neck is mobile.

Head. (1) Eyes are large and compound. In the common house-flies the sexes can be distinguished from the approximation of the eyes; these are close together in males but are divergent in females. The vertex or frons is the space between the eyes; (2) Simple eyes or ocelli, three in number; may or may not be present. (3) Antenna : 3-segmented; the last segment is the longest from which arises a long stylet called the arista; the arista may be simple or branched. (4) Proboscis (or labium) which is retractile in non-biting flies and stiff in biting flies. Maxillæ and mandibles are wanting in all flies though the maxillary palps are well developed. At the distal end of the proboscis lie the labellae. The proboscis of the house-fly consists of three parts named from above downwards, rostrum, haustellum, and oral disc. The rostrum can be voluntarily withdrawn or extended. The extension of the rostrum is brought about by the distension of the large air sacs which are contained within it. The rostrum is more dilated than the haustellum and consists of a thick membrane called the fulcrum which has a triangular outline, the cone of the triangle pointing downwards. The pharynx runs along the ventral wall of the rostrum. To the lower part of the rostrum are attached the two maxillary palps.

In a deep groove in the haustellum lie the labrum-epipharynx, and the hypopharynx. Through the hypopharynx runs the terminal part of the salivary duct. The labrum-epipharynx is dagger-shaped and lies dorsal to the hypopharynx. The food channel is formed by the apposition of the labrum-epipharynx and the hypopharynx.

The oral disc consists of 2 labellæ, the inner surfaces of which are grooved by a large number of chitinous channels, the pseudotracheæ, running transversely across the labellum almost parallel to each other. Each pseudotracheal ring is incomplete; one extremity is flattened and the other is forked or bifid. The rings are arranged in such a manner that along each side of the central fissure the bifid extremity of one ring alternates with the expanded extremity of the next ring. The area enclosed between the forks of the bifid extremity of a ring has been termed the "interbifid space" by Graham-Smith (1911). Food really enters the pseudotracheæ through the spaces between the bifid extremities of the rings (or interbifid spaces) and not through the continuous zigzag fissure which is extremely narrow and is probably closed during the act of feeding. The size of the solid particles which can pass into the mouth is regulated by the width of the interbifid spaces. Certain relatively large objects such as the ova of tapeworms, too large to pass through the filter, may occasionally be

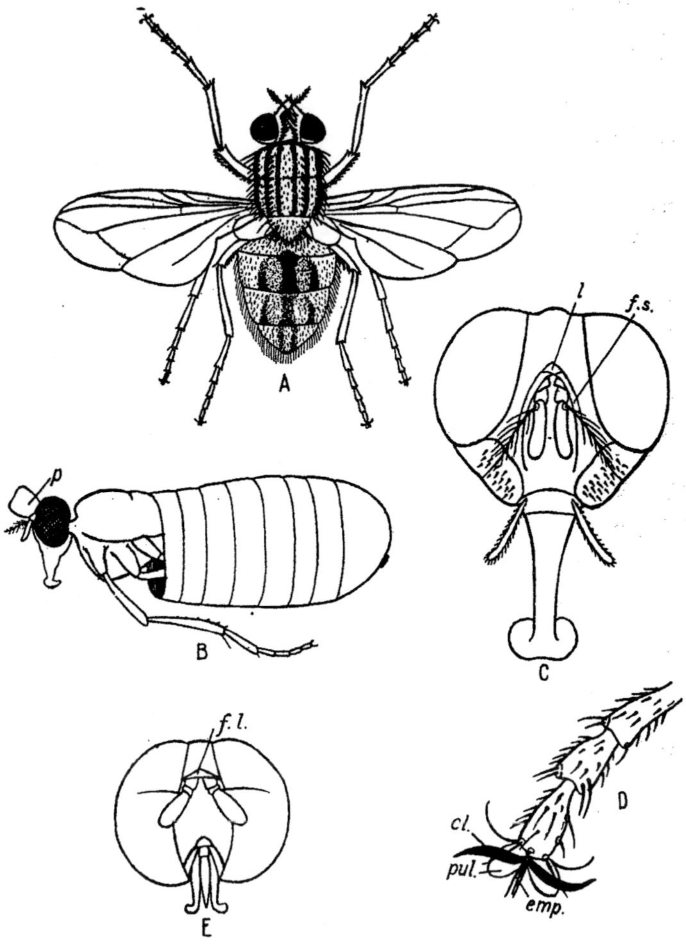

FIG. 87.

A. House-fly ; B. Emergence of fly from the puparium. p, ptilinum.
C. Head of house-fly. f.s., frontal suture ; l., lunule.
D. Foot of house-fly. cl., claw ; pul., pulvillus ; emp., empodium.
E. Head of Aschiza. f.l. frontal lunule.

swallowed. Such objects probably pass directly into the mouth. The liquid food is sucked into the pseudotracheæ and drawn through the collecting channels into the mouth. Each collecting channel opens into the corresponding gutter between the prestomal teeth in a remarkable manner.

The prestomal teeth are chitinous prong-like processes free at their distal bifid extremities but with their proximal portion embedded in the dental area of the pseudotracheal membrane. In the centre of the labellum are from within outwards four rows. The teeth are used for scraping solid food, e.g., sugar (Graham-Smith, 1930).

When at rest the oral surfaces of the labellar or oral lobes are in apposition, but during feeding they are spread out over the surface of the food so as to form an oval disc.

The labial salivary glands lie at the base of the labella. Each consists of a large number of glandular cells. They open externally by a number of small ducts on the labellar surface. It is thought that the function of the labial saliva is to seal the longitudinal fissures of the pseudotracheæ by its viscid secretion during the act of feeding.

Thorax. The thorax is mainly composed of the mesothorax; the pro and the meta-thorax are poorly developed. A transverse suture separates the mesonotum into prescutum and scutum. The scutum is further divided by a suture into scutum proper and the scutellum which lies posteriorly and overhangs the metathorax. There are numerous bristles on the dorsum of the thorax; the most important ones are: (1) a row of acrostichal bristles and (2) a row of dorsocentral bristles on each side of the middle line. On the pleuron of the mesothorax lie a number of sclerites, the important ones being sternopleura and hypopleura each bearing well developed bristles.

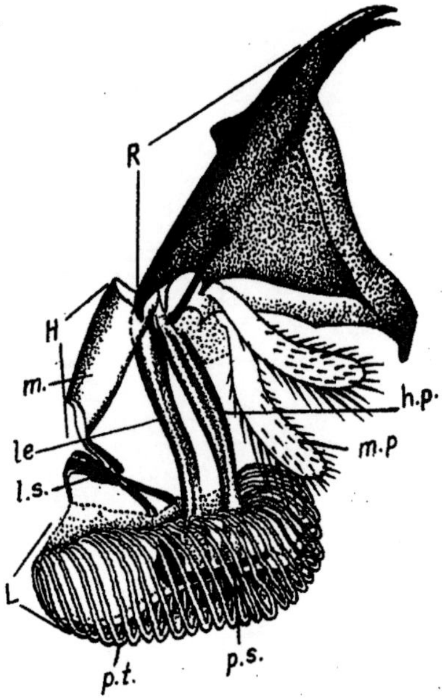

FIG. 88.
Mouth parts of a fly.
R, rostrum; H, haustellum; L, labella; m, mentum; le. labrum-epipharynx; hp, hypopharynx; mp, maxillary palp; l.s., labial sclerite; p.t., pseudotracheal tube; ps., prestomal sclerite.

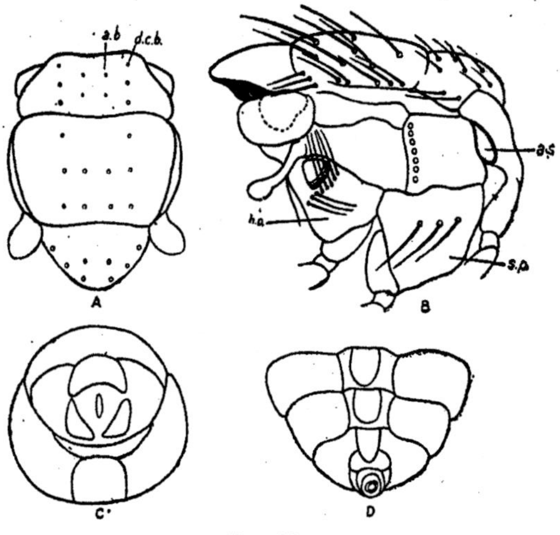

FIG. 89.
A. & B. Thoracic chaetotaxy of fly. C. & D. Ventral view of the last segment of the abdomen of *Chrysomyia megacephala.* C, male; D, female.
a.b., acrostichal bristles. d.c.b., dorsocentral bristles. a.s., anterior spiracle. s.p., sternopleura. h.p., hypopleura.

The thorax bears two pairs of spiracles, three pairs of legs, and a pair of wings.

Each leg is composed of (1) coxa, (2) trochanter, (3) femur, (4) tibia, and (5) five tarsal segments. The last tarsal segment ends in a pair of claws. On the inner side of each claw lies a foot-pad or cushion covered with hairs and called pulvilli. Between the pulvilli lies a long bristle called the empodium.

The venation of the wing of the house-fly is comparatively simple. There is no branched vein. The 4th longitudinal vein bends distally in a characteristic manner so as nearly to enclose the first posterior cell.

Near the base of the wing is a membranous expansion attached to the side of the thorax just posterior to the attachment of the wing. This is known as the squama. In house-flies it conceals the halteres when examined from the dorsal aspect.

Abdomen. There are 8 abdominal segments in the male and 9 in the female fly. Among them only four segments are visible. These are the second to fifth. The first segment is very much reduced. Segments 6 to 8 in the male are also reduced. In the female segments 5 to 9 form the tubular ovipositor, which, in repose, is telescoped within the abdomen. It is extended only at the time of oviposition. The tergal plates of the abdomen are well developed whereas the sternal plates are greatly reduced.

Alimentary canal. The alimentary canal is divided into (*a*) foregut, (*b*) midgut, and (*c*) hindgut. In the foregut is included the pharynx, œsophagus, œsophageal diverticulum and proventriculus. The midgut consists of the stomach and mid-intestine where active digestion of food takes place. The hind-intestine, rectum and malpighian tubules form the hindgut.

(1) The mouth lies between the labellae where the pseudotracheal collecting channels lead into the food channel. The common salivary duct which passes through the hypopharynx opens into the ventral surface of the buccal cavity. (2) Pharynx : its function is to pump liquid food and forms the first part of the alimentary canal. It extends upwards from the mouth along the posterior aspect of the fulcrum and terminates above at the proximal end of the latter. (3) Oesophagus : takes its origin at the proximal end of the fulcrum and passes through the cephalic ganglion and then the neck; it ends in the anterior thoracic region at the proventriculus. (4) Oesophageal diverticulum or crop. The reservoir or crop is a large distensible bilobed sac lying in the abdomen. When distended with liquid food, it becomes reniform. The duct of the crop begins and opens at the posterior end of the oesophagus where the latter terminates and opens into the ventral side of the proventriculus. The special function of the crop is to store liquid food. When the alimentary canal is empty, food is rapidly expelled from the crop. It is filled with great rapidity. All food materials ingested by the fly first pass through the œsophagus into the crop and from there through the proventriculus into the intestine. From time to time the fluid contained in the crop is regurgitated into the mouth and appears as vomit drop. (5) The proventriculus is formed by the invagination of the foregut into the midgut, and acts as a valve between the midgut and the foregut. (6) The midgut or stomach is a long coiled tube and lined with the peritrophic membrane which also imperfectly lines the inner wall of the hindgut. (7) The hindgut is comparatively short. (8) The rectum has 4 papillæ, their function being to absorb water from the rectum and probably also to excrete urea from the blood. (9) Malpighian tubules : two on each side and arise from a common stalk which soon divides. They are very long, convoluted and bound up with the adipose tissue containing the fat body. They function as renal organs especially serving to excrete urea and other waste products from the adipose tissue and the blood.

Salivary glands. To avoid confusion and in order to distinguish them from the labial glands these are designated as lingual salivary glands. They are two extremely long fine tubes of uniform length except at the distal end. They are considerably longer than the total length of the body. The small salivary valve is situated within the common duct a short distance from where it enters the hypopharynx.

Respiratory system. The respiratory system is characterised by the presence of air sacs developed in connection with tracheal trunk in the head, thorax and abdomen. These large tracheal trunks open externally on the thorax and abdomen. The thoracic spiracles are two on each side; the abdominal spiracles differ in number in the two sexes; in the male there are seven pairs, in the female five. It is extremely doubtful if the abdominal spiracles are functional like the thoracic ones.

The reason of the existence of air-sacs in insects has not been properly understood. It may be a means of lowering the specific gravity of the insect or the volume of the tidal air may be increased by providing regions in the tracheal system which can be readily compressed and distended.

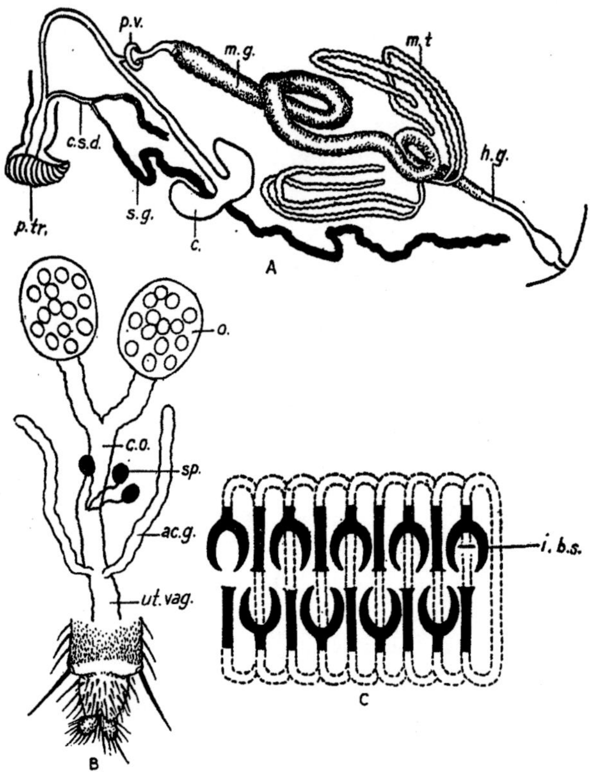

FIG. 90.

A. Alimentary system of house-fly.
c., crop; c.s.d.; common salvary duct; h.g. hindgut; m.g., midgut; m.t., malpighian tubules; p.tr., pseudotracheae p.v., proventriculus; s.g., salivary gland.

B. Generative organs of a female house-fly.
ac.g., accessory gland; c.o., common oviduct; o., ovary; s.p., spermathecae or vesiculae seminalis; ut.vag., utero-vaginal tube.

C. Pseudotracheal tube showing interbifid space (i.b.s.).

Reproductive system. Female: (*a*) a pair of ovaries containing a large number of ovarioles; (*b*) short oviducts; (*c*) common oviduct; (*d*) vagina; (*e*) spermathecæ, three in number; their common duct opens in the common oviduct; (*f*) one pair of glands; these lie on either side of the common spermathecal duct. *Male*: (*a*) a pair of brown pyriform testes; (*b*) vasa deferens; (*c*) a long ejaculatory duct opening into a small ejaculatory sac and continued into the aedoeagus; (*d*) a pair of accessory glands.

Egg. See in life history.

Larva. It is an apodous, hairless maggot; the body consists of head, three thoracic and 9 abdominal segments. The anterior end is tapering and the posterior is much broader and truncate. The head is not demarcated from the rest of the body. The head or pseudocephalon which represents the first segment is really a reduced cephalic segment. It is bent towards the undersurface of the larva. The pseudocephalon consists of two oral lobes separated from each other. On the ventral surface of the oral lobes are numerous fine grooves which resemble the pesudotracheæ found on the labella of the adult fly. These converge towards the mouth. It is along these channels that fluid material runs into the mouth. On the anterior portion of the oral lobes lie two papillæ, the antenna and the maxillary palp. The mouth is triangular in shape and within it the pseudocephalon is retracted. Sometimes projecting from the cephalic end of the body is a pair of highly chitinised hooks. The eyes are absent.

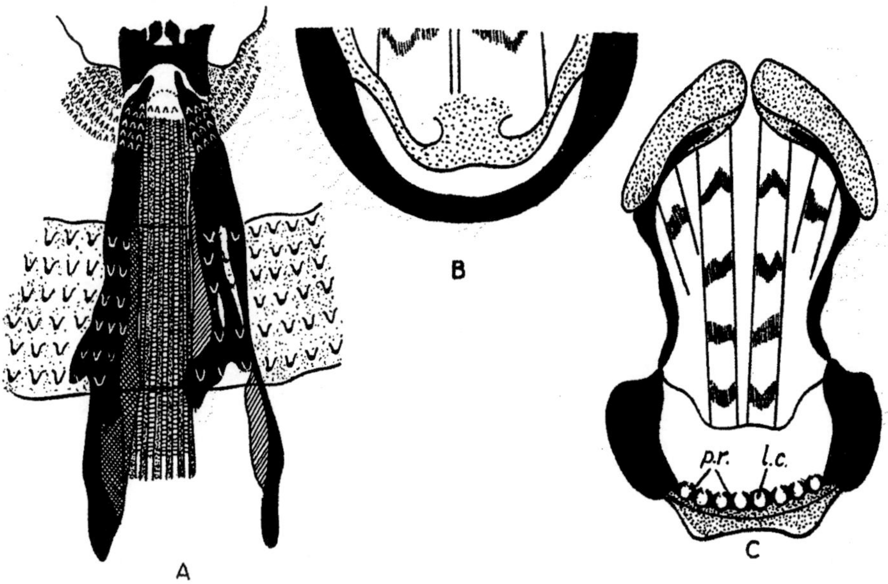

FIG. 91.

A. Calliphora erythrocephala, showing the pharyngeal ridges on the ventral side of the pharynx of the larva. (After Roy). B. Transverse section through the pharynx of *Gasterophilus equi* (*intestinalis*) larva showing the absence of ridges on the floor of the pharynx. (After Roy). C. Transverse section through the pharynx of the larva of *C. erythrocephala* showing the pharyngeal ridges (p.r.) and the longitudinal channels (l.c.) where the solid particles are usually caught. (After Roy).

Three pairs of setiferous papillæ present on the ventral surface of the thorax in connection with the imaginal discs of the legs have been described by Keilin (1915). This indicates that the atrophied legs of the maggot have been transformed into sensory organs.

On the second segment lie the anterior spiracles and the posterior spiracles are situated in the middle of the truncate end of the last segment. The anterior spiracles consist of a number of finger-like processes. The posterior spiracle is round and the openings are slit-like. On the ventral surface of the last six segments are pads which are used in locomotion.

The cephalopharyngeal skeleton lies well within the oral lobes of the larva. It consists of 3 pieces of sclerites; these are from before backwards, mandibular, hypostomal and pharyngeal. On the ventral surface of the last-named sclerite lies the pharynx the wall of which in some species may be ridged.

The alimentary canal of the larva is divisible into several well defined regions such as œsophagus to which is attached the œsophageal crop, proventriculus, ventriculus, midgut, hindgut and the rectum. The crop is homologous to œsophageal diverticula. The peritrophic membrane extends from the proventriculus to the rectum. It forms a tubular sheath. The salivary glands are disproportionately large.

Pupa. The pupa is barrel-shaped. The colour is dark-brown. The last larval skin is not cast off in flies but hardens to form the outer covering or puparium inside which the pupa develops. The puparium shows almost the same outward structures as the fly larva. Thus anteriorly the antennæ, the maxillary palps and the anterior spiracles are visible while the posterior spiracles are also noticed on the last segment.

Life history. The egg of the house-fly measures about 1 mm. in length, is creamy in colour, roughly oval, and presents a highly polished surface. The anterior end is narrow and the posterior is broader. On the dorsal surface of the chorion are two longitudinal ridges with a concave area between them. Temperature has a marked effect on the developmental cycle of the fly and a sudden change from heat to cold will materially prolong any of the stages. During the most favourable time in summer in this country eggs may hatch within 3 hours after they are deposited. In the winter this period is extended to more than 2 days. On an average eggs do not take longer than 48 hours to hatch. Eggs of house-flies are laid in any fermenting material of organic or vegetable origin, or in refuse that is likely to ferment, e.g., garbage, human faeces, horse-manure, kitchen refuse etc. Other flies select cow-dung for oviposition, or a carcass. Some have a predeliction for dead birds while there are some which oviposit in human and animal wounds. Thus it will be seen that different species have different breeding habits. The eggs are seldom scattered on the surface but are carefully deposited one on top of another. By means of its long ovipositor the fly generally deposits its eggs in dark places away from light and some distance from the surface. On an average a single *Musca vicina*, *M. nebulo* and *Chrysomyia megacephala* can deposit in summer, 97, 75 and 182 eggs in a batch respectively, whereas in December the respective numbers are reduced to 72, 64 and 166. The ovaries completely discharge their contents when a batch of eggs is laid, and after a single act of oviposition, further deposition of eggs does not take place within 3 days which is the normal time for the development of eggs and their maturation within the ovary.

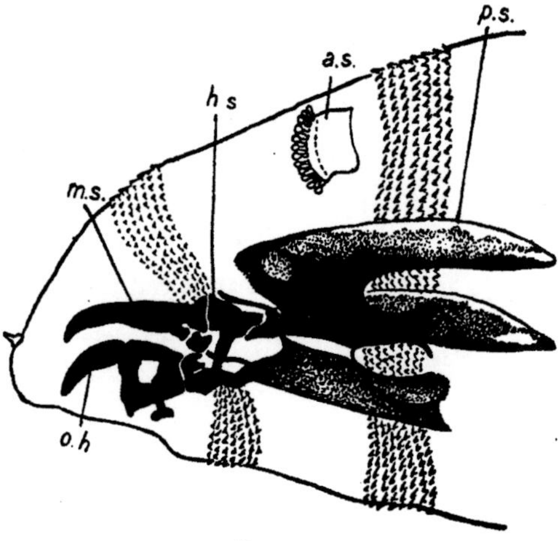

FIG. 92.
Anterior part of a fly larva showing cephalo-pharyngeal skeleton and anterior spiracle.
a.s., anterior spiracle; h.s., hypostomal sclerite; m.s., mandibular sclerite; o.h., oral hook; p.s., pharyngeal sclerite which is deeply incised posteriorly.

The larva is about half-an-inch in length; it is an apodous, hairless maggot and is very active. The fleshy pads which lie on the ventral surface of the body help the larva in its progression. It resents light, either natural or artificial. The larva is a fluid feeder and sometimes minute particles of solids are ingested with the liquid food.

The floor of the pharynx in certain Diptera larvæ is provided with longitudinal ridges. The nature of the ridges varies in different types of larvæ and Keilin (1912) considers this to be an adaptation to the manner of life. In saprophagous larvæ which feed on decayed vegetable matter or the excrement of herbivorous animals, the buccal armature is similar to that of *Calliphora* or *Lucilia.* These have a number of longitudinal ridges projecting in the lumen of the pharynx. Parasitic larvæ, on the other hand, have a smooth pharyngeal wall without any ridges. This type of pharynx is seen in larvæ of *Chrysomyia bezziana, Cochliomyia americana,* etc. The function of the ridges seems to be connected with filtration in the nature of separation of large particles of solids from the liquid food; the former are lodged in the spaces enclosed by the ridges and remain there during the rest of the larval period. The solids thus caught in the ridges are found in the same situation in the adult stage and these gradually pass out from the alimentary canal within 12 to 16 hours after emergence of the fly (Roy, 1937).

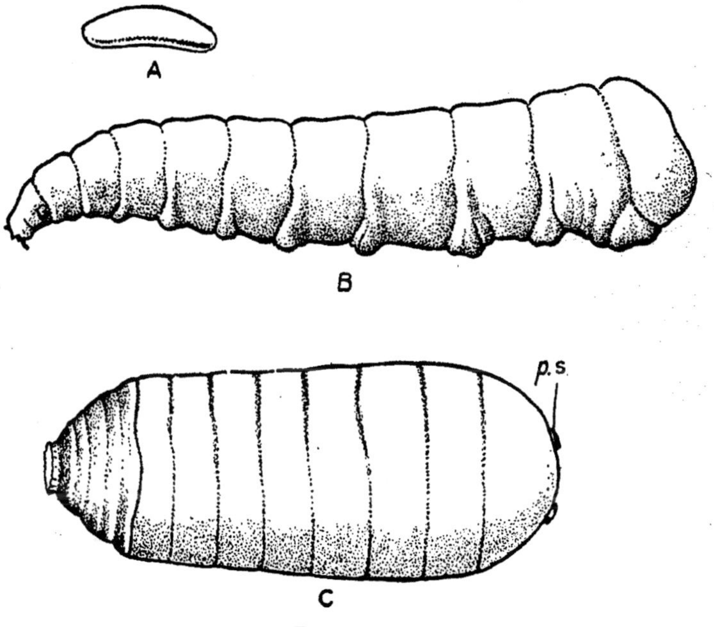

FIG. 93.

Egg, larva and pupa of house-fly.
A, egg ; B, larva ; C, pupa ; p.s., posterior spiracle.

The larva moults twice. The first stage larva has no anterior spiracles and the posterior pair is very small. The third stage larva has the habit of migrating. They may even burrow to a depth of 2 ft. in loose earth before they pupate. This migratory habit, which is so much in evidence during the prepupal state of the house-fly, has long been known. As pupation draws near, the larvæ pass from the very moist regions of a manure heap and seek the comparatively dry outer regions. If no such places are in the heap, they will leave it to pupate in the ground or in loose material of any kind even at some distance from the breeding place. This habit offers an important point of attack in the control of this pest. Farm refuse and manure may be placed in a barrel in the bottom of which several holes are made, with the result that on the following day thousands of maggots will be found in water placed in a tub beneath. In this way perhaps 70 per cent of the maggots can be destroyed. Larvæ kept in horse-manure, sterilised from time to time, will fail to grow. The length of life of the larva varies according to temperature. The earliest recorded in Calcutta is between one and a half and two days. Before pupation the larva does not feed, but empties its alimentary canal and ceases activity. It contracts and transforms itself into a barrel-shaped object; its colour gradually changes from white to brown and finally to chestnut colour. The last larval skin is not cast off but hardens to form the puparium.

Under the conditions present in the tropics the pupal stage does not last more than 4 to 6 days. In cold places this stage may be considerably prolonged.

When the fly is about to emerge air is forced into its anterior part and an air-cushion or ptilinum appears on its forehead. The anterior part or the cap of the puparium is pushed off, a circular slit appears through which the fly escapes leaving behind the nymphal sheath, which lines the inside of the puparium, behind. In course of time the ptilinum retracts giving rise to a rhomboidal scar called lunule and a frontal suture, the latter being absent in many flies, *e.g.* Aschiza. The ptilinum not only helps the fly in escaping from the puparium but also in forcing its way through soil when the pupating larvæ are buried in the earth.

In the tropics during summer a fly is able to complete its cycle within 5 to 6 days. At other times, this period is correspondingly increased depending primarily on the temperature and secondarily on the quantity of food available for its growth.

House-flies are usually more attracted to sugar than to other types of food although they are omnivorous in nature. The fly feeds intermittently and at the same time voraciously. It has the habit of sitting on filthy substances such as excreta, sputum, sores etc., and also on food. After feeding, the fly cleans its head and proboscis with its legs. From time to time it vomits or regurgitates its food from the crop, the regurgitation appearing as a minute drop of fluid hanging from the oral lobes. The vomit also contains a drop of saliva. Some times the vomit drops are deposited like faeces upon the surface on which the fly is sitting. It is not easy to distinguish between the two, vomit and faeces, although it has been said that their colour and characters are different. It is thought that the vomit drops are intended to soften solids such as sugar, but even after 24 hours' starvation when no vomit drops are seen on the labellæ the flies will still feed voraciously on sugar. Perhaps the saliva helps such solids to be converted into the liquid form. The presence of a large amount of fluid in the crop always leads to an increase in regurgitation. The fly has also the habit of constantly defæcating.

Under normal conditions, flies fed on coloured food will continue the defaecation of coloured excreta for at least 48 hours. If, however, the flies have been starved before ingesting the coloured food, this food may appear in the faeces within 2 hours and all the coloured food will be expelled within 6 hours. The emptying of the crop normally takes place within 48 hours, but if the fly is placed in cold surroundings it is possible to detect the coloured food in the crop even after 72 hours.

Copulation takes place within 24 hours after emergence. Though mating frequently occurs, yet one act of fertilization is probably sufficient for the female to produce fertile ova all its life. Protein food is necessary for the development and maturation of the egg. The adult flies are generally not capable of dispersing to a distance greater than a quarter of a mile.

It is doubtful if adult flies live longer in nature than inside the laboratory cage. They have a large number of enemies and they fall a ready prey to the latter when the flies rest at night. Spiders and lizards may be counted as their most deadly enemies. Flies do not generally live longer than 15 days in summer and 25 days during winter. Provided there is no dearth of the proper type of protein food a fly may deposit a batch of eggs every four days.

The house-fly cannot pass the winter in the adult state when exposed to outdoor conditions. Its larval and pupal stages are greatly extended and are passed in or under large manure heaps. It is possible that the method of overwintering by continued breeding is much more widespread than is commonly realised.

Parasites of Flies. (1) *Habronema muscae* Carter, and *H. megastoma* Rud. The embryos are passed with the alimentary dejecta of horses and find their way into the alimentary canal of fly larvae. In the adult flies the parasites are commonly found in the head and frequently in the proboscis and labella. They may escape from the proboscis of *Musca* when the fly feeds on the lips or wounds of horses. (2) *Allantonema stricklandi* Roy, and *A. muscae* Roy. These nematodes are found in the haemocoele of *Musca vicina* and *M. nebulo.* (3) *Herpetomonas muscae domestica.* (4) Different species of Hymenoptera. A very large number of Hymenoptera of the Superfamily Chalcidoidea are known to parasitise Muscoid puparia and act as natural enemies of the latter. A list of those found in India has been given by Roy and Siddons (1939). (5) A fungus, *Stigmatomyces baeri* is a common parasite of adult flies.

Flies and human diseases. Although the danger of house-flies towards the spread of human diseases had long been suspected, yet the first microscopical proof was demonstrated by Raimbert when in 1896 he isolated the causative organisms of anthrax from flies exposed to the cadavers of animals dead of anthrax. In 1886 Tizzoni and Cattini isolated the cholera organism from flies caught in hospitals. However, the real danger of flies was realised after the Spanish American war in 1888.

The habits which are particularly dangerous from the point of view of the carriage of disease germs are : (*a*) The fly is an intermittent feeder; it constantly defæcates and vomits every now and then. (*b*) It feeds indiscriminately on filth and articles of food intended for human consumption such as sugar, sweets, milk, etc. (*c*) It has a restless nature. Whether for the purpose of feeding or not it will alight on any article making no difference between food and filth. It has a tendency to settle on sputum, sores, stools, etc., on which it also feeds. (*d*) During the larval stage it may feed on pathogenic organisms which sometimes may be carried over and may be discovered in the excreta of the adult fly. Such habits of house-flies leave no room for doubt that they are highly conducive to the spread of pathogenic germs in general and not to any one in particular, though they play an important part in the spread of intestinal diseases of man. The different ways by which domestic flies can propagate disease-germs are : (*i*) by passing ingested pathogenic germs, e.g., of cholera, typhoid and dysentery etc., sucked up from the excreta of man, tubercle bacilli from sputum, or lepra bacilli from leprotic ulcers; (*ii*) by carrying them on the external surface of the body e.g., on hairs and bristles, wings, foot-pads, proboscis etc., and (*iii*) by vomiting the ingested materials. It will thus be clear that the fly spread disease organisms by acting merely as a mechanical agent. The evidence of the fly being responsible for the spread of disease organisms has been obtained mainly from three sources such as epidemiological, experimental and recovery of specific disease-producing organisms from wild flies. It may be pointed out that no proof has yet been obtained of the multiplication of disease germs in either the crop or the intestine of the fly.

The diseases with the spread of which flies are associated are typhoid fever, cholera, summer and infantile diarrhœa, dysentery, intestinal worms, poliomyelitis and other diseases, e.g., leprosy, anthrax, yaws, etc.

It is supposed that the virus of poliomyelitis gains entrance into the body from the naso-pharynx. Even then sufficient of it is swallowed so that in almost every case its presence can be detected in the stool. The virus, when administered appears in the blood in sufficient concentration within 4 to 6 days. There is an interval of 3 to 7 days before it is possible to demonstrate invasion of the nervous system.

Fly prevention. In combating flies and diseases carried by them our aim should be to eliminate the breeding sites altogether. As far as possible conditions should be made unfavourable for the flies to deposit their eggs. The eggs, larvæ and adult flies wherever found should be killed.

House-flies in India have a wide choice of breeding materials and in cities these are garbage, refuse of all sorts, ash-pits and human and animal excreta. Human excreta and conservancy trenches are favourite breeding sites for most of these flies. City refuse dumps and incinerators, if badly worked, are fertile breeding grounds. Offal and undigested materials from the stomach of ruminants obtainable in slaughter houses also yield a fair number. Cow dung, both in bulk and patches, cannot be ignored and some of the domestic flies are attracted to it. Other flies in addition to house-flies breed almost exclusively in human excreta, in bulk and in single deposits. The blue-bottle breeds in cesspits containing almost liquid faeces. This fly is perhaps least conservative in breeding habits, for it also readily breeds in putrefying meat and in cadavers. The flesh-fly breed to a large extent in putrefying animal remains. Thus all fermenting and excrementous organic matters will breed flies.

The broad measures of fly prevention are grouped under the following heads : (1) *Prevention of egg-laying* by proper disposal of refuse and garbage. This largely rests on municipal authorities. The ordinary householder can also contribute greatly towards carrying out these measures by maintaining cleanliness in the house. (2) *Chemical manure*, e.g., bleaching powder; it is useless as it destroys the valuable properties of the manure. (3) *Biothermic method* of allowing the manure to accumulate by close packing though not very satisfactory, yet can be profitably applied in rural areas. The object is to raise the inside temperature so high as to kill all eggs and larvæ. A large number will always escape the heat and migrate to the outer and cooler parts of the heap. These migrations may be prevented by an open channel constructed all around the manure heap in which they are trapped and destroyed in cresol or bleaching powder solution. This method applies only to horse manure. With cow-dung, the temperature does not rise to the same extent. (4) *Disposal of human excreta* by the best method suitable for that particular locality. (5) *Construction of bore-hole latrine* in rural areas. If the boring is carried to a depth of 2 ft. below the subsoil water, the latrine will last indefinitely otherwise it simply acts as a container for faeces and fills up relatively quickly. (6) Such measures as the *use of traps, poisoned baits, fly papers* etc. are not satisfactory. They are regarded as useful adjuncts to other ways of controlling flies. Sodium arsenite is an excellent fly poison ($\frac{1}{2}$lb. of arsenite, $2\frac{1}{2}$ lb. treacle and $2\frac{1}{2}$ gallons of water) but it is dangerous to use in the household. Household fly papers are made as follows : Resin 12 oz., ground nut oil 5 oz., crude vaseline 1 oz. The resin should be reduced to powder and the ingredients heated together without boiling until all the resin is dissolved. Strips of paper, tape, string or wire dipped in this preparation may all be used with advantage. If systematically used, it will kill a very large number of flies. (7) *Composting*: this was originally based upon the natural process of conversion of the waste products of agriculture whereby the residues of plants and animal life are converted into humus through the agency of fungi and bacteria. It was later adapted as a simple solution for sanitary disposal of the night soil and rubbish and other organic materials which compose the ordinary street refuse. The temperature reached is 160°F., and this high temperature is generally maintained in the compost heaps for at least 3 weeks; this serves to kill not only all pathogenic bacteria and helminth eggs but also fly maggots living in the cooler parts especially at the surface for which the compost has to be turned. Smell is also reduced to a minimum. (8) *Fly sprays*: A large number of fly-spraying fluid can knock down and stupefy but do not kill all the flies. They are of little use in the open especially in warm climates. The use of fly-sprays is therefore limited to hospitals, private houses, hotels, mess etc.

Amongst the spraying fluids DDT and Gammexane are widely used. DDT itself has not proved very successful in the control of house-flies. Blue-bottle flies are much more susceptible to it. For this reason a combination of DDT, pyrethrum and lethane should be used against house-flies. Though Benzene hexachloride is about 9 times as effective as DDT against the house-fly, it lacks the residual effects of DDT and its musty smell is very disagreeable.

For control of flies by chemicals and the possibility of the development of resistance against such insecticides, see later in Part II.

Breeding. Breeding flies in the laboratory is not difficult if the proper type of nidus is provided and the proper temperature is maintained. For egg laying by *Chrysomyia*, *Lucilia*, *Sarcophaga* and *Synthesomyia*, meat is the best medium while house-flies readily oviposit in fresh horse-dung, also on lint soaked in milk after the latter has turned sour. The right temperature for breeding is 85 to 90°F. The culture medium should not be disturbed till the larvae have pupated in sand placed at the bottom. The pupae should be protected from parasitic Hymenoptera.

Dissection of Fly Larva. After killing the fly larva in hot water, fix it in water on paraffin block by means of two pins placed at both ends of the body. Pinch off the skin from the dorsal surface with a pair of forceps and make an incision with scissors. Carry the incision forwards and backwards and reflect the cut ends of the skin laterally. Remove the fat body from the body cavity. The characteristic feature about the internal anatomy is the enormously enlarged salivary glands.

Dissection of Adult Fly. Remove wings and legs. Snip off the coxae with a pair of scissors. Hold the fly gently between the index finger and the thumb; with a knife-edged needle or preferably a safety razor blade make a small incision in the centre of the abdomen and carry it well forwards up to where the head joins the neck and backwards up to the anterior part of the last segment. The waist or the constricted 1st abdominal tergite must be carefully incised; this always gives some trouble. The thorax on either side is incised longitudinally. The incision is carried deep into the muscle and continued both forwards and backwards.

Pin the fly on a dissecting paraffin board in 4% formalin solution and remove the cuticle together with muscles from the abdominal and thoracic segments. The entire alimentary system stretching from the head to the last abdominal segment will come to view. When only the reproductive system has to be exposed, dissection may be confined to the abdomen only.

Preservation of Flies. Flies should be pinned while still in a fresh condition with the muscles and tissues pliable. The pin should not be passed through the fly on the median line since it may destroy the acrostichal bristles on both sides. It should be passed through the post-scutellum to one side of the middle line. The hypopygium of both sexes should be exerted and fully extended. This is particularly important in the case of male flies. The legs should be drawn away from the abdomen, i.e., they should be fully extended. Larvae should be killed in hot water and preserved in 80 per cent alcohol. Four per cent formalin is an unsuitable preservative in the case of fly larvae.

Catching. Flies should be caught in fly nets. A fly net is a bamboo ring set in an wooden handle about three feet long. To the ring is fixed a cone-shaped mosquito netting.

Acalyptrata Muscoidea

Acalyptrata Muscoidea which are generally of a small size are divided into many families; among them the following are important. (*a*) Drosophilidae (fruit-flies); (*b*) Sepsidae (flies with constricted waist); (*c*) Phoridae; (*d*) Chloropidae and (*e*) Scatophagidae (Cordyluridae). All are small and breed either in putrid animal or vegetable matters.

Family **Drosophilidae.** They are minute yellowish flies and often swarm over ripe fruits. The eggs are deposited in the same situation. The pupae which resemble minute seeds are sometimes encountered in human faeces and can be easily recognised by the long tubular spiracles. They may be swallowed with the food and are passed in a dead condition. The attention of the author was recently drawn to a case of urinary myiasis occurring in a hospital patient in which two living larvae of *Drosophila* were reported to have been found in the urine passed in a pot. *D. melanogaster* is an extremely common fly.

FIG. 94.
Pupa of *Drosophila melanogaster*.

D. melanogaster Meig. Abdomen shining black, each of the three segments with a basal yellowish band; mesonotum and scutellum yellowish red; on the inner dorsal surface of the basal joint of the front tarsi are short, stiff, black bristles which are confined to one diagonal row.

Family **Sepsidae.** They are small black flies and are characterised by marked constriction in the waist. The fourth vein runs straight without making an angle at the distal end. The wing has a dark spot near the tip. They breed in decaying animal and vegetable matters in which eggs are laid. *Piophila casei* breed in cheese and may occasionally find their way into human intestine. They have the peculiar habit of leaping. The lower part of the head, the antennae and parts of the legs are yellow. The larvae of Sepsidae can be easily recognised by the presence on the rim of the posterior spiracle of one or more bristle-like hairs best seen under a dissecting microscope.

Family **Phoridae.** They are all small or minute with a hunch-backed appearance. Antennae are of peculiar form; the third joint large and concealing the others, spheroid with an apical arista. The wings have two heavy veins anteriorly which reach only half way to the apex of the wing, and three or four much lighter ones which run obliquely across the disc of the wing. They can always be recognised by their rapid jerky movements.

The Phorids have remarkable life histories. Most of them breed in decaying animal and vegetable matters, in birds' nests and also in the nests of rats.

In the genus *Apiochaeta* spurs are present on four hind legs.

A. scalaris Mg. (*A. ferruginea* Brunetti). It is extremely common all over the plains of India, and has been recorded from most of the tropical and subtropical regions of the world. This species is of remarkable interest as it is thought to be capable of passing its entire life-cycle in the human colon. The fly is of extremely small size and the larva is also proportionately small. The adult is of an ochraceous or brownish yellow colour. The frons is furnished with 4 rows of 4 macrochaetae each. There are characteristic U-shaped marking on the dorsum of most of the abdominal segments. These markings are absent in *A. rufipes* and the bristles on the head are arranged differently. Tibiae with a distinct row of short bristles on outer side; the four posterior tibiae with two terminal spurs, the hind pair having a close row of very minute hairs on the outer side adjacent to the row of bristles. The larva is extremely small and the pupa looks like a small seed. The posterior larval stigmata are seen in the form of two small brownish papillae situated close together. It takes about 12 to 15 days for the first generation to emerge. Pupation takes place from 5 to 8 days after oviposition and the emergence of the adult occurs seven days later.

A. scalaris is a common fly breeding in human faeces. It also breeds in the excrement of laboratory animals, in both fresh and putrid meat, and in small dead animals, e.g., lizard, snail, caterpillar etc.

Family **Chloropidae** (**Oscinidae**). They are popularly known as eye-flies on account of their habit of hovering over the eyes. They are black in colour and of extremely minute size. They are all directly related with the carriage of the causative agents of conjunctivitis, which, when it is fly-borne, appears in an epidemic form and mostly affects babies and children. The fly merely acts as a mechanical carrier. These flies have also been incriminated with the dissemination of yaws. The spirochaete of yaws can

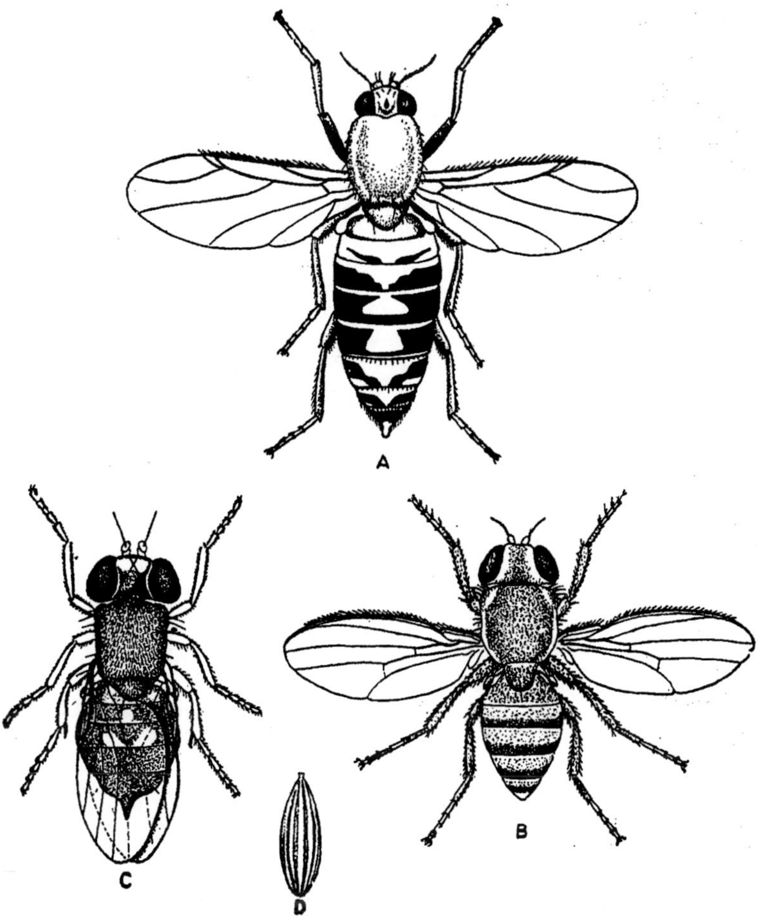

FIG. 95.

A. Female *Apiochaeta scalaris*.
B. *Drosophila melanogaster* (the fruit-fly).
C. & D. Adult and egg of *Hippelates pusio* (After Hall).

remain alive in the digestive tracts of *Hippelates pallipes* Loew.; they retain their motility for a much longer period in the oesophageal diverticulum than in the stomach. In the stomach the motility is lost very rapidly. The method of infection, therefore, has been suggested by Kumm *et al.* (1935), to be by the vomit drop.

In this family both male and female vertex are dichoptic; arista is bare or finely feathered which can be detected under the microscope; the third antennal segment is globular; males can be recognized by their terminalia.

Hippelates pusio Loew. (**H. flavipes**). This is found in various places in the United States and in the West Indies and acts as a true eye-fly. It does not feed on ulcers. Herms (1926) gave an account of conjunctivitis produced by this fly. Eggs are found on decaying animal excrement, decaying fruits and vegetable matter. In this species the mesonotum is black, as the scutellum is also black. There are more than seven rows of hairs on the mesonotum. The hind tibiae are unbanded and hind femora at least partly brown. The base of the abdomen and legs is yellow. Proboscis of *H. pusio* has six large pseudotracheae.

The eye-fly, *H. pusio*, is particularly abundant in South California. They cause considerable annoyance to man and are also responsible for the spread of conjunctivitis. The disease is especially prevalent among children. The greatest annoyance is caused by the adult females on warm days and their activities are correlated with heat and moisture. A man or animal wet with perspiration will always attract these flies.

H. pallipes Lamb. It is more common than *H. pusio* and prefers to feed on ulcers on the feet and lower extremities. The name eye-fly does not apply so well to *H. pallipes* as to H. *pusio*. *H. pallipes* are probably of more importance as far as yaws is concerned than the other species because of their positive tropism towards ulcers on men and animals. In the insectary human excreta are the most favourable media for rearing larvae. Faecal pollution of moist soil is an important, though not the only factor in the breeding of these flies. They also breed in decaying vegetable matter. It is common in many parts of West Indies. In this fly the antennae are reddish, blackened on the end of the third joint; the legs are wholly pale. The mesonotum which carries seven rows of hairs has a shiny surface devoid of any blackish dusting; cheeks are narrow and silvery.

Genus **Siphunculina.** These flies are extremely small or minute in size. The arista is microscopically feathered. The wings are shiny.

S. funicola de Meij.: In this species the tibiae and the tarsi of the front pair of legs and the tarsi only in the mid and hind pair are golden. The knee joints are golden in the mid and hind pair of legs. Males can be recognised by their massive terminalia.

This is the eye-fly of India, Ceylon and Java and is widely distributed in this country. It is a great pest in places where the climate is warm and moist. It is responsible for the spread of the severe form of ophthalmia found in this country especially in tea gardens and which usually occurs in an epidemic form in Assam. The fly is a serious pest from May to June. It feeds on warm human excretions, on the exudates of ulcers of man and animals, putrid sloughs of ulcers, the discharge from the nostrils or eyes and on cow and horse dung, either fresh or decomposed.

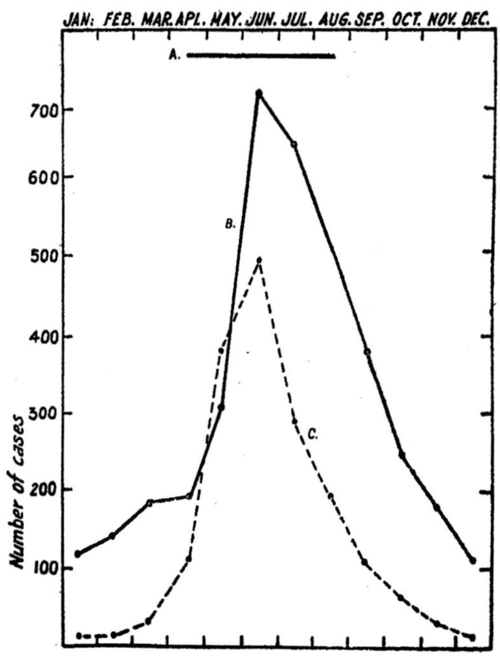

FIG. 96.

Chart showing the incidence of Naga sore (B), and epidemic conjunctivitis (C) in a tea estate in Assam during 1924-25. The horizonal line above (A) shows the period during which *Siphunculina funicola* is a serious pest.

Redrawn by Graham-Smith from the chart given by Roy.

The proboscis in *S. funicola* has undergone a considerable modification from the normal Muscid form though it is completely retractile. There is a great reduction in the number of pseudotracheal channels, some of the pseudotracheal rings are enormously elongated into spines. The fly uses the projecting free ends of the pseudotracheal rings as scarifying organs to remove dried crusts and reach the serous exudations beneath them. Its natural breeding places are : decomposing animal and vegetable matter (Hutson, 1928); earth sodden with dung and urine (Roy, 1928); ammoniacal soil (Syddiq, 1938); and house thatch from the roofs of dwellings (Hamilton, 1939). Females lay eggs in batches of 5 to 10; the duration of the life cycle under favourable conditions of climate is about 10 days according to Syddiq and 3 weeks according to Hamilton. Pyrethrum spray is the only remedy that can be suggested to cope with the nuisance.

Family **Scatophagidae (Cordyluridae).** In comparison to other acalyptrates they are of large size. They swarm over fresh excrement. *Scatophaga*, a common species, are slightly greenish or yellowish brown in colour and have a hairy body. They lay eggs in human and animal excreta.

CALYPTRATA MUSCOIDEA

Families			Genera	
	(1) **Muscidae**	1. **Non-biting**		**Musca.**
		2. **Biting**		**Stomoxys, Glossina, Lyperosia.**
	(2) **Calliphoridae**			
	Subfamily	**Calliphorinae**		**Lucilia, Chrysomyia, Cochliomyia, Calliphora, Auchmeromyia, Cordylobia.**
	(3) **Sarcophagidae**			
	Subfamily	**Sarcophaginae**		**Sarcophaga, Wohlfahrtia.**
	(4) **Oestridae**			**Oestrus, Hypoderma, Gasterophilus.**
	(5) **Tachinidae**			
	(6) **Anthomyidae**			**Fannia.**

Study of the External Characters of a Fly: Such studies are necessary for the determination of the genus and the species. Principally dry mounted specimens are used. Slide preparations especially of the hypopygium may also be required. It must be clearly realised that except in the case of the common house-frequenting flies specific identifications are difficult and it is desirable that the specimens should always be sent to an expert. No reliance can ever be placed on identifications made by non-experts. Ordinarily one should be satisfied with the genus which is not so difficult.

For the identification of flies certain terminology which are commonly used must be understood. The important ones are : (*a*) Acrostichal and dorsocentral bristles on mesonotum. (*b*) Eye length which is the longest axis of the eye viewed from the front. (*c*) Parafrontalia: these are the frontal portions of the genae from vertex to bases of antennae. (*d*) Vertex : the posterior edge of front. (*e*) Hypopleura : from the lateral walls of metathorax and lie just below the posterior spiracles. (*f*) Jowls: portion of genae between eyes and the lower edge of the cheek. (*g*) Humeral callus : a plate lying just above the propleura. (*h*) Gena : below the region lying between the eyes and the bases of the antennae.

The Calyptrata Muscoidea includes the house-fly and other flies which are known to play significant roles in the carriage of pathogenic organisms. This group may be divided into a number of families as follows:

Key to Families of Calyptrata Muscoidea
(Malloch)

1.	Hypopleura without a vertical series of strong bristles below spiracle ..	2
	Hypopleura with one or more vertical series of bristles below spiracle ..	3
2.	4th wing vein more or less bent forward near tip of wing	Muscidae.
	4th wing vein running straight to tip of wing	Anthomyidae.
3.	Mouth parts vestigial	Oestridae.
	Mouth parts well developed and functional	4
4.	Post-scutellum not conspicuously convexly developed; at least basal sternite overlapping tergite	Calliphoridae.
	Post-scutellum prominently convexly developed; sternite more or less hidden by overlapping tergites	Tachinidae.

(1) **Muscidae.** Hypopleura without any tuft of bristles; first posterior cell narrowed to any degree, in fact it may be almost closed; arista feathered to the tip either

on both sides or on one surface only. The genus *Musca*, the members of which are mostly non-biting, is included in this family. These flies are mostly oviparous though a few are larviparous. Among the biting Muscidæ there are three important genera; *Stomoxys* or the stable-fly, *Glossina* or tsetse-fly, and *Lyperosia* or horn-fly.

(2) **Calliphoridae.** Hypopleura with a tuft of bristles; first posterior cell as in Muscidæ. These flies are all metallic coloured, either blue or green. The arista is feathered on both surfaces along its entire length.

The subfamily (Calliphorinæ) contains 4 genera : (**a**) **Lucilia**. (**b**) **Chrysomyia**. (**c**) **Cochliomyia**. and (**d**) **Calliphora.** All are oviparous.

(3) **Sarcophagidae** are grey coloured flies with striped thorax and chequered abdomen. The hypopleura has a tuft of bristles and the 4th vein as in Muscidæ. The arista is differently plumose. They are all larviparous. The subfamily Sarcophaginæ contains two genera : (1) *Sarcophaga*; and (2) *Wohlfahrtia.* These are all flesh-flies; they have retractile proboscis. In *Sarcophaga* the arista is plumose up to the middle, whereas in *Wohlfahrtia* the arista is bare.

(4) **Anthomyidae:** They are not so important as the other two flies. The larvæ are parasites of plants especially in the garden and many of them parasitise other insect larvæ chiefly of Lepidoptera; one genus *Fannia* sometimes visits dwelling houses. The adult flies closely resemble house-flies from which they can easily be distinguished by the course of the 4th longitudinal vein which is more or less straight, the first posterior cell being thus widely open.

(5) **Oestridae:** Members of this family lead strict parasitic lives and infest cattle, horse and other animals. In the adult stage the mouth parts are entirely suppressed.

(6) **Tachinidae:** These are of great practical importance as they parasitise insect larvæ, chiefly of Lepidoptera. The body and legs are covered with stiff hairs or bristles.

Genus **Musca** Lin. The following are the main features of the genus *Musca.* (1) Generally small or moderate in size. (2) Greyish, blackish or bluish grey integument without any metallic lustre. (3) The arista is feathered on both sides up to the tip. (4) The proboscis is retractile. (5) The 4th longitudinal vein bends forwards at a sharp angle, so as almost to close the first posterior cell. (6) The sternopleural bristles are arranged as 1 : 2.

Patton has classified the genus *Musca* into three groups in accordance with their feeding habits. Group I includes the true house and bazar frequenting flies and though at times may be found on animals and show a haematophagous tendency, they are mainly associated with man, his dwellings and his food. They breed in garbage and excrement of all kinds. This group includes *M. domestica* L., *M. vicina* Macq., *M. nebulo* F., *M. sorbens* Wied., *M. vetustissima* Wlk. and *M. yerburyi* Patton. Flies in group II are non-biting, hæmatophagous species and suck blood from wounds made by true biting flies. They are only found on animals and in their vicinity, and breed almost exclusively in cow-dung. These include *M. ventrosa* Weid., *M. craggi* Patt. and *M. conducens*. Group III consists of the true biting species which are only found on animals and which breed exclusively in cow-dung. These are among others *M. inferior* Stein. and *M. crassirostris* Stein.

Medical and public health workers are primarily concerned with Group I, whereas the veterinarians should be familiar with those that bite cattle and are responsible for the deterioration of the health of these animals.

The research worker may feel the necessity of identifying the common house-frequenting flies and in this respect a key does not seem to be very helpful. It is for this reason that a more detailed description taken mainly from Patton has been given. It may be pointed out that the typical *Musca domestica*, the house-fly of temperate climate, does not occur in India and it is doubtful if it is found in Africa. Both are found in China. Its place in the tropics has been taken by *M. vicina*. The latter fly no doubt bears a close resemblance to *domestica*, the only difference being in the male which has an exceptionally wide frons in *domestica* than in *vicina* in which the frons is very narrow; in fact the two eyes nearly touch each other. It may be of interest to know that the *vicina* found in the hills above 5,000 ft. elevation is intermediate between the true *domestica* and *vicina* and interbreeding experiments between the hill and plain varieties of this species show a tendency for the frons to become narrower in males.

M. domestica L. The vital importance of horse-manure and pig-dung to the development of this species has been fully realised. There is no essential difference between horse-manure and pig-dung as regards their power to attract the ovipositing flies. Pig-dung is highly favoured as medium for egg-laying in many parts of Europe but when this is covered with cow-dung the oviposition is stopped.

Musca vicina (**Musca domestica vicina** Macq.). It is the commonest house-frequenting fly in the tropics. It breeds in domestic, stable and garden refuse, horse manure, night soil in patches but more extensively in trenches, and occasionally in piled up cow-dung mixed with organic and vegetable refuse matters especially during the rainy season. In fact it will breed in any fermenting material, but if covered with a thin layer of earth or cow-dung, it will at once fail to attract the ovipositing flies. All house-flies are oviparous.

In both sexes of *M. vicina* the mesonotum has four well defined dark stripes. In the male the abdominal terga have a median broad black stripe; the apparent first segment with a broad black band; tergum 3 on either side of the median band orange which is bordered on both sides by a silvery stripe; tergum 4 greyish yellow. In females the abdominal characters are almost the same as in males except that terga 1, 2 and 3, are more extensively orange; apparent first segment black or dark brown anteriorly.

Musca nebulo F. Roughly it comprises about 5 per cent of the catches made from dwelling houses though in certain houses it may be entirely absent, all being *M. vicina*. It has the same breeding habits as *vicina*. Like *vicina* both sexes possess 4 thoracic stripes. In male flies the median abdominal stripe expands anteriorly on the first apparent segment to form a dark band across the anterior half but it is narrower than in *vicina*; tergum 3, in addition to a silvery stripe, has silvery patches at the margins; on tergum 4 silvery stripe and silvery patches are more prominent. The female vertex and cheeks are creamy white; abdomen is light orange; silvery stripes and spots in remaining segments well marked. Both *Musca domestica* and *M. nebulo* harbour the flagellate parasite, *Herpetomonas muscae domesticae*. Flies become infected through the ingestion of the cystic stage of the parasite.

M. yerburyi Patt. It is not a house-frequenting fly but is essentially a bazaar fly though found in considerably smaller numbers than *vicina* and *nebulo*. It breeds in entrails of animals, decomposed fish etc. Eggs are also laid in animal excrement.

In both sexes there are four mesonotal stripes. In the male terga 1 and 2 yellow; tergum 3 orange with a broad median black stripe bordered with narrow to broad silvery stripes but without silvery marginal spots. In the female terga 1 and 2 are the same; tergum 3 with a broad, black median stripe bordered with broad silvery stripes, also well marked silvery marginal spots.

M. sorbens Wied. Eggs are laid principally in large accumulations of cow-dung. At times the larvae are found in isolated patches of cow-dung and small deposits of human faeces.

M. vetustissima Wlk. (**M. pumila** Macq). It is essentially a bazaar fly in India and is occasionally found in camps. It breeds in entrails of animals and fish and sometimes eggs are laid in animal excrement. In India it is mostly prevalent in South India. It also occurs in Ceylon but it is more an Australian than an Indian species.

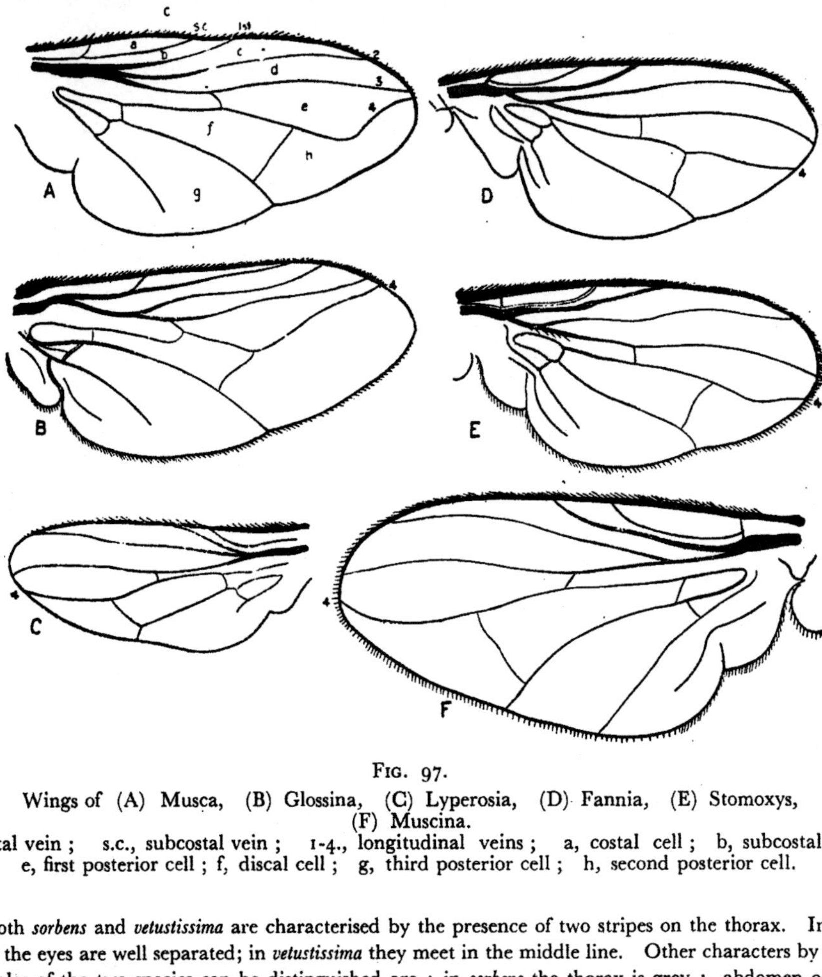

FIG. 97.
Wings of (A) Musca, (B) Glossina, (C) Lyperosia, (D) Fannia, (E) Stomoxys, (F) Muscina.
c, costal vein ; s.c., subcostal vein ; 1-4., longitudinal veins ; a, costal cell ; b, subcostal cell ; e, first posterior cell ; f, discal cell ; g, third posterior cell ; h, second posterior cell.

Both *sorbens* and *vetustissima* are characterised by the presence of two stripes on the thorax. In male *sorbens* the eyes are well separated; in *vetustissima* they meet in the middle line. Other characters by which the males of the two species can be distinguished are : in *sorbens* the thorax is grey : abdomen orange; apparent first segment either dark or light brown; apparent 3rd and 4th segments with a median black stripe and admedian and marginal silvery spots. In *vetustissima* the thorax is bluish grey, abdomen dark orange; apparent first segment black; other 2 segments similar to those of *sorbens*. The female *sorbens* has grey thorax with two black stripes separated before the suture; abdomen blackish grey with black bands and stripes. The female *vetustissima* has the same thoracic stripes but the abdomen is bluish grey with black bands and stripes.

As has been stated before the true house-frequenting flies in the tropics are mainly *M. vicina* and *M. nebulo*; where the dwelling house adjoins the cow-shed, *M. sorbens* are also found. *M. vetustissima* and *M. yerburyi* are really bazaar-flies though occasionally isolated specimens may be caught indoors.

The following two species, *M. crassirostris* and *M. inferior*, have the proboscis structurally adapted for scratching and tearing through skin and drawing blood.

M. crassirostris Stein. It is a widely distributed species. It is larviparous and breeds in cattle dung. It is readily recognised by its large size, olive green colour and by the large bulbous mentum.

M. inferior Stein. It is also widely distributed in the Oriental Region. It is oviparous and breeds in cattle dung. The eyes in the male are widely separated. It is a large species which is easily identified by the character of the proboscis and by the presence of hairs on squamae, being the only species which has such hairs.

Family **Calliphoridae.** Subfamily **Calliphorinae** (metallic coloured bluish or greenish flies).

Genus **Calliphora.** Lower lobe of squama covered completely or in part with long hairs; metallic bluish purple flies.

Genus **Lucilia.** Brilliantly shining greenish flies; lower lobe of squama quite bare.

Genus **Chrysomyia.** The small round protuberance below base of wing with erect hairs; lower calyptra subtruncate at apex, concave on outer margin, haired on part of upper surface; hairs on posterior surface of stem vein of wing; squama always hairy; only one pair of acrostichal bristles (post-scutellar).

Most of them breed primarily in animal materials and are of great veterinary and economic importance.

Genus **Lucilia** R. D. (Green-bottle flies) : It is characterised by setæ on a considerable part of the upper surface of the 3rd longitudinal vein, bare squama and presence of well developed dorsocentral and acrostichal bristles on both prescutum and scutum. The arista is plumose to the tip. The sternopleural bristles are 2 : 1. Hypopleural bristles 5 to 7 in number.

These flies are commonly called green-bottles and all have a bluish and bright metallic colouring. The females are oviparous and though larvæ develop mostly in decaying animal matter, some of them have the habits of blow-flies. Though these flies will occasionally blow meat placed outside but they seldom enter the house in search of food. *L. sericata* and *L. cuprina* are well known for their habit of blowing wool. The eggs are laid on living sheep on which the maggots feed and then make sores in which the larvæ may burrow. The production of wool is not only reduced but the quality of wool is greatly impaired. The health of the sheep suffers and it may even die. *L. caesar* L., is found in the Palaearctic Region. It has also been recorded from New Zealand and eastern Australia and South Africa. The subcostal sclerite has one or more upstanding bristles and two post-sutural acrostichal bristles on the thorax. In *L. sericata* Mg., and *L. cuprina* Wied., the subcostal sclerite has fine decumbent pubescence.

L. sericata and *L. cuprina* are cosmopolitan, both being common in the Oriental Region. In *sericata* the body is rounded, metallic green with sparse pollinosity on the mesonotum. In *cuprina* the body is elongate, metallic coppery with rather dense pollinosity on the back. In the female *sericata* the parafrontals have short, fine, decumbent stiff hairs among the frontal and fronto-orbital bristles whereas the female *cuprina* possesses bare parafrontals apart from frontal and fronto-orbital bristles. In the male the abdominal sternites bear tufts of long hairs which are absent in the male *sericata*.

In Australia, New Zealand and Great Britain sheep are infested with larvæ of *L. caesar* and *L. sericata*. The eggs are deposited in batches of 45 to 275 eggs in the soiled wool of the sheep. The wool of the hind quarters is most frequently blown by them. The maggots on hatching penetrate the skin. In this way extensive festering sores are caused.

L. cuprina which is an Ethiopian and Oriental species is widely prevalent in India;

it is a common bazaar fly; *L. sericata* is found in North Western India. Both breed in animal matter. *L., cuprina* is particularly attracted to fish and animal entrail in a decomposed state.

L. argyricephala Macq., is a common myiasis-producing fly in Africa. It is also found in India breeding in animal tissues.

Genus **Chrysomyia.** It is also like *Lucilia* a metallic coloured fly but differs from the latter in many important respects. It is bluish in colour and hence called blue-bottle. The prescutal and scutal acrostichal bristles are almost absent and the prescutal dorsocentral bristles are practically wanting. The sternopleural bristles are commonly 1 : 1 and in some cases 2 : 1. A row of 5 or more hypopleural bristles is present. The upper surface of the squama is hairy. The arista is plumose to the tip and in the male the eyes are close together whereas in the female they are wide apart. They generally breed in the carcass of both large and small animals. The spiracular slits of the larva are directed in one direction. i.e., towards the scar and the peritreme surrounds the slits on all sides.

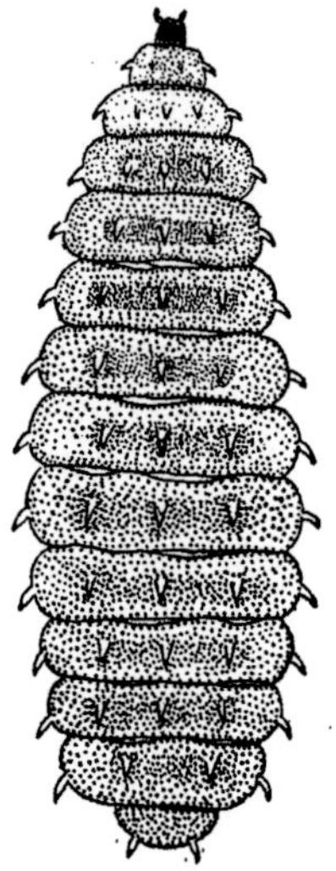

FIG. 98.
Chrysomyia rufifacies larva.

C. megacephala Fab. This fly is very common in Oriental and Australian Regions; also recorded from Mauritus. This species is very distinct on account of the much enlarged upper facets in the male and in both sexes by the black porthoracic spiracles and the dark squamæ which have on the upper surface a white spot exteriorly. Eggs are laid in (1) dead animals such as dogs, cattle, rats and big birds; (2) in human fæces after it has turned into the liquid form as a result of putrefaction; enormous numbers of their larvæ are commonly noticed on the floors and also in pots of service privies when they are not cleaned regularly; (3) larvæ have been recovered from isolated patches of cow-dung.

Next to *M. vicina* and *M. nebulo* it frequents dwelling houses. It is greatly attracted by mango, jack fruit and country toddy; it is quickly overwhelmed by the latter and dies. This fly is mostly attracted to baits of human fæces and is easily caught in traps in China. Next to it in China is *Musca vicina* and last of all *Musca sorbens* (Meng and Winfield, 1938). *C. megacephala* is also a common house-frequenting blue-bottle fly in China. It breeds almost entirely in liquid human excrement. It was found by Chow (1940) in Tsinan that 8 per cent of the flies from latrines carried *B. dysenterica* internally.

C. rufifacies Macq., and **C. albiceps** Wied. In nature they generally choose large birds for depositing their eggs although in the laboratory *C. rufifacies* will readily oviposit on any dead animal. The larvæ are hairy, dark in colour and predacious, feeding on other fly larvæ. Both larvæ possess fleshy lateral processes with a tuft of black spines on each process, first appearing in the second instar. Both *rufifacies* and *albiceps* have the anterior spiracles white and the wings entirely hyaline. In *albiceps* there is a strong bristle just below the prothoracic spiracle which is absent in *rufifacies*.

The method of reproduction in *C. rufifacies* is unique. Mating is necessary. A female produces eggs which turn out into adults of one sex not only in the first but also in subsequent batches. Some females therefore produce only males whereas others, only females (Roy and Siddons, 1939).

C. albiceps is a widely distributed blow-fly and one of the well known sheep maggot flies of Australia where it is known by the name of *rufifacies*. This is also one of the principal sheep maggots of South Africa.

C. rufifacies.: occur in Australia, New Zealand, Pacific Islands, Java, India and Ceylon. *C. albiceps.*: occur in the Palæarctic Region, South Africa, Southern Rhodesia, Asia Minor, Palestine, around the shores of the Mediterranean and in islands in the Indian Ocean.

Chrysomyia bezziana Vill. It is widely distributed in Africa, Ceylon, India and also further east including the Philippine Islands and is a true myiasis-producing fly. It deposits its eggs always in living tissues. In the human being the larvæ are found in any of the natural orifices of the body such as nose, ear, dental socket, and on rare occasion vagina and skin injury from burn. In animals the larvæ are commonly found in the scrotum after castration. The flies are attracted to an evil-smelling discharge which induces them to lay eggs. Patton thinks that the larva is capable of attacking living tissue such as muscle, tendons etc., causing an extension of the ulcer. This, according to him, may end fatally. No such bad effect has, however, been noticed by us and we are convinced that the larvæ feed on necrotic tissues only.

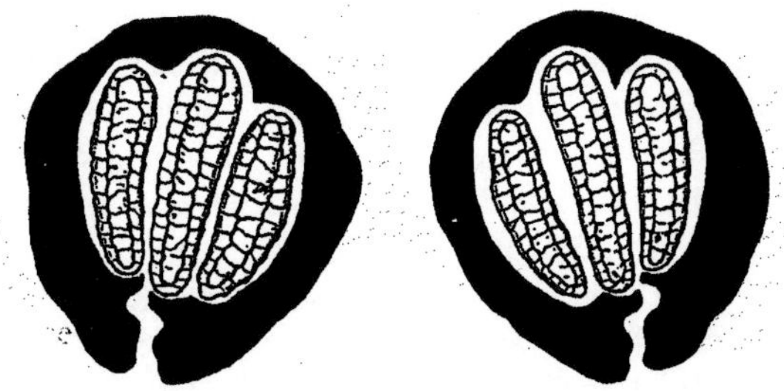

FIG. 99.
Posterior spiracles of the larva of *Chrysomyia rufifacies*.

The adult fly has not yet been encountered in nature. The mature larva presents a characteristic cork-screw-like appearance due to the presence of well developed intersegmental spines. Hence it is often called a screw-worm larva. The posterior wall of the pharynx of the larva is smooth suggesting that the larva is strictly parasitic. In Africa it causes myiasis only in the large animals, chiefly cattle. In India both man and cattle are affected. In cattle the predisposing cause is injury on the skin from any cause, castration being one of the commonest causes. The adult *C. bezziana* can easily be distinguished from *C. megacephala*. The squama in *bezziana* is waxy white, in *megacephala* it is greyish. In the male *megacephala* the upper facets of the eyes are large, which is not the case in *bezziana*. The intersegmental spines and the character of the pharynx will be helpful in distinguishing their larvæ. The spines are normal in *megacephala* and the posterior wall of the pharynx in this species is ridged. This is smooth in *bezziana*.

Genus **Cochliomyia** Town. The genus *Cochliomyia* may be immediately separated from related genera by the shape and size of the palpi, these being very short and slender in *Cochliomyia*; elongate and more club-shaped in others. The body colour is metallic blue. They possess bright yellow to orange red faces and genae, three distinct dark longitudinal stripes on the dorsum of the thorax.

C. hominivorax Coq. (*C. americana* C. and P.). It is restricted to the New

World. It is oviparous. It is the screw-worm fly of the New World. It is an obligatory myiasis-producing fly (Cushing and Patton, 1933), the posterior pharyngeal wall in the larva being without any ridge. According to Aubertin and Buxton (1934), it generally chooses one of the orifices of the body especially the nose for laying eggs. These authors believe that cases of myiasis in human being are more common than in cattle. This fly is very widely distributed in tropical and subtropical America and the West Indies.

The invasion of the wound by the larvæ may result in considerable destruction of tissues and death of the animal. Borgstrom (1938) found 8 larvæ to be the maximum which could be harboured without resulting in death. The death of the animal is due to the toxic effect following the presence of the larvæ in the wound. The predisposing factor is the discharge and a characteristic odour that is nearly always associated with wounds when the ears of animals are infested by the Gulf Coast tick, *Amblyomma maculatum.* The ticks generally attach to the inside of the ears. Such injury of the skin following the attachment of the ticks becomes very often infested by larvæ of this fly. The only way by which *C. hominivorax* and *C. macellaria* can be distinguished is by an examination of the male and female terminalia.

Benzol and pine tar oil are deadly toxic to the larvæ and can be used on animal wounds for treatment and control.

Cochliomyia (Chrysomyia) macellaria F. This blue-bottle fly, which is strictly a saprophyte, breeds principally in carcasses and was previously confused with *C. hominivorax* which is alone responsible for screw-worm infestations in man and warm-blooded animals in tropical and subtropical America.

Genus **Calliphora.** Squamæ with a white hind border and distinct hairs above at least towards the middle. Eyes of male separated; abdomen dusted; dorsocentral and acrostichal bristles both presutural and postsutural are well developed.

C. erythrocephala Mg., is the well-known blue-bottle fly. It is a large bluish-black species with a more or less whitish pollinose abdomen and black legs; palpi reddish; antennæ black and jowls (part of face below cheeks) red, with black hairs. *C. erythrocephala* oviposits on fresh or tainted meat, on carcasses, offal, and other animal refuse. It is a cosmopolitan species though it is essentially a hill species in India where it is found only during summer when it frequents dwelling houses.

Auchmeromyia luteola F. It represents a special group of Calliphorinæ characterised by their blood sucking habits and intermittent parasitism of the larvæ. They are popularly known as Congo-floor maggots.

This fly is distributed throughout the whole of the Belgian Congo where it is a common pest. This species is found in close association with man and is considered to be a specific human parasite as the larvæ are blood suckers and live on the blood of man. The fly resembles Tumbu-fly in colouration, the head and the body are pale yellow in both, and both have dark markings on the thorax and the distal half of the abdomen is blackish.

The Congo floor-maggot is devoid of the characteristic spines found in the Tumbu-fly larva. The stigmatal plates are extremely small and wide apart whereas in the Tumbu-fly maggot they are much larger and close together. The larva is found in huts in cracks in floors where urine is voided. The adult fly is commonly found in the huts. Numerous eggs are generally laid in the loose and slightly moist soil. There is a dark, quadrate, median longitudinal mark on the dorsum of the second abdominal segment of the larva. The maggots are unable to climb and are entirely dependant for their existence upon the habit of the local people of sleeping on floor mats. The larvæ crawl out by night to suck blood. The adults are also believed to be blood suckers and to live on blood; the mouth parts are, however, adapted for sucking only.

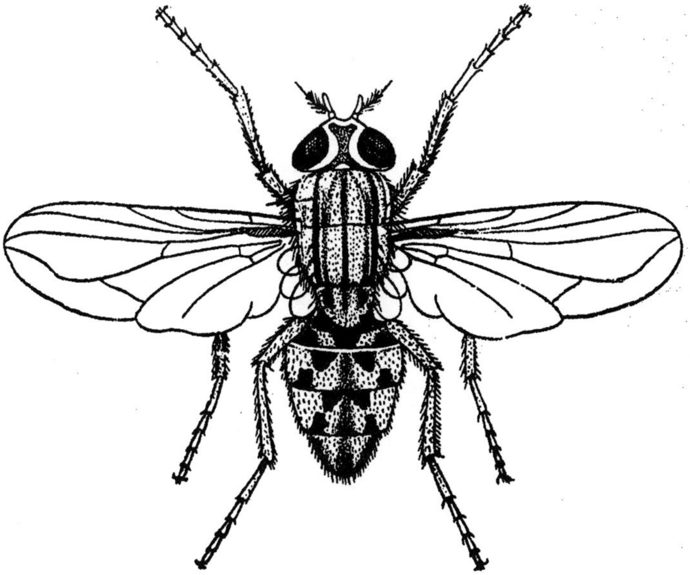

FIG. 100.
Female *Sarcophaga ruficornis.*

Genus **Cordylobia** Grun. **C. anthropophaga** Grun. This genus is peculiar to Africa and is known as Tumbu-fly. The larvæ are found in dogs, goats, cats, domestic rabbits, also in man, causing subcutaneous tumours. The life history has been worked out by Blacklock and Thomson (1923). The fly generally deposits its eggs on the ground where there is smell of animal or human perspiration and urine and other animal matters. Sometimes eggs and young larvæ are found on clothing impregnated by urine. In man tumours in connection with the presence of larvæ are always found in parts of the skin coming into intimate contact with the ground. In animals the larva shows a marked preference for the skin of the scrotum. The larva takes about 10 to 15 days to attain full development. If allowed to run its course the larva escapes through a small opening and falls to the ground to pupate. Pupation occupies about a fortnight. The larva is yellowish white in colour and half an inch in length. It is bluntly pointed at the anterior end and truncate behind. The segments of the body are thickly covered with minute recurved spines. The first instar larva is capable of penetrating unbroken skin by means of the cephalopharyngeal skeleton. The shape of the puparium is rather characteristic. It has the posterior end very squarely cut off.

Metazoal Immunity. Blacklock and Thomson (1923), working with *Cordylobia anthropophaga*, proved experimentally that an immunity against the larvae can be acquired by both men and animals. This immunity is confined to the area of the skin into which the parasite had previously penetrated, and from this area later the immunity spreads to other areas which have never been the site of invasion (Blacklock and Gordon, 1927 ; Blacklock, Gordon and Fine, 1930). Similar observations have been made in regard to the invasion of the human or animal body by other arthropods such as *Sarcoptes scabiei, Dermacentor variabilis,* larvae of *Cochliomyia americana* etc. (Trager, 1939; Lake and Smith, 1939; Mellanby, 1944).

Family **Sarcophagidae.** Subfamily **Sarcophaginae** (Flesh-flies). Genus **Sarcophaga.** The *Sarcophga* flies contain many species distributed throughout all the zoological regions of the world. These flies are generally of large size though some are quite small. They are grey in colour, the thorax is banded or striped and the abdomen is chequered or tessellated. The sexes cannot be definitely distinguished from the approximation of the eyes; this can be better done by examination of the terminal segments of the abdomen. The arista is plumose on both sides up to half its length, the terminal part being bare. All are larviparous ; first-stage larvæ are deposited. Posterior spiracles of the larva lie in a deep recess whose lips may be kept closed. The spiracular slits are not directed in one direction. The internal slit runs vertically and outwards, and the central one is practically parallel to the median sagittal plane. The peritreme is interrupted in its lower part and the scar is absent or very poorly developed.

They are called flesh-flies on account of their habit of breeding in flesh for which purpose carcasses of small animals like rats are generally chosen. Some species also choose freshly passed human faeces for breeding. Some deposit their larvæ in human and animal tissues. They are not known to produce eggs but always first-stage minute larvæ usually about 20 to 30 in number. Development on suitable pabulum is rapid, the third instar usually appearing in 3 to 4 days.

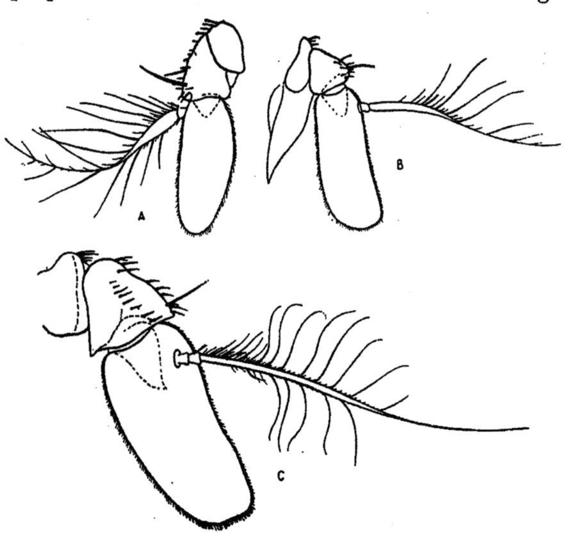

FIG. 101.
A. Arista of house-fly.
B. „ „ *Stomoxys.*
C. „ „ *Sarcophaga.*

It is not really a house-frequenting fly; its visit in the house is always associated with the presence of meat. It is not necessary for this fly to sit on the pabulum for depositing its larvæ. It can drop its larvæ from a distance while flying over the meat. Ordinary wire-meshed meat covers are no protection against them. For this purpose meat has to be covered with paper.

The female *Sarcophaga,* unless they are bred out or caught *in copulo,* are difficult to identify. The male flies can be readily identified by drawing back the anal cerci and exposing the terminalia, the characters of which are, as a rule, strikingly different in different species.

The two common species frequently encountered in connection with cutaneous myiasis in man and animals are *S. ruficornis* Fab., and *S. dux* Thom. The former is by far more common than the latter. Their larvæ feed entirely on necrotic tissue and their presence in the wound may be considered as beneficial, the process of healing being greatly accelerated.

Males of both *S. dux* and *S. ruficornis* have the frontal width two-thirds that of an eye while in females the width is equal to that of an eye.

S. ruficornis Fab. Prescutellar acrostichal bristles are either absent or only one is present. The tip of apparent 4th segment reddish. No spot on second abdominal segment. Antennæ and palpi yellow-orange. They breed in both large and small animals such as rats and big birds, also in freshly passed human fæces. It occurs in the Oriental Region and is the commonest among the *Sarcophaga*. It is also found in the Ethiopian Region.

S. haemorrhoidalis Fall. It is a Nearctic species. Larvæ are deposited on carrion, excrement and tainted meats. Antennæ and palpi black; only prescutellar acrostichals are present.

S. dux Thom. They generally breed in dead animals such as snail, snake, etc.

Genus **Wohlfahrtia.** The genus *Wohlfahrtia* includes medium-sized Sarcophagidæ. The third joint of the antenna is short. The arista is hairy and not armed with bristles. The abdomen is marked with well-defined black spots in three longitudinal series and there are no glistening reflection as in *Sarcophaga*. It has, however, the same breeding habits. It has been recorded from America, Central Asia, Africa, Europe and Southern Russia. The adult flies are found in the garden and never enter human habitations.

Wohlfahrtia magnifaca Schin. This is the only European myiasis-producing fly. It is widely distributed in the United States, South America, South Africa, South Russia, Asia Minor and Egypt. The antennæ are mostly black except at the apex of the second segment which may be yellowish or reddish; palpi black; the lateral abdominal spots rounded and well defined, and the fourth abdominal segment has three small spots at the apex. Cases of myiasis involving *W. magnifaca* have been dealt with by Salem (1935).

W. vigil Wlk. This fly is a frequent cause of human and animal myiasis in North America and Canada. It occurs in the Nearctic Region. This species differs from all other known members of the genus in that the usual black abdominal spots are large and confluent, the abdomen consequently appearing black with gray pollinose spots.

(For Wohlfahrtia infestation see Hill, C. G. (1933)—*Trans. Roy. Soc. Trop. Med. & Hyg.*, 27 : 93.)

Family **Oestridae** (Bot flies) All members of this family live parasitic lives. They are oviparous or larviparous and live either in the skin, nasal or gastro-intestinal tract and reach the skin by more or less complicated migrations. The mature larvæ always drop to the ground. Generally the imago does not feed but only lives for the propagation of the species. The eyes of both sexes are placed well apart. Ocelli are always present. The antennæ are placed in a deep groove; the palpi are generally not developed; the thorax is almost quadrangular ; the abdomen is as a rule short and is constricted where it joins the thorax. The body is generally covered with dense hairs. The female is recognised by its ovipositor. The following are the important genera: (*a*) *Oestrus*. (*b*) *Hypoderma*. (*c*) *Gasterophilus* and (*d*) *Dermatobia*.

Genus **Oestrus** L. Body well narrowed; scutellum distinctly incised on margin and showing two divisions; metatarsi much shorter than the following joints; female vertex more than half the head width.

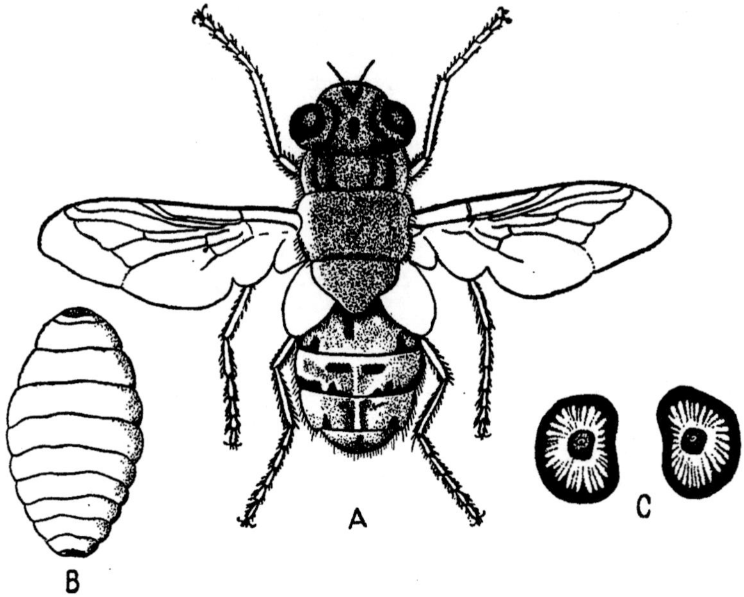

FIG. 102.
Oestrus ovis.
A, adult; B, larva; C, posterior spiracles.

O. ovis L. It is primarily a pest of sheep and occasionally of goats in India. In some rare cases it attacks man. In certain regions of the Sahara and Caucasus, the larvæ may be deposited on the conjunctival and nasal mucous membranes of man and cause much trouble. The first stage larva does not grow in man. The fly is larviparous. The larvæ are deposited just inside the nose when the female fly hovers over and strikes at the nostrils. The young larvæ crawl to the nasal sinuses or even to the base of the horns and attach themselves by hooks to the mucous membranes. Occasionally they may migrate to the cranial cavity or turbinate bones, where they may become imprisoned and die. The larvæ usually live in the host for 8 to 10 months but may take only 6 months to mature. They crawl out and are expelled by a violent fit of sneezing.

Genus **Hypoderma** Latr. Body stout; scutellum practically entire; metatarsi nearly as long as the following joints together; female vertex less than half head width.

All are primarily parasitic on cattle and goats. Only four species are important. These are : (*a*) *H. bovis*, (*b*) *H. lineatum*, (*c*) *H. crossi* and (*d*) *H. diana* which is confined to the Palaearctic Region. *Hypoderma* larvæ enter their hosts through the skin close to the spot where the eggs have been laid. The adult flies are rarely seen elsewhere than in the vicinity of their hosts.

Domestic cattle in Europe and North America are subject to attack by two species of *Hypoderma*, *H. bovis* and *H. lineatum.* Both species are widely distributed, but *H. bovis* is the prevalent species on cattle in Europe whereas *H. lineatum* in North America and in Asia. In India they occur practically in the Punjab and further north in the North Western Frontier Province. The flies appear during the summer weather. The favourite places for the fly to strike is the region of the hock in particular and the back of the knee, occasionally striking as high as the stifle. The effect of the parasites in cattle is very serious and the production of milk is appreciably reduced. The hide is also considerably damaged. *H. bovis* causes losses estimated at many millions of sterling in England alone. Calves and young cattle are more frequently affected than older ones.

H. bovis DeG. They are dark bee-like flies and are thick set. Front and cheeks are covered with light yellow hairs; presutural area of mesonotum with dense yellow hairs hiding longitudinal markings and contrasting strongly with black mesothoracic region; wing veins dark brown; legs smooth with fine hairs; middle third of the abdomen with orange tipped dark hairs forming a conspicuous dark band; apical part with yellow hairs.

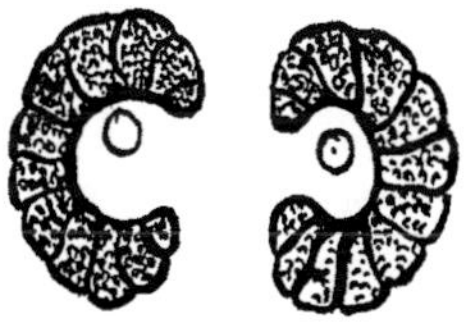

FIG. 103.

Posterior spiracles of larva of *H. lineatum*. *Hypoderma lineatum*, female.

The eggs are laid on the legs and more rarely on the flanks, apparently never on the back. The eggs hatch in about 4 days and give rise to a minute larva armed in every segment with small spines and furnished with a median piercing tooth, flanked by a pair of well developed mouth hooks.

H. lineatum Vill. It resembles *H. bovis* but is smaller. The mesonotum is uniformly covered with mixed black and white hairs; the mesonotal sutures are visible owing to sparse covering of hairs; the terminal part of the abdomen covered with orange red hairs; the middle part is dark brown to black; wing veins black whereas in *bovis* they are dark brown; the femora and tibia especially tarsi are rougher and more hairy.

H. lineatum is the prevailing species on cattle in North America and Asia. It attaches several eggs in a series, 1 to 14, to a single hair whereas *H. bovis* lays its eggs singly. Oviposition inflicts no pain on the animal and may be unnoticed but if the victim becomes aware of the presence of the fly, great uneasiness and terror is excited. By their persistence in oviposition they frequently stampede herds of cattle, which take refuge in water if available.

In Indian cattle the larvae are found in September deeply embedded in submucosa of the gullet; subsequently they are found in loose connective tissue outside the oesophagus. In the North Western Frontier Province the infestation rate in some places varies from 50 to 80 per cent.

Cases of infestation by *H. lineatum* larva is occasionally seen in people in this country who live in close proximity to cattle particularly in the northern parts of India. This condition is not, however, common.

Fig. 104.
Hypoderma bovis, female.

The larvae of *H. bovis* and *H. lineatum* can be easily distinguished. On the ventral convex side *H. bovis* has the two last segments entirely destitute of spines, while in *H. lineatum* the last segment alone is naked. The central boss of the posterior spiracle in *H. bovis* is depressed and funnel-shaped while in *H. lineatum* the whole surface is level.

Life history. Carpenter and his co-workers in Great Britain, Glaser in Germany and Hadwen in Canada worked out the method by which the larvæ reach their final destination. The eggs are laid on the hairs of the legs and the first stage larvæ on hatching out, penetrate the skin adjacent to the hair, causing a temporary dermal myiasis. The larvæ are negatively geotropic being impelled by an upward urge acting in opposition to gravity, which is responsible for their migrating habit. They now migrate upwards to appear later in the region of the throat as the second instar, in which stage they are very different to the third stage larva. After a time they leave the site and migrate along the muscles of the back eventually reaching the subcutaneous tissues of the back. Here a cyst is formed. The larva now pierces a hole to the exterior towards which it advances, the caudal end protruding through the opening. It now produces the well known warble. The movement of the larva causes some inflammation in the walls of the cyst. On account of contamination of the wound with putrefying organisms the exudation becomes dark and fœtid. The larva grows very rapidly after connection with the exterior has been effected. The orifice in the skin is widened by frequent protrusion of the posterior part of the larva through the opening. The larva is spindle-shaped, convex on the ventral surface but flattened dorsally and furnished on either side with three longitudinal rows of tubercular ridges and its spiracles are large and conspicuous and of a dark colour. Successive batches of tumours often arise in the same animal. As the orifice is proportionally smaller than the width of the body, exit is made with some effort and it leaves the skin to drop to the ground where it pupates. The pupal stage averages about five weeks while the larval life lasts 8 months and for about 7 months there are no external signs of the whereabouts of the larva.

Young maggots of *H. bovis* proceed comparatively straight up after penetrating the skin and do not occur so frequently in the œsophagus but frequently enter the spinal canal, whereas *H. lineata* rarely enter the spinal column but are more frequently found in the œsophagus.

In man the maggots of both *H. bovis* and *H. lineatum* traverse the subcutaneous tissue before they come to a halt often at the end of many months. Numerous cases of parasitism in man by *H. bovis* are on record in North America and N. Europe. Parasitism in man is accompanied by considerable pain and discomfort. The larva in course of its wanderings may invade the spinal cord or the interior of the eye. Infestation by *H. lineata* in man has been recorded by Pavlovskii and Glezer (1932) from Russia, Lawson (1929) from South Africa and Ghosh (1949) from India.

The first stage larva of *Hypoderma* is characterised by a very dense spiny armature on all the body segments. The second stage larva is smooth and found in the œsophagus. While still in the œsophagus this stage moults again and changes to the third stage, which is almost devoid of spines. This stage may also be found in the subcutaneous tissue in the back. This instar moults again producing the fourth stage which is really regarded as the third.

Creeping Eruption. This term was introduced by Lees in 1874. Parasites belonging to three different groups are known to be responsible for creeping eruption in man. The parasites generally follow the percutaneous route in their mode of entry. The infestation of man is accidental and comes to notice very rarely.

The following parasites are generally responsible for this peculiar condition: (*a*) Species of *Hypoderma.* (*b*) Species of *Gasterophilus.* (*c*) Immature worms of the genus *Gnathostoma, G. spinigerum.* (*d*) Infective or third stage larvae of certain hook-worms, *Ankylostoma caninum, A. braziliense* and *A. duodenale.*

Genus **Gasterophilus** Leach. All the larvæ of the known species are parasites in the stomach and intestine of Equidæ and are easily carried to distant places. They attach themselves to the mucous membrane by means of well developed oral hooks. The abdomen of the adult fly is sessile; eyes are widely separated in both sexes; arista bare; squamæ small and generally with long cilia not covering the halteres; the venation is like that of Anthomyid flies; mouth parts very small; palpi small; proboscis not extensible. The head is clothed with conspicuous pile, the thorax hairy but without bristles. The adult fly in size and appearance bears a strong resemblance to the honey bee and like the latter it makes a buzzing noise while flying.

The larva of *Gasterophilus* is reddish in colour, due to the presence of a pigment similar to haemoglobin, as is found in certain Chironomid larvae. The larva was believed by Buchner (1922) to feed on blood extracted from the mucous membrane of the stomach, to which it remains attached. It has been demonstrated that it feeds entirely on the stomach contents of the horse, as a consequence of which the health of the animal is not affected, even in the presence of a heavy infestation (Roy, 1937).

Gasterophilus haemorrhoidalis L. This is a European and North American species. The fly is very annoying to horses at the time the eggs are laid. The eggs are deposited singly in the pores of the minute hairs of the lips; they hatch in 5 to 10 days. The larvae are ingested with food and water, and attach themselves to the walls of the stomach, where they remain for a long time but later migrate to the rectum where they reattach themselves. Before leaving the host they usually attach themselves close to the anus and protrude from it. They drop off after a certain length of time. The pupal stage lasts 2 months. The length of life of the adult is from 1 to 7 days during which time they take no food but are very active, the female ovipositing throughout her existence.

G. nasalis L. (Throat bot-fly). It deposits its eggs on the hairs under the jaws and to some extent on the shoulders and other parts of the host. The larvae attach themselves to the wall of the pharynx, and later to the wall of the stomach and duodenum. This species is found throughout Europe, North America, Australia, French Guinea and Iraq.

G. intestinals De G. (*eqi* Clark). It is the common bot-fly of horses in Europe and North America. It has also been recorded from many parts of Africa. It is also found in Australia and New Zealand. The eggs are unstalked and are deposited on all parts of the body but preferably on the forelegs. The larvae which develop upon the application of moisture and friction supplied by the rubbing and licking of the horse, are ingested and attach themselves to any part of the stomach by means of powerful oral hooks.

In this species, which is also found in India, the transverse posterior vein is always present and placed immediately behind the small one. There are smoke coloured transversal bands and smoky spots on the wings. The abdomen is brownish yellow. The thorax has interrupted belts of black hairs behind the suture.

FIG. 105
Gasterophilus intestinalis.

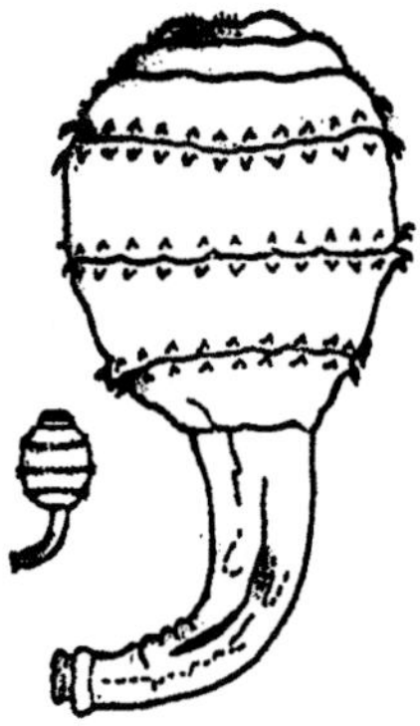

FIG. 106.
Larva of *Dermatobia.*
(After Blanchard.)

Genus **Dermatobia** Brauer. The flies which produce subdermal myiasis are *Cordylobia anthropophaga*, *Hypoderma bovis*, *H. lineatum*, *Dermatobia hominis* and other species of *Dermatobia.*

The larvæ of *Dermatobia* are parasitic in the skin of animals and occasionally in man in Central and South America and neighbouring islands. The eggs are never laid on the skin of the host but the female fly has the characteristic habit of making use of a mosquito or other blood-sucking insect for the transportation of the eggs to its host.

D. hominis L. (**D. cyaniventris**). It is as big as an ordinary Calliphora and bears some superficial resemblance to the blue-bottle fly. The head is mainly yellow. Wings are hyaline and the scutum not metallic; thorax is striated; abdomen metallic blue but lustreless. In living specimens the eyes are brick red. The arista is minutely feathery on the dorsal side. The first posterior marginal cell of the wing open. The legs are yellowish. The larva resembles *Hypoderma* larva.

D. hominis, the Tropical American warble-fly, a parasite of cattle, monkeys, pigs, and dogs in addition to man in Central and South America from Mexico to Argentina, is a frequent cause of subcutaneous myiasis in human beings. The larva is a bottle-shaped maggot surrounded by circlets of chitinous spines and when adult, often measuring about 1 inch in length. The effect of infestation is a sensation of sharp itching, irritating pricking and the oozing of blood and serum from the wounds. An individual generally harbours one larva at a time and it is generally found in the arm and leg, though occasionally the face may be attacked.

For the dissemination of *Dermatobia*, house-flies, *Stomoxys*, *Anthomyia* and *Synthesomyia* flies and mosquitoes like *Janthinosoma* and other Culicines are frequently chosen by the female *Dermatobia*. The egg-laying female fly is attracted to horses and cattle.

She pounces upon these mosquitoes and flies as they approach to within a centimeter or so of her, catching them in mid-air. Then, usually while hovering, she holds them head forward and back up against her thorax and plants the eggs, about 30 or more in number, rapidly on the abdomen of the captive. After this she releases the carrier. During the time the mosquito feeds, the eggs come in contact with the warm surface of the animal; they hatch and the maggots proceed to enter the host. The most common point of entry is the puncture made by the mosquito. The same use is made of other biting flies and in particular of the stable-fly. According to the observations made by Dunn (1930) in Panama, the larva leaves the body between 46 to 55 days later and the adult fly emerges after a pupal period of 22 to 24 days passed on the ground. No area of the body whether clothed or not, from the eyelids to the middle of the back, is apparently exempt. In children head and the neck seem to be affected more often than other regions. One fatal case of parasitism in a child resulting in invasion of the brain by the larva has been recorded by Dunn (1934). As the adult flies live in woodland regions especially in the margins of woods which border grazing land, people inhabiting these regions are attacked and become infested by the larvæ. These bots do an enormous amount of damage in South America to the hides of range cattle.

Family **Anthomyidae.** Members of this family resemble house-flies and are characterised by the 4th longitudinal vein running outwards in such a way as to leave the first posterior cell widely open. The maggots are generally parasitic either on plants or on insects.

Fannia canicularis L., also called the "lesser house-fly", and occurs in temperate climate, frequent dwelling houses though it is less attracted to food than is *Musca.* The larvae breed in decaying and fermenting vegetable and animal matters and also in animal excrements. Larvae can be readily recognised from the spine-like processes on the thoracic segments.

F. scalaris Fab. In size it is slightly larger than *F. canicularis.* The habits of this species are somewhat similar to those of *F. canicularis* but it prefers excrementous matter as a nidus for the eggs and is very commonly found breeding in human excrement. It is very common in European countries and in North America. On account of its most common breeding habits, it may be called the "latrine fly." Although *F. canicularis* and *F. scalaris* bear general resemblance to each other, the two can easily be distinguished by the mid-tibiae which in *F. scalaris* bear a distinct tubercle which is absent in *F. canicularis.* The basal segments of the abdomen of *F. canicularis* are particularly transluscent and the abdomen of *F. scalaris* is black. The larva closely resembles *F. canicularis*; the former is, however, provided with lateral processes on all the segments of the body, which are feathered.

Several cases of intestinal and urinary myiasis have been reported in which the larvae of the above two species have been thought to be responsible. The validity of such findings has, however, been doubted.

One species *Muscina stabulans* Fall., has been reported to cause intestinal myiasis in man. The female fly may enter houses and may oviposit on food. A number of cases of intestinal myiasis have been recorded by Portchinsky (1913) from Eastern Europe. This fly is known throughout the temperate zone. It breeds in decomposing animal waste matter and sometimes in human faeces. Larvae of *Muscina stabulans* feed on the larvae of other flies.

FIG. 107. Larva of *Fannia scalaris.* (After Hewitt).

Blood-sucking Muscidae

Genus **Stomoxys** Geof. (Stable-fly). It closely resembles house-flies and is found everywhere where house-flies occur. It can, however, be quickly distinguished from the latter by its stiff proboscis. It is a medium-sized fly of a greyish colour. In most of the species the sexes can be differentiated by the approximation of the eyes; in males the eyes are closely placed whereas in females the eyes are widely divergent. The arista is plumose only on its dorsal surface. The proboscis greatly differs from the common *Musca* type. It encloses the sucking tube. Palps are slender and much smaller than the proboscis. The abdomen is differently marked in different species. They commonly attack domestic animals such as pigs, dogs, cats, sheep and goats but

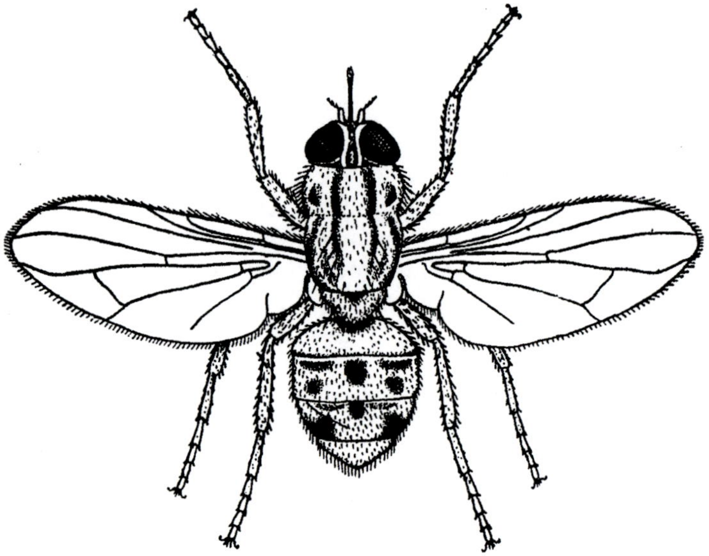

Fig. 108.
Stomoxys calcitrans.

horses and cattle are preferred. In the resting position the proboscis projects out like a bayonet. It is black, shiny and stiff. The labium is strongly chitinised and the labial gutter in which lie the hypopharynx and the labrum-epipharynx is deep. These latter structures are united posteriorly. The labellæ are greatly reduced and the pseudotracheæ are totally absent. On the inner wall of the labella is located the cutting and scratching apparatus. These consist of about 5 teeth on each side and an interdental armature which is of a membranous character. The maxillary palps are less than half the size of proboscis. Eggs are laid preferably in horse dung. The larva is smaller than that of the house-fly, the posterior spiracle presenting striking differences in form and in the position of the slits. Both males and females suck blood. In the tropics under the most favourable conditions the whole cycle may be completed in 12 days. A female may deposit between blood meals even 122 eggs and throughout life 632 (Bishopp, 1939).

S. calcitrans L : is a very common species and is readily distinguished from others by the spotting on the abdominal terga. The presence of 6 round dark spots is a characteristic feature of this species. It is more closely connected with the horse than with the cattle. Although it is a true blood-sucker, yet it can be fed in captivity on syrup, or the liquid from decomposing vegetable matter or dung.

The frons in both sexes is wide, but the females are invariably the larger and of a lighter colour.

Adult *S. calcitrans* will often bite man while working in stables. It will even attack through thin clothing. The bite is painful but the pain soon passes off.

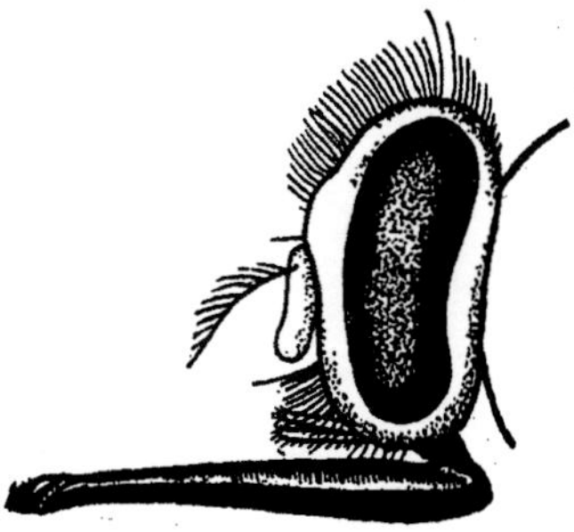

FIG. 109.
Head of *Stomoxys calcitrans*.

It has been proved that *Trypanosoma evansi* may be carried in a mechanical manner by certain biting flies, especially Tabanus. There is no evidence to show that any cyclic development of the organism takes place within the bodies of such insects. For successful transmission of surra by *Stomoxys* flies, the interval between the feeding on the infected and on the healthy animal must not be more than 10 minutes (Moutia, 1928). However, both Mitzmain (1912) and Kelser (1927) found *S. calcitrans* an inefficient transmitter of surra. *G. palpalis* is incapable of transmitting this disease (Duke, 1912). In addition to *T. evansi*, *T. soudanense*, which attacks horses and camels in North Africa, is spread by *Tabanus* and not by *S. calcitrans*. In the United States the disease is spread through the agency of *Tabanus* flies also. In deserts where *Tabanus* is absent, surra yet occurs. This is due to other biting flies, e.g., *Lyperosia*, being responsible for the transmission but seldom Stomoxys.

S. calcitrans is the intermediate host of *Habronema microstoma* Sch., a parasitic nematode which lives in the adult stage in the stomach of the horse. It also carries

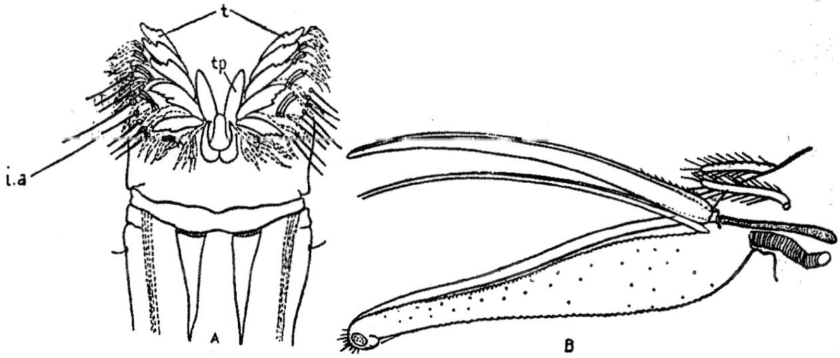

FIG. 110.
Mouth parts of *Stomoxys*. (after Newstead).
A. Labella of *Stomoxys*.
i.a., interdental armature ; t, teeth ; t.p., tooth plate.
B. Labium of Stomoxys in the gutter of which labrum-epipharynx and hypopharynx are lodged.

Setaria cervi to domestic animals and *Hymenolepis carioca* to fowls. *H. microstoma* reaches the final stage in the pupa of the stable fly. In the adult fly the larva lies in a free state in the body cavity and passes forward into the proboscis. The larval worm escapes into the wound at the time the fly bites and develops in the alimentary canal of horses. *Setaria cervi* occurs in cattle and various species of antelopes and deer. There is a strong suspicion that *Stephanofilaria dedoesi* is transmitted by blood-sucking arthropods, particularly *S. calcitrans*. The parasite is responsible for causing ulceration on the neck of cattle.

For laboratory use *S. calcitrans* are bred in horse-dung.

Genus **Glossina** Wied. The tsetse-flies or *Glossina* which are restricted to Africa, are the most important group of blood-sucking flies affecting man. They are the

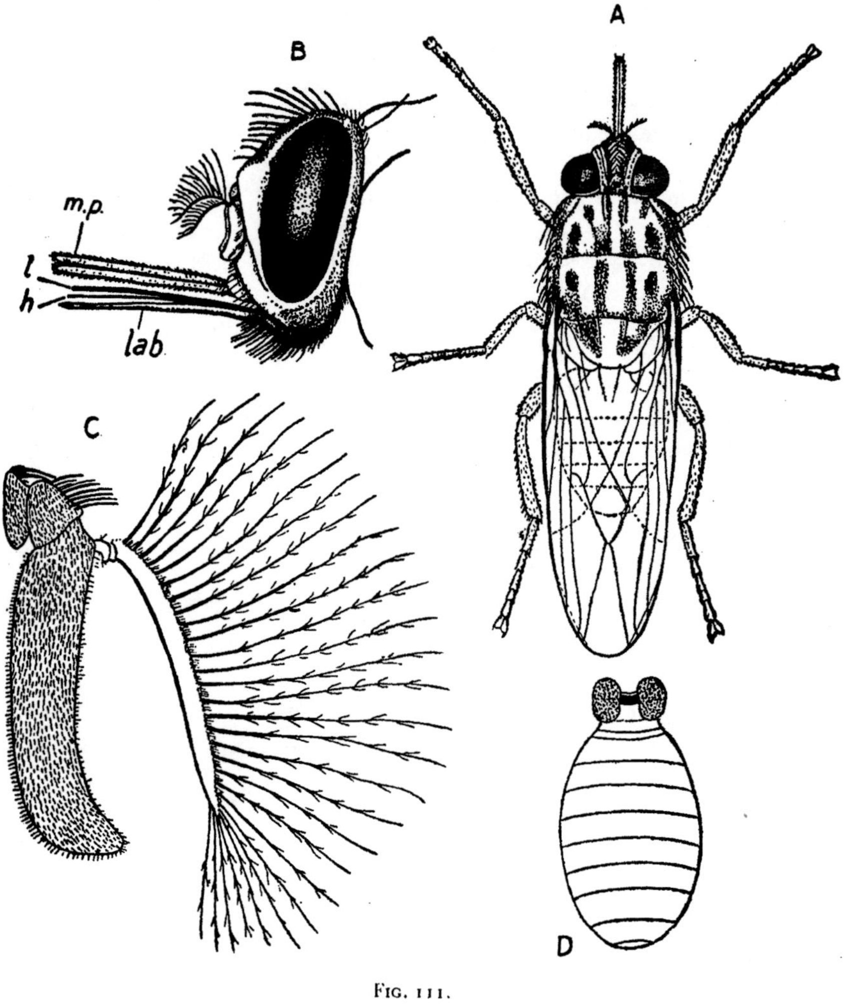

FIG. 111.

A. Resting attitude of *Glossina*. (After Austen).
B. Head of *Glossina*.
m.p., maxillary palp ; lab, labium ; l, labrum-epipharynx ; h, hypopharynx.
C. Antenna of *Glossina*. (After Austen).
D. Pupa of *Glossina*. (After Castellani and Chalmers). (Placed anterior end down.)

intermediate hosts of a fatal trypanosomal disease, the sleeping sickness in Africa, and also transmit *Trypanosoma brucei* which is the causative agent of Nagana in domestic animals.

They are readily distinguished from other flies by their characteristic antennæ, wing venation and the proboscis. The method of reproduction is unique. They are in reality larviparous, mature third stage larvæ being passed which readily pupate after their discharge from the uterus of the parent fly. The eyes are separated in both sexes. The antenna consists of 3 segments of which the third is greatly enlarged. It has three olfactory pits on the 2nd and 3rd segments and numerous sensillæ and spines. The lower anterior end of the third segment projects forwards. The arista bears on its dorsal surface a large number of long, curved and branching hairs. The proboscis or the labium is rigid and is nonretractile. It is considerably enlarged into a bulbous expansion at its proximal end where it is attached to the oral sclerite. The pharynx runs along the ventral wall of the latter. At the distal end of the labium are the much reduced labellæ. The proboscis is composed of the same parts as that of other blood-sucking Muscoidea. The sucking tube is composed of the dorsally placed labrum-epipharynx and the hypopharynx which lies ventrally. The common salivary duct passes through the entire length of the hypopharynx. Both epipharynx and hypopharynx are accomodated in the labial gutter. The labellæ are extreme distal parts of the proboscis and are fused ventrally. There are three rasps in each labellum and they bear rows of very sharp teeth. At the anterior end of each rasp there are four strongly sclerotised prestomal teeth. The cutting and scratching apparatus consists mainly of the prestomal teeth and rasps. Maxillary palps are long, even longer than the labium, well developed and covered over with spines and hairs. Normally the palps cannot be distinguished from the proboscis. When at rest the wings cross each other over the abdomen like the blades of scissors. The venation of the wing is in many respects different from *Musca* type. The veins are concentrated in the anterior part of the wing. The fourth vein ends at the costa at a point much anterior to the tip of the wing thus greatly narrowing the first posterior cell distally.

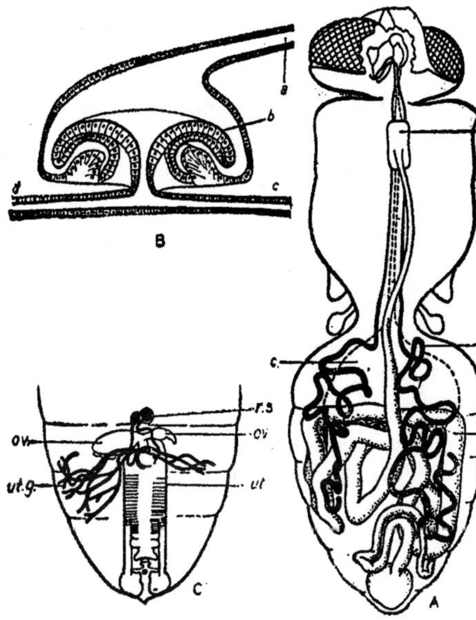

FIG. 112.

A. Alimentary canal of *Glossina*. (After Minchin).

c., crop; mg., midgut; p.v., proventriculus; s.g., salivary gland.

B. Diagram of proventriculus of *Glossina* as seen in sagittal section. (After Wigglesworth).

a., midgut; b., invaginated portion of midgut; c., duct of crop; d., oesophagus.

C. Generative organs of female *Glossina*. (After Minchin).

ov., ovarioles; r.s., receptaculum seminis; ut., uterus; ut.g., uterine glands.

The alimentary canal is of the usual Muscoid type. The crop contains air-bubbles. The salivary glands are two very long and very much coiled tubes. The posterior end of the gland extends as far as the sixth abdominal segment. The ducts of each side pass ventrally behind the pharynx. An anticoagulant function of the saliva was first suggested by Stuhlmann (1907), and its presence was demonstrated by Yorke and Macfie (1924) and subsequently by Lester and Lloyd (1928). The proventriculus is as well developed and has the same function as in other Muscidæ. The peritrophic membrane gradually grows downwards from the proventriculus. It forms a uniform membranous tube and extends as a chitinous sheath far into the midgut being interposed between the lumen of the gut and the epithelial lining. This is intact anteriorly but not so as it extends backwards. The early development of the trypanosomes of *Glossina* takes place in the posterior part of the midgut between the peritrophic membrane and the stomach wall. As they migrate forward they are found outside in the lumen of the gut. The gut of the tsetse-fly contains certain elongated strips of greatly hypertrophied cells, in whose cytoplasm occur certain rod-shaped bacteria-like organisms called "bacterioids." These are carried from one generation to another, and are also found in many other insects. Their exact function is at present unknown.

The hypopygium of the male is a large oval swelling which is anteriorly bisected by the anal slit. The reproductive system is different from the house-fly type. The ovaries are two in number and each ovary contains a single ovariole. The two ducts of the ovary open into a medially placed common oviduct. This is continuous with the uterus; the spermathecæ are two in number and their ducts also open into the uterus. The accessory glands are two in number and each consists of branched tubular glands. They lie on the uterus, and the single duct of the two glands opens into the uterus. They secrete a milky white fluid which is thought to nourish the growing larva.

Life history and habits. The tsetses are confined to Africa, where they have a very wide distribution in the tropical and subtropical zones in South Africa. Within this wide area these flies are not found uniformly everywhere but are restricted to "belts" or patches of forest, bush, usually on the margins of water courses, rivers and lakes and seldom far from water of any kind. These "fly belts" range from the Gambia in the west to Somaliland in the east, extending south from this line to as far as the northern portion of Zululand. Different species are distributed in different zones, the distribution of any particular species in a tract depending on several factors, e.g., presence of the particular type of wild animals on which it is principally dependant for its blood meal, physical characters, existence of its breeding facilities and the proper jungle cover. As a rule the tsetses live in hot, damp places either in the borders of rivers or lakes in bush or forest. The tsetses are not as a rule far away from shades and except for the purpose of procuring blood meal, they seldom move out from their habitations in bush or forest.

The activity of the fly is restricted to the day time and is greatest when the sun grows hot. In the morning and on cold days it is slow in its movements. Its flight is swift and it darts at the object from a distance being guided more by sight than by smell. The tsetse will travel on men and animals across strips of cleared country. It is not unusual to find a large number of males of *G. morsitans* resting on the grass beside native or game paths away from bush and awaiting the passing of men and animals on whom there may be females.

Both sexes are blood suckers; some prefer big animals to bite, others prefer reptiles and birds. Sometimes they are found attacking snails and Lepidopterous larvæ. The bite may be felt as a sharp prick but it is painless. At the time of feeding the palps stick out while the proboscis is lowered at right angles and introduced into the skin. The entire length of the proboscis is inserted in the flesh. During feeding it excretes a drop or two of dark fæces and sometimes some clear fluid. The weight of blood ingested by male *G. morsitans* is 0.034 g. (140 per cent of body wieght), 0.37 g. (160 per cent of the body weight) by female; and by *G. tachinoides* 0.03 g. (220 per cent of body weight) in male and 0.028 g. (140 per cent of body weight) in female flies. A female *G. palpalis* can imbibe 0.044 g., i.e., 1.6 times the body weight of blood and a male, 0.027 g., i.e., 1.3 times. The tsetse can and do bite through cotton and woollen clothing and sometimes even through a thicker texture. Tsetse-flies, like many biting Diptera, prefer dark surfaces and they can rarely fly higher than 6 to 8 feet from the ground. The female fly does not lay eggs but deposits mature larvæ. The larva is adapted to an intrauterine life, the uterus containing only one larva at a time. The larva on extrusion appears as a yellowish slow-moving creature which crawls and burrows by means of peristaltic movements and longitudinal contractions and possibly aided by its antennæ and maxillary appendages. The female fly deposits the larva in the bush in damp soil and is not guided by any selective instinct to choose loose ground easy for the offspring to burrow. In case the ground is unsuitable, the larva crawls for a little distance until it finds shelter to protect itself from enemies. It pupates within an hour of its deposition. Larvæ are produced quite regularly at an interval of 10 days. In the laboratory the female is not a prolific breeder as the number of progeny does not seem to exceed a dozen during her whole life time. Abortions are common in captivity by flies obtaining too little food. The larva moults twice before it is discharged. The newly-born larva is covered with a slimy secretion which protects it against insects, particularly ants. It also keeps the cuticle from being injured when the larva burrows into the soil.

The larva is yellowish white in colour, segmented and bears two well marked round protuberances posteriorly. These are separated by a depression in which the spiracles lie.

The colour of the pupa is dark brown. It is of large size about 7 mm. long and 3 mm. wide. The last segment is deep black and the posterior protuberances in the larva persist in the pupa. The pupal period lasts about 6 weeks. The minimum pupal period for *G. palpalis* and *G. tachinoides*, bred in the laboratory during the rainy season, has been recorded as 19 and the maximum, 46 days. It is not difficult to rear tsetse flies in captivity (Mellanby and Mellanby, 1937).

The adult tsetse-fly depends for its life and its activities on a high atmospheric humidity. Dry air is unfavourable to the continued life of the insect. The adult *Glossina* can be kept alive in the laboratory for several months. Males have a shorter length of life than female flies. Kleine and Fischer (1912) kept a female *palpalis* alive for 227 days, i.e., 7 months.

Among a large number of species of *Glossina*, those known to transmit sleeping sickness number only four, though the work of Bouet and Roubaud (1910) showed that in laboratory experiments all the local species of tsetse were able to transmit any given local trypanosome. In nature, on the other hand, one species plays this part to the exclusion, complete or partial, of the rest. Those that are known to carry the trypanosomes of sleeping sickness are : *G. palpalis*, the vector of *T. gambiense* in West Africa, Belgian Congo and Uganda; *G. tachinoides*, which conveys *T. gambiense* in Northern Nigeria; *G. morsitans*, the carrier of *T. rhodesiense* in Rhodesia and Nyasaland and *G. swynnertoni* which transmits a trypanosome of the *T. rhodesiense* type in the Tanganyika Territory. In this case the cyclical mode of transmission, as seen in the case of *G. palpalis* and *T. gambiense*, plays but a minor part, if any. Man, not game, is the reservoir. The parasite is transmitted from man to man without the intervention of the game.

G. morsitans West. It is the most widely distributed species of *Glossina* and next to *palpalis* is the most important species. It occurs in Rhodesia, Nyasaland, Southern Sudan, W. Kenya, Tanganyka, Portuguese East Africa, i.e., in Central and South Africa. In nature this fly feeds on wild games and there is evidence that the fly will disappear when the large game animals are considerably reduced. Experimentally it has been proved that in absence of big games the fly is able to breed on a diet of smaller mammals, birds and reptiles, though reptilian blood is not suitable to this species as a continued diet.

G. morsitans pupae are often found in hollows in trees even at a height of 6 ft. They are commonly found beneath fallen trees, their branches, logs, in accumulations of dead leaves and twigs and even in termite nests and in the burrows of animals.

This fly seeks its host entirely by sight and if the antennae are rendered functionless, the fly is ready to feed. It is greatly attracted to moving game and objects. It will follow moving objects such as motor vehicles, cyclists, pedestrians etc., for long distances, and having fed, it makes for the nearest shady woods and digests its meal. A meal is required every 5 or 6 days, hence the presence of a number of game animals is not essential. Though they like shade, they feed in open grounds. They have a marked preference for forest in contrast to their dislike for open plain and cleared spaces. They are particularly common in forests where there is thick cover and the existence of large game. It is seldom found near rivers and

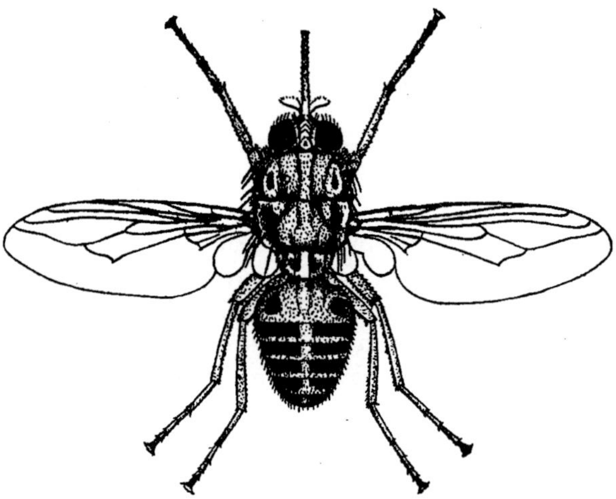

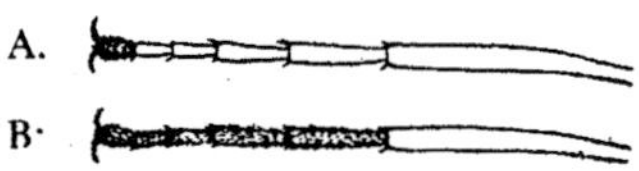

A. Hind tarsi of *G. morsitans*.
B. ,, ,, *G. palpalis*.

FIG. 113.

Glossina palpalis.

is more abundant in the savannah forests with mopani trees. It is very widespread. This species migrates by following herds of game for long distance. Domestic stock do not survive long in the presence of *G. morsitans*. It carries *T. rhodesiense* which is extremely fatal for animals. It also carries other trypanosomes such as *T. pecorum*, *T. nanum*, *T. vivax* and *T. brucei*. The latter causes nagana in horses, cattle, antelopes and other ruminants. The wild *G. morsitans* is generally heavily infected.

G. palpalis R. D. It occurs in West Africa, Belgian Congo, Uganda, West Tanganyka and West Coast. It is essentially a riverine species and is found along the banks of rivers where vegetation is dense. In the savannah country *G. palpalis* and *G. tachinoides* are the predominating species of tsetse. In the forest

G. palpalis has a very wide distribution. There is little evidence that the two tsetses depend on wild game though it is commonly believed that both *palpalis* and *tachinoides* feed on rodents, antelopes, monkeys, birds and reptiles. In two minutes' time *G. palpalis* can take its own weight of blood. This species is particularly restricted to the tall vegetation and fringing shades of rivers, lakes, and small streams, which seldom extends for a width of more than 20 yards along the banks. The highest concentration is found round rivers and lakes. Very dangerous places are roads crossing the river. It very rarely goes out in open sunshine. The fly selects certain breeding places for years at a time. Besides their hunting movements, there are seasonal movements from its permanent haunts at the end of the wet season. Pupae occur almost anywhere in the bush along lake shore and river. The pupae are found exactly in the same situations as those of *G. morsitans*.

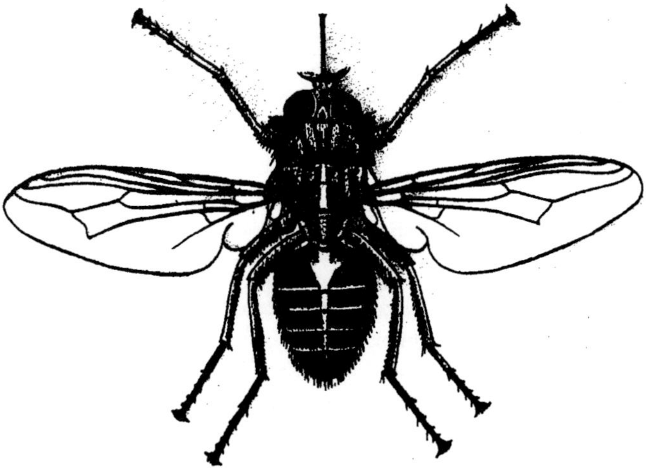

FIG. 114.
Glossina morsitans.

The fly prefers to remain in the shade and near the ground preferably in low shrubs and darts out in the open in full sunlight when in search of a meal. The maximum distance to which *G. palpalis* will venture into an area with no shade may be 150 yds. The distance this fly can ordinarily cover in open countries has a bearing on the width of clearing of shrubs and trees for the protection of the inhabitants. It is very active at 70 per cent humidity and at 45 per cent the fly disappears.

G. tachinoides West. It occurs in Northern Nigeria, Cameroon, Gambia and West Coast. It has the same breeding habits as *G. palpalis* and is found in or near bushes. The trypanosome infection rate in *G. tachinoides* is as a rule, low whereas in *G. morsitans* in the same locality, the rate is much higher. It appears to stand dry air better than *G. palpalis*. It refuses to feed at high temperatures. *G. tachinoides* takes blood from reptiles, bats as well as from ungulates. Flight experiments indicate that the greatest distance covered by a single female *G. tachinoides* is 4 miles and it is undoubtedly found further from its breeding places than *G. palpalis*.

G. palpalis will bite anywhere on the body but *tachinoides* generally attack low down, e.g., ankles, feet etc. While the favourite hunting grounds of *palpalis* are those where there are high trees and shrubs, *tachinoides* prefers high trees and no shrubs.

G. swynnertoni Aust. It occurs in Tanganyka Territory. For all practical purposes its habits are the same as of *G. morsitans*. This species dislikes dense bush and long grass and it may travel more than two thirds of a mile to its feeding grounds independent of artificial transport.

G. palpalis group : *G. palpalis*, *G. tachinoides*. Hind tarsi entirely dark or at least all the joints more or less dark. In the female *G. tachinoides* the basal half of the first joint and the extreme bases of the following joints are usually pale.

G. morsitans group. *G. pallidipes*, *G. longipalpis*, *G. morsitans*, *G. longipennis*, and *G. fusca*. Hind tarsi not entirely dark brown; last two joints alone dark; remainder pale and forms a strong contrast with remaining joints.

Glossina palpalis R.-D. It is larger than *G. tachinoides* being about 9 mm. in length. It is the darkest of all tsetse flies. The tarsi of the hind legs are entirely black and the 3rd joint of the antenna dusky brown. The abdomen is blackish brown with a narrow median pale stripe, the pale area in centre of second segment triangular, the apex is directed backwards and is continuous with a fine pale median line on the other terga. The abdomen is not conspicuously banded.

G. morsitans West. It is a relatively small species. This species and its Western representative *G. submorsitans* have the abdomen conspicuously banded with interrupted blackish bands on a yellowish ground. The thorax is grey with dark markings. The last two or three tarsal segments of hind legs are dark; last two segments of tarsus of front and middle legs with sharply defined dark brown or black tips; the third antennal segment without any distinct fringe of hair on front margin.

G. tachinoides West. It belongs to the *palpalis* group. It is a small species and slenderly built. There are dark brown transverse bands on the abdomen and sharply defined pale hind borders to the segments; second abdominal tergum buff with a large dark spot on either side and a very conspicuously square or oblong pale area in the centre of the second segment; all the segments of the tarsus of the hind legs entirely dark brown except at the base and tip of the first three dark segments of the tarsus which may be pale; the third segment of the antenna rounded at the end and has no long hairs along its anterior border.

G. swynnertoni Aust. It is a member of the *morsitans* group. The colour of the tarsi resembles that of *G. morsitans*. It is much darker than *morsitans* and resembles *G. pallidipes* and *G. longipalpis* in having the dark abdominal bands much more extensive than in *morsitans*. These bands unlike those of *pallidipes* and *longipalpis* are sharply defined and squarely truncate on their inner edges, leaving a narrow paler stripe on the 3rd to the 5th tergum. The markings on the tergum are much more distinct than those of *G. morsitans*. It is found in parts of Tanganyika Territory.

Disease relationship. The world owes to Bruce for the proper conception of the nature of nagana and its transmission by tsetse-flies. He showed in 1895 that the disease was not due to a poison injected by the tsetse-fly but to an actual parasite, a trypanosome, which the fly got from an infected animal and re-injected into healthy ones. He supposed this was a simple mechanical inoculation. In 1909 Kleine showed that a development of the trypanosome takes place in the fly analogous to that of the malarial parasite and the filaria in the mosquito and that infection is transferred with the saliva when the tsetse-fly bites. As a result of subsequent investigations carried out by Dutton, Low, Castellani, Christy, Bruce and others, the etiology of sleeping sickness was satisfactorily explained.

In West Africa trypanosomiasis is a fundamental barrier to the progress of man. It produces sleeping sickness which, if untreated, nearly always proves fatal. It has been estimated in the Gold Coast that a sleeping sickness incidence of 13 per cent causes rapid depopulation and that at 50 per cent extermination results. At 6 per cent the population remains stationary while an incidence of 3 per cent permits an increase of population (Davey, 1948).

In addition to man, trypanosomiasis affects nearly all the domestic stock. There is not only the consequent direct economical loss on account of the death of cattle but in addition to it, the presence of the fly indirectly affects agricultural progress and progress in other fields.

In 90 to 95 per cent of flies the ingested trypanosomes die in a few days and in only about 5 per cent do they establish themselves. In starved flies the proportion is greater. The persistence of the trypanosomes after the 5th day is an indication of their establishment in the gut. In the meantime they had another feed of clean blood subsequent to the infective feed and when established in the mid-intestine, they multiply in the posterior part. The early development of the trypanosomes of *Glossina* takes place inside the peritrophic membrane which is formed from the secretion of the glands in the proventriculus and which hangs like a sleeve up to the junction of the mid and hindgut. About the 5th or 6th day the trypanosomes pass below the lower free end of the peritrophic membrane into the space between the latter and the wall of the midgut. After the 10th day these forms become all sizes and shapes. They now increase in enormous numbers. From the 10th to 15th day after the infective blood feed they migrate forward in large numbers to the proventriculus. They are found in this situation from the 20th day onwards. They now penetrate the soft parts of the peritrophic membrane at its junction with the proventricular fold and come to lie within the proventriculus. They now make a forward movement into the buccal cavity. From here they pass through the hypopharynx into the salivary ducts from where the long, thin trypanosomal forms reach the salivary glands. There they undergo transformation into crithidial forms and then into short metacyclic forms, these being the infective forms. New specimens keep on arriving from the hypopharynx. The infection may take place directly from the buccal cavity or from the salivary secretion at the time of biting. The parasites reach the infective stage in about 30 days.

Mechanical transmission of trypanosomes of sleeping sickness cannot be totally ruled out though under ordinary conditions the chances are small. It is, however, thought that when the disease occurs in an epidemic form, the flies transmit the parasites mechanically. This can take place only if the transference of the flies from the infected to the healthy animal is instantaneous, *i.e.*, by interrupted feeding. Mechanical transmission of the disease has been effected when *Stomoxys calcitrans* and a variety of insects including fleas have been employed. On the other hand, transmission experiments with *T. brucei* from dog to dog by the agency of the common flea *C. canis*, and infection of oxen with *T. congolense* by injection of faeces of bovine lice *Haematopinus eurysternus* and *Linognathus vitule* as well as by engorged ticks have been reported as unsuccessful in the mechanical transmission of either *T. brucei* or *T. congolense* (Duke, Meltan and Wallace, 1934).

The method of transmission also varies widely in different species. Though the usual method is by the bite of the insect, faecal contamination or crushing and ingesting the crushed insect are the methods of infection in the case of *T. lewisi* and *T. cruzi*. *T. equiperdum*, on the other hand, is transferred from one horse to another by sexual intercourse without the intervention of any outside agency.

The sites where the various trypanosomes undergo development are as follows : Stomach and salivary glands—*T. brucei*, *T. gambiense*, *T. evansi*, *T. rhodesiense*, *T. equiperdum*. Stomach and labial cavity and hypopharynx of the proboscis—*T. congolense*, *T. simiae*, *T. pecorum*. Labium and hypopharynx—*T. vivax*, *T. uniformae*, *T. caprae*. Alimentary canal and mouth sheath—Trypanosomes of fish, reptilia and amphibia. Alimentary canal and rectum—*T. cruzi*, *T. lewisi*. Transmission through coitus—*T. equiperdum* (dourine).

T. brucei which produces Nagana in Central and South Africa and *T. rhodesiense* are extremely virulent to laboratory and domestic animals, *T. gambiense* being less so. Whether man is susceptible to *T. brucei* is still a matter of doubt. However, epidemic of sleeping sickness in Tanganyka Territory has been traced to *G. swynnertoni* and *T. brucei* was the parasite concerned.

T. equiperdum produces dourine in horses and donkeys and is widely distributed in Africa, Europe, India and North America. It is conveyed by direct inoculation during coitus. It has now been stamped out from India. *T. congolense* occurs particularly in cattle though sheep and goats are occasionally attacked. *T. vivax* affects equines and ruminants; *T. simiae* is extremely virulent to monkeys and domestic stock. *T. pecorum* is the most important trypanosome disease among bovines in Africa. Horses, donkeys, sheep, goats, pigs, dogs are all attacked. It is carried mainly by *G. morsitans*.

Control measures against tsetse-flies and sleeping sickness. These include : (1) Hand Catching. (2) Trapping. (3) Mass Treatment. (4) Clearing. (5) Mass Removal of Population. (6) Destruction of Big Game. (7) Grass Fire. (8) Application of Insecticides.

Hand Catching : Hand catching of the adult fly will generally cause only temporary reduction of the fly population though parts of bush heavily infested can often be cleared of Glossina by hand catching alone. The adult is caught by hand in blocks of bush along the streams after the blocks have been isolated by clearing.

Trapping : In order to produce any appreciable result, a very large number of traps have to be employed and even if the operation is a sustained one, this method is not of much practical use. There are many designs of tsetse traps and some of them give promising results.

Mass Treatment : It has been established that in human trypanosomiasis, as in kala-azar, the incidence of the disease can be considerably reduced and kept at a low level by proper survey and mass treatment alone. It should be borne in mind, however, that by this method, eradication of the disease is not generally possible; only its incidence is lowered.

Clearing : Clearing is of two kinds, "Aggressive" and "Defensive." By *Aggressive Clearing* is meant extirpating the fly by rendering the riverine and other areas unsuitable for its breeding. It may involve the wholesale clearing of bush, forests and undergrowth and to achieve the object training and diversion of the channel of the river or stream may also be necessary.

Defensive Clearing : It really aims at reducing the man-fly contact to negligible proportions. This is achieved by clearing the bush in blocks. It has been carried out with measurable success against *G. tachinoides* and *G. palpalis*.

Mass Removal of Population : This is necessary in places where the incidence of the disease is high in human beings and where clearing of bush is not likely to produce the desired effects on the invasion of the flies in the same areas.

Destruction of Big Game : *G. morsitans* depends on big game for its blood meals and if the big game is destroyed, this fly, being deprived of its natural food supply, will attack man and domestic stock with great ferocity, and its danger to infection with trypanosomes will considerably increase. However, in limited and selected areas wholesale destruction of big game has produced good results which are generally of a temporary character.

Grass Fire : Several investigators have advocated the use of late organised grass fire as a method of controlling tsetse-flies in various parts of Africa. There is a temporary reduction of flies which become concentrated along the shady water courses and other parts of the forest.

Natural enemies : Puparia of *Glossina* are attacked by at least 9 species of parasites. These belong mostly to the order Hymenopteron.

Insecticides : For control of tsetses by chemicals, *e.g.*, DDT and Gammexane, see in Part II.

Dissection of Tsetse-Fly. The technique of dissection of the proboscis and salivary glands has been described by Lloyd and Johnson (1923). The thorax is held steady by means of the left needle and with the twisting movement the head is separated from the thorax by using the right needle. As the head is separated, the salivary glands remain attached to the head. The head is placed on its dorsal surface and pressed by needle No. 2; this stretches the proboscis which is most carefully separated from the head together with the salivary glands. The right needle is thrust into the bulb. By means of the left needle the gutter-shaped labrum is released from the labium and with the help of the right needle the hypopharynx is dislodged.

Genus **Lyperosia** Rond. Lyperosia flies are cattle pests and are distributed all over the globe. They possess biting and piercing mouth organs as in *Stomoxys*. In size they are much smaller than the latter and are generally yellowish grey in colour. The proboscis is very slender, and projects downwards. The maxillary palps are very long. When at rest the position of the wings is scissor-like and they are kept folded over the abdomen. The thorax has some inconspicuous longitudinal markings. Fourth longitudinal vein has a more gentle upward curve than in *Stomoxys*. Eggs are reddish-brown in colour and are deposited in fresh animal dung, generally of the animal on which the adults are parasites.

L. irritans L. Occurs essentially in Europe and America. It is a comparatively large species; palps and legs in *L. irritans* are dark-brown to black. It feeds only on the blood of horned cattle on whose bodies it remains constantly even while resting, hence they are called horn flies.

L. exigua Meij. This fly is widely prevalent in Asia. It is a large species in fact one of the largest in this genus. It is slate grey in colour; thorax with 2 narrow well separated stripes; abdomen with a dark longitudinal median band on the first and second segments. The legs are yellow. It is widely distributed in India. The fly is commonly seen sitting on the shoulders of cattle with its head directed downwards.

L. minuta Bezzi. A small dark grey species which has five or six hairs on the arista. The thorax is brown with two brown lateral stripes and the abdomen is dark grey without any distinct markings. The legs are pale. The first posterior cell is very narrow distally. This fly is present throughout India, and resemble *L. exigua* in its breeding habits.

MYIASIS. Myiasis is the condition where any part of the body of man or animal is invaded by eggs or larvae of Dipterous flies. The larvae may be found either in any of the natural orifices, in wounds on the skin, in the alimentary canal or in the tissues of the body. Patton has placed myiasis-producing flies into three groups : (*a*) obligatory, (*b*) facultative, and (*c*) accidental.

Obligatory. Here the larvae will only develop on living tissues of animal and also of man. The following is a list of obligatory myiasis-producing flies.

(*a*) *Oestrus ovis* L.: widely distributed; affect nasal sinuses of sheep and on rare occasions the conjunctiva of shephards. (*b*) *Hypoderma bovis* De G.: the common warble-fly of cattle in Europe. (*c*) *H. lineatum* Vill.: the warble-fly of cattle in America and Asia. (*d*) *H. crossi* Patt.: the true warble-fly of goats and sometimes of cattle in the Punjab in India. (*e*) *Dermatobia hominis* L.: normally a subcutaneous parasite of cattle in Central and South America ; sometimes attacks man in the same manner. (*f*) *Auchmeromyia luteola* F.: restricted to Africa and its larvae are considered to be specific human parasites as they live on the blood of man. (*g*) *Cochliomyia hominivorax* Coq.: widely distributed in America and West Indies; it is the typical screw-worm fly of the New World. It attacks generally one of the natural orifices of the body of man. It causes external myiasis in animals. The infestation in animals especially in young ones may occasionally prove fatal. (*h*) *Cordylobia anthropophaga* Grun.: the larvae form subcutaneous tumours in domestic animals and also in man in Africa. (*i*) *Chrysomyia bezziana* Vill.: the screw-worm fly of Africa and India; in man its larvae are found in one of the natural orifices of the body. In animals they are found especially on scrotum after castration. (*j*) The larvae of *Gasterophilus nasalis* L., *G. haemorrhoidalis* L. and *G. intestinalis* De G., are found in the alimentary canal of horses.

Facultative. (*a*) *Sarcophaga* species : on external wounds of man and animals all over the globe. (*b*) *Wohlfahrtia magnifaca* Schin: extermal myiasis in animals in Southern Russia. (*c*) *Lucilia sericata* Meij. and *L. caesar* L., are the blow-flies of sheep in Europe and Australia. (*d*) *Chrysomyia rufifacies* Macq. and (*e*) *Pollenia stygia* F., also attack sheep in Australia. (*f*) *Phormia regina* Mg., is known as the black fly which is a well-known pest of sheep in many parts of Canada and the United States. (*g*) *Chrysomyia* (*Cochliomyia*) *macellaria* Fab., cause external myiasis in domestic animals in Mexico.

Accidental. Dipterous larvae may occasionally be noticed in human stools. They are generally ingested with the food and are passed in a dead condition. Anong them larvae of *Eristalis* may at times cause unpleasant intestinal symptoms especially in children and lunatics. *Apiochaeta scalaris* Mg., is believed capable of passing its life cycle in the human intestine, living larvae being passed from time to time. It is extremely doubtful if Muscoid larvae can survive in the intestinal canal of man and pass unchanged in the faeces as the experimental works of Causey (1938), and Strickland and Roy (1940) clearly indicate. Regarding living larvae of *Apiochaeta scalaris* passed in the human stool, it may be pointed out that such cases, that have come to our notice, occurred during the most favourable time, i.e., summer, when the developmental process was very rapid. The larvae were all of one size, immature and were seldom intimately mixed with the stool.

The following is a list of myiasis-producing flies in man in India recorded by Strickland and Roy (1941). Eye : *Anthomyidae* Sp.; Nose: *Chrysomyia bezziana* Vill., *Sarcophaga* Sp.; Ear and dental sockets: *C. bezziana* Vill; Mastoid region: *Sarcophaga dux* Thoms.; Skin: *S. ruficornis* Fb.; Intestinal: *S. ceylonensis* Park., *S. ruficornis* Fb., *S. dux* Thoms., *S. craggi* Park., *Drosophila* Sp., *Apiochaeta scalaris* Mg.

The following explanation will help to understand the significance of the finding of Muscoid larvae in the urine of man. The urinary bladder as well as the urine, whatever their conditions may be, are inhospitable environments for the sustenance and growth of any Dipterous larva. The passage of a foreign body like the larva of a fly through the urinary canal cannot pass unnoticed by the patient. Instead of first stage larvae being found, the larvae generally show to have undergone considerable development at the time they are detected in the urine.

Treatment of Myiasis. Protection of the wound with gauze will prevent flies from ovipositing on the wound. Where there is any foul-smelling discharge from the nostril, mouth, etc., adequate treatment should be carried out. Where infestation has actully taken place, the parts should be painted with a thick layer of castor oil with the idea of suffocating the larvæ and forcing them to appear on the surface when hand picking with forceps should be resorted to. In our opinion the myiasis-producing larvæ in India are harmless and cause no injury to living tissues. Their beneficial effects in removing the necrotic tissue should be borne in mind though their presence in the wound is repugnant to every human being.

DDT and pyrethrum are effective and they have been dealt with in Part II.

Maggot Therapy. Although it was known that maggot infested wounds heal quickly, it was not till Baer (1931) demonstrated the value of blow-fly larvae in the treatment of osteomyelitis and of other infected wounds, that maggot therapy was established as a useful surgical procedure. When placed in infected wounds, maggots decrease the bacterial population and remove necrotic tissue. Wounds treated with maggots become sufficiently alkaline to turn red litmus blue (Baer, 1931). The liquifaction of the necrotic tissue is thought to be effected by a trypsin-like protease found in both the digestive tract and excreta of the larva, and by collagenase in the excreta (Hobson, 1931). Larvae of *Lucilia cuprina* and *Chrysomyia rufifacies* are capable of liquifying and digesting protein without the intervention of bacteria (Mackerras and Freney, 1933). Aqueous solution of excreta of larvae of *L. sericata* and possibly of other species has been shown by Simmons (1935) to contain potent bactericidal properties.

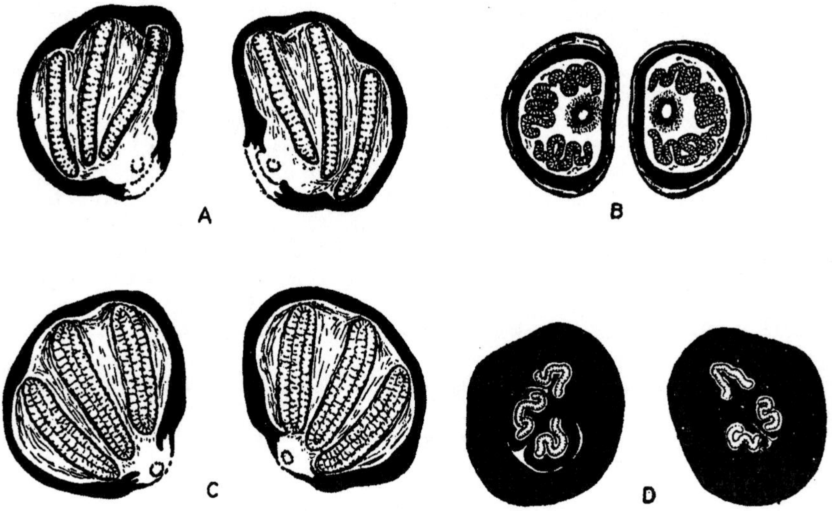

FIG. 115.

A. Posterior spiracles of *Sarcophaga ruficornis*.
B. „ „ *Musca vicina*.
C. „ „ *Chrysomyia megacephala*.
D. „ „ *Stomoxys calcitrans*.

Thus certain species whose maggots attack only dead tissues can be advantageously employed in the treatment of surgical conditions such as cellulitis, osteomyelitis etc. For this purpose the maggots should be bred out under very clean conditions and released on the wound as soon after their emergence from the eggs as possible. In India larvae of *Chrysomyia megacephala* can be used. Other flies which have been employed in maggot therapy elsewhere are *Lucilia sericata*, *L. caesar* and *Phormia regina*. They all feed on dead tissue and breed in animal matter.

Rearing Larvae From Cases of Myiasis. Inspite of the presence of a large number of larvae, it may take some time to extract a few as the larvae are very active and may burrow deeply. The larvae may be recovered by exposing the affected part to the vapour of chloroform or better by covering it with melted paraffin. The latter method is adopted with the purpose of partially asphyxiating the larvae, when they will readily appear on the surface. One by one they will have to be picked up with forceps. A few may be killed in hot water and preserved in 80 per cent spirit and the rest reared to the adult stage. For this purpose meat is used as the pabulum. It must be flamed or just placed in boiling water to ensure freedom from other larvae. The meat is now placed on a thick layer of sand in a deep vessel or a tin and the larvae are dropped on the meat. The opening of the vessel should be covered with two or more layers of muslin to prevent other flies from depositing their eggs or larvae on the meat. Pupation takes place in the sand. The pupae may be placed in test-tubes plugged with cotton wool till the flies hatch out. On no account should bred-out flies be killed within 24 hours after emergence. In the case of the specific myiasis-producing Diptera, it is useless trying to rear the immature larvae on meat. Only third stage larvae should be allowed to pupate in sand.

Identification of Larvae From Cases of Myiasis. Specific identification from examination of larvae alone is by no means easy. An attempt may be made to identify the genus from (1) the posterior spiracles, (2) intersegmental spines, (3) anterior spiracles, and (4) the pharyngeal ridges. The character of the cephalo-pharyngeal skeleton may also be of considerable help in the identification.

Diptera—Pupipara

All the known species are included in three families, *Hippoboscidae*, found on mammals and birds, *Nycteribiidae*, and *Streblidae*, on bats. Streblids resemble Hippoboscids but the fifth tarsal segment is much enlarged and wing veins have spines or bristles on them.

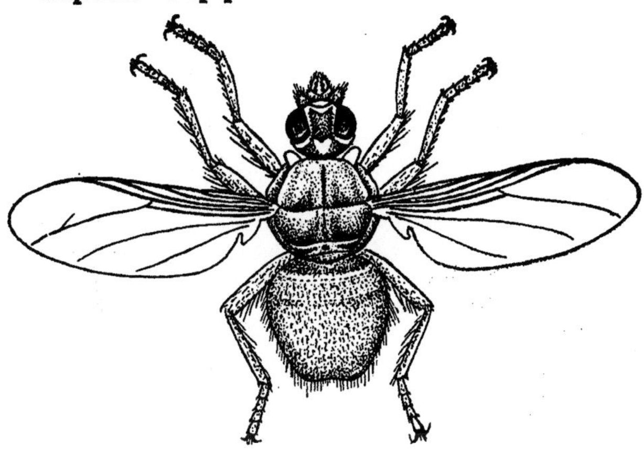

FIG. 116.
Lynchia maura, female.

The family Hippoboscidæ includes the three important genera which are of some interest to us: (1) *Hippobosca*, (2) *Melophagus*, and (3) *Pseudolynchia*.

Genus **Hippobosca** L. These are all degenerate Diptera which have adapted themselves to more or less continuous existence on the bodies of birds or mammals. They are sometimes called Pupipara on account of their peculiar mode of reproduction. They give birth to third stage or mature larvæ which soon after their extrusion are

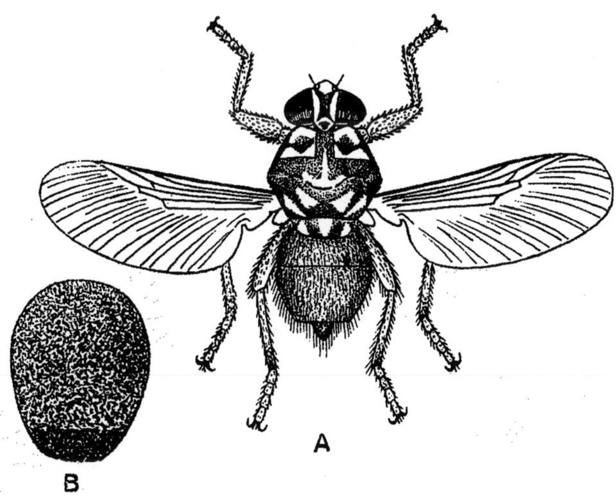

FIG. 117.
Hippobosca maculata. A, adult; B, pupa.

changed into pupæ. But it must be remembered that this method of reproduction is not confined to this group alone. These flies all possess a shining hard cuticle which is also ornamented in many species. The head is impacted in the thorax. The antennæ are greatly reduced. The segmentation of the abdomen may or may not be present. Wings may be present or absent. The legs are stout. The proboscis consists of the same parts as the proboscis of a blood-sucking fly with which there is a close similarity. In place of the psedotracheal teeth in Muscid flies there are prestomal teeth which are 10 in number. Each tooth terminates in two denticles (Jobling, 1926).

Hippobosca maculata Leach., is a common species found especially on horses, and less frequently on dogs, camels, cattle, etc. In this species the head does not impinge on the thorax; wings are always present; the squama and antesquama of wing fringed with short, white pubescence; femora thickly covered with long hairs; tibiae with scanty pubescence scattered over the surface. The male penis protrudes on pressing gently on the abdomen. A horse may even harbour several hundred flies. The larva is of white colour but it soon becomes shining black which is an indication that the larva has pupated. It has been stated that the pupae occur in cowdung, and horsedung. It is perhaps more correct to say that the flies deposit their larvae on the ground, and sometimes perhaps in holes of trees. Pupae are also found in fodder troughs, in cracks in the ground, etc. This species is believed to be capable of transmitting trypanosomiasis mechanically like other biting flies.

H. capensis Olf., is found on dogs and *H. camelina* Leach., on camels. *H. equina* L., is restricted to horses. It is smaller and more slender than *H. maculata.* The wings are brown and eyes black. The markings on the dorsum and thorax are characteristic.

Genus **Melophagus** Latr. **Melophagus ovinus.** L., the sheep tick-fly, also known as ked of the sheep, is especially injurious to young lambs, to which it migrates in large numbers when the older sheep in a flock are sheared. The fly does not possess any wing nor any haltere and does not leave its animal host. The full grown maggot which is roundish and yellowish white is extruded by the fly, which sticks to the wool being glued to it by a gelatinous material. In this situation the maggot changes to the pupa. The larvae or puparia are born one at a time and generally 3 to 4 are produced during the life time of the fly. The pupal stage lasts from 5 to 20 days. The fly lives 3 to 4 months.

Genus **Pseudolynchia** Beq. **Pesudolynchia maura** Bigot., are ectoparasites of the pigeon, and are the vectors of *Haemoproteus columbae.* This fly is viviparous and is cosmopolitan in its distribution. The pupae are deposited in birds' nests and are affected by kerosene-pyrethrum mixture. The pupal life is nearly a month. In this species the wings are of a peculiarly pointed form, though they are rounded at the tip.

REFERENCES (Muscoidea)

Aubertin, D. and **Buxton, P. A.** (1934). *Ann. Trop. Med.* & *Parasit.*, 28 : 245. **Baer, W. S.** (1931). *J. Bone and Joint Surgery*, 13 : 438. **Bishopp, F. C.** (1939). *Farmer's Bull. U.S. Dept. Agric.*, No. 1097. **Bishopp, F. C., Lake, Brundrett** and **Wells** (1926). *Ibid*, No. 1369. **Blacklock, D. B.** and **Gordon, R. M.** (1927). *Ann. Trop. Med.* & *Parasit.*, 21 : 181. **Blacklock, D. B.** and **Fine** (1930). *Ibid*, 24 : 5. **Blacklock, D. B.** and **Thompson, M. G.** (1923). *Ibid*, 17 : 443. **Borgstrom, F. A.** (1938). *Amer. J. Trop. Med.*, 18 : 395. **Bouet, G.** and **Roubaud, E.** (1910). *Ann. Inst. Past.*, 24 : 658. **Bouet, G.** (1910). *Bull. Soc. Path. Exot.*, 3 : 599, 722. **Bruce, N.** (1949). *Pests*, June. **Bruce, W. N.** and **Decker, G. C.** (1950). *Soap*, 26 : 122, 145. **Causey, O. R.** (1938). *Amer. J. Hyg.*, 28 : 481. **Chow, C. Y.** (1940). *Chinese Med. Jl.*, 57 : 145. **Cushing, E. C.** and **Patton, W. S.** (1933). *Ann. Trop. Med.* & *Parasit.*, 27 : 539. **Davey, T. H.** (1948). Trypanosomiasis in Br. West Africa. His Majesty's Stationery Office, London. **Duke, H. L.** (1912). *Proc. Roy. Soc.* Sec. B., 85 : 563. **Duke, H. L., Mettam** and **Wallace** (1934). *Trans. Roy. Soc. Trop. Med.* & *Hyg.*, 28 : 77. **Dunn, L. H.** (1930). *Psyche*, 37 : 327. **Fay, R. W.** (1949). *J. Nat. Mal. Soc.*, 8 : 137. **Fraenkel, G.** (1935). *Proc. Roy. Soc.* Sec. B., 118 : 1. **Ghosh, S. M.** (1949). *Ind. Med. Gaz.*, 85 : 96. **Graham-Smith, G. S.** (1911). *Jl. Hyg.*, 11 : 390. **Graham-Smith, G. S.** (1930). *Parasitology*, 22 : 47. **Guillaume, A.** (1950). *C. R. Acad. Agric.*, Fr., 36 : 728. **Hackman, R. H.** (1947). *Abs. in Rev. Appl. Ent.*, B., 36 : 103. **Hamilton, C. S. P.** (1939). *Ind. Med. Gaz.*, 74 : 210. **Herms, W. B.** (1925). *J. Parasit.*, 11 : 149. **Herms, W. B.** (1926). *J.*

Econ. Ent., 19 : 692. **Hutson, J. C.** (1928). *Rep. Insect. Pests Ceylon*, 1927. Dept. Agriculture. **Jobling, B.** (1926). *Parasitology*, 18 : 319. **Johnson, W. B.** and **Lloyd, Ll.** (1923). *Bull. Ent. Res.*, 13 : 373. **Keilin, D.** (1912). *C. R. Acad. Sci.*, 155 : 1548. **Keilin, D.** (1915). *Bull. Soc. Fr. et Belg.*, 49. **Kelser, R. A.** (1927). *Philip. J. Sc.*, sect. B., 34 : 115. **Kleine** and **Fisher** (1912). *Zeitschr. f. Hyg. u. Infection Krankheit*, 73 : 252. **Knipling, E. F.** (1942). *J. Econ. Entom.*, 35 : 63. **Kumm, H. W., Turner** and **Peat** (1935). *Amer. J. Trop. Med.*, 15 : 209. **Lake, E. W.** and **Smith, C. L.** (1939). *J. econ. Ent.*, 32 : 339. **Lawson, G. B.** (1929). *Southern Med. Jl.*, 22 : 1003. **Lester, H. M. O.** and **Lloyd, Ll.** (1928). *Bull. Ent. Res.*, 19 : 39. **Mackerras, M. J.** and **Freney, M. R.** (1933). *J. Exp. Biol.*, 10 : 237. **Madjinicolaou, J.** (1948). *Archieves of Hygiene*, 6 : 7. **Mellanby, K.** (1944). *Parasitology*, 35 : 197. **Mellanby, H.** and **Mellanby, K.** (1937). *Proc. Roy. Ent. Soc.*, Lond. (A), 12 : 1. **Mitzmain, M. B.** (1912). *Philip. J. Sc.*, B., 7 : 475. **Moutia, A.** (1928). *Bull. Ent. Res.*, 19 : 211. **Pavlovskii, E. N.** and **Glezer, B. M.** (1932). Abs. in *Rev. Appl. Ent.*, B., 23 : 51. **Perry, A. S.** and **Hoskins, W. H.** (1950). *Science*, 111 : 600. **Portchinsky, I. A.** (1913). Abs. in *Rev. Appl. Ent.*, B, 1 : 108. **Roy, D. N.** (1928). *Ind. Med. Gaz.*, 63 : 369. **Roy, D. N.** (1937). *Parasitology*, 29 : 150. **Roy, D. N.** and **Siddons, L. B.** (1939). *Rec. Ind. Mus.*, 41 : 223. **Roy, D. N.** and **Siddons, L. B.** (1939). *Parasitology*, 31 : 442. **Salem, H. H.** (1935). *J. Egyptian Med. Assoc.*, 18 : 238. **Simmons, S. W.** (1935). *Bull. Ent. Res.*, 26 : 559. **Smart, J.** (1939). *Parasitology*, 31 : 130. **Sternberg, J.** (1950). *J. Econ. Entom.*, 43, No. 2. **Strickland, C.** and **Roy, D. N.** (1940). *Ind. J. Med. Res.*, 28 : 593. **Strickland,** and **Roy, D. N.** (1941). *Ibid*, 29 : 863. **Stuhlmann, F.** (1907). *Arbeit a.d. Kaiserl. Gesundh.*, 26 : 301. **Syddiq, M. M.** (1938). *Ind. Med. Gaz.*, 73 : 17. **Trager, W.** (1930). *J. Parasit.*, 25 : 233.

CHAPTER X

ORDER SIPHONAPTERA

(Fleas)

Fleas belong to the order Siphonaptera which comprises a large number of insects; they are distributed all over the world. Some of them are of particular concern to us as they act as intermediate hosts of the parasite of bubonic plague. They are found on many warm blooded animals and they are parasitic only during the time they feed on the host. Although they are regarded as free parasites, there is one family, Sarcopsyllidae, in which the females after impregnation attach themselves permanently to their hosts. Besides the name Siphonaptera, the name *Aphaniptera* has been commonly applied to fleas.

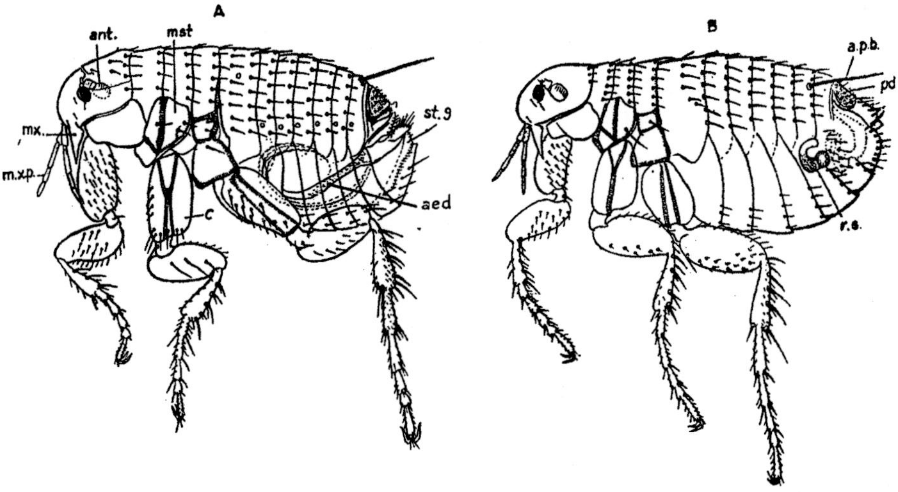

FIG. 118.

A. *Xenopsylla cheopis*, male. B. *Penopsylla cheopis*, female.
aed, aedoeagus ; ant., antenna ; apb., antepygidial bristle ; mst., mesosternite ; mx., maxilla ; mxp., maxillary palp ; pd., pygidium ; rs., receptaculum seminis ; st. 9, 9th sternite.

The chief characters of this order are : (1) they are small insects; (2) the body is laterally compressed so as to enable them to move freely in the hair of the host; (3) the cuticle is strongly chitinised; (4) the adult is wingless, hence it is incapable of flying; (5) both sexes lead parasitic lives on warm-blooded animals including birds; (6) the mouth parts are adapted for piercing and sucking blood which constitutes the only food of both sexes; and (7) metamorphosis is complete, the larva being caterpillar-like and the pupa enclosed in a cocoon.

The body is heavily chitinised and the thorax is attached to the head without the interposition of a neck.

The head is roughly conical in shape and is divided by an antennal groove into an anterior and a posterior part, known as the front and the occiput respectively. The front may be provided with a slightly raised projection, the frontal tubercle. There are rows of bristles on the frons which are usually three in number, prefrontal, frontal and ocular. Some of these rows may be absent. The lateral lower portion of the front beneath the eye is the genu or the cheek which is wide and in some species there is a row of strongly chitinised teeth or comb arranged in a symmetrical manner along its free edge. The eyes may be present, absent or rudimentary. The antenna at rest lies in an antennal groove which is situated above and behind the eyes. The antenna consists of three parts, the terminal one being club-shaped and may be completely or incompletely divided into a number of smaller segments. There are some bristles on the head which are of taxonomic importance.

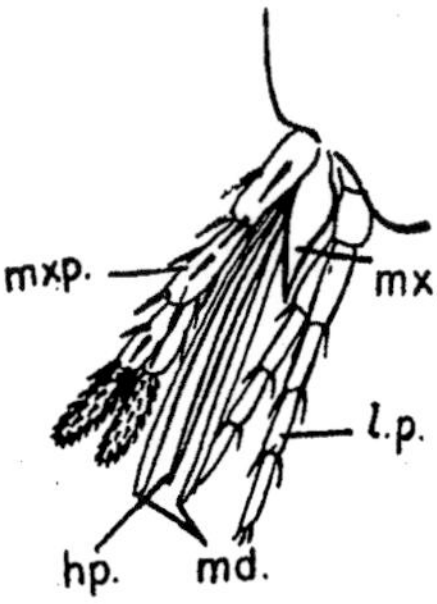

FIG. 119.
Mouth parts of *Xenopsylla astia.* hp, hypopharynx; lp, labial palp; md, mandible; mx, maxilla; mxp, maxillary palp.

The mouth parts are conspicuous structures and project from the head downwards. They consist of (1) a pair of maxillae, (2) a pair of maxillary palps, (3) labium, (4) a pair of labial palps, (5) a pair of mandibles, (6) the fused labrum-epipharynx, and (7) a rudimentary hypopharynx which is difficult to see. The maxillae are two triangular plates with an acute angle at the free end. They are attached to the lower end of the head. The maxillary palp projects like an antenna from the lower front of the head and is a four-segmented structure. The labium is small and insignificant, and to it is attached the labial palp which is also segmented. This is grooved on the inner surface and the two fitting together form a labial beak. The mandibles are long, needle-like, and finely serrated. The labrum-epipharynx is also needle-like, barbed and pointed at the end.

The actual cutting organs are the finely serrated mandibles and possibly the apex of the triangular maxillae. The sucking tube is formed by the apposition of the mandibles and the labrum-epipharynx.

The thorax is composed of the usual three segments. They are as a rule quite distinct but are greatly reduced in *Echidnophaga* and *Tunga*. Each thoracic segment is made up of a dorsal plate, the tergite, and a ventral plate, the sternite. In some species the pronotum has a row of chitinised teeth or comb along the posterior border. The mesosternite is divided by a ventral chitinous thickening into an anterior and a posterior portion. The anterior portion is divided by an internal incrassation into an upper part, the episternum and a lower part, the sternum.

The abdomen is really composed of 10 segments but only 9 are clearly discernible, the last three being modified for sexual purposes. The first abdominal segment is represented only by its tergite. The posterior region of the metathorax overlaps the abdomen and replaces the first abdominal sternite. Each of the abdominal tergite bears one or more rows of bristles. The ninth tergum bears a pitted and setose sensory plate known as the pygidium whose function is at present unknown. Near the posterior margin of the seventh tergum there is a pair of prominent stout bristles, one on either side of the middle line, which project over the pygidium; these bristles are known as antepygidial bristles. Sometimes there are more than one pair.

The male flea is characterised by the presence of the aedoeagus or penis which is a complex structure and lies coiled up inside the abdomen. There are two pairs of claspers. The 8th tergite in the male is small. The 9th tergite bears dorsally the pygidium. The 9th sternum lies for the most part inside the abdomen and only a small part projects out from it. In mounted specimens the 9th sternum appears to be covered by the sternum of the 8th segment.

In the female the receptaculum seminis is a striking object and is situated about the middle of the distal end of the abdomen. On pressure it is liable to be displaced. It is arbitrarily divided by a deep constriction into a highly chitinised round portion called the "head," and the rest being called the "tail." Its shape is of considerable significance in the identification of the flea.

There are three pairs of thoracic and seven pairs of abdominal spiracles. The first abdominal spiracle is wanting.

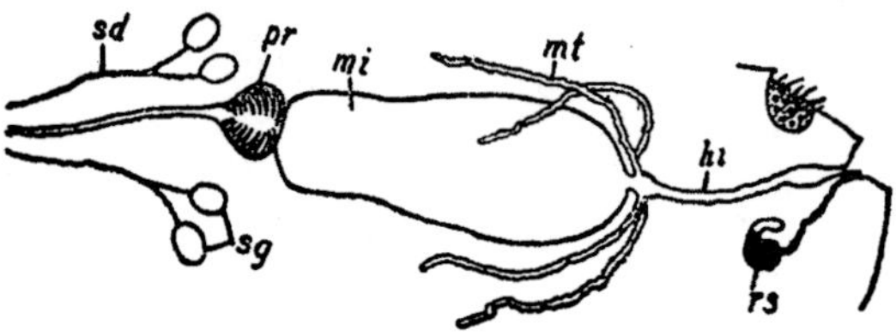

FIG. 120.
Alimentary canal of *X. astia*.
hi, hindgut ;
mi, midgut ;
mt, malpighian tubules ;
pr, proventriculus ;
rs, receptaculum seminis ;
sd, salivary duct ;
sg, salivary glands.

The pharynx is provided with a powerful pump for sucking blood from the wound. The proventriculus has on its inner surface a number of chitinous rods which project into the cavity, and in the ordinary circumstances they act as a valve which prevents the regurgitation of the ingested blood.

The midgut is a short saccular tube which when fully distended occupies almost the whole abdomen. The hind intestine is extremely narrow and short. At the junction of the stomach and intestine are four Malpighian tubules. The rectum contains six papillae. The anus opens externally below the 9th tergum.

There are two pairs of small round salivary glands and the common salivary duct runs in the head and enters the salivary pump.

In the male the reproductive system consists of (1) a pair of testes with the corresponding vas deferens which opens into a single tube; (2) the seminal vesicle from which (3) the ejaculatory duct leaves to enter the penis or aedoeagus; (4) there are in addition two pairs of accessory glands which open into the vas deferens of each side.

The female reproductive organs consist of (*a*) ovaries, each consisting of a number of independent tubes containing egg follicles and converging to form a single tube; (*b*) the common oviduct; (*c*) the vagina; and (*d*) the receptaculum seminis or spermatheca.

Life history. The flea has a complete metamorphosis, the successive stages in its life being egg, larva, pupa and adult. The egg is oval, transparent and on account of its softness it is liable to be damaged on handling. It has a tendency to adhere slightly to the material on which it is laid. Its length is about 0.6 mm., and though small, it is visible to the naked eye. The eggs are deposited in places where the host rests. They are never laid on the body of the host. For egg laying the female usually selects dark and dry places such as rubbish, debris and dust in dwelling houses, under carpets, in granaries etc. Rat-fleas often lay eggs in rat holes, bird fleas in birds' nests, and dog and cat-fleas where these animals live at night. The female generally lays a few eggs at a time and provided it has access to frequent blood meals, it will continue laying eggs throughout its life time. The number of eggs laid by a single individual during its life time is not known but it is thought that a single female can lay at least 200. *X. cheopis* or *X. astia* does not usually lay more than 6 to 8 eggs after a blood meal. Strickland (1914) working with *Nosopsyllus fasciatus* found that the eggs are laid even at a temperature as low as 50°F. During summer in the tropics the eggs do not take more than 4 to 5 days to hatch. In common with other insect eggs a low temperature proportionately prolongs the incubation time of the egg. In unfavourable conditions the eggs may even take 9 days. Ova of *X. cheopis* do not hatch at temperatures below 55°F., and the optimum is attained at 80°F., and saturation deficiency 0.23 inches.

The larva is an apodous hairy caterpillar-like maggot, extremely active, and is of a whitish colour. The body consists of a distinct head, three thoracic and ten

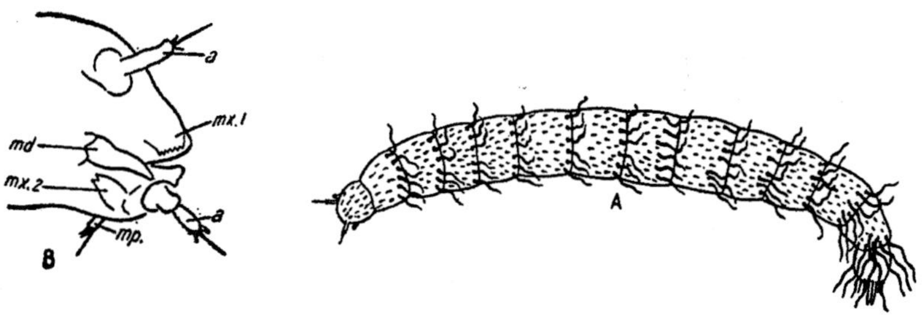

FIG. 121.
A. Larva of *X. astia*.
B. Mouth parts of larva of *X. astia*.
a, antenna ; mx. 1, first maxilla ; mx. 2, second maxilla ; md, mandible ; mp, maxillary palp.

abdominal segments. The head carries a pair of antennae, a pair of toothed mandibles and two pairs of maxillae with jointed palps. In *X. cheopis* there are two pairs of hairs, an anterior row of small hairs and a posterior row of larger ones on each side of the thoracic and abdominal segments. The last abdominal segment is very small and possesses a pair of chitinous processes which are of use to the larva at the time of locomotion. The anus is situated on a raised area of chitin which is sunk into the tenth segment. The larva, though it is devoid of eyes, is yet very sensitive to light, especially in its younger stages. It is rapidly killed when exposed to direct sunlight. The flea larva moults twice and there are three larval instars in the life history of the flea.

Bacot and Ridewood (1914) found that the chief food supply of larvae of most species is the excreta of the parents. It is believed that the larva of many species of fleas requires blood in its food and without it, growth will be retarded. There is always the appearance of a dark reddish line in the alimentary canal of very young larvae suggesting the ingestion of blood by it. In nature the blood is supplied by the adult fleas, which swallow more blood than they can digest, so that some of it is discharged along with faeces. But fleas will seldom deposit their faeces during feeding. Blood, although it may constitute the most important part of the nutritional requirements of the flea larvae, is by itself not sufficient for their normal and successful development.

The flea larvae require an additional food which in nature is supplied by the organic refuse present in the bed of the host of adult fleas. The larva cannot grow on blood alone (Sharif, 1948). It is, however, always possible to breed *Xenopsylla astia* and *X. cheopis* exclusively on dried animal faeces including rat excrement, cow-dung and warehouse and cattleshed sweepings without any blood.

The third stage larva spins a cocoon inside which the larva rests for some time and then changes into pupa.

The flea larva is killed not only by high temperatures but by low humidities at all temperatures. At high temperature in presence of suitable humidity the larva grows more quickly.

The normal length of life of a larva varies with the species and the environments in which it has been placed. The larval life of both *X. cheopis* and *X. astia* is approximately a fortnight.

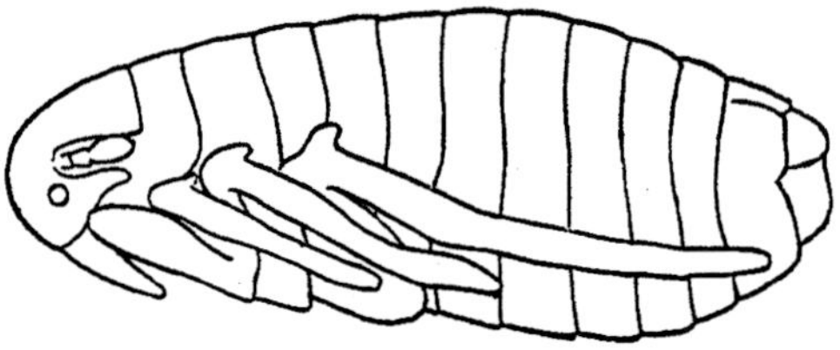

FIG. 122.
Pupa of *Xenopsylla astia*.

The cocoon is made of silken material to which particles of sand, grit etc. adhere. Inside the cocoon the pupa develops. The pupa closely resembles the adult. Sharif (1935) has recorded the presence of wing buds on the mesothorax in the pupa of *Nosopsyllus fasciatus*.

Bhattacharji (L. M.) has noticed that the ratio of the two sexes in *cheopis* is about 1.6 female to 1 male and in *astia* it is 2.6 for every male.

It is only in the adult stage that the flea is parasitic. The animals which harbour fleas are those that provide themselves with some nest or den. The flea is a temporary parasite and uses its host as a source of food and frequently leaves it between feeds. After the death of the host and as soon as the body becomes cold, it seeks a new host. The flea is able to jump about three inches from the ground when gorged, and about 4 inches when starved. It is also capable of walking up a vertical sheet of glass for about 8 inches after which it falls back to the ground. Horizontally it is able to cross a ditch 6 inches wide and can remain afloat on water for some hours.

Fleas of either sex live exclusively on blood which is an absolute necessity for their own nutrition, also for the nutrition of the ovum. If fleas are unable to obtain blood of the proper animal on which they are usually found, they may attack other animals. Under normal circumstances the rat-flea will always be able to find a rat to return to, but where the number of rodents has been practically decreased, which happens during an epizootic of plague among rats, it is probable that the flea may be driven by hunger to attack man. It will not refuse to bite caterpillars. This explains the occurrence of cat and dog-fleas on rats, and rat-fleas biting man. Nevertheless a flea is averse to biting any other animal which is not its natural host and hence even if it bites an unnatural host, the required quantity of blood necessary for egg formation may not be ingested.

Fleas are very sensitive to light and also to air currents. They always hide under dark objects, and when a number of fleas are blown upon, they at once become very agitated.

Oviposition takes place within 24 hours of copulation but copulation, according to Strickland (1914), does not take place in *N. fasciatus* unless the flea has fed on rat's blood.

The duration of life not only of the adult flea but also of its active stages is influenced greatly by the two important factors, temperature and humidity. An increase of humidity and also of temperature has a tendency to retard the growth of all these stages. Under the most favourable conditions Bacot succeeded in keeping fleas alive for 95 days. During summer in Calcutta *astia* and *cheopis* do not usually survive more than 4 weeks even when they have opportunities of feeding on rat's blood regularly.

In unfed fleas high temperatures and low humidities tend to shorten life and, conversely, low temperatures and high humidities with a possible optimum of 90 per cent, tend to prolong it. Webster and Chitre (1930) found under natural conditions in Bombay that during hot weather the majority of fleas died off during two and four days, a few living as long as a week. Fed fleas, on the other hand, live longer. Females live considerably longer than males.

Although according to Bacot and Martin (1924), and Hirst (1926), the duration of life of *X. cheopis* is directly proportional to saturation deficiency, Leeson (1932) pointed out that the life of the flea is determined mainly by temperature and that the effect of humidity, though definite, is relatively slight. The adult flea shows a remarkable resistance to low humidity.

Disease relationship. The following diseases are carried by fleas. (*a*) Rat trypanosomiasis (*T. lewisi*). (*b*) Endemic or murine typhus. (*c*) Cat and dog tapeworms. (*d*) Plague, bubonic in man and epizootic in rats. (*e*) Chiggerosis.

(*a*) **Rat trypanosomiasis.** *T. lewisi* is transmitted by fleas parasitic on the rat. Infection in rats can take place by eating infected fleas but commonly it is effected by way of the rats' mouth, the rat licking from its fur or skin the moist faeces of infected fleas containing the final propagative form of the cycle. The trypanosome is confined during its whole development to the digestive tract. The mouse-flea *Ct. musculi*, the rat-louse *Polyplax spinulosus* and even *Melophagus ovinus* can carry it and are the infective vectors from rat to rat (Mooser *et al.*, 1931; Dove and Shelmire, 1931). The flea becomes infective in 4 to 7 days.

(*b*) **Endemic typhus.** A number of typhus-like diseases whose aetiological backgrounds were then unknown, had been described under different names and endemic typhus has now replaced all those in whch the reservoir of the disease is the rat and the vector, the rat-flea. The louse-borne typhus is a different disease being known as epidemic typhus.

Since the discovery of a strain of typhus derived from rat-fleas collected from wild rats by Dyer *et al.* (1931), and from the brain of a wild cat by Mooser *et al* (1931), the importance of the rat and the oriental rat-flea, *X. cheopis*, in the transmission of murine typhus has become well established. Other fleas which have been found naturally infected are *N. fasciatus*, *N. anisus*, *Ct. felis and Lept. segnis*. The endemic typhus in nature is seldom carried by lice though transmission of murine typhus organisms by the louse has been determined by Castaneda and Zinsser (1930), Lepine and Bilfinger (1934), and Liu and Zia (1941). The disease is transmitted among rats through the skin soiled by the faeces or by the crushed bodies of fleas, the rickettsiae gaining admission through lesions, scratches etc., or through the small wound made by the flea when biting. Alternatively the organisms may enter through the mucous membranes by the digestive route, the fleas being swallowed and faeces licked off the skin and swallowed. The rickettsial bodies are present also in the urine and faeces of infected rats. The oral transmission of murine typhus in humans has been pointed out by Pollard *et. al.* (1946). The infection is not transmitted hereditarily. The rat-louse *Polyplax spinulosus*, and some mites as *Liponyssus bacoti* are the infection vectors from rat to rat.

Typhus of murine origin is much more prevalent in this country than the reported figures indicate. The distribution of the disease in different parts of the country has been given by Kalra (1947). Rao and Kalra (1949) claim to have isolated a strain of murine typhus rickettsia from ticks, *Boophilus annulatus*, collected from bushes in the suburbs of Jubbalpore (C.P.).

(*c*) **Cat and dog tapeworms.** Fleas act as the intermediate hosts of the developmental stages (cysticercoid stage) of certain cat and dog tapeworms, *Dipylidium caninum.* Cats and dogs become infected by the ingestion of infected fleas. It is not uncommon to find similar infection in man due to the accidental ingestion of fleas carrying the cysticercoids. The tapeworm embryos hatch in the ventriculus of the flea larva and immediately begin to migrate through the intestinal epithelium into the body cavity. The parasites increase rapidly in size in the pupa and adult flea.

Hymenolepis diminuta undergoes larval development in several arthropodal hosts including fleas.

(*d*) **Plague.** Human plague in India is essentially a disease contracted from domestic rats through rat-fleas of the species of *Xenopsylla.* Other fleas, e.g., *Ct. canis*, *Ct. felis* etc, may under experimental conditions transmit the infection from rat to rat but as they show a natural disinclination to bite any animal other than their true hosts, they do not play any significant role in the spread of this disease.

Fleas and Plague. The distribution of the three species of *Xenopsylla, astia, cheopis* and *brasiliensis,* has some bearing on the epidemiology of plague (Cragg, 1921). *X. brasiliensis* is confined to Peninsular India and Central Provinces; it is absent in the rest of India. *X. cheopis* and *X. astia* occur all over India, *astia* being more common on the east coast where the climate is tropical throughout the year and the humidity is always very high. While *X. cheopis* is found in places at high altitudes, such as the Simla Hills, *X. astia* does not appear to be well adapted to cool climate and tends to disappear at higher elevations.

X. astia and *X. brasiliensis* as a rule do not generally bite man whereas *cheopis* take feed on man more readily. In laboratory experiments it is noticed that fleas may not accept a feed on man's blood in the presence of a rat or guineapig, but if the fleas are starved for a few days, they are more readily attracted to the human skin. Experimentally *X. astia* does not exhibit any marked susceptibility to plague as is the case with *X. cheopis.*

It has been established by the Indian Plague Commission that in absence of rat-fleas plague cannot spread and that fleas are the real vectors of this disease. According to the oldest theory of plague transmission which held the field for many years, the bacilli contained in the droppings of the flea were inoculated into the animal on which the flea was feeding, either by the puncture of the fleas' proboscis or by the scratchings of the animal itself. Fleas seldom deposit their faeces while feeding. Contamination of the wound inflicted at the time of the bite may be possible but this should not be regarded as the normal mechanism of infection but an exception.

The normal mechanism for the transmission of plague is by regurgitation of the infected blood from the stomach (Bacot and Martin, 1914). The proventriculus of fleas is provided internally with chitinous processes which project into its lumen in such a manner that when the muscles surrounding the proventriculus contract, the entrance from the stomach is effectively blocked, whereby the regurgitation of blood from the stomach is prevented. On account of the rapid multiplication of the bacilli in the proventriculus, its lumen is completely blocked. When the flea becomes hungry, it makes repeated attempts to bite but the blood drawn in the pharynx cannot pass the blocked zone and is regurgitated back into the healthy host. An infected and "blocked" flea is capable of transmitting plague every time it inserts its proboscis.

Infected fleas which do not show signs of blockage never transmit plague. Even some with blockage of the stomach cannot transmit the infection. The danger of *cheopis* as the most dangerous plague-carrying flea is due to early and complete blocking of the proventriculus by plague bacilli. When the temperature is increased, plague may be transmitted as early as five days but ordinarily it takes about ten days (Eskey, 1938).

As different species of fleas exhibit different blocking potentialities even when placed under the same conditions, the blocking of the stomach cannot be regarded as a simple mechanical phenomenon.

The fate of an obstructed flea depends on the atmospheric conditions. In the dry tropical heat the fleas are quickly killed but in the presence of a cool and moist atmosphere, obstructed fleas may remain alive and are capable of conveying the infection for several weeks. It is also possible that after some time the obstruction is resolved.

The chief endemic centres of plague are India, Burma, Java, China, parts of Asiatic Russia, Madagascar, south, central and east Africa. Bubonic plague has spread through the channels of international traffic to nearly all the principal ports of the world. How this has taken place is not known. The disease may be introduced through the medium of infected rats carried in a ship or infected fleas without a host may serve as the reservoir of plague infection.

A rat after it has contracted plague dies within ten days. It has been demonstrated by Long (1935), Machiavello (1943) and George and Webster (1934) that plague is not always a fatal disease in the rat and the plague organisms may persist in the tissues producing no discoverable lesion in the rat though inoculation of guineapigs with emulsions of spleen and liver of such a rat produces septicaemic plague from which it dies. Leger and Baury (1922), Williams and Kemmerer (1923) and Bordas *et al* (1922) have drawn attention to apparently healthy human carriers of plague and rats showing no signs of the disease yet harbouring the bacillus in an attenuated form. Such inapparent and latent plague has also been noticed in ground squirrels in California (Meyer *et al*, 1943, 1947). Such a state of latent plague probably precedes the discovery of mild cases at the beginning of the epidemic. The capture of a rat suffering from resolving type of plague from a vessel at New Orleans and possibly carried from Calcutta has been recorded in the *Pub. Hlth. Rep.*, March 19, 1920, p. 722.

It is worth considering that the shrew which occasionally frequents dwelling houses may harbour *P. pestis* but actually does not die of plague.

The possibility of infected fleas being carried in ships in merchandise such as jute bags has been discussed by Long and Mostajo (1934) and Macchiavello (1947). Hart and Pelican (1934) have experimented with healthy *X. cheopis*, *X. astia*, and *C. canis* and *felis* and found that these would not ordinarily live longer than five days. Though it is the rule for *X. cheopis* to die within a week of the blocking of the gastrointestinal tract which makes it infective, successful transmission of the disease can take place even after four weeks after the infective feed (George and Webster, 1934; Swellengrebel, 1913).

In addition to *X. cheopis* which is the most important rat-flea in the tropics and which is responsible for the carriage of plague from rat to rat and from rat to man, other fleas which are mostly concerned with the spread of plague amongst these rodents are *N. fasciatus*, the European rat-flea ; *N. anisus*, the rat-flea in China, Indonesia, Japan, and Formosa; *N. montanus*, the squirrel-flea in the United States; *Hop. anomalus*, the squirrel-flea in California, and *Stivalius* (*Pygiopsylla*) *ahalae*, the (field) rat-flea in Java. However, *X. cheopis* is probably the only flea which is responsible for the spread of the disease to man.

How the infection is carried over from year to year is not known. It is possible rats and shrews play significant roles in this respect.

The mortality from bubonic plague in the Punjab during the period 1901-1924 is :—Jan.-March, 37353; April-June, 79542; July-Sept., 1076 and Oct.-Dec., 5183. The mortality can be regarded as an expression of the number of infected fleas at different times of the year and which attack man. The number of fleas depends on the cycle of reproduction of fleas especially *X. cheopis*, and perhaps on the number of rats also. In the laboratory, on the other hand, the breeding is intense from April-June and is abruptly reduced from July and may cease altogether in the winter.

Collection of Living Fleas. If for any special purpose, e.g., for breeding, living fleas are required, the muzzle of the rat is caught with ovum forceps and after holding the rat inside a battery jar, the animal is anaesthetised by giving bromethol intraperitoneally in doses of 0.012 c.c. It causes anaesthesia lasting for about an hour. Half the dose will render mice unconscious for about three hours. The drug has the special advantage of a quick action, so that when given to animals lightly anaesthetised by chloroform or ether, consciousness is not regained during the brief period it is needed for the production of its effect. There is also a good margin of safety. The other method is to lightly anaesthetise the rat with chloroform; this will stupefy the fleas. After a short while the fleas will revive.

Flea Index. The flea index represents the population of fleas on a single rat. This is determined by catching rats and collecting the fleas after killing both the rats and fleas. The Specific Flea Index signifies the population of the different species of fleas on a rat and this can be determined after the fleas have been identified. In order to prevent fleas from escaping, rats should be caught alive and the cage with the trapped rat should be placed inside a cloth bag and the two ends of the bag are tied with a string. The bags are placed in the killing chamber in order to kill both rats and fleas.

The danger of handling rats infected with plague should be borne in mind and in plague endemic areas all necessary precautions to guard oneself against being bitten by infected fleas and contracting plague under such circumstances should be taken.

It is held that *X. cheopis* index of 1 per rat is able to start an epidemic. During epidemics there is a significant rise of the *cheopis* index when compared with the pre-epidemic periods.

Indentification of Living Fleas. The flea is lightly chloroformed and then placed on its left side on the stage of the microscope, while a powerful light, preferably sunlight, is directed upon the hinder end of the abdomen. The characteristic shape of the receptaculum seminis can then be made out. Dead fleas are cleared in caustic potash and examined in carbolic acid.

Detection of Plague Bacilli in Fleas. Devignat (1938) advises dropping the fleas in Broquet's medium which is composed of calcium carbonate—2 gm., glycerine—20 c.c., and distilled water—80 c.c. It has been claimed that the plague bacilli will survive for several days in the flea and putrefaction will be prevented at the same time. Another method is to collect a large number of fleas from a single rodent or from several rodents of the same type, macerate them in normal salt solution and use this for bacteriological culture. About 1 c.c. of the emulsion should also be injected subcutaneously in a guineapig.

Classification. Fleas are grouped into a number of families and genera. A detailed classification is, however, not necessary.

The order Siphonaptera is divided into the following three families : (1) Pulicidae or the true fleas; (2) Sarcopsyllidae or chigger fleas; and (3) Ceratopsyllidae or bat-fleas. The last named family is unimportant.

The difference between the two families Pulicidae and Sarcopsyllidae is given below :

Family Pulicidae : (*a*) The head is conical. (*b*) The three thoracic segments are distinct and not foreshortened or telescoped into one another. (*c*) The coxae and femorae of all the legs are equally well developed.

Family Sarcopsyllidae : (*a*) The head is angulated. (*b*) The three thoracic segments are foreshortened. (*c*) The coxae and femorae of the first two pair of legs are very thin.

Family **Pulicidae.** The following represents the important genera of Pulicid fleas found on rats and their broad characters are given.

Eyes	Genal comb	Pronotal comb	Flea
Present	Absent	Absent	*Xenopsylla.*
,,	,,	,,	*Pulex.*
,,	,,	,,	*Hoplopsyllus.*
,,	,,	,,	*Ceratophyllus.*
,,	,,	,,	*Nosopsyllus.*
,,	,,	,,	*Ctenophyllus.*
,,	,,	,,	*Trichopsylla.*
,,	,,	Present	*Pygiopsylla.*
Rudimentary	,,	,,	*Stivalius.*

Present	Present	Present	*Chiastopsylla.*
Absent	,,	,,	*Hystrichopsylla.*
Present	,,	,,	*Ctenocephalides.*
Absent	,,	,,	*Ctenopthalmus.*
Rudimentary	,,	,,	*Leptopsylla.*
Absent	,,	,,	*Dinopsyllus.*
Rudimentary	,,	,,	*Neopsylla.*
Vestigial or absent	Absent	,,	*Ctenopsylla.*

Amongst them *Xenopsylla*, *Pulex*, *Nosopsyllus*, *Stivalius*, *Ctenocephalides* and *Leptopsylla* occur in India.

Genus **Hoplopsyllus** Baker. The pronotal comb is composed of more than 8 spines and not extending on either side to lower margin of pronotum; coxa III with a row of patch of spinelets on the inner surface; segment 5 of each tarsus with 4 pairs of lateral plantar bristles.

It is distinguished from *Nosopsyllus* (*Ceratophyllus*) from the segmentation of the third segment of the antenna and the antepygidial bristles. In the former the club of the antenna is incompletely segmented and only one antepygidial bristle is present on each side. In the latter, the antenna is completely segmented and there are more than one antepygidial bristle on each side.

It occurs on squirrels and rabbits in the United States, Canada and Russia. It is accidentally found on rats. *H. anomalus* Baker is suspected of carrying plague. The comb of the prothorax consists of 8 to 10 spines. It is found on ground squirrels in California.

Genus **Nosopsyllus** Jordan. This genus has been separated from *Ceratophyllus* which is restricted to birds and fowls.

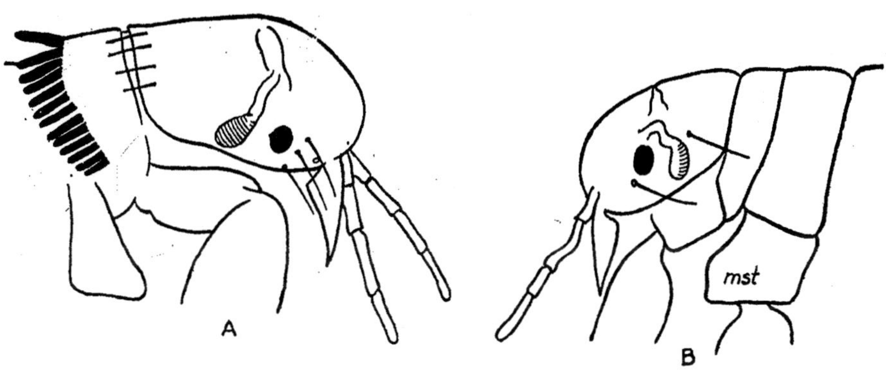

FIG. 123.

A. *Nosopsyllus* (*Ceratophyllus*) *fasciatus*. B. *Pulex irritans*. mst., mesosternum.

Pronotal comb of less than 24 spines on the two sides together; inner surface of coxae *ii* and *iii* with their setae at most on distal half but without a row of spinelets; labial palps scarcely reaching apex of trochanter I; pronotal comb with usual long, heavily pigmented spines; femur I with several setae on outer surface; male sternum VIII vestigial without setae and entirely contained inside sternal plate VII; 2 antepygidial bristles in male and 3 in female; spermatheca with globular head and long cylindrical tail.

The genus *Nosopsyllus* is large and widespread. *N. fasciatus* Bosc., is the common rat-flea of temperate climate and is a dangerous carrier of plague in Europe. The distribution of the fleas of this genus in India is limited to the hills.

Among the *Nosopsyllus* recorded from India, the important ones are *N. fasciatus*, *N. nilgiriensis*, *N. punjabensis*, *N. simla* and *N. tamilanus.*

N. fasciatus Bosc. It differs from others in having the 8th tergite of the male rounded off and does not extend backwardly beyond the last dorsal bristle. It is essentially a palaearctic species with a wide distribution. It is found on rats.

N. nilgiriensis J. and R. It is fairly common in the Nilgiri Hills above 3,000 ft. and is found on rats. The hind femur has 3 to 4 bristles on its inner surface and the hind tibia with 6 to 7 bristles on outer surface, the first bristle being placed near the second dorsal pair; the first hind tarsal segment without any bristle on its outer surface.

N. punjabensis J. and R. The hind coxa on its inner surface has a row of 4 to 7 bristles and a subapical bristle; hind tibia with a row of 8 to 12 bristles on its outer surface. It is restricted to the Punjab and is also found on rats. This species was formerly identified as *N. fasciatus*.

N. simla J. and R. Its distinction from *N. punjabensis* is difficult. It is found in N. W. India, rat being its usual host.

N. tamilanus J. and R. Recorded from Ceylon and S. India. A row of 3 to 4 bristles present on the inner surface of hind femur; on the outer surface of hind tibia there is a row of 6 to 8 bristles; longest apical bristle of segment 2 over-reaching apex of segment III.

N. anisus Roths. This species is confined to Japan, Manchuria, Formosa and China and takes the place of the most common *C. fasciatus* which is essentially the European rat-flea, but has been extensively dispersed in the temperate regions of the world. Thus *N. fasciatus* is found on the rats in Japanese and Australian sea coasts and is widespread in America.

Genus **Stivalius** Roths. **Pygiopsylla** Roths. In both *Stivalius and Pygiopsylla* the frontal tubercle is absent; pronotum with 2 or more rows of bristles; pygidium very strongly convex projecting backwards.

In *Stivalius* the frons is strongly curved; between the eye and anterior row of bristles there are 2 or 3 long bristles placed nearer to the antennal groove than to the ventral margin of the groove; eyes placed far backward.

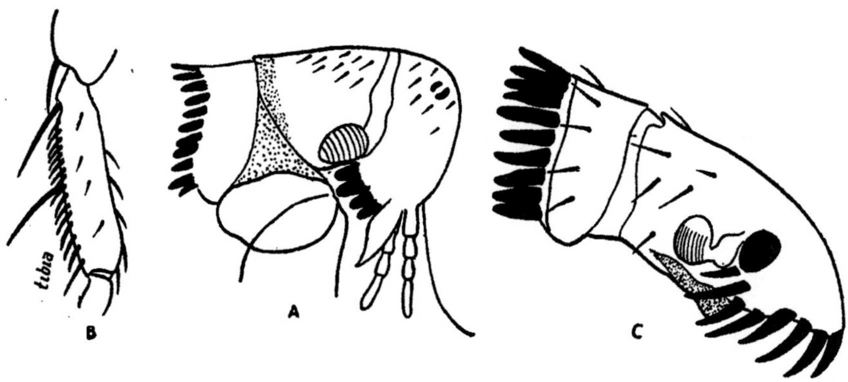

FIG. 124.
A. *Leptopsylla musculi.*
B. Tibia of *Leptopsylla musculi.*
C. *Ctenocephalides felis* var. *orientis.*

It is a large species widely distributed in the tropics and is found in N. Australia, New Guinea, Malaya Archipelago on marsupials and rodents. From the Indian Region 4 species only have been recorded, the commonest being *S. ahalae* Roths., found on rats in S. India and Ceylon, particularly in the hills. In Java *S. cognatus* J. and R. (previously diagnosed as *Pygiopsylla ahalae*, also *S. ahalae*, is a rat-flea and is suspected of playing some part in the transmission of plague.

Pygiopsylla Roths. Frons much less curved; the bristles much further away from antennal groove, the one near the eye below the level of the middle of the eye. *P. hilli* Roths., is commonly found in Australia on rats. In the Malayan Archipelago these fleas are found on marsupials and rodents.

Genus **Ctenocephalides** Stiles and Collins. In this genus the frons and anterior margin are rounded; genal comb is present and extends anteriorly along ventral margin of gena to fronto-genal angle; eye present; pronotal comb is composed of 16 to 18 spines which are heavily pigmented; segment 5 of each tarsus with 4 pairs of lateral plantar bristles.

These are principally the cat and dog-fleas and are distributed all over the globe. They are known to attack rats and sometimes man. In India this genus is represented by *C. canis* Curtis, *C. felis* Bouche and *C. felis* subsp. *orientis* Jordon. These three are very closely related to one another and the subspecies *orientis* is undoubtedly intermediate between *felis* and *canis*. *C. canis* is confined to the hilly regions in India and is very scarce in the plains. In India the subspecies *orientis* appears to be the commonest flea of the genus *Ctenocephalides* (*Ctenocephalus*) and replaces *C. canis*. Its chief host is the dog but it is also found on cat, goat, rat and squirrel. Both cat and dog-fleas are capable of transmitting plague under laboratory conditions.

C. canis is almost cosmopolitan. It is distinguished from *C. felis* in having a much shorter forehead and in having spine I of the head comb very distinctly shorter than the posterior ones, whereas in *C. felis* spine I is nearly as strong. In *felis* form *typica* the female is without a row of short hairs posterior to antennal groove and in form *orientis* there is a row of 1 to 8 short hairs posterior to antennal groove. In the typical form the frons is elongate and pointed anteriorly whereas in the subsp. *orientis* the frons is short and broadly rounded anteriorly.

Genus **Leptoposylla** J. & R. (**Ctenopsylla** Kolenati). Head angulate; pronotum with one row of bristles; labial palp consisting of 5 segments extending much beyond apex of maxilla; genal process narrow and short not reaching further backwards than uppermost spine of genal comb; the genal comb consists of 4 spines.

This genus is represented in India by two species, *L. segnis* Schonh. (*C. musculi* Duge), and *L. himalaica* Roths. They are easily differentiated by the number of spines on the genal comb and on the anterior angle of the head. In *segnis* there are 4 teeth on the genal comb and there are two short spines close to the anterior angle of the head. In *himalaica* the genal comb consists of only 2 spines and there are no spines on the anterior angle of the head.

While *L. segnis* is found principally in the plains of India, *L himalaica* occurs in the hills. Both are mouse-fleas. *L. segnis* is cosmopolitan and may occasionally be found on rats. This flea has been suspected of playing a minor role in the transmission of plague.

Genus **Ctenopthalmus** Kolenati. Eyes vestigial; frontal tubercle large and prominent, pronotum with a comb of 16 spines; genal comb consisting of 3 spines, the first two being much shorter than the third; labial palpi 5-segmented; coxa III without a patch of spinelets on inner surface; 3 antepygidial bristles in both sexes.

They are widely distributed in Europe, Russia, South Africa and America. *Ct. agyrtes* Hell., are found on brown rats, field mice and bank voles in England and also in Western and Central Europe. *Ct. pseudagyrtes* Baker, are found on prairie dogs, ground squirrels in the eastern parts of the United States. They seldom bite man and are chiefly implicated in the transmission of plague from rodent to rodent.

Genus **Ctenopsylla** Wagner. Eyes vestigial or absent; hind legs of tibiae with about 12 short and 3 long bristles; front with anterior margin broadly rounded; only anterior submarginal row of frontal bristles pigmented and spine-like; these are rodent-fleas and are mostly prevalent in Asia.

Genus **Chiastopsylla** Roths. Frontal comb extending at most down to two-third the margin; the number of teeth on pronotal comb is about 12.

It is similar to *Nosopsyllus* except that it has two teeth at the angle of the head behind the maxilla. It is a rare flea and has been described from rat and mice in South Africa.

Genus **Hystrichopsylla** Tasch. Genal comb of at least 5 long slender spines; eyes absent; pronotal comb consisting of many long slender spines which are obliquely arranged; abdominal tergal plates II, III, IV with rows of apical spines; females with two receptaculum seminis. Fleas are of large size and hairy. Occur in America on meadow mice, weasel etc. *H. lalapae* Curtis, are found in Central and North Europe on moles and shrews and are only accidentally found on rats.

Genus **Neopsylla** Wagner. Eyes vestigial or absent; pronotal comb extend down to spiracle; at genal angle two overlapping spines; number of teeth on pronotal comb 18; 5th segment in fore tarsi with 5, and in hind tarsi with 4 lateral bristles; males with two and females with 3 antepygidial bristles. They occur on mice and are common in Asiatic Russia.

Genus **Dinopsyllus.** Spines or comb on gena restricted to anterior margin of anterior groove; a prominent lobe below the genal comb and the ventral margin without spines. These are large hairy species and are restricted to Tropical Africa where they occur on small rodents.

Genus **Trichopsylla** Kolenati. Pronotal comb usually with less than 12 spines. Parasitic on chipmunk, meadow mouse, squirrel in the United States and Canada.

Genus **Ceratophyllus** Curtis. This genus has been split up in recent years into *Nosophyllus*, *Trichopsylla*, *Ceratophyllus* and *Ctenophyllus*.

Ceratophyllus are all bird-fleas and are restricted to the northern temperate zone in Europe. *C. gallinae*

Schr., is found on fowls in Europe and N. America. *C. wickhami* Baker, is the common flea on squirrel, weasel and porcupine in England.

In *Ceratophyllus* the pronotal comb consists of 24 spines or more, inner surface of mid and hind coxae with longish and thin bristles only on apical half; 3 antepygidial bristles in females and head of spermatheca concave above; labial palp not reaching the apex of fore coxae. All are bird-fleas and restricted to the northern temperate zone in Europe.

Genus **Pulex** L. In *Pulex* the sternite of the mesothorax is very narrow, without any internal rod-like incrassation or thickening from the insertion of the coxa upwards. In *Xenopsylla* the sternite of the mesothorax is wider and contains a strong internal chitinous rod besides an oblique one. The internal rod runs from the insertion of the coxa vertically upwards to the upper border of the mesosternite. *Pulex* is essentially a human flea, whereas *Xenopsylla*, a rat-flea.

This genus contains one species only, *P. irritans* L., which is the human flea and is practically cosmopolitan. It also bites other animals including the rat. In the tropical countries of the eastern hemisphere *P. irritans* occurs only in the hills, though it is found in the plains in North, South and Western Africa. In Europe and Central Asia it is essentially a parasite of man and is not found on rats or other animals.

In *P. irritans* the bristle situated in front of the eye is absent and is replaced by one below the eye. The hind coxa bears a number of hairs situated on the inner surface of the posterior portion.

Genus **Xenopsylla** Glink. This genus is represented by a large number of species. It is primarily a rodent-flea and is widely distributed. At least one species

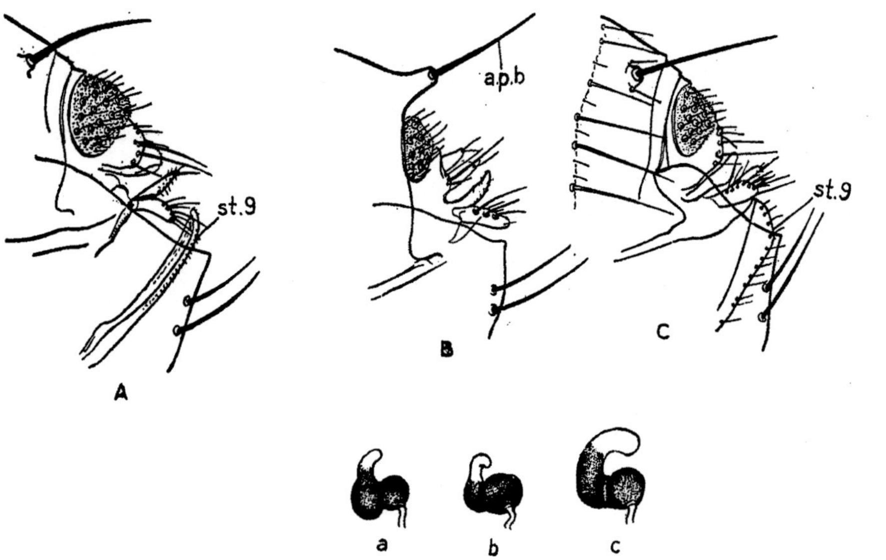

FIG. 125.

A. *Xenopsylla astia*, male. B. *Xenopsylla brasiliensis*, male. C. *Xenopsylla cheopis*, male. a, Receptaculum seminis of X. *astia* ; b, Receptaculum seminis of *X.brasiliensis* ; c, Receptaculum seminis of *X. cheopis*; ap.b, antepygidial bristle ; st. 9, 9th sternite.

X. cheopis, is principally concerned in the transmission of plague. Rothschild in 1911 separated the three species of fleas of the genus *Xenopsylla* found on rats. These are *astia*, *cheopis* and *brasiliensis*. In addition to these, two rare species *sewelli* and *hussaini*, both taken from *Gerbillus indicus*, have been reported from India. Other important species of *Xenopsylla* that occur on rats are *X. eridos* Roths., which is a rodent-flea in Africa and *X. hirsuta* found on striped mouse in Cape Regions.

Xenopsylla cheopis Roths. It is widely distributed in the tropics and is the principal plague flea in India, Ceylon, Africa and Hongkong. It is thought to be originally a native of Tropical Africa. It is very widely prevalent in the plains and also in the hilly regions in India.

In the male *cheopis* the antepygidial bristle is situated on a short pedestal, which is placed at some distance from the apical edge of the 7th segment. The 9th sternite which usually projects but slightly from the interior of the 8th segment, is widened towards the apex, having more or less the shape of a club.

In females the "tail" of the receptaculum seminis is very long and near the constriction it is a little wider than the "head.". The abdominal segments III to VI bear ventrally on each side a row of three or four, rarely five, bristles and the 8th segment has less than 30 bristles, usually 20 to 25.

Xenopsylla brasiliensis Baker. It is found in India and Africa (Natal, Nigeria, Belgian Congo, Kenya, etc,), and has recently been introduced into South America from Africa. Its distribution in India is very restricted.

In the male the antepygidial bristle is placed on a long pedestal which projects beyond the apex of the 7th segment. In the female the "head" of the receptaculum seminis is very much wider than the "tail." The abdominal segments III to VI bear ventrally on each side a row of 4 bristles, and the 8th segment has on the outer surface less than 20 bristles.

Xenopsylla astia Roths. Like *X. cheopis* it is prevalent all over India. Before the separation of the three species by Rothschild it was confused with *cheopis*. Outside India it is found in Burma, Ceylon, Hongkong and Iran. *X. nubicus* Roths., represents *astia* in Africa.

The antepygidial bristle in male *astia* is similar to that of *cheopis*. The 9th sternite is characteristic. This sternite instead of being club-shaped has the appearance of a ribbon viewed from a point on its edge, which is due to the ventral margin being strongly chitinised, whereas the sides and upper margin are very thin and transparent.

In the female the "tail" of the receptaculum seminis is so strongly widened near the constriction that it is here very much wider that the "head;" the "tail" moreover is shorter than in *X. cheopis*. The abdominal segments III to VI bear ventrally a row of 7 to 8 bristles on each side, and segment VIII has more than 30 bristles on the outer surface.

Key to the Identification of the Species of Xenopsylla (Sinton, 1925).

Male

1. With antepygidial bristle on a long pedestal, which projects beyond the apex of the 7th segment. .. .. *brasiliensis.*
2. With antepygidial bristle on a short pedestal, which does not project beyond the apex of the 7th segment. .. .. ..
   - (*a*) IXth sternite club-shaped. .. .. .. *cheopis.*
   - (*b*) IXth sternite ribbon-shaped. .. .. .. *astia.*

Female

1. The "head" of the receptaculum seminis very much wider than the "tail." *brasiliensis.*
2. The "tail" near the constriction is wider than the "head."
   (*a*) with the "tail" near the constriction very much wider than the "head;" the "tail" comparatively short. .. .. *astia.*
   (*b*) with the "tail" near the constriction distinctly wider than the "head;" the tail comparatively long. .. .. .. *cheopis.*

[An account of the Indian species of Siphonaptera has been given by Iyengar, R. (1950)—Thesis for the M. Sc. of the University of Calcutta.

Rearing Fleas. The fleas are kept with a mouse in a glass jar. The mouse is provided with autoclaved oats, a small quantity is added daily, and drinking water. The water is contained in a specimen tube closed with a cork bored to take a small length of glass tubing, the end of the tubing which projects about 1 [illegible] has been previously rounded off in a Bunsen flame, so that the opening is less than half its original diameter. The whole is then suspended by a wire, mouth downwards inside a jar. The jar is covered with a fine muslin cover over which is fitted a perforated zinc cap. When fleas are required, the debris is tipped into one side and the other side is shaded; the adult fleas hop into the shade and are collected with the aid of a suction pump attached to a water tap (Leeson, 1931).

Fleas may also be bred on a rat kept in a cage inside a kerosene tin. Dried rat's or rabbit's faeces and powdered blood may be given as food for larvae. From the time a rat with living fleas is introduced the culture becomes ready within about 4 weeks. In this way *astia* and *cheopis* may be reared separately.

For oviposition by individual fleas small glass jars should be selected. If some rough paper of a dark colour and folded in a longitudinal manner is placed inside, fleas may be induced to lay eggs in the folds.

Dissection of Fleas. On account of the presence of tough chitin on the body, dissection of fleas is difficult. The needles must be strong; one must be blunt pointed for steadying and the other sharpened to have a knife edge for cutting. The type of instrument used in this country by barbers for cutting finger

FIG. 126.
Special knife for dissection of fleas and bed-bugs.

nails has been found to be very useful. This is a rectangular steel rod as wide as the handle on which dissecting needles are mounted and has a sloping, sharpened edge at one end. This is strong and at the same time the edge is sufficiently sharp for chipping off the chitinous cuticle of fleas.

The legs are removed with a knife-edged needle from the attachment of the coxa with the body. Dissection is done in salt solution on a glass slide. The insect is steadied with a needle by pressing the blunt end against the ventral part of the abdomen. This will push the alimentary canal towards the dorsal side. The abdominal sternites are now cut along the ventral margin. In the same way the cuticle along the entire length of the dorsal margin of the abdomen is chipped off. The sternites and tergites of the different abdominal segments are then gradually freed from their tracheal and other attachments and removed with the help of forceps. In this way a large part of the alimentary canal becomes free. Though the salivary glands generally project out from the metathorax, some further dissections may be necessary in order to have a good view of these organs.

The other method is to steady the insect and gradually and carefully push a knife-edged needle between two segments about the middle of the abdomen. The nick is extended and carefully widened. The posterior part of the abdomen is cut transversely and the alimentary canal is exposed by pulling the head anteriorly.

Rats. The essential part in the prevention of plague is the destruction of rats especially those (*R. rattus*) which come into intimate contact with man in his dwelling houses. A reduction of the rat population by 50 to 75 per cent should ensure at least a year's freedom from plague.

In connection with the spread of plague, the two domestic rats, *R. rattus* and *R. norvegicus*, are of primary importance. They have followed in the wake of civilised man. The control of plague is absolutely dependent on the control of these two rats, especially *R. rattus*, which all the time lives in comradeship of man. As a rule rats like cats and dogs have a great homing instinct. They show little or no inclination to leave their haunts. Wholesale migration of rats is occasionally noticed. During epizootics and when poison is used to kill them, there is a tendency on their part to leave the locality and disperse to neighbouring areas. In this way plague may spread from one area to another in a city.

The common rats which are encountered in this country are the following.

Rattus rattus (R. alexandrinus). It is a domestic rat and is a good climber. It generally lives in rooms particularly in tiled houses. The tail is often about one and a quarter times as long as the head and body combined; the eyes are large and prominent; the ears are large; muzzle narrow; tail uniformly dark, feet slender, white but sometimes dark. The colour of the Indian species is greyish brown and the belly yellowish white. It is widely distributed in India.

Rattus norvegicus. The tail is about 90 per cent the length of the head and the body. It is a heavy bodied rat with large, heavy tail which is generally white or distinctly light in the lower half; ears short, rounded and nearly naked; eyes small; the feet are large, heavy and flesh coloured; muzzle broad, heavy and cheeks full; no long piles or bristles on the body, this being the chief characteristic in *Nesokia*; colour light yellowish brown, and silvery grey on the ventral surface. It does not spit or bristle when caged. It is a drain rat and visits houses at night. It is also a burrowing rat and a good swimmer.

Bandicoota indica (Nesokia bandicoota). Greyish brown in colour, light grey on the sides; the tail is equal to the length of the head and the body; colour uniform; rings on the tail well marked; tip of tail covered with hair; ears long and large; muzzle like *R. norvegicus* but bigger in size. It has long piles on its body which stand out when enraged.

Gunomys varius (Nesokia bengalensis). Greyish in colour; the tail about 80 per cent of the length of the head and the body; colour uniformly black, tail almost naked; broad at the base and tapers suddenly; tip not covered with hair; eyes large; ears large, flesh coloured and when pulled over the forehead almost cover the eyes; muzzle short and pig-like. It makes a characteristic grunting noise, erects its bristles and spits when enraged. It is essentially a field rat and nests in burrows.

Crocidura coerulea (musk rat). Greyish in colour; head and body combined is about 4-5 inches; tail very short and pointed; smells strongly. Eats smaller rats, mice and insects. Seldom found inside the dwelling house and is never affected by plague.

The identification of all types of rats in a locality is an important part in the study of epidemiology of plague particularly those from which plague fleas have been recovered. The salient points which will help in the identification of these rats are given below.

Chief Ectoparasites of Rats in India :

Fleas. *X. cheopis, X. astia, X. brasiliensis, N. punjabensis, N. fasciatus, N. simla, Stivalius ahalae, Echidnophaga gallinacea, Pulex irritans, Ct. canis* and *felis, Lept. segnis.*

Mites. *Dermanyssus muris, D. gallinae, Laelaps nuttalli, L. echidninus, Liponyssus bacoti, L. bursa, Notoedres muris, Trombicula deliensis, T. indica, T. oudamansi, T. acuscutellaris, T. bhimtalensis, T. squamosa, T. gliricolans, T. parapalpalis, T. intermedia, Scongastiella lingula, Aschongastia indica, schongastiella bengalensis.*

Lice. *Polyplax spinulosus.*

Ticks. *Hyalomma sp., Boophilus* sp., *Rhip. sanguineus, Haem. leachi.*

Diseases Carried exclusively through Rats: (*a*) Leptospirosis, (*b*) Trichinosis, (*c*) Spirillosis (rat-bite fever), (*d*) Salmonellosis, and (*e*) infection with *Streoptobacillus moniliformis* (*Streptothrix muris*).

Examination of Rat for Plague. The rat is fixed with belly upwards on a board by means of tapes tied round its legs. After reflecting the skin from the jaw to the pubis, cervical, axillary and inguinal lymph glands are examined. It should be remembered that the sex glands in front of the pubis normally contain numerous small bipolar bacilli. In rats dead from plague the glands are swollen, of purple colour and surrounded with gelatinous oedema. There is subcutaneous haemorrhage and congestion. Liver and spleen are enlarged, congested and show signs of haemorrhage and necrosis. Diagnosis is made from the general post-mortem characters, culture from the spleen tissue and inoculation into a guineapig subcutaneously at a shaved spot on the abdomen.

Rats dead from plague seldom show the typical picture given above. Reliance has to be placed on the bacteriological diagnosis only which, however, becomes extremely difficult when putrefaction has set in. In such cases the bone marrow should be bacteriologically examined. The marrow should be wetted with a minute drop of salt solution before it is drawn into a capillary pipette.

Salient points in the Identification of the Common Rats of India

Rat	R. rattus	R. norvegicus	G. varius	B. indica
Body	Slim & slender.	Heavy, rather large.	Large and heavy.	Extremely large.
Muzzle.	long and sharp.	broad, blunt & heavy with cheeks full.	short & pig-like.	greyhound-like.
Length of head & body taken together	6-8 inches.	10 inches.	less than 9 inches.	9-10 inches.
Bristles or long p les on the body.	absent.	absent.	present; fur thin.	very long & black
Tail	125 per cent of the length of the head & body taken together.	80-90%	Very small about 70.1.	90.1.
Number of teats	10 (2+3) × 2	12 (3+3) × 2	16-18 (4+4) × 2	12 (3+3) × 2
Ears	large, almost naked, when drawn forward, they reach the eyes; they also touch each other over the head.	small.	large like *R. rattus*.	long & large
Make sound	a series of squeaks.	squeaks.	grunts; erects its bristles and spits when disturbed.	grunts, erects its bristles.
Palatine foramen.	8.6 mm.	8.6 mm.	7.6 mm.	10.6 mm.

Anti-rat measures. Extermination of rats is always a matter of considerable difficulty on account of their habit of hiding in holes in inaccessible places and their capacity to increase their progeny at an enormous rate. Anti-rat measures in a city should include : (1) Stopping access of food to rats which is not a practicable measure. (2) Destroying rats by (*a*) trapping; (*b*) poisoning; and (*c*) fumigation. These are the most important methods of dealing with rats in the house and in preventing and controlling an epidemic of plague. Any plan of action which aims at destruction of rats must be

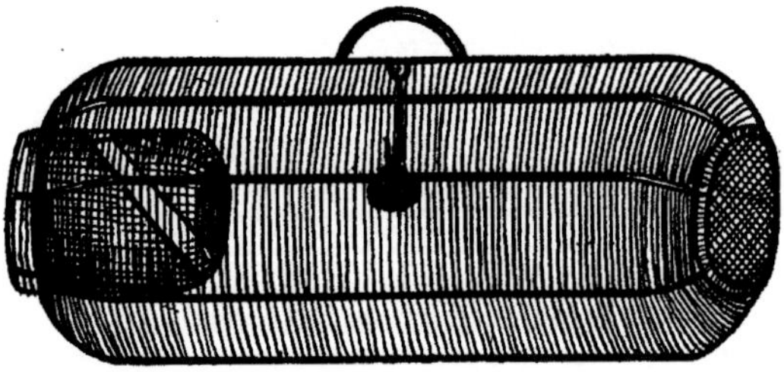

FIG. 127.
'Wonder' variety of rat-trap.

systematic, and well planned. (3) Preventing the accumulation of garbage in and near the house.

The control of grain trade and the proper storage of grain are almost synonymous with efficient plague protective measures. All godowns where food grains are stored should be properly planned both regarding their sites and construction. The godowns must be not only rat proof but also equipped with arrangements for fumigation.

Various types of traps are sold in the market. The "wonder" type which can trap more than one rat at a time is widely used. In dealing with a large colony of rats, traps should never be set unless a very large number, at least 40 to 50 per 1,000 inhabitants are set simultaneously.

The exclusive use of poison for destruction of rats in the control of plague is a more or less temporary expedient. A large-scale anti-rat campaign by poisoning should be conducted in three different stages. During the first stage the locality should be properly surveyed in order to estimate the personnel and the quantities of material required. The second stage should be devoted to prebaiting for at least 4 to 5 days and thereafter poison baiting which is the third stage of the operation. The bait should be frequently changed. The required objective will be difficult to achieve when there is plenty of waste food in the house. Poisons should always be handled carefully and also in the matter of their use extraordinary precautions must be exercised.

The meal is prepared with flour, biscuit, maize, liver, fish, cheese etc. The poisoned meal must be coloured with a dye; Prussian blue, lamp black etc., may be used.

The poisons commonly employed are: (*a*) Arsenic; (*b*) Phosphorus; (*c*) Strychnine; (*d*) Barium carbonate; (*e*) Zinc phosphide; (*f*) Squill; and (*g*) Some newer poisons.

Though phosphorus has a peculiar smell, yet rats and mice readily take it. Dissolve 2 per cent of yellow phosphorus in lard and spread over bread. Strychnine is not a suitable bait for killing rats. Squill in the freshly chopped up condition or in the powder form may be used. In the case of barium carbonate, 1 oz. in 1 pound of the meal will form an effective poisoned bait particularly for use in the household. $1\frac{1}{2}$ gr. of barium carbonate is required to kill a large-sized rat. One pound of barium carbonate in 3 pounds of flour can be made into 4,8000 pills, each pill containing about $1\frac{1}{2}$ grain each. It is only fatal to man when taken in a dose of 60 grains. Rats seldom die in the hole after taking these poisons. Zinc phosphide should always be used with dry baits. In contact with water a poisonous gas is given off. The poison bait should be made with vegetable oil in proportion of 2 lb. of the former in 1 qt. of the latter.

[For details of these poisons and for the newer poisons, see later in Part II.]

Amongst the fumigants hydrocyanic acid gas is widely used. Though the gas is extremely poisonous, the risk is actually not so great as the gas is always handled by well trained and experienced persons. One half to three quarters of an ounce of Potassium or sodium cyanide will suffice for 100 c. ft. of space and the gas should be allowed to act for 4 to 6 hours. A concentration of 0.2 per cent of the gas is necessary to kill rats and also fleas.

Calcium cyanide dust may be blown into the rat hole with a special dusting machine and the moisture of the soil acting on the powder generates hydrocyanic acid gas. The dust generally reaches 15 ft. at the most and is most effective up to 5 ft.

For both cyanide and cyanogas special apparatus has to be used. Imperial Chemical Industries' preparation 'Cymag' is also largely used for gassing rats in rat holes.

Anti-flea measures. For control of fleas by chemicals, see Part II. However, amongst the chemicals, DDT is the one of most commonly used not only for checking the spread of plague but also for controlling murine typhus.

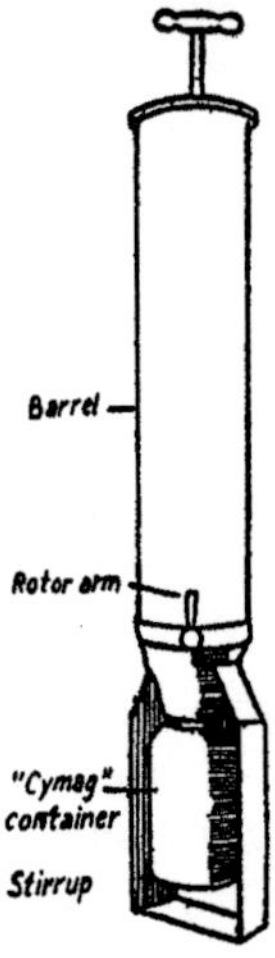

FIG. 128. 'Cymag' blowing foot pump.

It has been demonstrated that while a single application of 10 per cent DDT is generally sufficient to keep the flea index below 1 for about a year; three or four applications of cynogas will be necessary for the same place during the same period. The cost of DDT application is also markedly cheaper.

Sylvatic Plague. Sylvatic plague is a term first proposed by Ricardo Jorge in 1928 to indicate a new type of plague infection which occurs among wild rodents as distinct from the domestc rodents or rats. The epizootics of sylvatic plague occur in places far outside the normal line of communication with man, e.g., steppes and deserts. At least 71 species of rodents have been known to suffer and no less than 50 species of fleas are found to infest these rodents. The epizootics generally spread slowly and after some time they apparently die out to reappear in future. The outbrakes as a rule remain localised amongst a particular type of animals and the epizootic is maintained by a particular species of flea peculiar to these animals. It is presumed that contact of healthy wild domestic rodents suffering from plague is essential for the former to contract the plague infection for which a type of flea which bites both these animals is essential for the transfer of the infection from rat to the wild rodent. The particular flea which infests the wild rodent now serves to spread the infection among these rodents.

The wild rodents which serve as reservoirs of plague are different in different countries. They may include ground squirrels, wood rats, prairie dogs, gerbille, arboreal rodents, marmots, tarabagans, susliks etc.

One distinct feature of sylvatic plague is the absence of any obvious symptoms of infection or gross pathology except presence of lesions in the lymph nodes. The bacilli have a great tendency to affect the lungs much more frequently than what takes place in rats.

The disease is enzootic over a wide area in the Union of South Africa among the rodent population and human cases have occurred sporadically. Two groups of veldt rodents are responsible, the gerbille *Tateroma* and *Sesmodillus*, and the multimammate mouse *Mastomys*. The latter becomes infected from the former, and frequent human dwellings and outbuildings. Man becomes thus infected from the gerbille. The fleas associated with the sylvatic plague in S. Africa are *Xenopsylla brasiliensis*, *X. ellobius*, *Dinopsyllus* and *Chiastopsylla* spp. of fleas.

In North America plague infection in the ground squirrel, *Citellus beecheyi*, forms an important reservoir of this disease and other rodents *e.g.*, chipmunk, marmot, prairie dog etc., are of secondary importance. The principal fleas concerned and found on *C. beecheyi* are *Diamanus montanus* and *Hop. anomalus*.

In South East Russia another member of the gerbille family is the reservoir, and in Manchuria, ground squirrels of the genus *Citellus* are occasionally found infected with plague.

The killing off not only the rodents concerned but their fleas as well is always difficult. Methyl bromide, hydrogen sulphide etc. are used. Methyl bromide is colourless and odourless and extremely poisonous to man. Chloropicrin is ineffective in killing fleas.

Family **Sarcopsyllidae.** In this family the mouth parts are directed forwards. The legs are not nearly so well developed as in Pulicidae on account of their parasitic mode of life.

The labial palp consists of 3 or fewer, very feeble, chitinised segments; the genal edge of the head is in all cases produced downwards into a triangular lobe situated behind the mouth parts. The three segments of the thorax are always shorter than the first abdominal tergite.

Key to the Genera of Sarcopsyllidae

1. Head small and blunted; hind coxa with a patch of small spines; abdominal segments II and III each with a pair of spiracles. Occur chiefly in tropical and subtropical regions of the world. *Echidnophaga.*

2. Head larger and angulated; patch of small spines on the hind coxa absent; abdominal segments II and III of female without spiracles. Restricted to the warmer regions of Asia and Africa and to tropical America. *Tunga.*

Genus **Echidnophaga** Olliff. *E. gallinacea* Westwood: It is an Old World species and has now been introduced into North America and Western Australia. It is widely distributed in the tropics. Its chief host is poultry but it attacks other animals as well. On poultry it is found especially on the naked parts of the head and the neck.

Kelsall (1927) reported an interesting case of sweating blood in a human female. A large number of fleas of this species were extracted from little black specks in the neighbourhood of each nipple of the breast.

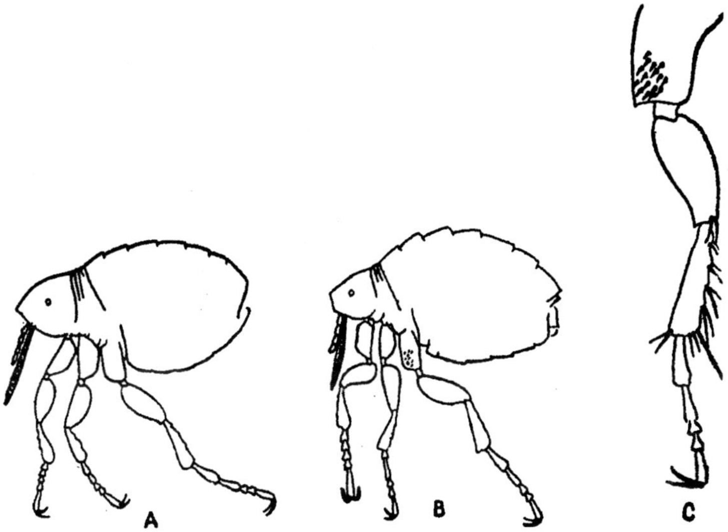

FIG. 129.

A. *Tunga penetrans.* B. *E. gallinaceus.* (After Turkhud) C. Hind leg of *E. gallinaceus.*

This species can be easily trapped by letting a guineapig loose in the infested place and then chloroforming it (Turkhud, 1928). The life history has been worked out by Parman (1923).

Fifth tarsal segment in this species has two ventral apical bristles; hind edge of the head with a lateral lobe.

Genus **Tunga** Jarocki (**Dermatophyllus**). Of the known species of this genus only two are found in the Old World. *T. caecata* Ender., in which the eyes are vestigial, occur in Brazil on rats. In *T. penetrans* L., the eyes are distinct. This is the flea which causes chiggerosis.

T. penetrans Linn. has occasionally been recorded from the western ports of India. This is originally a South American species and has been introduced into other tropical countries. In Africa it has propagated with astonishing rapidity. In East Africa it is very common. Although it might have been introduced into this country, yet it has not been able to gain any foothold. It flourishes best in sandy soil. Its hosts are usually domestic animals from which man becomes infested,

Detailed descriptions of the early stages have been given by Hicks (1930). The flea has only two larval stages. It takes from 3 to 4 weeks for the imago to escape after the egg is laid.

Chiggerosis. Chiggerosis is caused in man and domestic animals by *T. penetrans*. After fertilization, the female attaches itself firmly to the host. It imbeds itself in the skin generally of the foot and less commonly of the hand. The toes and fingers are affected, as also the palm of the hand and the sole of the feet.

In this condition the insect grows from a minute size to that of a pea on account of the distension of the abdomen by the ovaries containing eggs. There is considerable inflammation of the surrounding skin which is raised; the insect lies in a depression in the centre and the terminal segments of the abdomen project outside for the discharge of the eggs. The eggs fall to the ground where further development takes place. After the eggs have been discharged, the female shrivels up and dies.

Treatment is mainly surgical. The insects are first killed with 8 per cent pyrethrum ointment applied over the affected parts for 24 hours and thereafter the dead fleas are removed with forceps. The wearing of boots may be regarded as the only prevention against infestation by *Tunga.*

[For references, see end of Chapter XI.]

CHAPTER XI

ORDERS HYMENOPTERA, COLEOPTERA & LEPIDOPTERA

Order **Hymenoptera.** This order represents ants, bees and wasps, and if it were not for the pathological effects they cause to man, this would not have come within our purview. In certain ways their activities are beneficial to man, e.g. they produce honey and wax. A large number of them parasitise flies in their early stages and act as their natural enemies.

The following constitutes the chief characters of this order. (1) Mouth parts are adapted both for biting as well as for sucking liquid food; the mandibles are well developed. (2) Wings may or may not be present but in typical Hymenoptera there are two pairs of small wings. (3) In the female the abdomen ends in an ovipositor which may be modified to form the sting; the sting generally lies concealed in a sheath. (4) Metamorphosis is complete. Bees, wasps and ants occasionally inflict painful sting to man. These may also attack in swarms. There are records where death of the individual has occurred.

The honey bee is a well known type of the order Hymenoptera. It exhibits the social life among insects in its most fully developed form. It provides us with honey and the beehive is the source of wax. The stinging apparatus consists of three parts : (*a*) the sheath, (*b*) the sharp, barbed darts, and (*c*) the venom sac consisting of three glands, two of which secrete formic acid and the other a fluid which is alkaline in reaction. The venom of the honey bee induces a rapid fall of blood pressure and a dilatation of the splanchnic vessels causing haemorrhage in those areas. It also affects the heart. The venom is markedly haemolytic and causes a reaction identical with that caused by rattle-snake venom and by histamine (Essex, *et. al.*, 1930). As is well known the bee sting affects certain individuals much more seriously than others. When a person is stung by the honey bee several times, a certain amount of tolerance develops. After being stung once or twice, a person may, on the other hand, develop idiosyncrasy to bee stings and in subsequent stings may show exaggerated symptoms of an allergic type.

As the poisonous fluid is acid, the pain of a sting can be immediately relieved by application of strong ammonia or bathing in a solution of washing soda. It should be remembered that the sting should be extracted first and thereafter alkaline solution should be applied. The poison of the wasp produces a similar reaction in man.

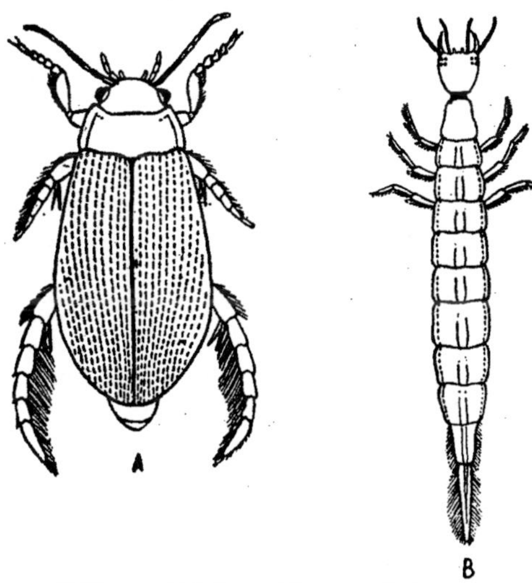

FIG. 130.
A & B. Adult and larva of *Cybister*.
(aquatic beetle)

Order **Coleoptera** (Beetles). A very large number of insects are represented in this order which is primarily important to the forester, agriculturist, stockist of grain and the householder. This order is characterised by : (*a*) Strongly built body ; (*b*) Mouth parts adapted for biting and chewing solid food; (*c*) Presence of double pair of wings; the anterior pair is the elytra and acts as a protective covering for the membranous hind pair which is used for flight; (*d*) Complete metamorphosis. All are ovigerous. The larva is provided with three pairs of legs.

The character of the antenna is chiefly used for classifying this order into different families. Only three families may be considered to have some connection with diseases in man. These are : (*a*) Staphylinidae; (*b*) Cantharidae; and (*c*) Scarabaeidae. Others which cause damage to stored grains, rice, wheat, wheat products, dried fruits etc., are many. Among them a few important ones are the following: *Silvanaus surinamensis*, *Tenebrio molitor*, *Tribolium confusum*, *Trogoderma granarium* and *Calandra* species.

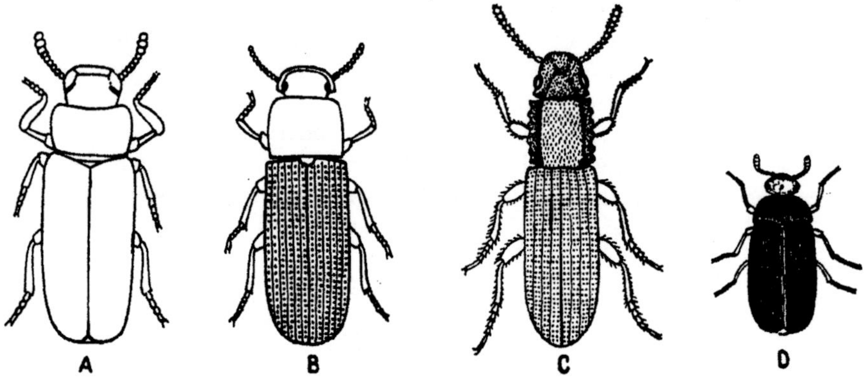

FIG. 131.

A. *Tenebrio castaneum*. B. *Tribolium confusum*. C. *Silvanus surinamensis*. D. *Trogoderma granarium*.

Although a large number of arthropods may harbour cysticercoids of *Hymenolepis diminuta*, *Tenebrio molitor*, the meal worm, is believed to be one of the commonest hosts. The cestode occurs frequently among rats, mice and other murine species and is an occasional human parasite. Human parasitization probably results from the accidental ingestion of infected insects living in cereals such as flour and meal. Chandler (1949) has reported 23 cases in India. The cycle of development of the cestode has been described by Joyeux (1920).

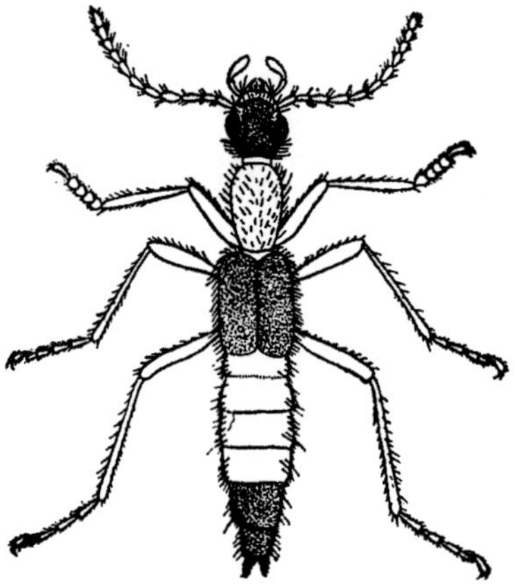

FIG. 132.
Paederus fuscipes.

The intermediate hosts of the poultry tape worm, *Hymenolepis carioca*, are *Tribolium castaneum* Hirst. and *T. confusum* Duv.

Family **Staphylinidae.** They possess a long and slender body; the elytra is extremely short and often brightly coloured; antennae are clubbed at the end but not flattened distally; tarsal joints are always more than three.

Genus **Paederus.** Antennae posteriorly flexible, not strongly geniculate, basal segment not very elongate; anterior coxal cavities entirely open behind; fourth segment of maxillary palpus compressed, truncate and pubescent. As many as 8 species have been recorded as causing dermatitis such as redness and blister. They include *P. irritans* Chapin, from Eucador; *P. amazonicus* Sharp, from Brazil; *P. brasiliensis* Er., from Argentina; *P. fuscipes* Curt., from India, Central Europe and Russia; *P. peregrinus* from the Oriental Regions and *P. crebripunctatus* Epp., from East Africa.

The most severe reaction follows vigorous rubbing of the beetle. The blistering fluid is not emitted voluntarily but exudes from any break in the integument. The fluid is essentially the haemolymph and is contained in all parts of the body. An alcoholic extract of the insect will also produce an erythema on the skin (Strickland, 1924). The vesicating properties of *Paederus* are due to cantharidin.

Breeding takes place in damp, porous soil, rich in decaying matter, along the banks of rivers, irrigation channels, drains and pools, a few inches above the edge of the water. Egg stage lasts 2 to 3 days, larval 7 days, and pupal 3 to 4 days. The adult is attracted by artificial light. The insect becomes active during May, June and July. They generally annoy man after heavy rains (Isaac, 1934).

Treatment. Magnesium sulphate compresses and application of lotio calamine give immediate relief.

Family **Meloidae or Cantharidae.** All possess vesicant properties in the blood. Some possess wings while others are wingless. The adult lives on nectar of flowers also, on leaves. The important poisonous ones belong to four genera : *Mylabris, Epicauta, Cantharis* and *Lytta.*

Blister beetles are found all over India but chiefly in the Central Provinces. The beetles appear in the fields in July and can be collected on cloths spread below the plants. Indian Spanish beetles yield on an average 0.7 per cent of cantharidin and in other beetles the yield is from 0.7 to 1.92 per cent. *C. hirticornis* from the Muree Hills contains as much as 2.02 per cent (Dutt, 1922).

In India there are nearly 70 species of blister beetles. Among them the following are the common ones. *Mylabris pustulata* Thunb.; *Epicauta hirticornis* Haag.; *Lytta actaeon* Cast.; *Lytta tenuicollis* Pall; *Gnathospastoidis rouxi* and *L. sybaromorpha.*

The presence of these beetles in the house is made evident by the powerful disagreeable odour which is due to a volatile principle present in the oily secretion exuded from the joints of the legs. In the living state when disturbed, it discharges a fluid which contains cantharidin from the mouth and joints of the legs. Cantharidin poisoning by mixing pulverised beetles with food is always of a homicidal character and follows its use as a "love philter." The beetles commonly used for this purpose are *Epicauta* and *Mylabris.* The symptoms are gastro-intestinal such as salivation, vomiting, diarrhoea and tenesmus.

Family **Scarabaeidae** (Dung beetles). These beetles, though they vary enormously in size, can be easily recognised by the shape of the body. They are commonly known as dung beetles and are familiar objects on account of their habit of rolling balls of dung much larger than their size with the aid of the front legs. The balls are

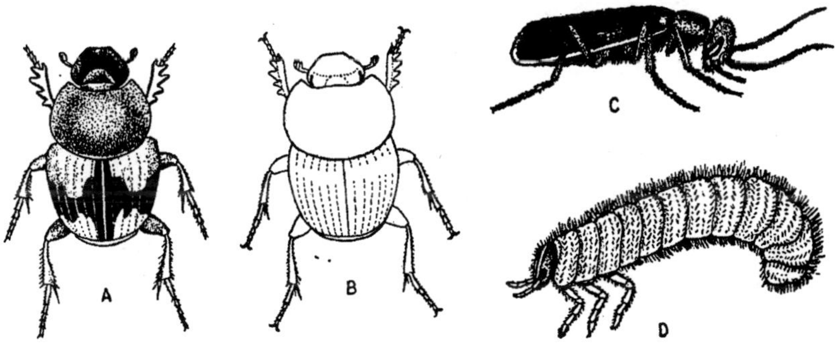

FIG. 133.

A. *Onthophagus unifasciatus.* B. *Caccobius unicornis.* C. *Cantharis* (blister beetle.). D. Larva of Coleoptera.

used for their own food and that of the larvae, the eggs being laid inside the ball. The antennae in this family are broad and leaf-like distally. Their chief importance lies in their ability to invade the alimentary canal particularly of children from 2 to 5 years of age. The occurrence of dung-beetles in the stools of persons is a condition that has been called "scarabiasis" or "beetle-disease." From time to time cases have been reported mostly from Bengal and Ceylon. De Meillon (1937) has drawn attention to the same condition in Natal. The insects are passed at intervals which may extend over some months. It is believed that adult beetles live in the intestine of patients. This condition has not yet been reported in sucklings and rarely in adults. There is loss of appetite, diarrhoea, long-continued pyrexia and progressive emaciation. The cause of the trouble is detected only when the beetle passed in the stool gradually looms up to the surface, emerges and flies away. As a rule a large number of beetles are passed over a period of months. The exact mode of entry is not known. It has been suggested that they gain entrance through the anus during sleep especially as four different species of beetles were reported to have been passed by a single individual (Strickland and Roy, 1939). But against this is the fact that the disease has not been noticed in sucklings.

The species most concerned are *Onthophagus bifasciatus* and *Caccobius unicornis* though other species of *Onthophagus* and *Caccobius* have been reported to produce the same condition. The treatment which has met with some success is change of environment as soon as the condition is detected.

Coprophagous beetles serve as intermediate hosts of a number of parasites, such as *Ascarops strongylina*, and *Physocephalus sexalatus* (both occur in the gastro-intestinal tract of pigs and other animals), and *Spirocera lupi* (found in the stomach and intestine of the dog, fox, etc).

Order **Lepidoptera.** This order includes the two familiar insects, moths and butterflies. They have a graceful appearance and possess sufficiently distinctive characters by which they can be easily recognised. The mouth part consists of a long tubular proboscis which is kept coiled up and is extended only during feeding. All are fluid feeders.

FIG. 135.
A caterpillar with bristly hairs found on mango trees during the rainy season.

FIG. 134.
Larva of *Lepidoptera*.

The metamorphosis of all Lepidoptera is complete. Eggs are laid on leaves of plants on which larvae are also found. The larva has a distinct head, the three thoracic segments are also distinct. There are three pair of legs. On the ventral surface of many abdominal segments are present pseudopods or sucker feet which help the larva in crawling on leaves. The larva lives entirely on leaves of plants and different species have particular attractions for different types of plants. When the larva attains maturity, it spins a cocoon inside which pupation takes place. The pupa is highly chitinised. The abdominal hooks of the pupa assist the imago in escaping from the cocoon.

This order though of great importance to agriculturists, florists and horticulturists is not so to us. It is on rare occasions that they come to the notice of the medical man. There are some caterpillars which are either uniformly hairy or with erect dorsal or lateral tufts. These hairy caterpillars are responsible for a sort of distressing dermatitis which is commonly met with in this country among gardeners during the rainy season. Sometimes the hairs may be blown into the eye and may set up conjunctivitis. The skin lesion generally results from direct contact with the caterpillar. It may also result indirectly by touching any object to which the caterpillar hairs may be adherent. If clothing is hung up out of doors in the vicinity of bushes, trees or creepers especially during the rainy season when these caterpillars are extremely common, an extensive rash all over the body may result when the garments are worn. These hairs are generally hollowed at the base and barbed at the end: the base contains a gland composed of one or two cells. The larvae which have urticating hairs on their body are of two different types (Weidner, 1936).

Treatment of caterpillar dermatitis consists in the application of lotio calamine or slaked lime mixed with water.

Some caterpillars infest stored grains and they not only live on the grains but make the latter unsuitable for human consumption. Among the insects which generally invade stored articles of food such as rice, cereals, biscuits, dry fruits, proprietary food stored in paper bags, tins and bottles etc., the following are the important ones :

Lepidoptera : *Ephestia kuhniella.*

Coleoptera : *Rhizopertha dominica* (*pusilla*). *Silvanus surinamensis. Trogoderma* sp. *Tribolium castaneum* (*ferrugineum*). *T. confusum. Calandra oryzae. C. granaria. Ptinus tectus, etc.*

A deposit of DDT is toxic to most of the adult beetles which infest rice, grains and flour though not to the larvae of *Ephestia.* It has been demonstrated by Jones (1947) that after spraying the bags containing the infested rice and flour with an emulsified solution to give an external deposit of DDT, most of the insects were killed in course of a week and though a small number were able to migrate to the bag surface and pass to untreated bags, they did not survive.

REFERENCES (Siphonaptera, Hymenoptera, Coleoptera and Lepidoptera)

Baccot, A. W. (1941). *Jl. Hyg.*, Plague Supplement, 3: 477. **Baccot, A. W.** and **Martin, C. J.** (1914). *Ibid*, 3: 423. **Bacot, A. W.** and **Martin, C. J.** (1924). *Jl. Hyg.*, 23: 98. **Bacot, A. W.** and **Ridewood, W. D.** (1914). *Parasitology*, 7: 157. **Bordas, F., Dubief** and **Tannon** (1922). *Presse Med.*, 30: 831. **Castaneda, M. R.** and **Zinsser, Z. H.** (1930). *J. Exp. Med.*, 52: 661. **Chandler, A. C.** (1949). Introduction to Human Parasitology, New York., 8th. Ed. **Cragg, F. W.** (1921). *Ind. J. Med. Res.*, 9: 374. **De Meillon, B.** (1937). *South African Med. Jl.*, II, 11: 479. **Devignat, R.** (1938). *Ann. Soc. Belge Med. Trop.*, 18: 215. **Dove, W. E.** and **Shelmire, B.** (1931). *J. Amer. Med. Assoc.*, 97: 1506. **Dutt, B. N.** (1922). *Ind. Med. Gaz.*, 57: 92. **Dyer, R. E., Ceder, Rumreich** and **Badger** (1931). *Pub. Hlth. Rep.*, 46: 1869. **Dyer, R. E., Rumereich** and **Badger** (1931). *Ibid*, 46: 334. **Eskey, C. R.** (1938). *Amer. J. Pub. Hlth.*, 28: 1305. **Eskey, C. R.** (1938). *Pub. Hlth. Rep.*, 53: 49. **Essex, H. E., Markowitz, J.** and **Mann, F. C.** (1930). *Amer. J. Physiology*, 94: 209. **George, P. V.** and **Webster, W. J.** (1934). *Ind. J. Med. Res.*, 22: 77. **Griffiths, J. J.** and **Tauber, O. E.** (1943). *J. Econ. Ent.*, 36: 536. **Hart, R. W.** and **Pelikan, E. R.** (1934). *Pub. Hlth. Rep.*, 49: 699. **Hirst, L. F.** (1926). *Ceyl. Jl. Sc.*, Sec. D., 1: 155. **Isaac, P. V.** (1934). *Ind. J. Agric. Sci.*, 4: 200. **Jones, B. M.** (1947). *Bull. Ent. Res.*, 38. **Joyeux, Ch.** (1920). *Bull. Soc. Franca Belga.*, Suppl., 2: 219. **Kalra, S. L.** (1947). *J. Ind. Army. Med. Corps.*, 3: 10. **Kalra, S. L.** (1951). *Ibid*, 7: 166. **Kelsall, R.** (1927). *Ind. Med. Gaz.*, 62: 565. **Klein, H. Z.** (1933). *Z. Wiss. Zool.*, CXLIV, 102. **Leeson, H. S.** (1931). *Trans. Roy. Soc. Trop. Med. & Hyg.*, 25: 3. **Lesson, H. S.** (1932). *Bull. Ent. Res.*, 23: 25. **Leger, M.** and **Baury, A.** (1922). *C. R. Acad. Sci.*, 175: 734. **Lepine, P.** and **Bilfinger, M. F.** (1934).

Bull. Soc. Path. Exot., 27 : 298. **Liu, W. T.** and **Zia, S. H.** (1941). *Amer. J. Trop. Med.*, 21 : 507. **Long, J. D.** (1935). *Pub. Hlth. Rep.*, 50 : 923. **Long, J. D.** and **Mostajo, B.** (1934). *Bol. Officiana Sanit. Panamericana*, 13 : 1016. **Macchiavello, A.** (1943). *Amer. J. Pub. Hlth.*, 33 : 807. **Macchiavello, A.** (1947). *Bol. Officiana Sanit. Panamericana*, 26 : 225. **Meyer, R. F., Cannor, Amith** and **Eddie** (1947). (1947). *Arch. Inter. Med.*, 59 : 967. **Meyer, K. F., Holdenreid, Burroughs** and **Jawetz** (1943). *J. Infect. Dis.*, 73 : 144. **Moser, H., Castaneda, Zinsser** (1931). *J. Amer. Med. Assoc.*, 97 : 231. **Mooser, H.** (1931). *J. Exp. Med.*, 54 : 567. **Parman, D. C.** (1923). *Jour. Agric. Res.*, 23 : 1007. **Pollard, M., Wilson, Livesay** and **Woodland** (1946). *Texas Repts. on Biol. & Med.*, 4 : 446. **Rao, K. N. A.** and **Kalra, S. L.** (1949). *Ind. J. Med. Res.*, 37 : 385. **Seal, S. C.** (1949). *Cal. Med. Jour.*, 46 : 167. **Sharif, M.** (1935). *Parasitology*, 27 : 461. **Sharif, M.** (1948). *Ibid*, 38 : 253. **Sinton, J. A.** (1925). *Ind. J. Med. Res.*, 12 : 471. **Strickland, C.** (1914). *J. Hyg.*, 14 : 129. **Strickland, C.** (1924). *Ind. Med. Gaz.*, 59 : 385. **Strickland, C.** and **Roy, D. N.** (1939). *Ibid.*, 74 : 416. **Swellengrebel, N. H.** (1913). *Geneesk. Tijdschr. Ned. Ind.*, 53 : 53. **Turkhud, D. A.** (1928). *Ind. Med. Gaz.*, 63 : 51. **Webster, J. W. Chitre, G. D.** (1930). *Ind. J. Med. Res.*, 18 : 407. **Weidner, H.** (1936). *Z. angew. Ent.*, 23 : 432. **Williams, C. L.** and **Kemmerer, T. W.** (1923). *Pub. Hlth. Rep.*, 38 : 1875.

CHAPTER XII

ORDER HEMIPTERA (RHYNCHOTA)

The members of this order are popularly known as bugs which represent a fairly large group of insects of varying sizes. Insects such as scale insects, aphids, cochineal bugs etc. are no doubt primarily important to the agriculturist, but the medical man is interested only in the bed bug and the cone-nosed bug, the former belonging to the family Cimicidae and the latter Reduviidae.

This order can be readily separated from others by the characteristic proboscis. The proboscis or the rostrum is a segmented structure and in repose is kept withdrawn under the head and thorax and is half or fully extended during feeding. Wings may or may not be present; when present, they may be two pair or one pair. Both may be membranous or the basal half of the front pair may be thickened. The mouth is of the piercing and sucking type, and all live on fluid food mostly of vegetable origin. The metamorphosis is incomplete and all are oviparous.

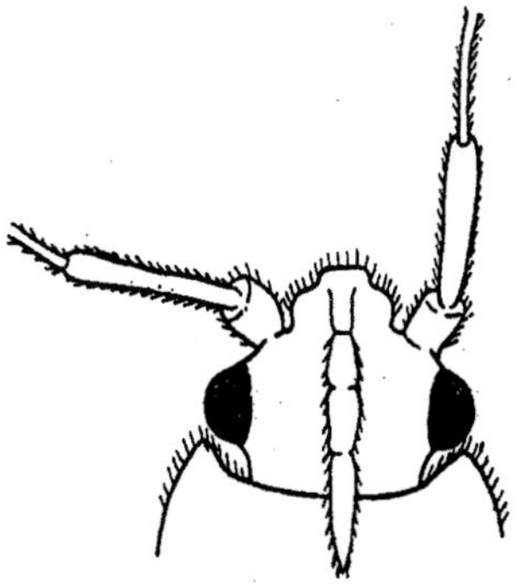

FIG. 136.
Ventral aspect of the head of *C. rotundatus* showing the segmented proboscis lying in a groove under the head and thorax.

While a large majority of bugs are terrestrial, thére are others which are aquatic. Among the latter some belonging to the famiy Hydrometridae prey upon mosquito larvae and other aquatic insects.

Of the many families only two will be mentioned; these are Cimicidae and Reduviidae.

Family **Cimicidae** (Bed bugs). These are small, dark-brown in colour and are flattened dorsoventrally. All are parasitic and the active stages live on blood alone. Metamorphosis is incomplete. There are usually 1 larval and 4 nymphal stages.

Genus **Cimex. Head :** antennae four-jointed; eyes simple; rostrum (beak or proboscis) three-jointed. The proboscis or the labium encloses the true biting and sucking organs which lie in a deep groove on the dorsal surface. The mouth organs in bugs are reduced to two pair of stylets, mandibles and maxillae. The two mandibles are grooved on their inner surfaces and when they appose, two tubes are formed, the dorsal one for the passage of blood into the mouth, and through the ventral tube the saliva flows into the wound.

Thorax : consists of the usual three segments. The prothorax is the largest and semilunar. The metathorax is inconspicuous and bears on its dorsal surface the rudimentary wings in the form of oval pads or elytra one on either side. The legs bear the usual number of joints except that the tarsi are three-jointed.

Abdomen : The tip of the abdomen is broad in females and narrow in males. The curved penis is accomodated in a notch on the left side of the 8th segment.

The midgut is a long tube, pear-shaped and occupies the largest portion of the alimentary tract. It is divided into an anterior and a posterior part. The hind intestine is globular and consists mainly of the rectum. The junction between the mid and hindgut is marked by four Malpighian tubules.

Stink glands. The offensive smell emitted by a large majority of the order Rhynchota is due to a fluid secreted by a pair of glands, one on each side of the thorax. They open by means of small apertures lying on the metasternum between the second and third pair of legs.

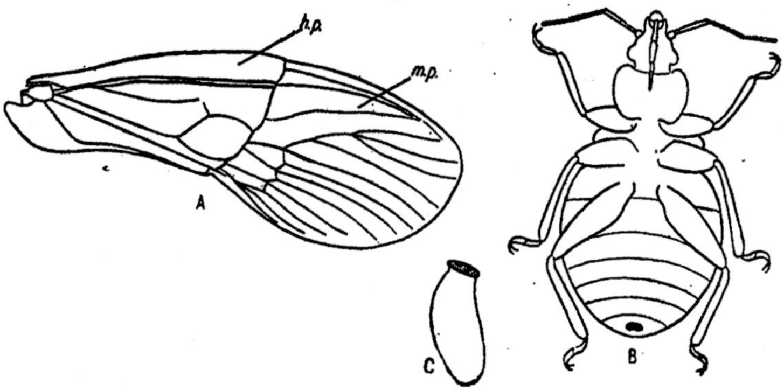

Fig. 137.
A. Wing of bug. *h.p.* hardened part. *m.p.* membranous part. B. Nymph of bed-bug. C. Egg of bed-bug.

Salivary glands. There are two pair of salivary glands, one being the gland proper and the other is thought to serve as a reservoir for storing saliva.

The small oval glands of *Triatoma rubrofasciata* possess an anticoagulant which prevents the coagulation of blood (Cornwall and Patton, 1914). The same properties also exist in the cardiac glands of *Cimex lectularius* (Puri, 1924).

Respiratory system. There are two pair of thoracic and seven pair of abdominal spiracles. The latter are situated on the under surface near to the outer edge.

Reproductive system. *Male* : A pair of testes with vasa deferens, vesiculae seminalae, accessory glands with their reservoirs, ductus ejaculatorius and the penis. *Female* : Each ovary consists of 7 egg tubes. The common oviduct terminates at the posterior end of the body between the modified ventral plates so arranged as to permit the expansion of the genital orifice at the moment of oviposition. The pouch-like spermathecae lie on either side of the common oviduct.

"*Organs of Berlese.*" This organ is peculiar to bugs and is found only in females. It consists of a spherical mass, whitish in colour, and lies in the body cavity on the ventral surface of segment 4 and to the right of the middle line. In the right lateral half of the ventral plate of the fourth segment there is a longitudinal slit. This is known as Ribaga's organ. The organ of Berlese is a solid mass of cells with no lumen and no duct. It contains very numerous spermatozoa. Also striking is the presence of spermatozoa free in the body cavity of the abdomen. During copulation the penis is inserted through the cleft in the right side of the fourth abdominal sternum into the organ of Berlese into which the spermatozoa are injected.

Life history. Metamorphosis is incomplete, the various stages in its life cycle being egg, larva, 4 nymphal stages and adult.

In the bed bug the ovarian development never takes place until the female has been fertilised. The eggs are white and flask-shaped. The eggs are laid singly in cracks, crevices, furniture, mattress, pillows, hats etc., in fact they select places well sheltered from light where the bug also hides in the day time. In the tropics the minimum time in which the eggs will hatch is 4 to 5 days and in the winter it is extended to 10 days. Eggs are sensitive to cold and if exposed to low temperature, they fail to hatch. Omori (1938) has conducted experimental studies on the influence of low temperatures upon the two bugs, *C. hemipterus* and *C. lectularius*, and Hase (1930) on the egg laying of bed bugs under different conditions of temperature. Ordinarily the eggs can withstand immersion in water at ordinary temperature for at least 4 days (Bacot, 1914), and a degree of partial vacuum for at least 2 days.

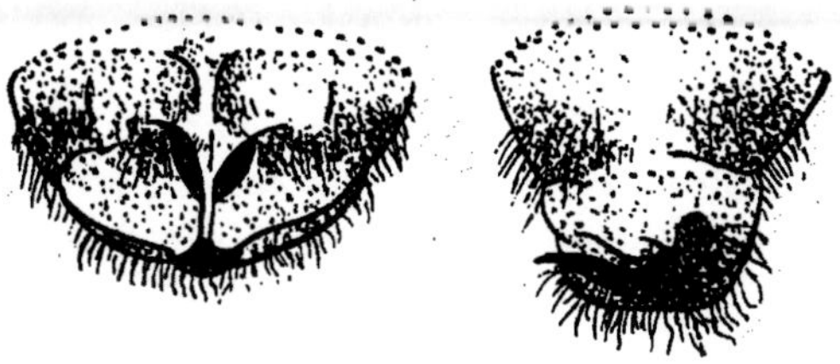

A B

FIG. 138.

Ventral view of the posterior part of the abdomen of *Cimex rotundatus*. (A) Female, (B) Male.

A female may lay about 50 eggs in course of 4 months. The irregularity as regards oviposition is very striking. The eggs do not begin to develop until they are fertilized and only a few are laid after the female has had a meal of blood. Females fertilized by unfed males do not produce so many eggs as those fertilized by fully nourished ones (Cragg, 1915, 1923).

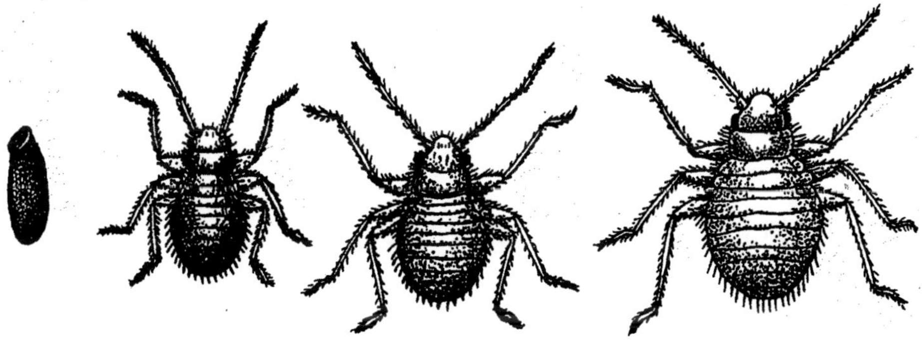

A B C D

FIG. 139.

Cimex rotundatus. A. egg ; B, C, and D. larva and two nymphal stages.

Within an hour or two after emergence from the eggs the larva feeds. The bug unlike many other insects feeds intermittently in all stages of its life. The wing pads make their appearance in the first nymphal stage. In all there are four nymphal stages, each stage lasting from 5 to 7 days, at the end of which a skin is cast off.

In the tropics adult bugs can subsist even for three months without food. Thus they may remain alive and active for months in uninhabited houses.

The weight of blood in mgm. sucked by *C. hemiptera* during one blood meal is 7.6 for the adult female and 4.9 for the male. *C. lectularius* males and females ingest

24

the same amount of blood during one blood meal (Johnson, 1940). A nymph sucks a total of 15.675 mgm. According to Omori (1935), on an average an adult female lives 118 days, lays 360 eggs, takes 37 blood meals and ingests 152.52 mgm. of blood.

All persons are bitten indiscriminately and its bite is extremely irritating to the skin. It will feed readily on any laboratory animal but in nature it shows a marked preference to bite man. Bed bugs often breed in association with rats, rabbits and poultry.

Heat is an important factor in stimulating bed bugs to obtain food and sebum and cerumen exhibit the greatest attraction (Rivnay, 1932).

Bed bugs are disseminated mainly through the agency of personal clothing to which the hungry bug is readily transferred from railway trains, tram cars, street buses, furniture, etc. Bedding and furniture are also responsible for the carriage of the bed bug from one place to another.

Diseases. A large number of diseases, e.g., relapsing fever, typhus, kala-azar, plague, leprosy etc. have been suspected of being carried by bed bugs, but none has yet been demonstrated. Castaneda and Zinsser (1930) indicated that *C. lectularius* was of importance in the transmission of murine typhus. This was later confirmed by Nauck and Zumpt (1940).

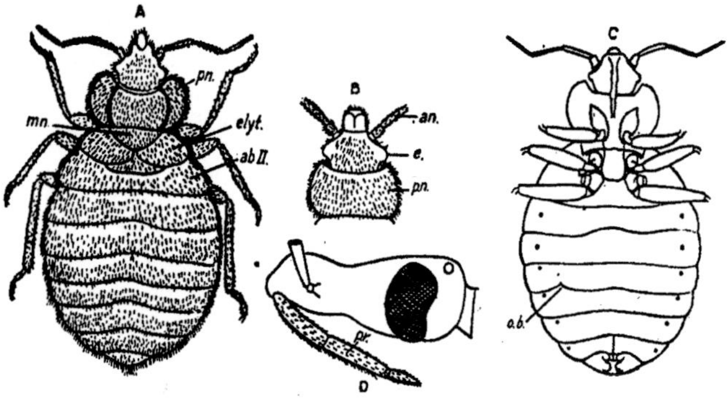

FIG. 140.

A. *Cimex lectularius,* female. B. Head of *C. rotundatus.* C. *C. rotundatus,* ventral surface. D. Lateral view of the head of *Triatoma robrofasciata.*

ab II. second abdominal segment ; *an*, antenna ; *e*, eye ; *elyt.* elytra ; *mn*, mesonotum ; *o.b.* opening into organ of Berlese ; *pn*, pronotum.

Species of Cimex. The two bed bugs are *C. lectularius* and *C. hemipterus* (*rotundatus*). *C. lectularius* L. is most commonly found in temperate regions while *C. hemipterus* Fab. is peculiar to hot climate being found in tropical and subtropical regions. *C. lectularius* is a little larger but otherwise the two are indistinguishable by naked eye. The main distinguishing feature is the character of the prothorax. In *lectularius* it is semilunar in shape, its lateral margins are flat and thin, and the anterior margin has a rounded lobe on each side of the head which projects lateral to the eyes. In *hemipterus* it is narrower and shorter; the margins are rounded and much less expanded laterally; the anterolateral margin, unlike in *lectularius,* just reaches the level of the eye margins.

C. lectularius has spread to America, Australia and South Africa, being carried by man. Omori (1941) has worked out the physiological behaviour of the two species when exposed to different temperature conditions.

C. rotundatus (Indian bed bug). India, Burma, Assam, Malay, Aden, Islands of Mauritius and Reunion, Perak, Ceylon, Thailand, East Indies, Borneo, Sumatra, Java, etc., South America.

C. lectularius (Temperate climate bed bug). Europe, North America, Suez, Egypt, the Sudan, N. W. Frontier Province, China, Japan, Siberia, Malta, Cairo, Khartoum, etc.

C. pipistrelli Jenys, is found on bats in England, India and South Africa. It is very hairy and the posterior part of the abdomen is tapering. *C. columbarius* Jen., has been found infesting rat cages in Edinburgh. It infests pigeons. The abdomen is almost circular in outline and the antennae are much shorter than in *C. lectularius*. *C. hirudinis* Jen., are found in nests of swallows. The body is hairy, the eyes are very small as also are the antennae.

Family **Reduviidae** (Cone-nosed bugs). This family of bugs is indeed very large. They are all terrestrial insects, and in size are much larger than bed bugs. They are found in all parts of the world. The body is elongate; the head is also elongate. The proboscis or rostrum has the first segment shorter than the second; antennae are larger than in Cimicidae and are placed at the side of the head midway between eyes and apex; the eyes are placed well forward, the abdominal connexivum or the lateral edge of the abdomen distinctly flattened; posterior tibia longer than the fibula. Only a few are blood sucking, others living on vegetable sap. All are fluid feeders.

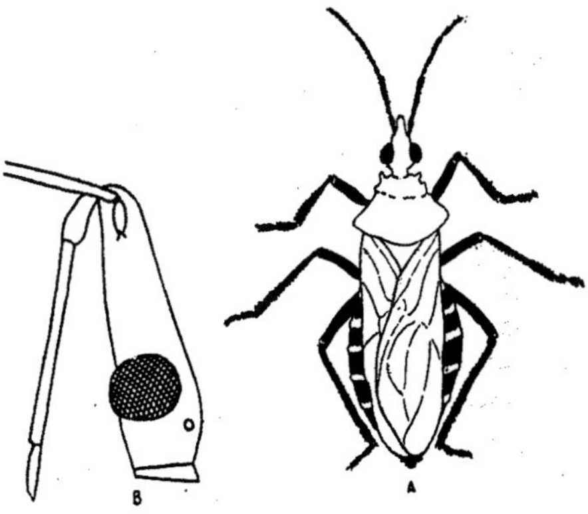

FIG. 141.
A. *Triatoma rubrofasciata.*
B. Head of *Rhodnius prolixus.*

Triatominae. The subfamily Triatominae comprises those members of the family Reduviidae which feed exclusively on the blood of vertebrate animals. The Triatomas apart from the ill effects that may be produced directly as a result of the bite, are principally concerned in the transmission of *Trypanosoma cruzi*, the parasite responsible for Chagas' disease in South America. The virus of equine encephalitis has recently been detected in this bug found inefcted in nature (Kitselman and Grundmann, 1940).

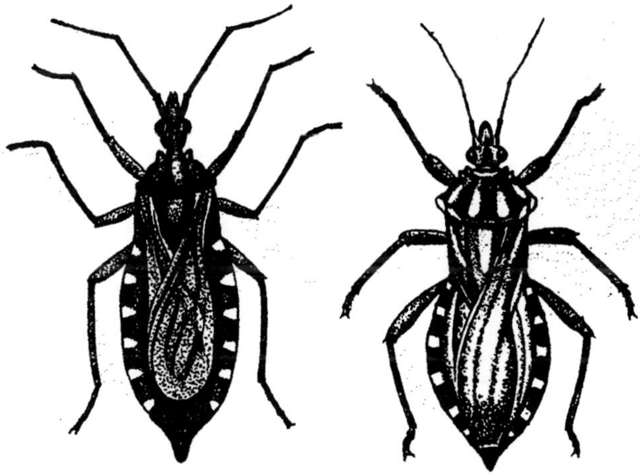

A B

FIG. 142.

A. *Triatoma megistus*, male. B. *Triatoma rubrofasciatus*, male.

Egg of *Triatoma rubrofasciatus*: the larva has hatched out.

Genus **Triatoma** Lap. All are of large size about an inch or more in length. Different species are marked differently on the thorax and abdomen. They are widely distributed in the Oriental Region, Africa and South America. At all stages they are blood suckers and man is also attacked. A sharp prick at the time of the bite is felt and after some time some sort of itching sensation is experienced.

T. rubrofasciatus De G., is distributed in the warm countries in the Old as well as in the New World. It is the commonest blood-sucking reduviid that are attracted by light and enters houses at night in this country. The adults probably live in tree holes and rodent holes and feed on mammalian blood. The tip of the abdomen in the female is broader than the male. When captured it will go on laying eggs provided it is allowed to have blood meals at frequent intervals. The bug moults 5 times and only in the adult stage the wings are developed. It will readily take blood feed on any laboratory animal. The lateral edge of the pronotum has a reddish linear tinge and connexivum has reddish lateral spots. The sexes are easy to distinguish. In the male the terminal part of the abdomen is pointed whereas in the female it is broad,

Though once suspected of playing a role in the transmission of kala-azar in India, it takes no part in the spread of any disease except in South America where this species in addition to a large number of others are responsible for the transmission of *Trypanosoma cruzi*, the causative organism of Chagas' disease. Those which are well known as vectors of this disease are *Rhodnius prolixus* Stal., in Venezuela and Mexico, *Panstrongylus geniculatus* Latr., and *T. rubrofasciata* in Brazil and Panama, *T. dimidiata* Latr., in Guatemala and Mexico, *T. sanguisuga* Lec., in Texas and *T. longipes* Barbo., in Arizona. *Rhodnius* are believed to be more dangerous than *Triatoma* as they defaecate immediately after withdrawing their rostrum from the puncture. *Triatoma*, on the other hand, do not defaecate until some seconds or minutes after feeding. The development of the parasite is confined to the intestinal tract and the metacyclic or infective forms are found in the rectum, these being voided with the excreta. The transformation of the trypanosomes into crithidia takes place in the stomach. Infection is acquired from contamination of the wound and of the mucous membrane of the mouth, nose or eye with infective excreta, and also by eating infected bugs. *T. cruzi* cannot pass through the unbroken skin. A large number of animals, e.g., dog, cat, armadillo, opossum, monkey, rat, besides man, serve as reservoirs of the trypanosome. In addition to Triatomas others which can also act as vectors are species of *Cimex* and *Ornithodorus* ticks (Brumpt, 1912), also *Melophagus ovinus* (Rodhain and Brutsaert, 1935).

Triatomas, if allowed to feed on the blood when the parasites are present, become readily infected. This is recognised as a sensitive "Xenodiagnostic test" of Brumpt for diagnosing the disease during the initial stage when parasites are scanty in the peripheral blood, and a large amount of blood is necessary for inoculation into experimental animals for establishing the diagnosis. In vertebrates the trypanosomes are found in the peripheral blood for about 2 weeks and thereafter they lodge in the muscles.

FIG. 143.
Armadilloes are reservoirs of *T. cruzi* in South America.

Genus **Rhodnius** Stal. The antennae are inserted well forward near the end of the head and the first and the third segments of the rostrum short, subequal, middle segment long, whereas in *Triatoma* the antennae are inserted further posteriorly and the first segment of the rostrum distinctly shorter than the second but longer than the third. It occurs in South America.

[For information on the Triatominae of North and Central America and the West Indies and their public health significance, see Public Health Bulletin No. 288 by Usinger of the U. S. Public Health Service, 1944; and Wood, 1941, 1943.]

Control. The measures that may be adopted for eradicating bed bugs are : (1) DDT 5 per cent. powder. (2) DDT and pyrethrum powder. (3) DDT and kerosene spray. (4) Heat, both dry and moist. (5) Chlordane : extremely toxic to bugs. (6) For disinfesting tram and railway coaches, gaseous insecticides, e.g., hydrocyanic acid; ethylene dichloride with carbon tetrachloride, ethylene oxide with carbon dioxide, chloropicrin and methyl bromide etc., may be used. (7) For hospital mattresses, supersaturated steam.

[For control of Reduviid bugs with pyrethrum and Gammexane, see in Part II.]

Dissection of Bug. Allow the bug sufficient time to digest its blood meal before it is dissected. After the legs have been removed the insect is placed in a drop of salt solution on a glass slide and firmly held in position with a blunt needle applied on the lateral side of the abdomen. With a sharp-edged needle the lateral margin of the abdomen on both sides except the last two segments is chipped off. The thoracic segments should be treated in a similar manner. It will now be possible to remove the skin either from the dorsal or ventral surface of the body and expose the alimentary canal which is left attached to the head in front and the abdominal segment posteriorly.

Preservation. All stages of bed-bugs can be preserved in 70 to 80 per cent spirit. They should never be preserved in a dry state. Reduviid bugs should be mounted on pins.

CHAPTER XIII

ORDER ANOPLURA

(Sucking lice)

The main features of this order are : (1) the body is flattened dorsoventrally; (2) the integument is highly chitinised; (3) wings are absent; (4) they are ectoparasites of warm-blooded animals; (5) mouth parts of the active stages are adapted for piercing the skin and for sucking blood; (6) metamorphosis is incomplete; (7) the eggs are laid on hairs, feathers or on clothes. This order is grouped in two suborders, Siphunculata and Mallophaga.

Suborder **Siphuculata.** The members of this suborder feed on blood and the piercing apparatus when at rest are not visible from outside as they are concealed within a diverticulum opening into the floor of the pharynx anteriorly; thorax is undivided. They are parasites of mammals.

Suborder **Mallophaga.** They possess chewing mouth parts and feed on solids, e.g., feathers and epidermis of their hosts. They are parasites mostly of birds.

Families. There are two important families of Siphuculata, Pediculidae and Haematopinidae.

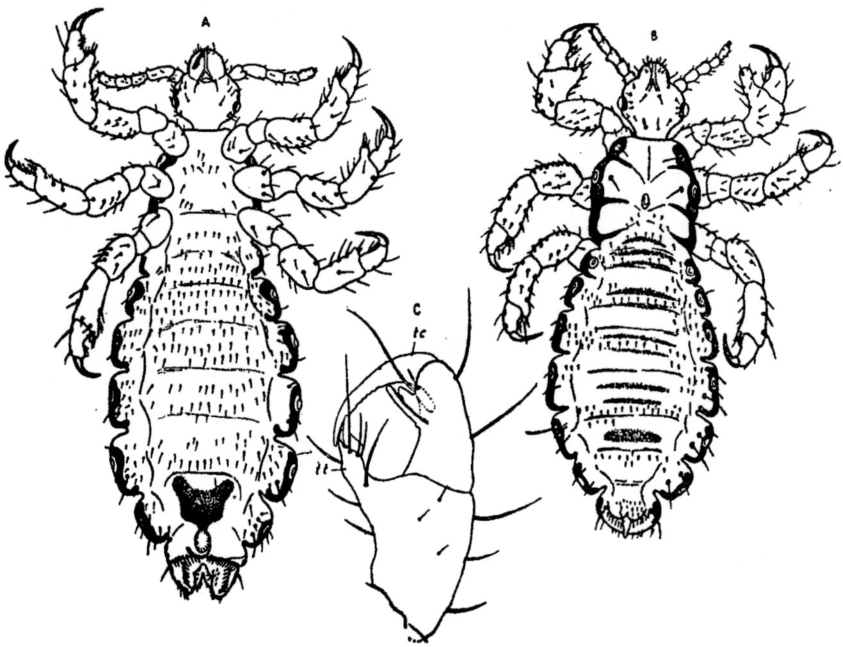

FIG. 144.

A. *Pediculus capitis*, female. B. *Pediculus capitis*, male. C. Terminal portion of the first leg of the male. (After Keilin and Nuttall). tc, tarsal claw ; tt, tibial thumb.

(1) **Pediculidae.** These are with pigmented eyes; head not retracted into thorax; antennae of 3 to 5 segments; tibia with thumb-like process and single incurved claw. All are parasites of primates. It contains two subfamilies, Pediculinae and Pedicininae.

In Pediculinae the larval antenna is 3-jointed and in the nymphal and adult stages the antenna is 5-jointed. It contains two genera, *Pediculus*, the human head and body louse, being found also on monkeys and apes, and *Phthirus*, the pubic louse which are found on man alone.

In Pedicininae the antennae are 3-jointed and all are parasites of old world monkeys and apes. It contains the genus *Pedicinus*.

(2) **Haematopinidae.** This family is characterised by the eyes being absent or very rudimentary and head retracted into thorax; antennae are 5-jointed. They are found on ungulates, e.g., pigs, buffalo, cattle, sheep and goat. The common species are *Haematopinus*, *Polyplax*, *Linognathus* and *Solenopotes*.

Genus **Pediculus. External characters.** Body is elongate and divided into head, thorax and abdomen. *Head* : The head is constricted proximally in a neck-like manner. The head is very small in relation to the size of the insect and is pointed anteriorly. The clypeus is prominent. The antennae consist of 5-articles. The eyes are small and simple. The actual mouth organs are kept retracted within the head; what is visible from outside is a short tube with a circle of chitinous hooks or teeth. These teeth are used by the insect for anchoring itself to the host at the time of feeding but while at rest, they are kept inverted within the tube. The actual cutting instruments are in the form of three chitinous structures, 2 lying dorsally and the third one ventrally, the three being enclosed in a sheath and lying ventral to the buccal cavity. They are normally kept retracted but are extended at the time of feeding.

The *thorax* is fused into a square shaped mass. The legs are strong, and fairly uniform in all stages, but the first pair of legs in the male are stouter and with larger thumb than those of the other two pairs. The third pair of legs in the female have a blunt baso-ventral femoral spur while the other two pairs are normal. There is only one tarsal claw. There is also a thumb-like projection from the lower part of the tibia. There is only one thoracic spiracle.

The *abdomen* is segmented. The abdominal spiracles are 6 in number and lie on the lateral plate of the abdominal segments. The sexes can be easily distinguished. In the female the last abdominal segment is bilobed; in the male it is pointed and from which the aedoeagus often projects. The gonopods lie on the ventral surface of the last segment in the female, and consist of two flattened and convergent hooks concave internally and serving to encircle and grasp hair or fibre during oviposition.

Internal anatomy. (1) Mouth, around the distal end of which lie some chitinous hooks or teeth. (2) Buccal cavity or the sucking tube. (3) Ventral to the buccal cavity lie three stylets enclosed in a sheath; these are the actual cutting and perhaps also sucking apparatus. The salivary duct passes between the two dorsal and the single ventral stylets. (4) The pharynx is the chief pumping organ and consists of two chambers, an anterior and a posterior. (5) This is followed by a very narrow oesophagus. (6) The anterior part of the midintestine is wide and is provided with lateral pouches. (7) A small hindintestine. (8) 4 Malpighian tubules, and (9) Rectum, which is ampullar.

Salivary glands. There are 2 pair of glands, one oval and the other tubular; the latter consists of two long tubules. The oval glands lie external to the oesophagus, whereas the tubular ones lie on a level with the upper part of the midgut. Each gland is provided with a separate duct which follows very closely to each other of the same side. The two ducts of each side join to form a very short common duct which after it has joined with its fellow of the opposite side opens into the buccal cavity between the dorsal and ventral stylets.

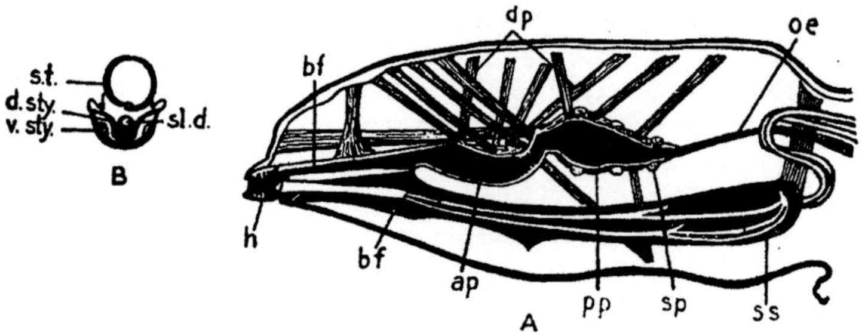

Fig. 145.

A. Pediculus humanus, longitudinal section of head. (From Buxton). ap, anterior part of pharynx ; bf, buccal funnel; dp, dilator muscles of pharynx ; h, haustellum ; oe, oesophagus ; pp, posterior part of pharynx ; sp, sphincters ; ss, stylet sacs. *B.* Transverse section through the piercing apparatus. (After Sikora and Patton and Evans). d.sty. dorsal stylet ; sl.d., main salivary duct ; s.t., sucking tube ; v.sty, ventral stylet.

Reproductive system. Female : (1) ovaries; each ovary consisting of five ovarian tubes containing eggs in all stages of development; (2) oviducts; (3) common oviduct; (4) accessory gland; (5) receptaculum seminis, and (6) vagina. Male : (1) A pair of testes on each side; (2) vas deferens; (3) seminal vesicle with accessory glands; (4) ejaculatory duct ending in aedoeagus.

Symbionts. Symbionts of the nature of mycetome which was first observed by Sikora (1919) lie in a free state in the abdominal cavity of the louse. According to Aschner (1934), these symbionts play an essential role in its life.

Life history and bionomics. Metamorphosis is incomplete, the different stages being, egg, three larval stages and adult.

Oviposition usually begins 24 to 36 hours after the adult female emerges from the third larval skin. The egg is oval; in its early stage, it is pearly white and proportionately very large. At the anterior end of the egg lies the cap which is perforated with minute holes through which air can reach the growing embryo. The eggs are attached to the hair by means of an adhesive substance. Only one egg is laid at a time. The pointed end is extruded first and at the time of laying each egg is carefully oriented, so that the operculum is directed towards the distal end of the hair. By means of its legs and gonopods the insect holds firmly to the hair before the egg is deposited which at once becomes firmly fixed to the hair. Oviposition is directly related to the temperature at which lice are maintained. The lower limit of temperature is 20°C., and above 37°C., when oviposition is stopped. The average number of eggs laid by a female in course of 24 hours varies from 3.7 (Bacot, 1917) to 9 to 10 (Nuttall, 1917; Buxton, 1940). The maximum total of 300 eggs laid by a single female has been recorded by Nuttall. The rate of egg production is increased when lice are kept close to the skin even when they are not feeding (Haddow, 1941). Under any circumstances eggs do not hatch below 4 days. When the larva is ready to emerge it swallows amniotic fluid and at a later stage, air. This causes the operculum to spring open. The young louse escapes from the egg-shell in 2 to 5 minutes. The empty shell usually remains attached to the hair for even more than 6 weeks.

There are three larval stages in the developmental cycle of the Pediculus.

The frequency with which lice feed is dependent on the rate of digestion which in turn is influenced by temperature. The amount of blood ingested by a louse varies from 0.7 to 1.2 mg. in weight of blood. In common with insects in general there is expulsion of blood in the act of feeding. Although *P. humanus* in nature thrives only on human blood, Culpepper (1948) has shown that it is possible to rear and maintain a colony of body lice by feeding on rabbits.

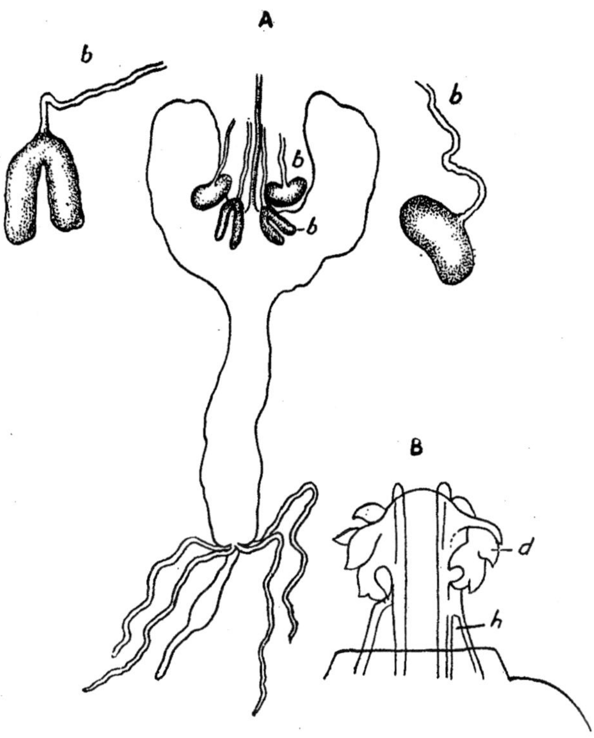

FIG. 146.
A. Intestinal canal and salivary glands (b) of head louse.
B. Mouth parts of louse. (After Peacock).
d, denticle ; *h*, haustellum.

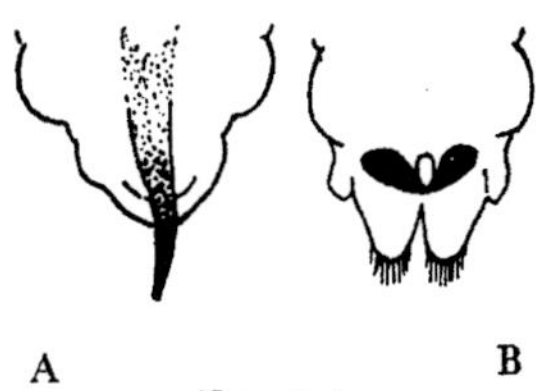

FIG. 147.
Ventral surface of the posterior end of the abdomen of *Pediculus capitis*. (A) Male showing the aedoeagus. (B) Female with the gonopods.

The time occupied by different stages of *P. corporis* for development to the next stage in temperate climate has been worked out by a number of investigators (Warburton, 1910; Bacot, 1917; Nuttall, 1917; Buxton, 1938, 1940). In India a total period of 15 to 16 days is necessary between the hatching of the egg and the appearance of the adult stage (Patton and Cragg, 1913). An increase in temperature causes a shortening of this period in *corporis* from 16 to 10 days, whereas in Calcutta the present author working with *capitis* noted an increase to 24 days during December and

January. The oviposition may take place from the second to eightth days after the final moult; most commonly they become sexually mature on the 4th or 5th days after ecdysis.

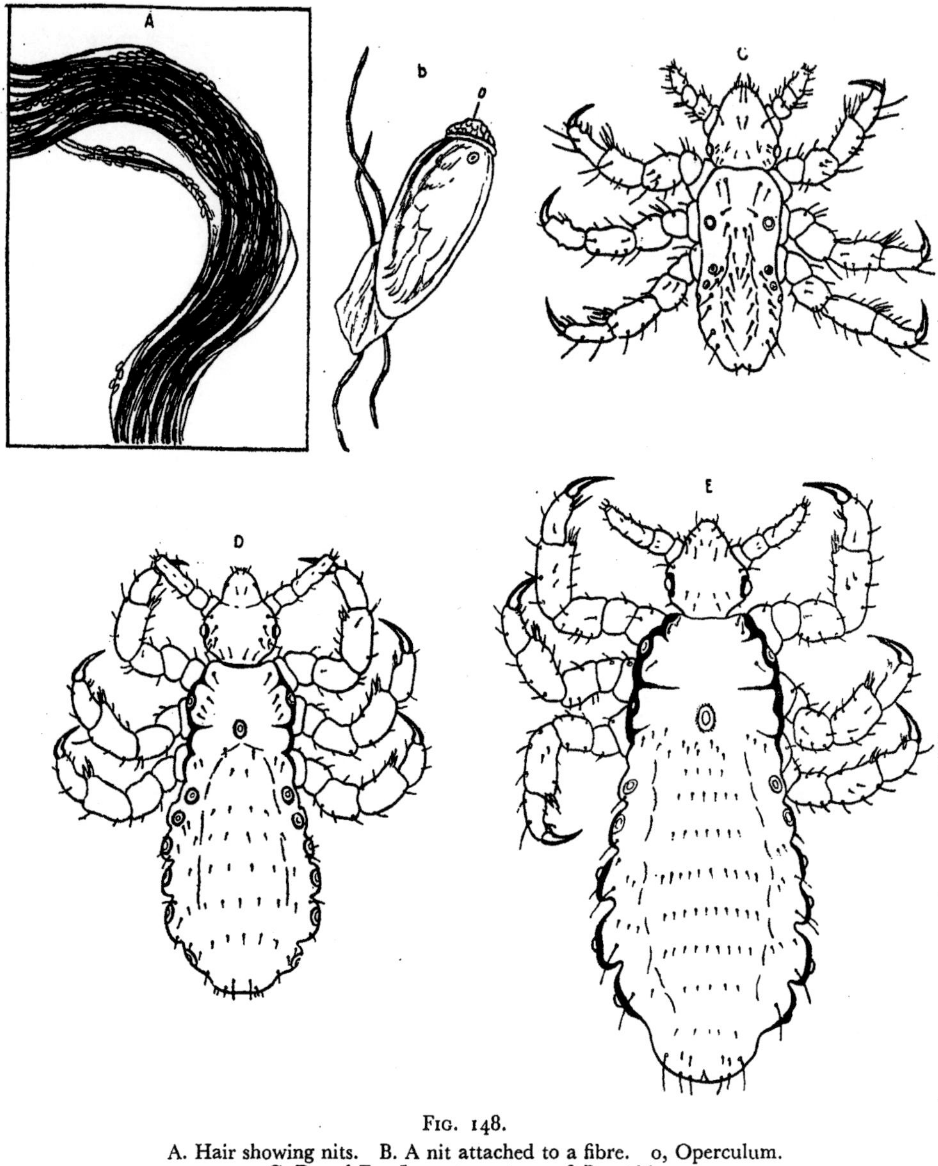

FIG. 148.
A. Hair showing nits. B. A nit attached to a fibre. o, Operculum.
C, D and E. Immature stages of *P. capitis*.

Under starving conditions Nuttall (1917) found the active stages of *P. corporis* living for 10 days at longest. Cragg (1922) recorded 93 hours as the maximum longevity of *corporis* but this must be regarded as exceptional and it can be taken for granted that most of them cannot resist starvation longer than 48 to 57 hours. *Corporis* is as a rule more hardy than *capitis*. The average length of life under the most favourable conditions in the case of *corporis* is 32 to 58 days (Buxton, 1940), and in *capitis* 16 to 27 days (Bacot, 1927).

Long-haired individuals as a rule show a greater infestation rate than those possessing short hair. On the other hand, Roy and Ghosh (1944) reported the highest count of lice, i.e., 5253 (adult 1434, larvae, 3819) from a short-haired female head. In another case a total of 9020 lice (adult 1670, larvae 7350) was noticed. The last patient had long hairs. It is by no means true that regular washing, and combing the hair combined with the use of oil will completely prevent louse infestation in women. In such cases the population of lice is generally at a low level.

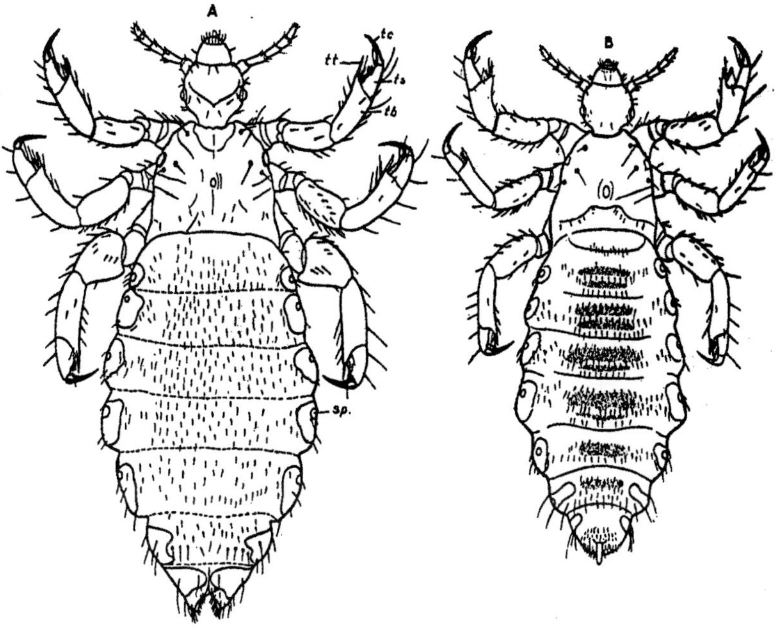

FIG. 149.
A. Pediculus humanus (corporis), female. B. Pediculus humanus (corporis), male. (After Keilin and Nuttall).
sp, spiracle ; tb, tibia ; tc, tarsal claw ; ts, tarsus ; tt, tibial thumb.

In nature a disproportion in the distribution of the two sexes is a marked feature. A preponderance of females over males has been noticed.

The activity of lice is greatly dependent upon the temperature of the surroundings. They are very active at 30°C., and at 37°C. they are extraordinarily active creatures. The temperature sense is very well developed in this insect. It can detect even small differences (Wigglesworth, 1941).

Lice are usually killed by 30 minutes' submersion in water but nits may survive immersion in water for 96 hours.

Nuttall found that the colour of lice has a direct bearing on their environments.

Lice may be fed through a membrane prepared from a piece of skin from a baby chick and experimental infection with *Rickettsia prowazeki*, *R. mooseri* and *Borrelia novyi* has been described by Fuller, Murray and Snyder (1949).

Dissemination. The following factors are concerned in the dissemination of lice. After dissemination the multiplication of the parasites depends mainly on the neglect of personal sanitation, and the wearing of underclothing for days together.

(1) Close contact with lousy individuals in the household, barracks, camps, underground shelters etc., especially when a large number of individuals have congregated together. Hence infestation spreads rapidly during war, in famine conditions and among refugees. (2) Indirect contact, e.g., through bedding, clothes, blankets, towels, hats, combs, brushes, etc. (3) Wind may be responsible to a small extent by blowing the nits. (4) It is thought that lice have a tendency to leave the host when the temperature of the body rises due to high fever. Such migration of head lice under conditions of malarial fever has not been observed by us.

P. capitis and P. corporis (humanus). As a rule *capitis* occurs on the head, mostly about the occiput and about the ears; *corporis*, on the other hand, is confined to the clothing worn next to the skin being found especially in the seams on the fibres of which eggs are laid. Nuttall (1917) has collected evidence showing that *corporis* may live upon the body and deposit its eggs upon human hair on the breast and axilla. These are no doubt rare instances of their breeding in unusual situations. It may be pointed out that *P. capitis* occur on the head in all seasons though they are found in large numbers during the winter, whereas *P. corporis* are seldom found in the summer except in very small numbers in countries with either very moist or very dry climate. They are no doubt found at all times of the year in temperate climates. In nature the population of *corporis* increases more rapidly than that of *capitis* for the reason that the females have a relatively greater egg-carrying capacity than the head louse and the number of eggs laid is also greater.

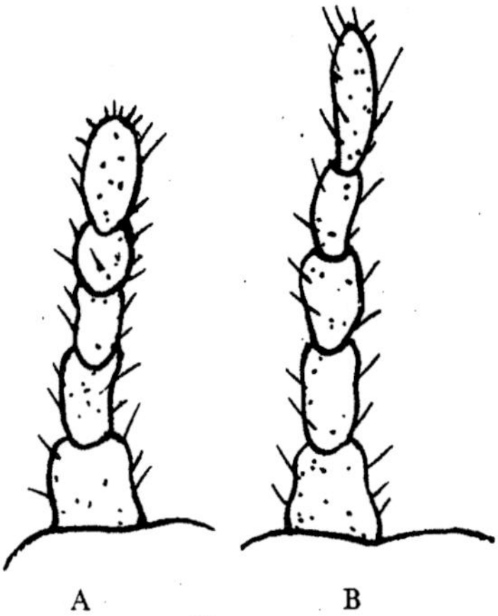

FIG. 150.
Antennae of (A) *P. capitis*, and (B) *P. corporis*.

It has been demonstrated that there are no constant differences between the head louse and the body louse and it has also been demonstrated that head lice under certain experimental conditions may acquire all the morphological characters of the body louse after four or more generations (Keilin and Nuttall, 1919). However, certain morphological differences are commonly found between them.

P. capitis	**P. corporis (P. humanus L.)**
Size smaller.	Size larger.
Deeply pigmented body.	Pigmentation absent.
Clearly marked indentations between successive abdominal segments (due to chitinous pleural plates).	Abdominal segments are rounded with shallower intersegmental indentations.
Antenna short and broad.	Antenna long and slender, the 3rd joint being relatively broader.
Legs shorter and stouter.	Legs relatively slenderer and longer than *capitis*.

Disease relationship. Both head and body lice have been proved to be arthropod vectors of (1) Relapsing fever, (2) Epidemic typhus fever, and (3) Trench fever. The first two are very important and both have a wide distribution. In addition to these diseases, lice cause considerable inconvenience to man. The bite especially at night often interferes with sleep. Their bites are associated with local pigmentation of the skin; at one time it was known as vagabond's disease. The adenitis met with in the posterior triangle of the neck in children is generally associated with the presence of lice on the head. The irritation of the skin due to their bites may result in considerable scratching and formation of sores.

Relapsing fever. The louse-borne variety which is liable to occur in epidemics has a widespread incidence and epidemics have been reported from Eastern Europe, Africa, Syria, Iraq, India and China. Great epidemics occurred during and after the war and it is stated that in Russia in 1920 there were over one million cases. The tick-borne variety, which never occurs in epidemics but always sporadically, has a tropical and subtropical distribution. It is found in the northern half of Africa, Iran and in the neighbouring countries and Central Asia. Both types may co-exist in the same region. It is undoubtedly prevalent in an endemic state in Northern India.

Mackie (1907) was the first to find body lice infected with spirochaete in nature; these lice were taken from boys and girls at Nasik who were found to be suffering from relapsing fever. Later, Sergent and Foley (1910) were able to communicate this disease to monkeys by using infected lice. The life cycle of *Treponema* (*Spirochaeta*) *recurrentis*, the causative agent of relapsing fever, within the insect host is different from that of *R. prowazeki*. Within two hours after the spirochaeta have reached the midgut, the motility of the latter completely ceases and they quickly degenerate. Their presence in the body cannot be detected by any means. They reappear, however, between the 8th and the 12th day, an enormous numbers of active spirochaeta being now easily found in the haemocoele of the louse. At first they are small but later they attain the normal size found in the human peripheral blood. The infectivity of licc does not depend on the presence of visible spirochaetes. They are most infective about the 6th day just before the spirochaetes reappear in the insect. Their infectivity subsequently diminishes. Lice may occasionally be infective as early as the third day and as late as the 15th day after the infective blood meal. The usual method of infection is by crushing the insect at the time it bites which causes the spirochaeta to escape on the human body. Infection cannot be produced by their bites alone.

Typhus fever. The following are the principal types of typhus fevers which have different aetiological entities and which have also different modes of transmission.

1. **Tick-borne typhus.** It occurs in the United States and in the Mediterranean littoral and is caused by *Dermacentroxenus rickettsiae*. Ticks are the transmitting agents. Cases from South Africa and India have also been reported.

2. **Mite-borne typhus.** It is a disease connected with shrubs and jungle and occurs in Eastern Asia as far south as North Australia. The rickettsia, *R. orientalis*, is carried by larval trombiculid mites. The virus of Japanese River fever can be transmitted experimentally by lice or fleas although the normal vectors are the mites, *T. akamushi* and *T. deliensis*. According to Zinsser and Mooser these experimental variations do not produce fundamental changes in the viruses. When the insect host is

of the unusual type, the rickettsial parasite becomes considerably attenuated in its pathogenicity.

3. **Q. fever.** This disease has been reported from Australia and United States. The causative agent is *R. burneti* and is transmitted directly and through ticks.

4. **Endemic typhus.** It has a world wide distribution and the causative agent is *R. mooseri.* Rat-fleas are vectors of this disease.

5. **Epidemic typhus.** (Classical typhus). It principally occurs in South Eastern and Eastern Europe and China, and is caused by *Rickettsia prowazeki.* It is carried principally by the body louse, *Pediculus humanus corporis*, and the head louse probably plays a much less important role. The lice of the monkey and the lice of susceptible animals such as guineapigs or rats may transmit the virus, *R. prowazeki*, between the respective animals (Burnet, 1937). The infection never takes place by the bite of the infected louse.

It is primarily a disease of man and occurs in epidemics. The home of epidemic typhus is Eastern Poland, Russia and the Balkans in Europe and in Asia, China. In India it is the most important epidemic disease in the mountain regions of Kashmir, the North Western Frontier Province and in the mountains on the Pakistan-Iranian border. In India classical typhus has never occurred in the plains. The basic factors in the spread of this disease are: (*a*) great movement of population, non-immunes being brought into infected areas; (*b*) great increase of the louse population on man which occurs in the winter when personal sanitation is generally neglected; and (*c*) overcrowding, famine, political upheavals, war, etc. Anything that causes overcrowding in a limited space will lead to a more intimate contact amongst man thus favouring the transfer of the louse from one individual to another. During the last world-war it occurred in Algeria, Egypt, Morocco, Tunisia, Germany, Hungary, Poland, Rumania, Italy and Spain. The fuller reports have been published in Public Health Reports, Washington 1942, Vol. 57, p. 630 and p. 682.

Although Nicolle *et al.* (1909) had demonstrated the transmission of typhus fever from monkey to monkey by means of the bites of infected lice, the causative agent was not known till it was discovered by Rocha Lima (1916) both in the peripheral blood of patients suffering from the disease and also from infected lice. The causative organisms, *Rickettsia prowazeki*, are found in clusters or pairs within the intestinal cells of the louse and can readily be stained with either Giemsa's or Romanowsky's stain. The rickettsia, *R. prowazeki*, connected with typhus and which is pathogenic, is different from the normal rickettsia, *R. pediculi*, harboured by healthy lice, and which is non-pathogenic. In lice *R. prowazeki* undergo multiplication in the epithelial cells of the intestine whereas *R. pediculi* lie free in the lumen. In infected lice *R. prowazeki* are passed in large numbers in the faeces and the organisms remain alive in the faeces even when dried. It has been suggested that the infective agent contained in the excreta may be carried with dust by wind and enter through the conjunctival mucous membrane. The infection is not, however, carried by the bite, the salivary glands being free from rickettsia, but by the faeces of the insect (Atkin and Bacot, 1922). Hereditary infection of the disease does not take place in the louse.

Lice can be artificially infected by injecting infecting meal of platelet material through the anus. Under such conditions rickettsia bodies are found in the faeces from the 5th day onwards (Bacot and Segal, 1922). In about 10 days the infected intestines normally sterile except for the rickettsiae, are dissected out and emulsified in carbolised saline. Weigl's prophylactic louse vaccine is based on this principle (1930). The vaccine, however, does not confer absolute protection against infection.

The rickettsial bodies with which lice become infected, are present in the peripheral blood of the patient up to about the tenth day of the disease. Infected lice begin to pass the rickettsiae in their faeces from 5 to 9 days after the infective feed.

Though both *corporis* and *capitis* can equally convey the disease to monkeys, epidemics of the disease are associated rather with the body louse, and generally when the infestation rate among the people and the gross infestation rate on the individual are both high. The great prevalence of body lice in winter is due to suitable combinations of temperature and humidity existing in the confined space between the inner garment and the skin. This environment is extremely favourable to the rapid multiplication of lice. In summer the air in this space is nearly saturated with moisture due to perspiration, a condition in which lice cannot thrive.

Rickettsia infection in louse is determined by teasing the gut and staining the teased-up bits with Giemsa's. Freshly passed faeces may also be examined but the former method is the better.

Trench fever. Trench fever, like typhus fever, is carried by *P. humanus* var. *corporis.* This disease first appeared in the British Expeditionary Force in 1915. The method of infection is by faecal contamination of the wound caused by rubbing and scratching. The louse remains infective for at least 23 days after feeding on the infected patients. The man suffering from trench fever is not infectious to lice for the first three days of his illness; after that he infects them throughout the fever and convalescence.

Breeding of Lice. Both head and body lice may be conveniently raised in a pill box. A hole is punched in the paste board lid and closed by a piece of fine chiffon tied with a silk thread around the rim of the box. The inside is filled with a mass of hair. It is placed inside a leather strap and worn around the wrist (Nuttall, 1917). The pill box may also be conveniently placed between the skin and the socks (Buxton, 1940). Warburton wrapped up his lice with cotton wool and kept them in his pocket.

Dissection. The insect is first starved and then rapidly killed in water warmed to 80° C. The insect is then put in 30 per cent alcohol. The louse is placed on a smooth glass slide with the ventral side upwards and is held in position by exerting a little pressure on the thorax with the head of a blunt needle. Then a fine longitudinal incision is made with the help of a sharpened triangular needle on each side of the thorax and the entire abdomen near their lateral margins. After the insect has been pinned on a paraffin tray at both ends, the sternites are removed and the tracheal and other attachments are cut. The alimentary tract can thus be easily exposed.

Control of head louse. (1) Shaving the hair. (2) 5 per cent DDT dust and repeated every night for one week. No child with scalp abrasion should be treated with DDT, but instead with 5 per cent pyrethrum powder. (3) Pyrethrum in kerosene mixture diluted to contain 0.12 mg. percentage pyrethrins and applied to every part of the hair. The action is quick and it kills both adult lice and eggs. (4) A mixture of Phenyle cellosolve—(40%), Ethenol (3%), water (25%) and Methyl salicylate (5%) kills both eggs and adult lice quickly. During application, eyes must be particularly protected. (5) NBIN : Only one application is necessary.

[For details, see in Part II.]

Control of body louse. This essentially comprises the disinfestation of garments, blankets etc. The disinfestation may be effected in various ways such as: (1) Cold : Articles like fur, feather, etc., which are likely to be destroyed by heat, either moist or dry, should be left in a refrigerator for 10 days. According to Chung (1937), a temperature of 17°C. (1°F) is fatal to nits in two hours. The eggs are more resistant to cold than the active stages. (2) Dry atmosphere—Warm clothing may be disinfested by storing the garments in a dry atmosphere for two weeks. Dry storage for 2 to 3 weeks at a temperature of 12 to 20°C. should suffice to dispose of the nits. Sunning clothes for disinfestation is not effective. (3) Dry heat—A temperature of 60°C. (140 F.) for 20 minutes will kill both eggs and adults. Hot ironing—Hot flat iron is especially useful for removing infestation of woolen goods. (4) Moist heat—Both nits and active stages of lice are rapidly killed in hot or boiling water. This is the most economical and useful way of dealing with *P. corporis*. The articles may also be exposed directly to a jet of steam from a kettle. Disinfestation of undergarments, clothing,

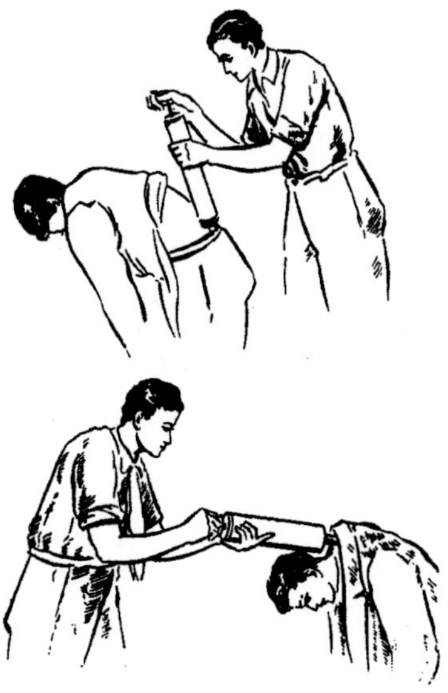

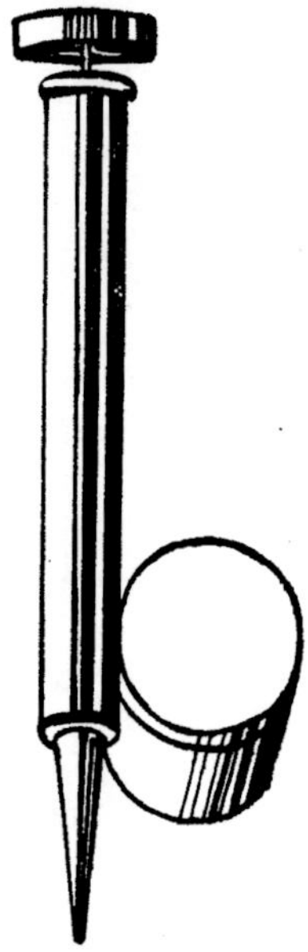

Fig. 151.

Disinfestation of garments while worn on the body, with DDT.

DDT dust blower.

bedding, blankets etc., may be effected by means of superheated steam. This is a very suitable war time measure. Any modification of the Serbian wine barrel type of disinfestor may be used. (5) Immersion in water in order to be effective must be continued for at least 5 days. At least 5 minutes' immersion in kerosene is necessary. Even after 3 minutes' immersion many of them not only survive but may even lay eggs (Joffick, 1924). (6) Naphthalene—It acts essentially as a contact poison and on the adult louse only but not on eggs. When rubbed into the shirt it causes itching and dermatitis.

[For disinfestation of garments and blankets by chemical insecticides, see in Part II.

Genus **Phthirus. Phth. pubis** L. This genus contains the only species *Phth. pubis*. This is popularly called crab louse on account of its general appearance and its powerful legs and claws. The general features of this insect are the following : (*a*) body is as broad as long; (*b*) head impacted on the thorax; (*c*) first pair of legs extremely slender; (*d*) thumb-like process from the tibia is not developed; (*e*) the abdominal segments have lateral projections; (*f*) the second abdominal segment bears three spiracles on each side lying almost in a line. The male is usually smaller than the female. The lateral processes on the abdominal segments are longer and thicker in the female. The posterior end of the abdomen is rounded in the male and bilobed in the female.

This species is less frequently encountered on man and is not known to convey any infective disease. It leads to considerable discomfort on account of the irritation of the skin caused by it. The parasites are usually confined to the pubic and perineal regions and even spread to the abdomen, thighs and forearms. They have never been found on the beard. In young children the louse may be localised on the eyelashes; this condition is extremely troublesome and leads to blepharitis and conjunctivitis. On rare occasions they are noticed on the temple. The infestation is at times a venereal one but it may be readily acquired from lavatory seats. The louse may also be picked up by direct contact.

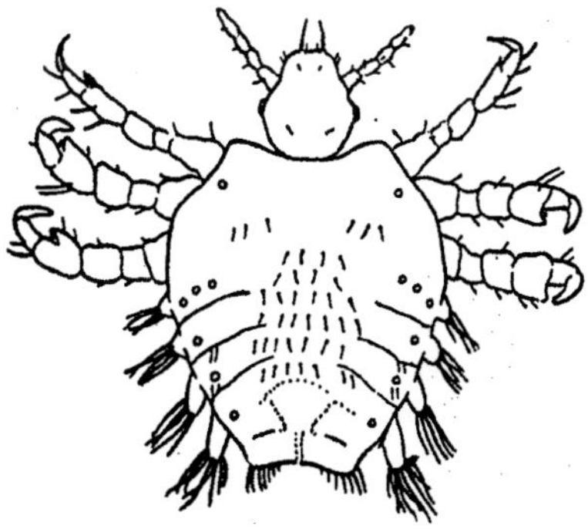

FIG. 152.
Phthirus pubis, female.

Treatment. (*a*) Shaving or cutting the hair. (*b*) Spraying with pyrethrum extract. (*c*) Use of 10 per cent DDT powder; the application should be repeated at night for seven nights in succession. (*d*) For eyelashes, pyrethrum ointment prepared with powdered pyrethrum flowers and vaseline in 8 per cent strength. The ointment should be repeated for seven nights in succession. Pyrethrum does not cause any irritation to the eyes. (*e*) Against the pubic louse the formula used in the Indian army is as follows : Benzyl benzoate—10 g., Ethyl p-aminobenzoate (benzocaine)—2 g., DDT—1 g., Ethyl alcohol—100 cc. The mixture is applied by means of a spray on the hairy parts of the body particularly the beard, moustache, chest, arm, pubic and anal regions and the thighs. No bath should be allowed for 24 hours. (*f*) For chemical pediculicides, see in Part II.

Family **Haematopinidae.** Genus **Haematopinus.** They are found on ungulates. They are of large size. The vestigial eyes are placed on projecting tubercles; there is a small triangular platelet between tibia and tarsus. Antenna is 5-jointed. *H. suis* L., is found on pigs; *H. tuberculatus* Burm., on the buffalo, and *H. eurysternus* Nitz., on cattle. The last one is the cosmopolitan short-nosed ox louse.

Genus **Linognathus.** They are found on dog, sheep and goat. Though *Linognathus* resembles *Polyplax*, the abdominal segments are soft and not transversely subdivided. Each segment has 2 or 3 rows of long hairs. *L. vituli* L., is the cosmopolitan long-nosed ox-louse, *L. pedalis* Osb., found on sheep, *L. stenopsis* Burn., on goats and *L. setosus* Olf., on dogs and foxes.

Genus **Polyplax.** These are found on rodents. A large number of species have been described. This genus is characterised by the presence of not more than two rows of transverse bristles on any segment of the abdomen. Antenna 5-jointed; eyes absent; front legs slenderer than others; no pretarsal platelets; abdominal segments are well chitinised and transversely subdivided. They are found on rodents.

Genus **Bovicola** (*Damalinia*). They are parasitic on cattle (*B. bovis* L., formerly *Tvichodectes scalaris*), equines (*B. equi* L.), sheep (*B. ovis* L.), and goats (*B. caprae* Gur).

Genus **Solenopotes.** Abdominal segments with one row of hairs; abdominal spiracles very large with a postero-lateral spicule. Host-ox.

Dairy cattle, sheep and pigs are liable to suffer from mange mites and lice. Pigs often suffer from heavy infestation with *H. suis* which is responsible for the spread of swine fever.

[For control, see in Part II.]

Order **Mallophaga.** (Biting lice). This is a small order and though the insects belonging to this order are popularly called bird lice, in reality they are more related to Orthoptera. In general appearance they are flattened, wingless, highly chitinised insects which do not undergo any metamorphosis. The head is usually large and the mouth parts are represented by a pair of mandibles used for chewing. They feed on epidermis and sometimes on feather. The eggs are laid on the under surface of the feathers and frequently at the base. The eggs have a characteristic shape for each species, and are deposited in little groups. They are generally parasites of birds and mammals. One genus, *Trichodectes,* is found on goats, dogs,, sheep etc.

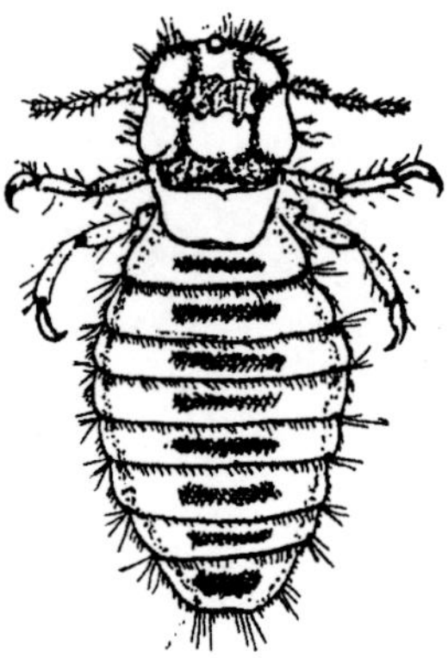

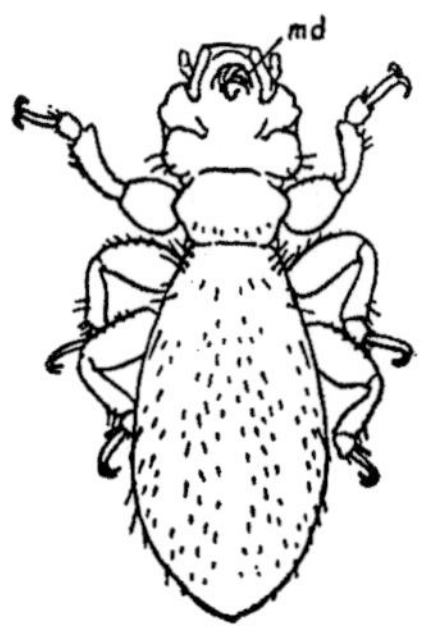

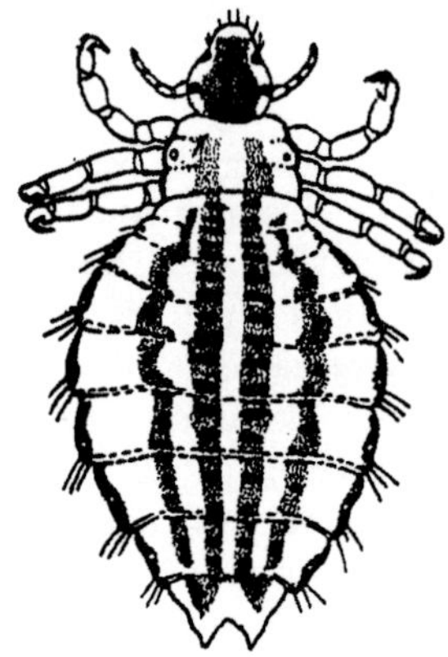

FIG. 153.

Trichodectes canis.

Mallophaga.
md, mandible.

FIG. 154.
Haematopinus sp.

They can be easily distinguished from sucking lice by their large head, mandibles and highly chitinised body.

The eggs of *Dipylidium caninum* passed in the faeces of dogs and cats undergo development to the "cysticercoid" stage in the dog-louse, *Trichodectes canis* Deg. (*latus*), and the dog flea, *Ct. canis.* The mechanism of transmission by the dog-louse is the same as by the dog and cat-flea of the genus *Ctenocephalides.* Infection follows the accidental ingestion of the flea by the definitive host. In *Trichodectes* the antenna is 3-segmented and the tarsi end in a single claw.

REFERENCES (Anoplura and Hemiptera)

Aschner, M. (1934). *Parasitology,* 25 : 309. **Atkin, E. E.** and **Bacot, A. W.** (1922). *Brit. Jl. Exp. Path.,* 3 : 196. **Bacot, A. W.** (1914). *Bull. Ent. Res.,* 5, 111; **Bacot, A. W.** (1917). *Parasitology,* 9 : 228. **Bacot, A. W.** and **Segal., J.** (1922). *Brit. Jl. Exp. Path.,* 3 : 125. **Brumpt, E.** (1912). *Bull. Soc. Path. Exot.,* 5 : 360. **Burnet, E.** (1937). *Arch. Inst. Pasteur,* Tunis, 26 : 391. **Bushland, R. C., McAlister, Eddy, Jones** and **Knfppling, E. F.** (1944). *J. Parasit.,* 30 : 388. **Buxton, P. A.** (1938). *Parasitology,* 30 : 65. **Buxton, P. A.** (1940). *Ibid,* 32 : 303. **Buxton, P. A.** (1936). The Louse, 2nd Ed., Edwards Arnold & Co., London. **Castaneda, M. R.** and **Zinsser, H.** (1930). *J. Expt. Med.,* 52 : 661. **Chung, H.** (1937). *Proc. Soc. Exp. Biol. & Med.,* 36 : 324. **Cornwall, J. W.** and **Patton, W. S.** (1914). *Ind. J. Med. Res.,* 2 : 569. **Cragg, F. W.** (1915). *Ibid.* 2 : 698. **Cragg, F. W.** (1922). *Ibid,* 10 : 78. **Cragg, F. W.** (1923). *Ibid,* 11 : 449. **Culpepper, G. H.** (1947). *Amre. J. Trop. Med.,* 28 : 499. **Da Rocha Lima, H.** (1916). *Arch. f. Schiffs-u. Tropen-Hyg.,* 20 : 17. **David, W. A. L.** (1944). *Amer. J. Hyg.,* 39 : 177. **Dias ,E., Laranja,** and **Nobrega** (1945). *Mem. Inst. Osw. Cruz.,* 43 : 495. **Fuller, H. S., Murray and Snyder** (1949). *Pub. Hlth. Rep.,* 64 : 1287. **Haddow, A. J.** (1941). *Parasitology,* 33 : 40. **Hase, A.** (1930). *Z. Parasitenk.,* 2 : 368. **Johnson, C. G.** (1940). *Parasitology,* 32 : 127. **Kei-**

lin, D. and **Nuttall, G. H. F.** (1919). *Ibid,* 11 : 279. **Kitselman, C. H.** and **Grundmann, A. W.** (1940). *Kansas Agric. Exp. Stat. Tech. Bull.,* 50. **Knipling, E. F.** (1949). *Agric. Chemis.,* 4: 43. **Mackie, F. P.** (1907). *Brit. Med. Jl.,* ii, 1706. **Musgrave, A. J.** (1946). *Bull. Ent. Res.,* 37 : 43. **Nauck, E. G.** and **Zumpt, F.** (1940). *Zentralbl. Bakt., Abt. I* (originale), 146 : 97. **Nicolle, C., Comte** and **Conseil** (1909). *C. R. Acad. Sci.,* 158 : 1815. **Nuttall, G. H. F.** (1917). *Parasitology,* 10 : 80. **Omori, N.** (1935). *Bot. & Zool.,* 589. **Omori, N.** (1938). *Jl. Med. Assoc. Formosa.* 37 : 1004. **Omori, N.** (1939). *Verb. 7. int. Kongr. Ent.,* Berlin, 1938 : 895. **Omori, N.** (1941). *Jl. Med. Assoc. Formosa,* 40 : 555. **Patton, W. S.** and **Cragg, F. W.** (1913). A Text Book of Medical Entomology, Christian Literature Society for India, London. **Puri, I. M.** (1924). *Parasitology,* 16 : 84. **Rivnay, E.** (1932). *Ibid,* 24 : 121. **Rodhain, J.** and **Brutsaert, P.** (1935). *C. R. Soc. Biol.,* 118 : 1228. **Roy, D. N.** and **Ghosh, S. M.** (1944). *Bull. Ent. Res.,* 35 : 231. **Sergent, E.** and **Foley, H.** (1910). *Ann. Inst. Pasteur,* 24 : 337. **Sikora, H.** (1919). *Biol. Zool.,* 39. **Smith, M. I.** and **Stohlmann, E. F.** (1944). *Pub. Hlth. Rep.,* 59 : 984. **Warburton, C.** (1910). *Rep. Local Govt. Board on Pub. Hlth. & Med. Subj.,* London, No. 27 : 23. **Weigl, R.** (1930). *Bull. Int. Acad. Cracorie Classe de Med.,* 4 : 25. **Wigglesworth, V. B.** (1941). *Parasitology,* 33 : 67. **Wilson, S. G.** (1948). *Bull. Ent. Res.,* 39 : 423. **Wood, F. D.** and **Wood, S. F.** (1941). *Amer. J. Trop. Med.,* 21 : 335. **Wood, S. E.** (1943). *Ibid,* 23 : 315. **Zinsser** and **Mooser.** See Burnet (1937) above.

CHAPTER XIV

Class : **ARACHNIDA**

The class Arachnida includes such creatures as ticks, mites, scorpions, spiders, king-crabs, tongue-worms, etc. A majority of them are terrestrial and only a small proportion live in water. They possess certain well-defined characteristics. These are : (*a*) The body, as distinct from insects, is not regionally demarcated, the proper segmentation being obliterated. Two distinct parts are seen, the anterior one being the cephalothorax and the posterior, the abdomen. (*b*) Antennae are absent. By this character alone this group can be easily distinguished from others. (*c*) There are two pairs of mouth appendages; these are : (*i*) one pair of chelicerae placed in front of the mouth, and (*ii*) a pair of pedipalps which lie one on either side of the mouth. With the aid of these mouth organs the insect is able to suck fluid food. (*d*) In addition to the two pairs of mouth appendages, chelicerae and pedipalps, there are four pairs of walking legs. (*e*) Metamorphosis is incomplete. Though a large majority is oviparous, there are some which are viviparous. (*f*) In this class the larva is characterised by the presence of three pairs of walking legs and the nymph, four pairs. The nymph can, however, be easily distinguished from the adult as it is only in the latter stage that the genital aperture makes its appearance. (*g*) Sexes are separate.

The respiratory system differs widely among the several orders. Thus scorpions and spiders breathe by means of special organs called lung books; these open externally on the abdomen. The king crab (*Limulus*) breathes by means of thin lamellae which are attached to the abdominal legs. Ticks and a majority of mites breathe by means of tracheae which ramify throughout the body and open externally on the cuticle. Itch-mites and follicle-mites are devoid of tracheae and they breathe through the integument.

Arachnida are grouped in 11 orders and among them only a few are parasites, e.g., ticks, mites, and tongue-worms. The important orders are classified as follows.

Class : **Arachnida**

Orders	Acarina (Ticks and mites.)	Scorpionidea (Scorpions)	Araneida (Spiders)	Pentastomida (Tongue worms)

Suborders (of Acarina):

Suborders	Prostigmata (Velvet or harvest mites).	Mesostigmata (Hard ticks).	Metastigmata (Soft ticks; also insect and rat mites).	Astigmata (Itch mites; also cheese mites).	Heterostigmata (Soft-bodied mites).	Vermiformia (Follicle mites).

Order Acarina. (Ticks and mites). The order Acarina represents both ticks and mites. The ticks are all ectoparasites of vertebrate animals, both warm and cold blooded, and all live on blood. The sexual differentiation is often marked. Many of them act as vectors of numerous protozoal and a few rickettsial diseases. Some species cause a dangerous paralysis in animals and also in man.

In the tick the cephalothorax and the abdomen are fused and though there may be a vestige of constriction in the body demarcating the false head from the latter, there is no true segmentation as in insects. The adults and nymphs are all octopods, larvae hexapods. The mouth parts consist of mandibles and pedipalps. All are terrestrial and parasitic.

Though the general outline of the mite resembles that of the tick yet its life history presents many points of difference. Only a few are parasitic and a small number suck blood, the majority being free living. They are both terrestrial and aquatic. A few among them are concerned in the transmission of human diseases.

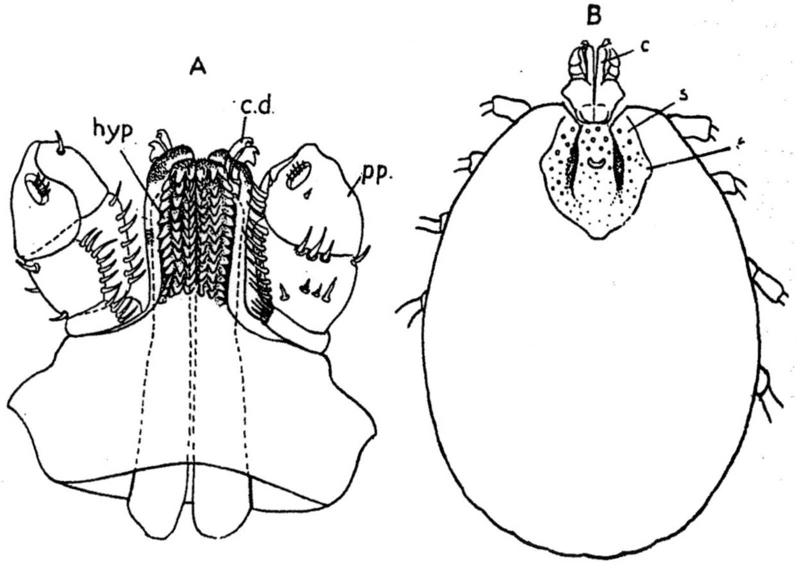

FIG. 155.
A. Head of *R. sanguineus*, (ventral aspect). B. *R. sanguineus*, female, (dorsal aspect). *c*, chelicera ; *cd*, cheliceral digits ; *e*, eye ; *pp*, pedipalp ; *s*, scutum ; *hyp*, hypostome.

Superfamily **Ixodoidea.** The superfamily Ixodoidea includes ticks only. They are distinguished from other Acarina by their relatively large size, by the position of the spiracle (which is absent in the larva) in the posterior half of the body behind the last pair or between the third and fourth pair of legs and absence of prominent hairs on the body. These are oval in shape, of varying colours, dorsoventrally compressed and slow in their movements. They are represented in almost every part of the globe. Ticks live strictly parasitic lives and are responsible for animal diseases due to *Treponema*, *Babesia*, *Rickettsia*, *Theileria* and *Anaplasma*. A spirochaeta of relapsing fever is carried by ticks which also play some part in the spread of typhus-like diseases in various parts of the world. The bite of some ticks is often followed by paralysis and this is particularly noticed in sheep and occasionally in man. The range of hosts is wide which includes not only warm-blooded animals, such as mammals, birds etc., but also cold-blooded creatures such as toads and reptiles. During their parasitic phase, they attach themselves to the host, and while they are free-living they pass a part of their life on ground. There are four distinct stages in their life cycle, egg, larva, nymph and adult. All exist strictly on blood.

Body. The body is oval, segmentation being obliterated. At the anterior end lies the false head or capitulum, and posteriorly the abdomen. The capitulum has a rectangular or hexagonal shape which fits into an excavation of the body. From its anterior part projects the proboscis or beak; this is composed of a pair of chelicerae and an unpaired hypostome. The pedipalps lie lateral to the chelicerae and are inserted at the antero-lateral angle of the basis capituli.

The chelicerae are the actual cutting instruments. They are capable of retraction. Each consists of a chitinous bar to the end of which is articulated two stout digits or teeth adapted for cutting the skin. Each chelicera is enclosed in a separate sheath both dorsally and laterally and which is continous with the basis capituli. At its distal end the sheath is so attached to the chelicera that the two move together and apparently the two chelicerae look united to each other. On the ventral surface of the chelicerae lies the hypostome which is dart-like and is a median prolongation of the basis capituli; it also is enclosed in a sheath. The hypostome though it appears to be a single organ, is essentially composed of two similar organs attached to each other. The distal half of its ventral surface is armed with longitudinal rows of strong recurved teeth giving it the appearance of a file. The character and the number of teeth vary in the different genera and species. Proximally it forms the floor of the buccal cavity. The sucking tube is formed by the apposition of the chelicerae above and the hypostome below. Sen (1935), on the other hand, claims to have discovered a separate chitinous tube through which blood passes into the buccal cavity.

Abdomen. Dorsal surface : (*i*) A chitinous shield or scutum may or may not be present. When present it is larger in males occupying nearly the whole back, and much smaller in females being confined to the anterior portion only. The scutum is generally pitted and contains minute hairs. In soft ticks where the scutum is absent, the integument may be mamillose, pitted or wrinkled and may be furnished with minute hairs. (2) Eyes may or may not be present. In species of Ixodidae or hard ticks which possess eyes, they are present as far as is known in all stages, but eyes do not occur at all in the genera (of hard ticks) such as *Haemaphysalis*, *Ixodes* and *Aponomma.* On hard ticks there is only one pair of eyes. In some soft ticks there may be more than one pair. (3) In hard ticks certain grooves are noticeable upon the scutum, especially a pair of marginal grooves. In soft ticks also there are certain well defined grooves. (4) The posterior border of the scutum is festooned conspicuously in the male and in the fasting female.

Ventral surface : (*i*) Genital orifice between the second pair of coxae; (2) Anal orifice lying in the middle line further posteriorly and often surrounded by the anal groove; (3) In the males of certain species adanal plates are found as projecting chitinous ridges on either side of the anus; (4) Spiracles : their position is different in hard and soft ticks; in the former they lie posterior to coxa IV and in the latter between coxae III and IV; (5) 4 pair of legs. Leg : Each leg is composed of (*i*) coxa, (2) trochanter, (3) 2-jointed femur, (4) tibia, (5) protarsus, and (6) one tarsal segment ending in a pair of claws. On the outer border of the tarsus of the first pair of legs there exists a small pitted structure which is lined with sensory hairs; it is known as Haller's organ and has been proved to be olfactory in function (Hindle and Merriman, 1912).

Alimentary canal. The alimentary canal in ticks is different from that found in insects. It is a straight tube which begins at the mouth formed by the apposition of chelicerae and hypostome and terminates in the anus. The pharynx and the oesophagus are small and narrow. The mid-intestine is wide and is provided at its two extremities with caecal appendages which are considerably distended with blood after feeding. As ticks can withstand starvation for a long time, these diverticulae serve as the store-house of food. The hind-intestine is small and very narrow. It ends in the rectal sac. The two Malpighian tubules are extremely long in ticks and open into the rectum; thus the excretory matters are probably carried directly into the rectum. The mid-intestine of *Argas persicus* contains an anticoagulin but no haemolysin. The salivary glands, one on each side, are large, elongated, grape-like bodies. The common salivary duct opens on the floor of the mouth near the base of the hypostome. The secretion of the salivary glands is strongly haemolytic (Nuttall and Strickland, 1908), and the amount of anticoagulin depends on the time which has elapsed since the time the tick ingested blood. (Cornwall and Patton, 1914).

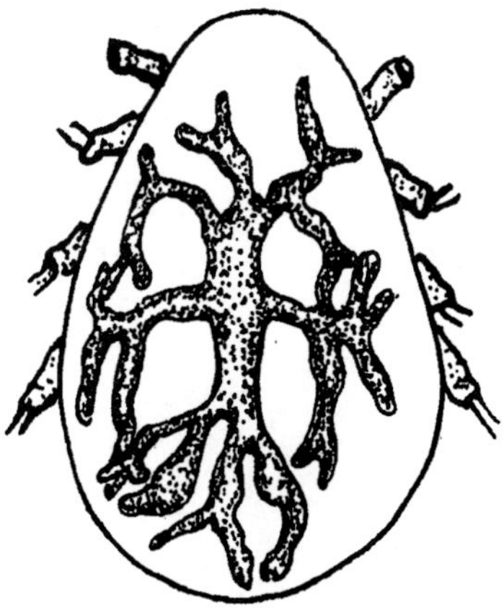

FIG.. 156.
Mid-intestine of *A. persicus* showing the intestinal diverticulae. The malpighian tubules and the salivary glands have been removed.

Coxal glands. The coxal glands are peculiar to ticks. They are two in number, one on either side of the body and are present in the adult and nymphal stages. They are comparatively small and are recognised with difficulty in dissections of the entire animal. They are readily seen in sections. They lie above the second pair of coxae. Each gland consists of the gland proper and a reservoir and the main duct passes between the first and the second coxa and opens externally behind the posterior border of the coxa of the first leg, the slit-like opening being concealed by a fold of the coxa. For some reason, which is not known, its secretion is increased while the operation of feeding is in progress; at that time the ventral surface of the body becomes bathed with the fluid. This is particularly marked in *Ornithodorus*. According to Christophers (1906), the secretion is slightly alkaline and prevents the coagulation of blood. The function of these glands may also have some relation to the excretion of nitrogenous waste products. These two tiny glands are capable of secreting a large amount of fluid either during feeding or soon after the feeding has been accomplished.

Reproductive system. In the female the uterus is a more or less globular organ placed in the middle of the body. The oviducts are convoluted. They open at the two antero-lateral angles of the uterus. The ovary is a large single tube placed across the body. The ovules are grape-like and contain eggs in different stages of maturation. The vagina is short and opens externally at the genital opening. A pair of accessory glands open into the vagina one on either side.

The *gene's organs*, so named by Nuttall after its discoverer, are also peculiar to ticks and are found only in the females in the anterior part of the body, hence they are called cephalic glands. They are concerned in the act of oviposition. They are visible as soon as the anterior end of the dorsal cuticle is reflected forwards. They are primarily concerned with egg-laying when they become considerably enlarged. The egg on being extruded out of the genital canal is rolled on the side of the tick by means of the everted lips of the genital aperture and thus it comes into contact with the secretion of these glands. In this way the egg is carried to the dorsal surface of the tick. This explains why eggs are found over the anterior part of the body of the female tick where the ducts open externally in a common opening, which is a broad concentric slit with everted lips and which is situated in the lower part of the camerostomal depression immediately anterior to the fold formed by the fusion of the dorsal surface of the basis capituli with the camerostome. It occupies the entire width of the camerostome. The camerostome is the cavity enclosing the capitulum.

FIG. 157.
Egg-laying by an Ixodine tick. Eggs are always noticed on the anterior part of the dorsal surface of the tick. After oviposition is complete, the female Ixodine tick dies.

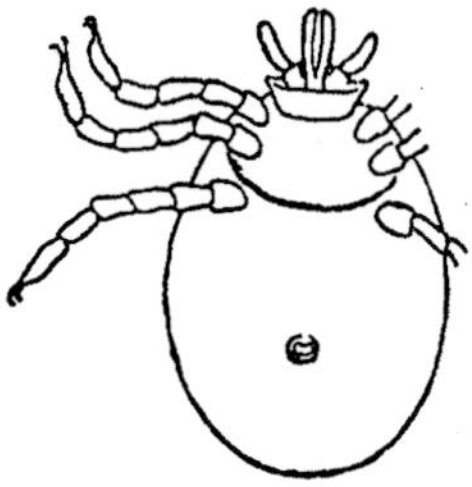

FIG. 158.
Larva of *Rhipicephalus*.

Life history. The metamorphosis is incomplete. The life cycle consists of four stages, (*a*) egg, (*b*) larva, (*c*) nymph and (*d*) adult. Mating may take place either on the host or on the ground. Eggs are laid in cracks, crevices or in grass adjacent to where the female has dropped off from the host. A very large number of eggs are laid. A female Ixodine tick may deposit as many as 15,000 eggs. Few arthropods possess a higher reproductive potential than do ticks. The oviposition often occupies more than a week. The egg is proportionately large, round and is brown in colour when ripe. Hard ticks deposit all their eggs in a single act of oviposition after which the female dies. The egg seldom hatches before three weeks. From the egg emerges the hexapod larva. The larva does not feed till a week has elapsed after its emergence. It climbs on grass, small shrubs etc., and waits for the chance arrival of its prey. It has ordinarily little tendency to attack any and every animal it finds, but only the proper type of host is selected. On finding one, it immediately attaches itself and begins to feed for some days. When gorged, it drops to the ground and remains quiescent during the time the blood is being digested. At the end of this period moulting takes place. The nymph is provided with eight legs.

In the same way as the larva does, the nymph seeks a new host, feeds and falls to the ground, where it moults for the last time to be transformed into an adult. After fertilization and feeding have been completed, the life history begins again.

This is an instance of a three-host tick where larval and nymphal ticks abandon their hosts after feeding. Metamorphosis to the next stage takes place on the ground. Thus they require access to a host on three occasions when after feeding in the larval, nymphal and adult stages they abandon their hosts. *Haemaphysalis leachi* is a 3-host tick. Other examples are *Rhipicephalus appendiculatus*, *R. sanguineus*, *R. simus*, *Amblyomma hebraeum*, *Dermacentor reticulatus*, *Ixodes ricinus*. *Boophilus* species are examples of one-host tick, for they remain upon a single host throughout their development from larva to replete adult. *R. evertsi*, which is a 2-host tick, undergoes metamorphosis from larva to nymph upon the host.

Both sexes are parasitic on vertebrate animals such as mammals, birds, snakes and other vertebrates, and they feed entirely on blood. The adult female *Dermacentor andersoni* can withdraw from 1.7 to 2 gms. of blood in the act of engorgement, whereas *D. venustus* about 1.95 gms. In many instances copulation will have a marked effect on the subsequent engorgement of the female and the presence of the male may also greatly influence the blood feed in regard to the time occupied during the process and also the quantity of blood ingested.

At the time of feeding, the tick holds its prey firmly with the help of tarsal claws and then makes a rent in the skin by the cheliceral digits. Through this opening the proboscis is inserted, the palps always remaining outside. The teeth pointing backwards and arranged in definite rows on the hypostome help the creature to anchor itself to the host at the time of feeding. The wounds caused by the insertion of the mouth parts of the ticks are scarcely noticeable, and little or no sensation is experienced at the time of attachment, during the feeding or at any time following the detachment of the tick.

The duration of the life cycle therefore must depend on the ease with which the tick in its different stages can find its suitable prey on which it can thrive. When it fails to find its host, it dies. In the case of the dog tick the cycle from the egg to the adult occupies a little over two months. In *Ornithodorus* it occupies about 7 to 8 months.

The longevity of ticks in their different stages varies in different species. As a rule the soft ticks survive for a longer period than the hard ticks. An unfed tick generally lives longer than a fed one. The longest periods that an adult hard tick has been known to survive is 413 days as in the case of *Dermacentor andersoni* reported by Hunter and Bishopp (1911), 1053 days in the case of ungorged females of *D. variabilis* as reported by Bishopp and Smith (1938).

Under the ordinary conditions it is always difficult to induce ticks to feed through a rabbit's skin used as a membrane but after the removal of the tips of the first pair of legs in which the olfactory organs are situated, they not only readily attach themselves to the membrane but also feed on various other fluids in addition to blood.

Ticks apparently have the power to regenerate lost parts such as amputated legs and the ability to repair mutilated mouth parts.

Parasites. Internal parasites belonging to the order Hymenopteron recorded from ticks in various parts of the world are ; *Ixodiphagus caucurtei* du Buys ser, *I.* [illegible]

How., and *Hunterellus hookeri* Hos. Amongst the three the last one has been recorded in India. *I. caucurtei* has been employed as natural enemies of *D. venustus*, but the rate of multiplication of the parasite is slow and the rate of parasitism low.

The most predacious enemies of ticks are various birds and domestic fowls.

The position of the respiratory openings and the shape of the body may be used in classifying the order Acarina into different groups or suborders.

Suborder	Family
1. **Prostigmata.** (Spiracles placed anteriorly on the basis capituli).	**Trombidiidae.** (Velvet or harvest mites). The pedipalps with an additional, subterminal finger-like appendix.
2. **Mesostigmata.** (Spiracles lie behind the 4th coxae).	**Ixodidae** (Hard ticks). Dorsal scutum present and varies in size in the two sexes.
3. **Metastigmata.** (Spiracles lie between the coxae of the 3rd and 4th pair of legs).	(1) **Argasidae** (Soft ticks). Dorsal scutum absent. (2) **Gamasidae** (Rat mites). The chelicerae are strongly exsertile; pedipalps long and 5-segmented.
4. **Astigmata.** (Spiracles absent).	(1) **Sarcoptidae** (Itch mites). Integument with fine parallel striae; legs stumpy, and they terminate in a long stalk; at the end of some of them is a sucker. (2) **Tyroglyphidae** (Forage mites). Integument not striated; chelicerae chelate and projecting like a rostrum; pedipalps small. Attack provisions and groceries.
5. **Heterostigmata.** (Spiracles placed otherwise).	**Tarsonemidae** (Plant and insect mites). The chelicerae are needle-like; the pedipalps are almost fused with the head.
6. **Vermiformia.** (Worm-like; spiracles absent).	**Demodicidae** (Follicle mites). Shape of the body worm-like; 4 pair of very stumpy legs; parasitic in sebacious glands of mammals.

The superfamily Ixodoidea is divided into two families, the Argasidae and Ixodidae, which differ not only in structure but also in habits. The Argasidae are generally known as soft ticks and the Ixodidae, hard ticks. The Argasidae are mostly inhabitants of warm climates whereas the Ixodidae have a world-wide distribution. Their distinguishing features are given below.

Difference between the two families, Ixodidae and Argasidae

Family **Ixodidae** (Hard ticks)	Family **Argasidae** (Soft ticks)
1. Scutum covers the whole back in the male and only a small part anteriorly in the female.	1. Scutum is entirely absent; hence they are known as soft ticks.
2. Capitulum is anterior.	2. Capitulum lies ventrally and mouth parts are not visible from above.
3. Pedipalps are not capable of movement like a leg.	3. Pedipalps are leg-like with subequal artile.sc

Family **Ixodidae** (Hard ticks)	Family **Argasidae** (Soft ticks)
4. Spiracles lie behind coxa IV.	4. Spiracles lie between coxae III and IV.
5. Sexual dimorphism is well marked.	5. Sexual dimorphism is absent.
6. Pulvilli always present.	6. Pulvilli absent or rudimentary.
7. Only one nymphal stage. After egg laying the female dies. The eggs may number several thousands.	7. More than one nymphal stage. The females lay eggs in batches after successive feeds and the total number of eggs laid by a female in a batch is comparatively small, say about 20 to 100.
8. They feed both in the day time and at night and take several days to become gorged. The female feeds only once.	8. They feed only at night and usually become replete with blood within 20 to 30 minutes. The female feeds more than once.
9. They are ectoparasites and remain on the host for a long time.	9. They hide like bed bugs and are parasites only during the time of feeding except in the larval stage when they are fixed parasites; they emerge from their hiding places at night.
10. They cannot resist starvation for a long time.	10. These ticks are able to live a year or even much longer without any food.

Family **Ixodidae.** The Ixodid or hard ticks are readily recognised by their scutum and also by the capitulum which lies anteriorly. The two sexes can easily be differentiated. The scutum or shield is derived from the integument and becomes highly sclerotised. In the male it covers almost the whole back and thus prevents the tick from increasing to any appreciable extent in size, whereas in the female it covers only a small part of the back, thus allowing the tick to increase considerably; the female is therefore much larger. The capitulum is always terminal in all stages. Eyes, if present, lie at the sides of the scutum.

The nymphs and the adults remain for a week or more upon the host. The females lay several thousands of eggs, feed but once and die after oviposition is completed. They generally do not live long when fed, but when unfed they may remain alive for a long time provided the conditions are favourable. The metamorphosis in 3-host and 2-host ticks takes place on the ground.

There are in all 9 genera of Ixodidae and they have been grouped by Nuttall as follows :

Ixodidae

- Prostriata — Genus *Ixodes*
- Metastriata
  - Brevirostrata
    - Group I — Genera. *Haemaphysalis.*
    - Group II — *Dermacentor. Rhipicentor. Rhipicephalus. Margaropus. Boophilus.*
  - Longirostrata
    - Group I — *Hyalomma.*
    - Group II — *Amblyomma.*

Among the 9 genera, two *Margaropus* and *Rhipicentor*, do not occur in India. In Prostriata the anal groove surrounds the anus in front, while in Metastriata it lies behind it or is obsolete.

Genus **Ixodes** Latr. Anal grooves surrounding the anus in front; scutum inornate; eyes and festoons absent; spiracles round or oval; sexual dimorphism pronounced in respect to the capitulum; males with ventral surface of the body covered by hard non-salient plates.

The adults are found on a large variety of hosts, e.g., sheep, cattle, dogs, birds etc. They have been recorded from Europe, Japan, North America etc. On rare occasions man may be attacked.

I. ricinus L, is the common sheep tick in Europe. On sheep pastures in infested farms larvae, nymphs and adults are exceedingly plentiful at certain seasons. It is suspected of conveying to sheep in Scotland the virus of an encephalomyelitis called "louping ill." It is a three-host tick with a long list of wild and domestic hosts, the important being sheep and cattle. It is the common European cattle tick and is responsible for piroplasmosis of cattle caused by *Babesia bovis*. It is suspected of carrying *B. canis* in France. In Europe *I. ricinus* commonly attacks man. This tick is also distributed in North Africa, W. America and W. Asia. It has no eyes. The shield is horseshoe shaped. When engorged, the female resembles a castor bean slightly flattened, and becomes a dark bluish colour.

In *I. holocyclus* Neum., which has been recorded from Western Australia and India, the limbs of the anal groove converge posteriorly. In Australia it is a common parasite of dogs, marsupials, bandicoots and even birds, bandicoots being its most important hosts. Man is occasionally attacked. *I. pilosus* Koch, is primarily a sheep tick in South Africa though found on other animals. *Ixodes hexagonus* Leach., is found on dogs in the northern parts of France. *I. ricinus* and *I. acutitarsus* have been recorded from India attacking sheep, squirrel, etc. They are known only as coming from the hilly regions. *I. ricinus* is known to feed on a variety of animals, e.g., cattle, horse, tiger, mouse, etc.

Genus **Haemaphysalis** Koch. Inornate without eyes; capitulum transversely oblong about twice as wide as long; second article of palp projects laterally beyond the base, this feature being of diagnostic significance; coxa of leg I not bifid in either sex with a blade-like dorsal retrograde spur on the first trochanter; sexual dimorphism is slight and no adanal or ventral shields in males; coxa IV of normal size; spiracles usually ovoid or short comma-shaped.

These are usually of small size and of a yellow or brown colour but the engorged female is bluish. It is an interrupted feeder. While the generic identification is not difficult on account of the peculiar feature of article 2, the specific identification is not easy. The diagnosis may have to be based on numerous minor points.

H. bispinosa Neum., is common on a wide variety of domestic and wild animals in India. It is essentially an Asian species which has spread to other countries such as Africa, being carried on cattle. In this species the palps though salient laterally yet article II has no ventral retrograde process; there is a moderate sharp spur on coxa I; the male scutum is finely punctate; in female the dorsal spine on palpal article III moderate and more or less erect; trochanter spurs feeble or absent. *H. leachi* Aud., is the principal cattle and horse tick in Australia. The palps protrude greatly in this species. In South Africa it is the chief carrier of canine piroplasmosis and is especially noticed upon the dog, the younger stages being found on rodents and in their nests. The transmission is effected by the adult tick which acquired the infection in its previous sexual forms, the infection having passed from the mother tick into the eggs, remaining latent in the larvae, in the nymphae and appearing in the adult. The tick's offspring are not capable of conveying the disease until it attains sexual maturity. In other words the infected larvae are not infective whereas the adults are (Lounsbury, 1904). It does not attack man. In India it has been recorded from tigers. *H. montgomeryi* Nuttall, found on horses and sheep and distributed in India in the hilly regions in the northern part of the United Provinces but mostly in Kashmir. It is common in the U. S. In this species the trochanters are all spurred; palps very slightly salient laterally and article 2 with dorsal and ventral retrograde process on the lateral salience; scutum in both sexes narrow. *H. leporis-palustris* Pack., is the North American rabbit tick and is responsible for maintaining a continuous reservoir of the rickettsia of Rocky Mountain spotted fever. Its importance as a vector of tularaemia has been proved. It does not bite man. It chiefly infests rabbits, both domestic and wild, and hares.

Birds are also important hosts of the immature stages; larvae and nymphae have also been collected from various other rodents, cats etc. The cotton-tail rabbit is an important source where this tick may be readily found. *H. punctata* is common on sheep in England. It is ellipsoidal in shape, marginal grooves deep, punctations on the scutum scattered and few in number, anal gooves bifurcate and Y-shaped. *Haemaphysalis humerosa* Warb-Nutt., is a 3-host Australian tick dropping to the ground after engorgement in each stage. All stages are frequently found on bandicoots, *Isodon torosus,* and generally in large numbers.

Genus **Dermacentor** Koch. Usually ornate; with eyes and festoons; basis capituli rectangular dorsally, wider than long; palpi short and thick, the second and third segments hardly, if at all, longer than wide; spiracle comma-shaped; coxa of leg I bifid in both sexes; male without adanal plates but with coxa of leg IV greatly enlarged.

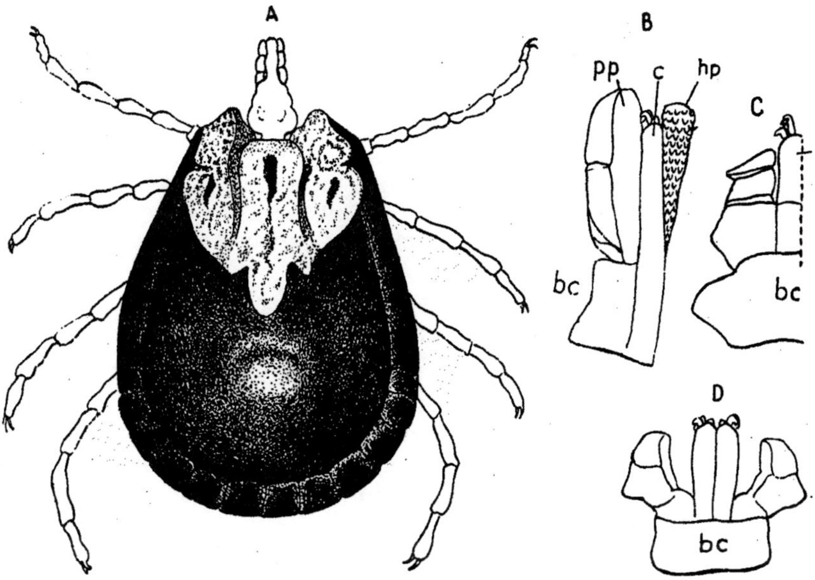

FIG. 159.

A. *Dermacentor andersoni,* female.
B. Head of *Hyalomma aegyptium.*
C. Head of *Boophilus australis.*
D. Head of *Haemaphysalis bispinosa.*
c, chelicera ; bc, basis capitulum ; pp, pedipalp ; hp, hypostome.

The only one of this genus which has attracted considerable attention is *D. venustus* Banks, the wood tick, which commonly attacks man and is the infective agent of the particular type of typhus prevalent in the Rocky Mountains in North America. It is a 3-host tick, the adult being found on domestic animals. It causes painless bites to man. The other species which is also responsible for human cases is *D. variabilis* Say, the common dog-tick in the central and eastern portions of the United States. Ungorged females may live as long as 1053 (Bishopp and Smith, 1938). Larvae and nymphae occur on mice and other small mammals.

Acquired immunity against ticks has been demonstrated by Trager (1939). One infestation of guinea-pigs or rabbits with larvae of the American dog-tick, *D. variabilis* Say., induces an acquired immunity which effectively prevents subsequent batches of larvae from engorging. In guineapigs the immunity develops fully within two weeks after the start of the first infestation and lasts at least three months.

Piroplasmosis in dogs due to *B. canis* in France, Russia and other European countries is transmitted by *D. reticulatus* Fab., which is found on dogs and other domestic animals. The immature stages are reluctant to attach themselves to dogs. In *D. reticulatus* the palpi are longer than hypostome and the second segment expanded distally and curved above with a backwardly directed spike; coxa of leg I strongly bidentate. It is widely distributed in Europe, Asia and America. It is readily found on horses and transmits equine piroplasmosis. It is a 3-host tick.

In British Columbia, *D. venustus* Banks (*D. andersoni* Stiles), popularly known as the spotted fever tick, and the commonest tick that attacks man, is also responsible for tick-paralysis which affects man, sheep and probably other animals. Its eggs have toxic properties (Gregson, 1941). Colorado fever is also caused by its bite. In Montana *Bacterium tularense* has been recovered in *D. venustus* collected under natural conditions. The bacteria are carried to the progeny (Parker, 1924).

The larvae and the nymphae of *D. venustus* feed on a wide range of small rodents especially the ground squirrel *Citellus* sp., which serves as hosts for the tick during its larval and nymphal stages. The nymphs hibernate and the adult stage is reached the following summer. *Dermacentor niveus* Neum, is widely distributed on domestic and field animals including rodents in Russia and sometimes attacks man. *D. nitens* Neum., is the horse-tick in Tropical America though rare in the U.S.A. It also occurs on mules. The inside of ears is the preferred place of its attachment. *D. occidentalis* Neum., the Pacific Coast tick, is the most abundant Ixodid in California and though all stages have been taken on cattle, horses, mules and dogs, the younger ones are particularly common on ground squirrels and deer mice, wood rats, rabbits, chipmunks etc. It has been found naturally infected in the adult stage with *Bact. tularense*. It is the common wood-tick of the Pacific Coast region and is often called the western dog-tick.

Only one species *D. auratus* Sup., has been recorded from India from wild animals and is of little economic importance.

Genus **Rhipicephalus** Koch. Inornate; with eyes and festoons, with short palps and basis capituli usually hexagonal dorsally; the third segment of palpi armed below with a backwardly directed spine; coxa I bifid; the male possesses a pair of accessory adanal shields; spiracles comma-shaped.

The genus *Rhipicephalus* contains many important species, e.g., *R. sanguineus* which conveys piroplasmosis in dogs; *R. appendiculatus*, the brown tick and the principal transmitter of East Coast fever; and *R. evertsi*, the red-legged tick, which is the transmitter of piroplasmosis of bovines and equines.

The *Rhipicephalus* is an interrupted feeder and requires three hosts during its development. Most of the members are brown in colour but the body and legs of *R. evertsi* are distinguished by being deep orange yellow.

R. sanguineus Latr. Its characters are : scutum with numerous punctations of variable size but evenly distributed, a few large ones being placed anterior to the lateral groove; the scutum in the male does not cover the entire dorsal surface of the body leaving the lateral portions exposed; eyes larger and more posterior than *R. bursa*. When fully fed, the male may show a slight caudal protrusion. Coxa I is visible dorsally.

R. sanguineus occurs in many parts of the world in tropical and warm climates being limited to 50°N. and 30°S. It is common upon the dog but also occurs on other animals. In many places it has been taken off horses, rodents and domestic animals. In Congo Free State it is common on cattle and occurs in association with *R. simus*, *R. longus* and occasionally *R. capensis*. The life history is similar to most other ticks. It is a three-host tick and has been proved by Christophers (1907) to be the transmitting agent

---

Chipmunk : It is the common name in North America and Canada for the ground squirrel. It has distinctive black and white stripes.

Marmot : It is a large burrowing Alpine rodent allied to the squirrels. There are numerous species ranging from the Alps through Asia north of the Himalaya. They are also found in North America. They are of stout built and have thickly haired tails. They are characterised by the absence of cheek pouches. They hibernate in winter.

Prairie dog : It is a North American burrowing rodent allied to the marmot. There are several species in the Western United States. The common prairie dog is a plump, short-tailed, squirrel-like animal of social habits. It lives in a burrow and feeds chiefly upon grass. They are called dogs on account of their barking habits.

of canine piroplasmosis in India and it doubtless occasionally carries the disease in Africa and elsewhere. In this tick, the behaviour of the parasite is different from what it does in *Haemaphysalis leachi*. The disease is heritable in the tick for nymphs descended from infected females and adults emerging from infected nymphs are capable of producing the disease in a new dog. The infection, however, remains in the larval stage.

It also carries the rickettsia of 'Boutonneuse fever' of the Mediterranean littoral. All stages of *R. sanguineus* are found on the dog. The larva remains attached to the host for 3 to 8 days and the nymph for the same period. At the different stages they remain off the host for about 2 to 3 weeks. The adult feeds on the dog for 4 to 7 days.

For keeping dogs clean, handpicking, particularly of the female ticks, should be done once a week. It is usual for dogs to contract infestation by the younger stages of this tick from the ground adjoining human habitations.

A mixture of 5 per cent. DDT and Gammexane powder, if regularly dusted in the hair, can be depended upon as an effective remedy against re-infestation. Derris powder with a rotenone content of 4 per cent. is also effective.

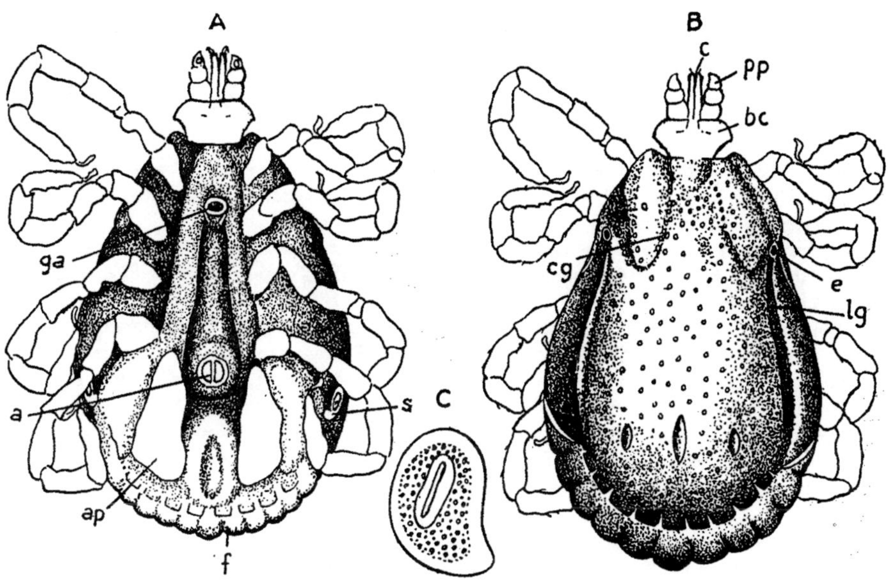

Fig. 160.

A & B. Ventral and dorsal aspects of *Rhipicephalus sanguineus*. C. Spiracle.
a, anus ; ap, adanal plate ; bc, basis capituli ; c, chelicera ; cg, cervical groove ; e, eye ; f, festoon ; ga, genital aperture ; lg, lateral groove ; pp, pedipalp ; s, spiracle.

R. appendiculatus Neum., is found mainly on cattle and is common in South and Central Africa. It is responsible for the Rhodesian East Coast fever caused by *Theileria parva*. The parasites are not transmitted hereditarily in the tick but only from one stage to another. This species is common in South Africa. It resembles *R. sanguineus*. The tail-like projection at the posterior end of the body is very distinct; coxa I is visible dorsally; punctations are very numerous and evenly distributed, a few large ones being placed anterior to the lateral groove; the marginal groove is deep, long and commencing near the eyes; eyes are flat and not prominent.

R. evertsi Neum., a 2-host tick, is common in South Africa. It is readily recognised by its eyes which are rounded and convex and lie in distinct orbits; the legs are bright orange red. It carries equine piroplasmosis in South Africa. It is not known whether it carries *B. caballi* or *B. (Nuttallia) equi*.

R. simus Koch. Dorsal scutum with large, equal punctations arranged in longitudinal lines with fine and hardly visible punctations; a 3-host tick; common in Africa especially on sheep, goats and dogs and is responsible for the carriage of *T. parva* to cattle. It is a stage to stage infection as in *R. appendiculatus*

Neum., which is the usual carrier in Africa. Ovine piroplasmosis is conveyed by *R. bursa* in Persia and Rumania. *R. bursa* is chiefly distributed along the shores of the Mediterranean.

R. bursa Can.-Fan. Punctations on dorsal plate very small but equal and close set; palpi with postero-external edges of segments 2 and 3 angular and outer border of segment I concave giving a constricted appearance to the palpi; coxae of last three pairs of legs with a small anterior-external spine and a small blunt posterior process.

R. nitens is confined to Africa being present on cattle, goats, and dogs and this tick and *R. capensis* are often associated with East Coast fever. *R. capensis* is also a 3-host tick.

This genus is represented in India by two species, *R. sanguineus* Lat., and *R. haemaphysaloides* Sup. The former is the most common Indian dog tick and the latter attacks both wild and domestic animals. The female *sanguineus* is characterised by punctations which are numerous, close-set, irregularly arranged and equal, larger ones being absent in the posterior portion of the median field, whereas in the female *haemaphysaloides*, the punctations are free, sparsely scattered, and strongly unequal; the larger ones are arranged in longitudinal rows and found in the posterior portion of the median field, finer ones being hardly visible.

Genus **Boophilus** Curt. *Boophilus* is essentially a cattle tick though found on other animals also. It is widely distributed and carries bovine piroplasmosis to cattle. This tick is a continuous feeder and remains on the host for about 3 weeks. It is a small tick and on account of the engorged female being bluish in colour, it is known as the blue tick.

B. annulatus Say. Festoons scarcely marked; eyes inconspicuous in males, the ventral surface is hairy and is furnished with two very large subequal, subsimilar adanal plates; coxae of leg I with a blunt anterior process and posteriorly bifid; remaining coxae with weak anterior spine.

In India this genus is represented by two species, *australis* Fuller and *annulatus* Say., the former being more common than the latter and represents the Indian form of the Texas fever tick. *B. australis* is a major pest of cattle and is widely distributed in India.

The life history of *B. australis* has been worked out by Sapre (1940) in India. Oviposition may extend over $1\frac{1}{2}$ months. The average number of eggs laid are 2552. The average duration of pre-oviposition, and post-oviposition periods and the egg stage, calculated from the beginning of oviposition to the hatching of the first larva, are 4, 18, 4 and 39 days. The larvae show a preference for the skin of the neck, dewlap and escutcheon. Feeding and metamorphosis last 3 and 4 days respectively in the larval stage, and 4 and 3 to 5 days in the nymphal stage. Unfertilised females do not engorge to the same extent as the fertilised ones. The adult female *B. australis* must be fertilised before it is able to reach a stage of repletion.

B. annulatus is prevalent on cattle in North America and Asia such as China, Formosa, Japan, etc., *B. australis* in South America, Australia, India and Africa and *B. decoloratus* Koch., in Africa, all being concerned in the transmission of piroplasma of cattle. There are a large number of varieties of *B. annulatus* described which have localised distribution, the difference being based on relatively characters of small importance. It is a one-host tick.

The remarkable investigation of Theobald Smith and Kilborne and published in the report of the Bureau of Animal Industry of the United States Department of Agriculture in 1893 supplied the first clear demonstration of the nature, causation and prevention of Texas or Southern cattle fever. The causative organism of Texas fever is *Babesia bovis* (*B. bigeminum*) and is transmitted by *B. annulatus* Say., *B. decoloratus* Koch., and other ticks. This organism infects the cow as the seed tick attaches itself for its first meal. Development of the protozoon parasite takes place in the tick's digestive tract. The developed organism passes through the wall of the gut from where it passes to the midgut and thence into the egg. Other changes in the organism occur during the development of the egg.

The parent tick is not therefore a direct carrier but as the infection passes through the eggs, its chances of reaching a susceptible animal are enormously multiplied.

In North America the chief carriers of Texas cattle fever are *Boophilus australis*, *B. annulatus decoloratus* and *Margaropus* (*Boophilus*) *annulatus*, whereas in South Africa the principal carrier is the blue-tick, *B. decoloratus*. In parts of Europe an identical type of the disease is transmitted by *Ixodes ricinus*.

In regard to *B. decoloratus* acting as an important transmitter of *B. bigeminum* in Africa, it has been observed that the tick retains the infection even when it has been reared for several generations on hosts that are not susceptible. Hereditary transmission has also been discovered in *Rhipicephalus bursa* C. & P.

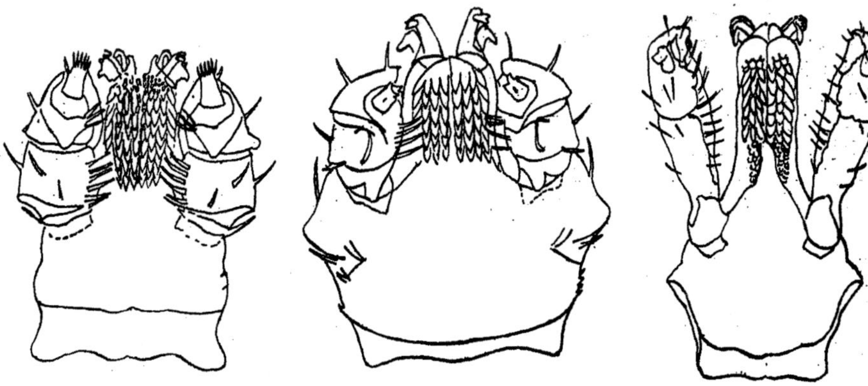

FIG. 161.
Capitulum of (A) *Haemaphysalis bispinosa*, (B) *Boophilus australis* and (C) *Hyalomma aegyptium*

B. decoloratus Koch., a one-host tick, occurs throughout tropical and southern Africa on cattle. This tick does not live long unfed and it has been most successfully eradicated by dipping processes and changing of the pasturage. It is commonly called the blue tick on account of its characteristic appearance.

The Texas fever organism is represented in India by *Babesia bigeminum* and is transmitted mainly by *B. australis*. The incubation period in cattle is 10 to 20 days.

Genus **Hyalomma** Koch. Pedipalps long; eyes distinct; festoons present; stigmata comma-shaped; legs banded; the male with adanal shields and paired chitinised points borne on posterior abdominal protrusion.

The commonest species is *H. aegyptium* L., which is more or less cosmopolitan. It is widely distributed in Africa, also in this country and is found on cattle and horses. It is common on camels in Northern India and attaches itself principally to the under parts of the body. During the earlier stages it is often a parasite on birds. This tick requires two and sometimes three hosts. It is very prolific and difficult to eradicate. The female can be easily recognised by its dark brown scutum with prominent eyes placed well inside its margin. Coxa I in both sexes is deeply cleft. The adanal plates of the male are much longer than broad and are provided with chitinised points posterior to them. The legs are marked white at the joints. The female when replete attains a very large size.

In Africa *H. aegyptium* is called the bont-legged tick owing to the variegated colour of the legs. It is suspected of being the arthropod host of the piroplasma-like parasites causing biliary fever in horse. It is the dog tick in Australia. It is a two and sometimes a three-host tick. In India it has been experimentally proved by Ray (1945) to be the vector of theileriasis (due to *T. annulata* infection) amongst calves in this country.

The number of eggs laid by a female *aegyptium* is from 6,000 to 8,000 during about 20 days, the maximum number during twenty-four hours being 800. The female dies before the larvae hatch. The incubation period of eggs is from 23 to 26 days. The larvae feed after a resting period of nearly 20 days. They are found on the ears and other soft parts of the horse, calf, camel, buffalo, etc. The larva feeds for 3 to 5 days and then drops to the ground; the nymphs emerge after 5 to 7 days and are ready to attach themselves to a host after a rest period of 5 or 6 days. They feed for 5 to 7 days and leave the host, the adults emerging after 9 to 15 days rest. After another 7 to 8 days they are ready to reattach themselves (Sharif, 1924).

Genus **Amblyomma** Koch. Scutum ornate with eyes; pedipalps long; festoons present; stigmata broadly triangular with rounded angles; males without any ventral armature but the scutum is more beautifully ornamented than in the female.

They are widely distributed and are found almost on any type of vertebrate, their host relationship being very loose. This genus is represented by 7 species being mostly found on reptiles, tigers etc.

The commonest bont-tick, *A. variegatum* Fabr., is responsible for the heart-water of sheep in Africa. The virus does not pass through the egg. This tick is found in Egypt, Abyssinia and East Africa. *A. americanum* Lin., the lone star tick, so called because of the prominent star-shaped marking on the back in the female, is widely distributed in both North and South America, occurring on a wide variety of animals particularly on the head and ears of wild deer and grouse. Next to *Boophilus annulatus*, it is the most important animal tick in the United States.

This species was found to be the vector of a typhus-like fever discovered in South Texas similar to Colorado tick fever, and to which the name of Bullis fever was given from camp Bullis in Texas where the disease was first observed by Woodland etc. (1943).

A. cajennense Fab., is a notorious pest to travellers in the bush during the dry season in Panama; they are very abundant in Texas and all stages attack man and animals indiscriminately. It occurs in Central and Tropical America, where it is known as the silver tick. This tick does not extend its range north of Mexican border. Horses and mules and other domestic and wild animals are commonly attacked. It is a 3-host tick and a vicious biter. Eyes flat; the first coxa has 2 short spines; scutum of male has about 9 elongate yellowish spots. In *A. hebraeum* Koch., the bont-tick, the festoons are whitish, the two spots at the sides are yellowish, the greater part of the light areas are pale violet or pinkish. In South Africa it carries "Heart Water" caused by *Rickettsia ruminantium*, a frequently fatal disease in sheep, goats and cattle. The disease is transmitted by the nymphal and adult ticks, which have been feeding on sick animals as larvae and nymphae. It is a stage to stage infection, i.e., infected larva transmits the disease in the nymphal and infected nymph in the adult stage. The virus does not pass through the egg. In the unfed condition, the tick may survive for nearly a year. It is a rare species in India.

A. maculatum Koch, the Gulf Coast tick, is of considerable local importance in the United States. The injury caused to cattle and other domestic animals due to their attachment predisposes to the invasion by the parasitic fly, *Chrysomyia hominivorax* for oviposition. It occurs in the areas adjacent to the Gulf of Mexico and on the Atlantic coast in Florida, Georgia and in South Carolina. It has also been recorded from the east coast of Mexico, West Indies and various countries in South America. It usually attaches itself to the inner surface of the external ear. The adult tick attacks the larger domestic and wild animals whereas the larvae are found on birds, rodents, lamb, and nymphs on sheep and calf. It is a 3-host tick. The disease and the tick are confined to Africa. *A. variegatum* Fab., another South African tick, and found on cattle and wart-hog, is also responsible for the carriage of the same disease (Heart water). It remains 4 to 5 days in its larval and nymphal stages on its first two hosts and about 12 days in the adult stage on its third host. *A. striatum* is a dog tick in Brazil.

Genus **Aponomma.** This is very closely related to *Amblyomma* except that eyes are absent. The palpi are long and slender and the second segment much longer than wide as in *Amblyomma.* This species are parasitic on tropical reptiles. *A. gervaisi* var. *lucasi* Warb., is widely distributed in India and attacks snakes.

Genus **Rhipicentor** N.-W. Inornate with eyes and festoons with short palps; basis capituli hexagonal and having prominent lateral angles; coxa I bifid; the male resembles *Dermacentor* ventrally and *Rhipicephalus* dorsally.

Genus **Margaropus** Karsch. (*Boophilus* of American authors) : Inornate with eyes, but without festoons; with short palps and capitulum intermediate between that of *Rhipicephalus* and *Boophilus*; highly chitinised; the articulations of the legs greatly swollen especially of leg 4; tarsi with a long spur extending far beyond the claws. *M. annulatus*, found on cattle, is of considerable economic importance in North America.

The following ticks have been reported by Strickland and Roy (1939) as biting man in India : *Amblyomma* sp. (nymph); *Rhip. sanguineus* Lat.; *Dermacentor* sp. (nymph); *Ixodes acutitarsus* Karsch.; *Haemaphysalis montgomeryi* Nuttall ; *Rhip. haemaphysaloides* Sup.; *Hyalomma aegyptium* L.; *Haemaphysalis* sp. (nymph); *Haem. bispinosa* Neum.; *Haem. aculeata* Lav. In addition to the above, Kirwan (1935) found *D. auratus* Sup., on upper eyelid of man.

The common ticks which infest domesticated animals in India are: *Boophilus australis* Fuller; *Rhip. sanguineus* Lat.; *Rhip. haemaphysaloides* Sup.; *Hyalomma aegyptium* L.; *Haemaphysalis bispinosa* Neum.

Hard ticks and human diseases. Tick-borne infections in man are mainly classified under the following heads : (1) Tick paralysis; (2) Tularaemia and (3) Rickettsiosis.

(1) **Tick paralysis.** A paralysis of man and animals, associated with the presence of ticks on the body, has long been known to occur in Australia, British Columbia and in other countries. Fatal cases commonly occur in dogs, pigs and other animals along the tropical and subtropical belts of eastern Australia and in sheep in Canada, Africa and Scotland. The onset is insidious and though the course of the disease is slow, it may end fatally unless the tick is removed. The first symptom noticed in sheep is lagging behind the rest of the flock followed by paralysis. The period elapsing between the attachment of the tick and the onset of the symptoms is important and it corresponds to the stage of tick's engorgement. The paralysis does not appear usually before the fifth day of its attachment and only adult females and nymphs are involved.

Children are especially susceptible to tick paralysis. Ticks attach themselves chiefly to the upper part of the body particularly the nape of the neck.

In Canada, Hadwen and Nuttall (1913) experimentally established the connection between tick, and tick paralysis in sheep, with *Dermacentor venustus*. This tick is also considered to be responsible for tick paralysis in man in Montana. Later, Dodd (1921) experimentally produced tick paralysis in the dog with *Ixodes holocyclus* Neum., and Ferguson (1924) recorded a fatal case of paralysis in a child due to this same tick. Other ticks, e.g., *Ixodes pilosus* Koch., and *I. ricinus* Lin., are suspected of causing tick paralysis in human being in addition to sheep in South Africa and Crete respectively.

The disease, however, cannot be transmitted by direct blood inoculation from affected to healthy animals. Ross (1926) has demonstrated the presence of two types of alveoli in the salivary glands of *I. holocyclus* and the appearance of the gland varies greatly according to the stage of engorgement of the tick. It is interesting that only the female tick is able to cause the disease.

Even a single tick is able to cause fatal paralysis. It has been suggested that the causal factor in the production of the disease is a toxin derived from the salivary secretion (Ross, 1926), the poison in the eggs being identical with that secreted by their salivary glands. The toxic properties contained in the eggs and ovaries of ticks when injected in animals produce symptoms of paralysis together with degenerative changes in the spinal cord (Regendanz and Reichenow, 1932; Hoeppli and Feng, 1933). It has later been shown that the eggs of certain South African ticks are also toxic to guineapigs (De Meillon, 1942)

The only treatment is to remove the tick as soon as the paralysis is detected.

(2) **Tularaemia in man.** Tularaemia, the so-called "rabbit disease" of the United States, is widespread in the United States and Canada but the percentage of infected animals is very small. It is only when the disease approaches epidemic proportions among rodents that it may become a serious menace to man. It has now been definitely identified in Japan and in the U.S.S.R. In the United States the disease was first described in the ground squirrel *Citellus beecheyi*. In the U.S.S.R., the disease has been found in an epizootic form in hamster *Cricetus frumentarius*. The water rat *Arvicola*, has played an important part in the U.S.S.R. epidemics. In Russia ticks of the genus *Ixodes* are among the chief vectors of the disease (Olsufev, 1939).

Transmission of the virus is effected through the bite of a fly or tick or direct through broken skin and conjunctiva. In the case of man, *Chrysops discalis* or a tick *Dermacentor andersoni* are vectors. In the case of animals, transmission is effected by means of fleas, lice, bugs, flies and other blood-sucking insects, especially ticks. The common rabbit tick, *Haemaphysalis leporis-palustris* is an important transmitter of the disease from rabbit to rabbit. Experimentally the rat louse *Polyplax* is able to carry the disease to healthy rats. *D. variabilis* has been found to be the vector in a large number of human cases occurring in several southern states of the Union. B. *tularense* has also been recovered from *D. occidentalis* in California. It has been demonstrated by Parker (1937) that *Bact. tularense* is able to survive from larvae to adults and to their offspring in *D. variabilis*, *Haem. leporis-palustris* and in other ticks (Parker, 1924).

(3) **Rickettsiosis.** Rickettsiae are normally parasitic in the alimentary tracts of arthropods, both of blood-sucking and non-blood-sucking. A very large number have been described but only 7 have pathogenic importance for man. These are *R. prowazeki* (epidemic typhus), *R. quintana* (trench fever), *R. rickettsi* (the Rocky Mountain spotted fever group), *R. mooseri* (murine typhus), *R. orientalis* (scrub typhus), *R. burneti* (Q. fever) and *R. acarina* (rickettsialpox).

(*a*) **Rocky Mountain spotted fever.** In 1906 Ricketts first transmitted the rickettsia of the Rocky Mountain spotted fever by the bite of ticks. This disease which is confined to certain parts of the United States, such as Idaho, Montana, Wyoming, Utah and neighbouring provinces, is a typhus-like fever and is carried to man by *Dermacentor venustus* Banks (*andersoni*) in the western parts, and *D. variabilis* Say., and *D. occidentalis* Neum., in the eastern and central portions of that country. Spontaneous infection has been recorded in *Amblyomma americanum*. Larvae, nymphae and adults of *A. americanum* all bite man. A large percentage of the immature ticks engorge on horses and cattle which are not susceptible to spotted fever.

Rocky Mountain spotted fever has been experimentally transmitted by at least 8 species of ticks. Generation to generation tramsmission from larva to larva has been demonstrated with three of these, *D. occidentalis*, *R. sanguineus* and *Amblyomma americanum*. The infection may be transmitted by the larva, nymph and both adult female and male ticks and also it is a stage to stage infection within one generation. Only the adult *D. andersoni* and *D. variabilis* attack man but the disease is transmitted to animals such as the mountain goat, sheep and rodents in which the virus is kept alive by larvae and nymphs. There is definite host-parasite relationship in this disease. Rabbits harbour the rickettsia of Rocky Mountain spotted fever, which infection is transmitted from rabbit to rabbit by the rabbit tick *Haemaphysalis leporis-palustris* (Parker, 1923). This tick is transported through certain ground frequenting migratory bird hosts such as the meadow lark which serve as agents in the extensive distribution of the virus. This tick does not bite man. The infection is transmitted from rabbit to man mainly by *D. venustus* and to a less extent by *D. variabilis* and possibly by a few other ticks. These ticks bite both man and rabbits.

Da Rocha Lima (1916) first described the causative agent of Rocky Mountain fever constantly found in the tissues of human cases as well as in the tissues and eggs of infected ticks. The causative organism, *Dermacentroxenus rickettsii*, but more commonly known as *Rickettsia rickettsi*, lives intracellularly in the tick and is passed with the faeces of infected ticks.

Four recognised diseased conditions are associated with the bite of the Rocky Mountain wood tick, *D. andersoni*: (*a*) Rocky Mountain spotted fever, (*b*) Tularaemia, (*c*) Colorodo tick fever and (*d*) Tick paralysis. Colorodo tick fever is a form of rickettsiosis of the same type as the Rocky Mountain spotted fever.

Though the virus of Rocky Mountain spotted fever is normally transmitted by the ticks *D. andersoni* and *D. variabilis*, other acarines can also serve as vectors. Weigl (1930) has shown that the virus can develop in the louse.

There are three measures of prevention of this disease : (1) To avoid ticks. (2) To examine the clothes very carefully after return from jungles and remove the ticks from the person as early as possible. (3) Prophylactic vaccination. Spencer and Parker (1925, 1930) and Parker (1935) have demonstrated the protection afforded against the tick-borne rickettsial disease of a killed vaccine prepared from ground-up virus-laden tissues of infected adults of *D. venustus*. Adult ticks infected in the larval and nymphal stages show a marked increase in rickettsial content of their tissues after feeding which is greater than that found when the ticks are infected as adults.

(*b*) **Fievre boutonneuse** (*Tick typhus*). This was first described by Megaw in 1916 in India and then by Conor and Bruch in 1919 in Tunis. It is now found widely distributed in countries bordering on the Mediterranean. This disease has now been shown to be carried by *R. sanguineus* though there are doubts that dog is the reservoir of the disease. Megaw in 1916 described a typhus-like fever in Kumaon hills in India and suggested transmission by *R. sanguineus*. Later Pijper and Dau (1933) proved in South Africa that this sort of eruptive fever could be transmitted by bite of the larvae of *Amblyomma hebraeum*, *R. sanguineus*, *R. appendiculatus* and *Boophilus decolcratus*. Reservoir of the disease of South Africa is probably the small rodents of the veldt. Tick-borne typhus also occurs in Abyssinia and in Kenya where the dog tick *R. sanguineus* acts as a vector. A typhus-like disease variously designated as tick-bite fever and tick typhus in East Africa and transmitted by ticks, *R. simus* and *Haem. leachi*, has been described by Dick and Lewis (1947).

Quite recently Kalra (1947) has drawn attention to the wide prevalence in India of the three types of typhus fever which are popularly known as scrub typhus, murine typhus and tick typhus. Tick strains related to *Dermacentroxenus rickettsii* have been isolated by him from *R. sanguineus*, *Boophilus sp.*, *Ixodes ricinus* and *Haemaphysalis leachi* in the Kumaon hills and in Imphal. He claims to have obtained definite proof of the occurrence of tick typhus in Srinagar, Almora, Jubbalpore, Lucknow, Secunderabad, Mysore, Imphal, Poona and other places. An outbreak of tick typhus in Central India of the boutonneuse fever type has been described by Audy (1947). Topping,

Heilig and Naidu (1943) came across some cases of typhus in Mysore which they regarded to have been caused by tick bite.

The immunological relationship between Rocky Mountain spotted fever and boutonneuse fever has been studied by Badger (1933).

(*c*) **Sao Paulo fever** in Brazil is transmitted by *Amblyomma cajennense.* It would appear to be established that the Rocky Mountain spotted fever of U.S.A. and Sao Paulo typhus are identical and can be carried by a common vector such as *D. venustus.* Other ticks which are able to transmit the rickettsiae are *A. striatum* and *R. sanguineus.* Possible reservoirs of infection are rodents and the opossum. Against Sao Paulo tick typhus Monteiro (1933) has successfully employed a vaccine prepared with the South American tick *A. cajennense.*

(*d*) **"Bullis fever"** in South Texas is carried by *Amblyomma americanum.*

(*e*) **Q. fever.** Q. fever in Australia particularly observed in abattoir workers and dairy farmers is carried by *Ixodes holocyclus,* which commonly attacks man in Australia and which is also responsible for tick paralysis in man and domestic animals, its principal host in Australia being the bandicoots, *Isodon torosus* and *Perameles nasuta. I. torosus* is a proved reservoir of Q. fever. Natural infection has been found in nymphs and adults of *I. holocyclus.* The infection in ticks passes from larvae to nymphs, from nymphs to adults but not from female to their progeny. Infection in bandicoots can take place by feeding on infected ticks. The rickettsia concerned is *Coxiella* (*R*) *burneti,* found in the lumen of the gut and in the cytoplasm of the epithelial cells.

The two commonest ticks on bandicoots, *I. torosus,* are *Haemaphysalis humerosa* War. & Nutt., and *I. holocyclus* Neum. *H. bispinosa* Neum., and *Ornithodorus* sp. have also been artificially infected and it is possible that the former may also carry the disease to cattle from bush animals. Excreta of ticks are infective. *H. humerosa* is a proved vector which carries the disease from cattle to cattle (Smith, 1940; 1942). Experimentally *R. sangüineus* and other ticks are also capable of transmitting the disease. *O. moubata* and *O. hermsi* infected with *Coxiella burneti* as nymphs, are able to transmit the infection by bite.

Human infection with Q. fever in Australia may result from : (*i*) attacks from the local ticks *I. holocyclus* in the absence of cattle; (*ii*) direct infection from the body tissue and fluids of infected cattle; and (*iii*) indirect infection from infected cattle by the excreta of ticks especially *Boophilus annulatus* and *Haemaphysalis bispinosa.* The third is probably the most important. Inhalation of dried faecal dust of the ticks is a possibility.

Q. fever in America is carried by *Amblyomma americanum.*

Different Forms of Typhus Fever and Their Vectors

Name of disease	Geographical distribution	Causative virus	Vectors	Reservoir of infection
1. Classical epidemic typhus	Old and New Worlds (Cosmopolitan)	*R. P. prowazeki*	Head and body lice	Man
2. Murine typhus	World-wide, common in the tropics.	*R. P. mooseri*	Rat-fleas	Rats

Name of disease	Geographical distribution	Causative virus	Vectors	Reservoir of infection
3. Tsutsugamushi fever (scrub typhus)	Sumatra, Malaya, Indonesia, Japan, India (Kashmir included).	*R. orientalis*	Larval mites : (1) *T. deliensis* in Malaya, Indonesia and India. (2) *T. akamushi* in Japan.	Rats, bandicoots, tree shrew.
4. Spotted fever group	Western and Eastern U.S., Europe, Africa, India and other Asian countries.	(*i*) *R. rickettsi* (Rocky Mountain spotted fever).	(*i*) Ticks	Various rodents, ruminants; dogs are also suspected.
		(*ii*) *R. conori* (Boutonneuse fever).	(*ii*) Ticks	
		(*iii*) *R. acari* (Rickettsial-pox)	(*iii*) Mite (*Allodermanyssus sanguinensis*).	Domestic mice
5. 'Q' fever	Queensland and North America	*R. burneti*	Tick	Cattle, bandicoots, dogs.

Blood Parasites of Domestic Animals Carried by Hard Ticks

Causative organisms :	*Tick carrier* :	*Host* :	*Distribution* :
Babesia bovis	*Ixodes ricinus*	Cattle	Europe.
(Red-water in cattle)	*Boophilus annulatus*	,,	North America.
	B. australis	,,	Australia.
B. bigeminum	*Dermacentor reticulatus*	,,	Africa.
	Boophilus decoloratus		,,
	Rhipicephalus bursa		,,
	Boophilus annulata		India and in other warm
	Boophilus Sp.		countries.
B. caballi	*R. evertsi*	Horse	France.
	D. reticulatus		Russia.
B. canis	*Ixodes ricinus*	Dog	South Europe.
	Haemaphysalis leachi	,,	South Africa.
	Rhipicephalus sanguineus		India.
	D. reticulatus		South Russia.
B. ovis	*Rhipicephalus bursa*	Sheep	Persia and Rumania.
	B. australis and *B. annulata*		India.
B. argentinum	*Boophilus Sp.*	Cattle	South America.
Theileriasis (*T. parva*)	*R. appendiculatus*	Cattle	Africa.
(Rhodesian East Coast Fever)	*R. simus*	,,	,,
	Hyalomma aegyptium	,,	,,
	R. nitens, *R. capensis*,	,,	,,
	Hyl. aegyptium var. *planum*	,,	,,
(*T. annulata*)	*H. aegyptium*	,,	India.
	Hyl. mauritanicum		Africa.
	H. dromederi		Central Asia.

Causative organisms :	*Tick carrier :*	*Host :*	*Distribution :*
	H. marginatum		South Kazakshistan.
	H. detritum		,,
	H. dromederi		,,
T. ovis	*Ornithodorus lahorensis*		Azerbaijan.
Nuttalla equi :	*R. sanguineus*	Horse	Tropics, Sub-tropics :
(Equine piroplasmosis)	*R. evertsi*		France, Italy, U.S.A., Russia.
	R. bursa		other parts of Europe,
	Hyalomma marginatum		,,
	H. volgense and other species of		,,
	Hyalomma.		,,
	In India the transmitting agent is probably *Hyalomma savignyi* (*aegyptium*).		,,
B. gibsoni	*Haemaphysalis bispinosa*	Dog	India.
(*Pattonella gibsoni*)	*R. sanguineus*		,,
(*Rickettsia ruminantium*)	*Amblyomma hebraeum*		South Africa.
"Heart water" of cattle	*A. variegatum*		,,
Rickettsia bovis	*Hyalomma Sp.*	Dog	Africa.
Rickettsia ovina	*R. bursa*	Sheep	Mediterranean Asia.
Encephalitis in sheep (Louping ill)	*I. ricinus*		Europe.
Anaplasmosis of cattle	*Amblyomma marginale*		America, Africa.
(*Anaplasma marginale*)	*D. occidentalis*		,,
	D. andersoni		,,
	B. microplus.		,,
	Ixodes ricinus		,,
	B. decoloratus		,,
	R. bursa and other ticks		,,
	Cattle ticks (*Boophilus* Sp.)		India.
Leucocytozon canis	*R. sanguineus*	Dog	,,

Haemoglobinuria. All members of the genus *Babesia* do not produce haemoglobinuria. For instance, in cattle *B. bigemina* and *B. bovis* do produce haemoglobinuria but this is not the case with *B. argentina* and *B. major* infections; in dogs *B. canis* do but not *B.* (*P*) *gibsoni.* With *B. motasi* and *B. ovis* jaundice and haemoglobinuria may occur but this is not a constant feature.

The following **key** prepared by Warburton has reference only to the ticks commonly found on domestic animals in this country.

Females

1. The whole rostrum twice as long as broad .. .. *Hyalomma aegyptium.*
   The whole rostrum quite or nearly as broad as long .. 2

2. Rostral base rectangular and narrower than the width across the palps. No eyes .. *Haemaphysalis bispinosa.*
   Rostral base hexagonal, pointed laterally. Eyes present 3

3. Peritreme nearly circular, scutum elongate and smooth .. *Boophilus australis.*
   Peritreme sub-triangular, scutum as broad as long and punctate *Rhipicephalus.*

Males

1. The whole rostrum twice as long as broad	..	*Hyalomma aegyptium.*
The whole rostrum quite or nearly as broad as long	..	2
2. Rostral base rectangular and narrower than the width across the palps. No eyes or anal plates ..	.. ..	*Haemaphysalis bispinosa.*
Rostral base hexagonal, pointed laterally	.. ..	2
3. All four anal plates well developed and linear, abdomen ending in a sharp point	.. .. ..	*Boophilus australis.*
External anal plates mere chitinous points	.. ..	4
4. Internal anal plates rounded	.. .. ..	*Rhipicephalus sanguineus.*
Internal anal plates hooked towards one another	..	*Rhipicephalus haemaphysaloides.*

Family **Argasidae**. All members of this family like those of Ixodidae, are parasitic and live on blood alone. The nymphs and adults are rapid and frequent feeders. Both in *Argas* and *Ornithodorus* the larvae generally attach themselves to the host. In *O. moubata* and *O. savignyi* the larva develops within the egg shell and is inactive and incapable of attacking a host. Like bed bugs, soft ticks hide in cracks and crevices in the day time and emerge at night to feed on the host. Sexual differentiation except from the character of the genital aperture is not possible. The females lay a moderate number of eggs in batches after each meal of blood. These ticks can fast for 3 to 4 years. Longevity is particularly marked in soft ticks. The longest adult life span has been reported by Smith (1944) in fed *Ornithodorus coriaceus* as more than 5 years.

It has been reported by Wheeler (1938) and this has been confirmed by later workers that a hungry *Ornithodorus papillipes* does not hesitate to suck the blood from freshly engorged specimens which may cause the death of the victim.

The Argasidae comprise two genera, *Argas* Latr., and *Ornithodorus* Koch., and their difference is shown below.

Difference Between the Genera of Argasidae

Genus **Argas**	Genus **Ornithodorus**
1. Body is oval.	1. Body is much rounder.
2. Margin of the body is flat.	2. Margin is round; a lateral notch between cephalothorax and abdomen present.
3. Chitinous plates arranged in radial lines are found on both surfaces.	3. Chitinous plates are absent, integument mamillose.
4. Legs are smooth.	4. Distal segments of legs provided with tubercles on the anterior border.
5. Eyes are absent.	5. Eyes may or may not be present.

Genus **Argas.** Body oval and covered with discs of all sizes on both surfaces.

Argas persicus Oken. It is cosmopolitan in warm countries and is found especially in poultry houses but seldom on fowls themselves. In exceptional cases it may bite man. Occasionally ducks may be attacked. It is commonly known as the fowl tick. It transmits the spirochaete, *Treponema anserina*, of fowls and ducks which causes a heavy mortality among them. It is one of the most troublesome pests which affect poultry, killing enormous numbers of birds from the loss of blood. It attacks poultry at night and is found on the walls and roofs of poultry sheds in the day time. It also hides in cracks and crevices even in the timber itself and lays its eggs in these situations.

A. persicus is also the vector of *Egyptionella pullorum* and is transmitted by bite.

The adult tick is brown in colour and the body is elliptical. Its lateral margin is distinctly flat and is formed by rectangular cells.

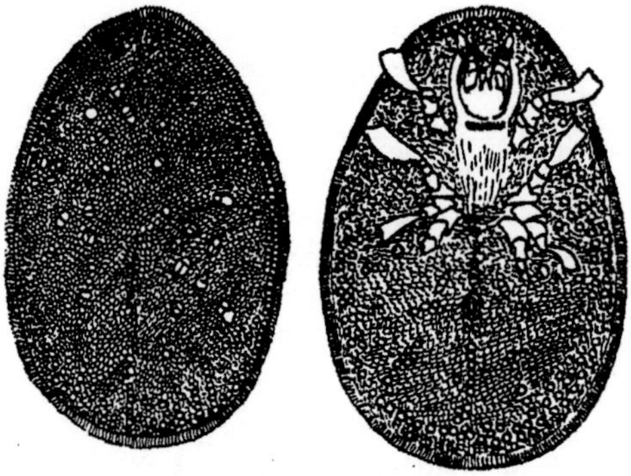

FIG. 162.
Argas persicus, dorsal and ventral aspects.

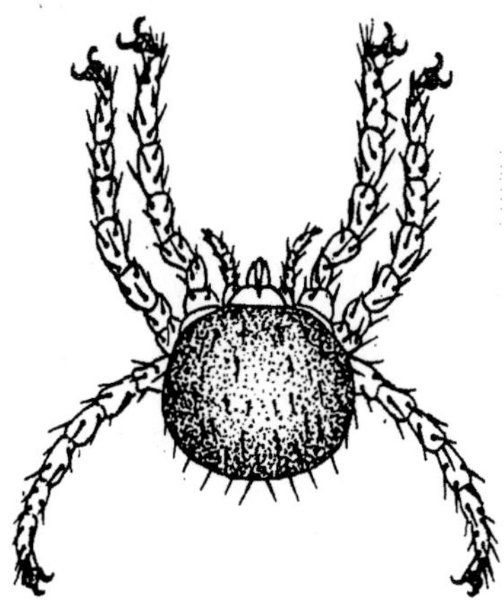

FIG. 163.
Larva of *Argas persicus*.

The larval forms are very minute, often colourless. These larval ticks climb the perches and walls of the poultry house and so reach the roosting fowls, to which they remain firmly attached usually under the wing until gorged with blood. In the larval stage they remain on the fowl for about 10 days and are likely to be dispersed to fresh coops and poultry yards. Both young and adult stages of these ticks can remain alive for long periods without food. The adult feeds once a month and appears to require that period for the digestion of the blood. The adult lays eggs in small batches of 20 to 100. From egg stage to egg stage it may take as long as 10 months. Adult ticks can travel great distances and in this way they migrate to new fowl houses. This species infests native dwellings in Persia and their bites are painful. Adults can live for 18 months without food.

Spirochaetosis of fowls caused by *Spirochaeta anserina* is carried by *A. persicus*. Infection is normally transmitted by the saliva and bite of the tick but occasionally the secretion of the coxal glands is also infective. Infection of gland may take place as early as the sixth day and the bite of the tick is infective on the seventh day. Eggs also become infected in the ovary and hereditary transmission of the spirochaete is known to occur. Fowls become infected also when infected ticks or eggs laid by them are ingested (Knowles, Das Gupta and Basu, 1932).

Pasteurella avicida, the causal organism of chicken cholera or fowl plague, is also carried by *Argas persicus*. The excreta are infective and are a cause of contagion to healthy birds. Basu (1930) refers to the possibility of the infection being acquired by ingestion though there is marked loss of virulence of *P. avicida* in the gut of the fowl.

Argas reflexus Fabr. It is the pigeon tick of Europe, North Africa and North and South America and kills by massive infestation. It closely resembles *A. persicus* in appearance and habits but differs from the latter in having the anterior end of the body more sharply narrowed and the margin being finely striate.

It is only when man lives in close association with pigeons or when the pigeon house has been abandoned, that this tick is attracted to a new host, and thus man gets bitten. Besides pigeons, they are known to attack ducks, geese, sparrows, etc.

Females lay about 50 eggs and continue to oviposit at intervals after each engorgement. The larvae hatch in about one month and attach themselves to pigeons on which they engorge for about a week. They then seek shelter in cracks, in pigeon lofts and transform into nymphs some months later. The nymphs moult four times before reaching the adult stage and engorge before each moult.

To eradicate this tick all cracks must be sealed. The ticks are resistant to lime-wash, insecticidal sprays and sulphur smoke. One may have to use boiling water or boiling decoction of tobacco leaves.

In addition to *A. persicus*, *I. vespertilionis* Koch has been reported from the Punjab as attacking birds.

Genus **Ornithodorus** Koch. The body is more or less round; in most of them eyes are absent. There is a depression or notch in the lateral margin of the body. The tibiae and tarsi are provided with protuberences. Both sexes live on blood of warm-blooded animals. A few carry the spirochaeta of relapsing fever from animals to man or from man to man. The reproductive potential is poor and only a small number of eggs are laid at each oviposition which takes place after a blood meal. They can withstand starvation for a long time. Sexual dimorphism is slight.

O. savignyi Aud. Body oval, rounded at both ends; integument mamillated, the supracoxal folds on the ventral surface are well marked and bear a pair of eyes on each side, the first pair facing coxa I and the other, the interval between coxae II and III; spiracles above the supracoxal folds; protarsal and tarsal humps of leg I on both, 3 large dorsal humps, the middle one being the largest, and placed equidistantly; leg IV protarsus 3 humps, the dorsal ones placed relatively near to each other; on tarsus the first two placed relatively close together.

In certain parts of North Africa such as in Somaliland, this tick is found buried in the soil in or around the huts on the outskirts of coastal towns especially in the more squalid insanitary areas. The ticks are common in the dusty soil and burrows to a depth of $\frac{1}{2}$ to 1 inch lying dormant for months until they are disturbed by man or beast.

It is chiefly parasitic on camels in Africa. It will attack human beings, camels, cattle, ponies, mules, donkeys and even sheep and goats with the same avidity. It is a rapid feeder and engorges itself within half an hour. As soon as its engorgement is complete and just previous to withdrawing its rostrum, it discharges a fluid excreta per rectum. There is also an abundant exudation from the coxal glands. The site of puncture is thus bathed in the excreta and the coxal fluid. These fluids may contain the spirochaeta of relapsing fever. These ticks may be transported in luggage of travellers.

In India this tick is found in the southern parts of the Peninsula. It generally attacks cattle and other domestic animals and occasionally man. When it does, the younger stages are generally involved.

Larvae of both *O. moubata* and *O. savignyi* are inactive, motionless and unable to feed ; they have no respiratory openings. Both these ticks feed for the first time in the first nymphal stage. In *O. moubata* the larva is found within the egg shell inside which the metamorphosis from larva to nymph takes place, the nymph emerging from the egg-shell. In *O. savignyi*, on the other hand, the larva escapes with the splitting of the egg-shell and remains near the empty egg-shell till metamorphosis takes place and the larva is transformed into nymph. In both cases after 3 to 4 days the larva undergoes the first ecdysis.

There are two pairs of eyes on each side and in this respect it differs from *O. moubata* which is eyeless. In other respects the two are indistinguishable.

On account of these ticks being scattered, they are difficult to eradicate. None of the measures ordinarily suggested for control of ticks is of much practical value. In confined areas where they swarm, such as the immediate vicinity of wells, the best and the cheapest method is to cover the infested areas with dry grass and brush wood after harrowing the surface, and then setting fire to the grass. Spraying the ground with insecticidal fluids is not only costly but ineffective. However, in limited areas DDT and Gammexane powders mixed together may be sprinkled over the ground. This, if repeated, will eradicate ticks provided the ground remains dry.

O. moubata Murray. Like *O. savignyi* the body is ovate, a little wider behind than in front; lateral margin of the body slightly constricted behind the level of the third leg; integument mamillose and studded with small tubercles best noticed with a magnifying lens; dorsal surface exhibiting some lateral grooves; there is a supracoxal fold on which the eyes when present are placed; tibia and tarsi of first three pair of legs with three humps or protuberances on outer border. The humps on the outer border of protarsus of leg I are subequal and tooth-like, while the tarsus of leg IV is shorter and thicker than in *O. savignyi* and its humps are nearly equidistant.

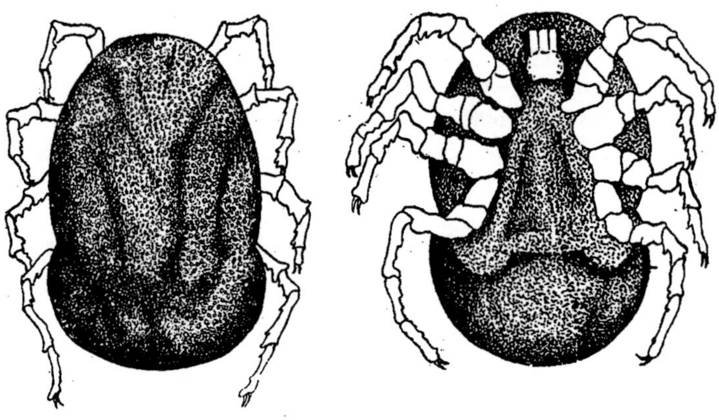

FIG. 164.
Ornithodorus savignyi, dorsal and ventral aspects.

This tick though it closely resembles *O. savignyi* can be readily distinguished by the absence of eyes, the details of its leg structures and being less hairy. In *O. savignyi* there are numerous hairs between the mamillae.

This species is widely distributed in Africa. It is especially found in Uganda, Congo Free State, Central and West Africa, Abyssinia and South Africa. In Morocco both *O. savignyi* and *O. moubata* occur. It infests huts and rest houses. It hides in the day time in cracks in the mud floors and in crevices in the walls and emerge at night to feed. It is said to be carried long distances in beds and other belongings by travellers. Man appears to be the chief host of *moubata.* Apart from man, pigs, sheep, goats, dogs etc., may be attacked. When replete, it reaches the size of a pea.

Generally the male is smaller than the female. The genital orifice in the female is broad, slit-like and situated more posteriorly than the male orifice. The lips of the vulva protrude from the oval depression of the integument. In the male the aperture is smaller. It has the form of a strongly arched crescentic slit.

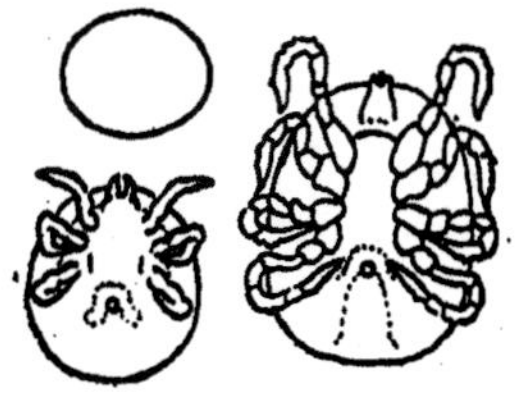

FIG. 165.
Egg, larva and nymph of *O. savignyi*

The life history is the same as in *O. savignyi.* Egg stage lasts 7 to 8 days. Nearly 150 eggs are laid in the first batch. Larvae are motionless and unable to feed. They have no respiratory openings on the skin. Three or four days later the larva undergoes the first ecdysis. In the first nymph stage it is very active and readily feeds as soon as the mouth parts are sufficiently hard. The female has five nymph stages and the male four. The intervals between the different ecdyses are not uniform and are dependent on complete digestion of the preceding feed of blood.

A hungry tick will soon start to gorge itself with blood; the body swells up, after which the discharge of the transparent secretion from the coxal glands commences. The coxal fluid is secreted by all the nymph stages as well as by the adult. The secretion of the fluid ceases several minutes after the engorged tick falls off. It generally chooses the fingers and toes for biting.

This is the well-known relapsing fever tick in certain parts of Africa. It is firmly believed by the natives that this tick causes fever in man by its bite and in order to counteract this condition, they get bitten by healthy ticks from time to time with the object of acquiring immunity against the tick-borne infection.

The productivity of the female is increased by large feeding, and several batches of eggs are laid provided opportunities exist for repeated blood feeds. Oviposition goes on slowly and generally lasts for 10 days.

The adult can remain alive for at least one year in captivity without blood feed.

O. turicata Duges. This tick is found in Columbia, Mexico, Florida, Argentine and Texas but not on Gulf Coast. It breeds in wild rodent nests or holes. It also infests burrows of tortoise.

Its chief characters are : The body with sides almost straight and parallel, the anterior extremity being narrowed to a rounded point; integument with brilliant contiguous granulations, larger posteriorly; eyes absent; two hairs at the base of the hypostome; palps long; the tibiae and tarsi of the legs furnished dorsally with three tubercles each, much smaller than those of *O. moubata* and more marked the more anterior the leg and scarcely visible on leg IV.

This species attacks pigs; man is occasionally attacked and very exceptionally domestic animals other than pigs. Larvae and nymphs have been recorded from rabbits, *Citellus* species, in California. In man the bite is painful and causes severe irritation. In this tick there is an abundant secretion of the coxal fluid during feeding.

O. megnini Duges. It is also called "spinose ear-tick," so named on account of the body in the nymphal stage being covered with numerous spines and is the only tick known in which this occurs. Its original home is America from where it has dispersed to South Africa, and other countries, being carried in the ears of horses and mules, which are its real hosts. In addition to horses, it attacks cattle, sheep, and other domesticated animals in South Africa. The adult tick lives in walls, cracks etc., particularly near cattle sheds.

It appears highly specialised in its adaptation to a life of parasitism. Unlike the larva of Argasid ticks, the larva of *O. megnini* remains attached to the host for about a week when the first moult takes place inside the ear. The nymph stage being reached, it remains fastened to the host from three weeks to three months or longer at the end of which it leaves the ear and falls to the ground, seeks cover and the second moult takes place. The adult tick does not seek a host and does not feed as the nymph stores up sufficient nourishment to carry the adults through. After mating, oviposition takes place in chinks, fissures and under the bark of trees, where the larva awaits the coming of a host. Infestation takes place at the time the horse and the mule rub their ears against infested trees. Unfertilised females may live even over three years. In the adult tick the hypostome is unarmed and the structure of the mouth parts suggests that the adult tick is scarcely able to suck blood.

The body of the nymph has a characteristic shape being like that of a kite but broad anteriorly and markedly narrow posteriorly. The integument is covered with posteriorly directed spines. The nymph has been recorded as infesting ears of man.

In India the infestation, so far recorded, is limited to horses imported from America or South Africa.

O. talaje Guerin-Men. It resembles *O. turicata* in general form. The sides of the body are almost straight and parallel, and pointed anteriorly; integument mamillated but the dorsum bears large discs; capitulum lies in a deep camerostome or pit with prominent flap-like borders situated at its antero-external border which is capable of covering the capitulum more or less completely; eyes absent; tarsi bear no humps.

It is regarded as a human tick as it is a great pest to mankind infesting native huts in Mexico, Venezuella and Chile. Man seems to be the selected host of the adult stage. It feeds on a variety of hosts including mammals, birds, reptiles, rodents and man. In the larval stage it has been found infesting marmoset, monkeys, opossums and rats; occasionally it has been found on chickens and snakes.

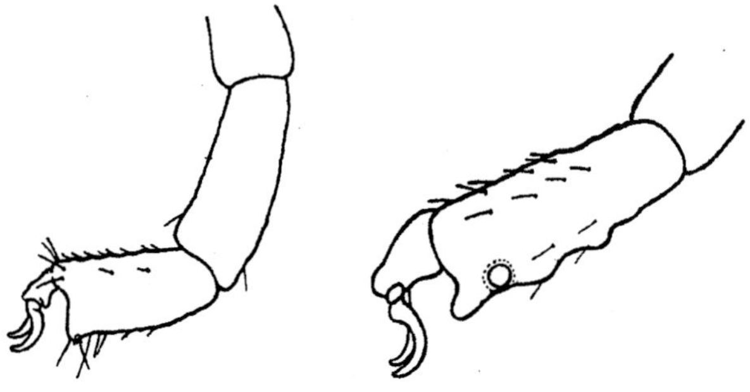

FIG. 166.
Showing the differences in tarsal protuberances of leg I between (A) *O. papillipes* and (B) *O. lahorensis*.

O. papillipes Bir. (**O. tholozani** Lab.-Meg.). This tick is a native of Persia and occurs in neighbouring countries such as Palestine, Syria, etc. Its natural host in Persia is chicken and it hides in chicken coops. It occasionally attacks camels. It has also been recorded in Uzbekistan, Caucasus, Turkistan and Central Asia. It has been described by Russian authors as *O. tholozani*. In India it is found in Muree Hills and Kashmir on dog, goat and cattle and sometimes on sheep.

The sides of its body are nearly as straight and parallel as *O. talaje*, while anteriorly they taper to a rounded process; the integument is covered with small granules; discs on the integument present, small and arranged on the anterior median prominence in three rows 3 : 2 : 2; there are other similar discs; eyes absent; tarsi 2 to 4 with terminal dorsal protuberance prominent, pointed and directed distally.

O. lahorensis Neum. It is common in Baku, Uzbekistan, Palestine, Armenia, Turkmenistan, Central Asia and North West India. So far as India is concerned, this and *O. papillipes* Bir., are the most important among the soft ticks found in Kashmir.

It is essentially a tick of domestic animals, sheep being its most favoured host. The adult tick is found in the mud walls of sheep enclosures and less frequently in human dwelling huts. Whenever— opportunities are available, man is freely attacked. Larvae and nymphs may be found on camels, sheep, dogs and other animals chiefly at some particular time of the year.

It closely resembles *O. papillipes* (*O. tholozani*). The body is elongated and the lateral border almost parallel, the anterior end extremely narrow but not pointed like *O. turicata* and *O. talaje*, nor rounded as in *O. moubata* and *O. savignyi*. The constriction in the middle of the body is not deep; a small number of discs present on the dorsum; the integument between the discs without granules but finely wrinkled; eyes absent; tarsi I with three blunt dorsal prominences; tarsi II, III, IV with very short proximal false articulation, prominent dorsally and with progressively longer blunt spurs near the distal end of the false articulation. It differs from *O. papillipes* in the structure of the integument and tarsi.

O. erraticus Lucus. It is like *O. talaje* but does not possess the lateral expansions of the camerostome. Integument with closely studded granulations; large symmetrical discs present of which there are four in a transverse series across the middle of the dorsum; tarsi only slightly humped. It is identical with *O. marocensis* Velu.

This tick is found everywhere in Morocco in places where pigs shelter at night in rough stone enclosures and in porcupine and rodent burrows. It is also found in Spain, Algeria and Egypt. The Spanish African relapsing fever can be transmitted by its bite. In Morocco *O. erraticus* takes the place of *O. moubata* as the transmitting agent of this disease. Natural infection has been recorded in this species collected from rat burrows in Tunisia.

O. coriaceus Koch. Recorded in Mexico. It is found on small rodents. It also infests cattle and other domestic animals. It is doubtful if it bites man.

O. hermsi Wheeler. It has only recently been described and it has been proved to be a vector of relapsing fever in California.

The hypostome is very small; the capitulum is located in a deep camerostome with rounded integumental folds along the lateral and posterior margins; distal end of article IV of palpi with two clumps of bristles; integument fairly and irregularly rugose and under the microscope mamillae are seen; eyes absent : tarsi with well defined protuberances; both ends of each anal valve bearing 3 long hairs and one terminal bristle on either side of the anal opening.

It differs from *O. talaje*, a close relative, in the absence of large discs on the dorsum, of the characteristic sculpturing of the integument and of flap-like borders at the margins of the capitulum.

O. hermsi differs from *O. turicata* in the absence of clubbed hairs between the mamillae, a decidedly smaller hypostome and the arrangement and number of protuberances on tarsi I and IV in particular and the smaller size of the adults. It differs from *O. venezuelensis* in the sculpturing of the dorsum and the protuberances of the tarsi.

Relapsing fever. In 1857 Livingstone, the famous explorer, first suggested that the relapsing fever of Africa was tick-borne. Ross and Milne in 1904 followed by Dutton and Todd in 1904 from Uganda and Congo and Robert Koch from German East Africa proved that the tick was the vector of the spirochaete which was responsible for the fever. Relapsing fever which is due to *Treponema* (*Spirochaeta*) *recurrentis* is widely distributed and is carried both by ticks and lice. The louse-borne disease occurs in an epidemic form and is particularly noticed in war time and during famine. It is also endemic in certain countries. The tick-borne relapsing fever does not present any feature distinct from the louse-borne disease except that it does not occur in an epidemic form which is a characteristic feature of the louse-borne variety. This form is distributed in Africa, Palestine, Syria, Persia, Caucasus, Central Asia, and North Western Pakistan and Kashmir. It also occurs in Central and South America. The animal reservoir of this disease varies in different countries. The spirochaetal infection is maintained in nature in such animals as chipmunks, rats, squirrels, rabbits, porcupine and other animals and is transmitted by species of soft ticks from animal to animal.

The relapsing fever in Tropical Africa is carried from man to man by the human tick, *O. moubata* (Dutton and Todd, 1905). The tick serves as the true intermediate host for this disease and the infection is transmitted to their offspring (Koch, 1905). The incubation period in man after the tick's bite is about 10 days.

The morphological differences between the various spirochaetes are insignificant and they are almost entirely distinguished by their immunity reactions. Manteufel (1908) and Neumann (1909) have shown that *Spirochaeta recurrentis* and *S. novyi* as well as *S. duttoni* may be transmitted by *O. moubata*.

The life history of the spirochaeta in the tick is almost identical with that of African relapsing fever which develops in the louse. The spirochaetes of relapsing fever which are plainly visible in the cells of the intestinal epithelium an hour after ingestion, completely vanish at the end of three days. They seldom appear in the blood of the tick earlier than three weeks. The spirochaetes can also be detected in some of the eggs. The disappearance of the spirochaetes from the intestine of the tick is probably due to their breaking up into minute fragments in which state they enter the coelomic cavity where further development takes place. Sometimes the spirochaetes remain unaltered in the digestive tract even at the end of five days.

The acutal method by which *Spirochaeta duttoni* is transmitted by *O. moubata* is much disputed. There are three possible ways, namely, they may appear in the secretion of the salivary glands, in the secretion of the *Malpighian* tubes which escape from the anus (faecal material) or in the coxal fluid, which is exuded from the coxal orifice during the feeding of the tick. In order for the spirochaetes to appear in the secretion of the coxal glands, they must traverse the walls of the glands and accumulate in the secreted fluid in order to be excreted with it. It has been suggested that the coxal fluid is haemocoelic fluid filtered through a thin membrane of chitin and that the spirochaetes pass through the membrane with the fluid. Bone (1933) found that red corpuscles, micro-organisms or spirochaetes when injected into the coelomic cavity of ticks and then immediately fed on healthy mice did not lead to the excretion of the bodies introduced into the body cavity just before feeding. Marchoux and Couvy (1913) thought that the spirochaetes of chicken cholera passed from the coelome in *Argas persicus* into the acini of the salivary glands and gathered together in their excretory duct. The spirochaetes were transmitted by bite through the salivary secretion of the tick at the time of feeding, a view shared also by Knowles, Das Gupta and Basu (1932).

It is therefore thought that infection can take place by the bite alone. Brumpt (1933) pointed out that *O. turicata* allowed to feed and removed before coxal fluid was excreted, produced the infection by the act of biting. *O. papillipes* does not pass any coxal fluid while feeding and infection takes place by the bite alone (Pavlosky and Chodukin, 1929; Moskwin, 1929; Adler, Theodor and Schieber, 1937).

When infected in the nymphal stage, the infection is transmitted in the adult stage only to a small proportion of experimental animals. Only the first generation of ticks which are the offspring of an infected female becomes infected. The transmission of infection from the first to the second generation, although liable to occur, is very rare.

Francis (1938) reported that *O. turicata* survived for five years under condition of starvation and that these ticks, originally infected, were still able to transmit relapsing fever to a monkey.

Although *O. savignyi* has been suspected of playing the same role as *O. moubata* in the carriage of relapsing fever spirochaete, Kirk (1938) failed in transmitting this disease through *O. savignyi*.

The infectivity of an individual tick is irregular as the spirochaetes are not transmitted by every bite (Davis, 1941).

Relapsing Fever Ticks.

O. moubata. In Africa the important carrier is *O.* moubata which is widely distributed all over Africa, north, central and south, including Madagascar. No case of relapsing fever has occurred through ticks in South Africa.

Among the 25 species of ticks recorded in both North, Central and South America, at least 5, e.g., *O. venezuelensis, O. hermsi, O. talaje, O. turicata* and *O. parkeri* are associated with relapsing fever. (*O. parkeri* is a synonym of *O. wheeleri*, and *O. venezuelensis* of *O. rudis*).

O. venezuelensis Brumpt. It transmits relapsing fever in Panama, Columbia and Venezuela and not *O. turicata* as was previously believed. *O. venezuelensis* is however, not found in Mexico where *O. talaje* is the prevailing species and is also the loca carrier of relapsing fever.

O. parkeri (*O wheeleri*). This tick extends over a wide area in U.S.A. and is responsible for relapsing fever outbreaks in Utah and in places along the Colorado river. It feeds readily on man. When engorgement is nearly complete, there is a copious discharge of coxal fluid but the fluid has not been shown to play any part in transmitting relapsing fever. It has been found naturally infected with relapsing fever spirochaete. Its habitat is sandy floor of caves or rodent burrows.

O. talaje and **O. turicata.** In Mexico, Texas and California no case of relapsing fever has been recorded from the areas of distribution of *O. turicata.* In these parts *O. talaje* appears to be the sole vector of relapsing fever spirochaete. Moreover the distribution of the latter tick corresponds with the distribution of the disease. Although tick-borne character of the relapsing fever in the United States had been suspected in the past, definite evidence in this direction was brought forward for the first time by Weller and Graham in 1930 by discovering infection in nature in *O. turicata.* This tick, however, is not a carrier of major importance. Darling suspected rats as disseminators of relapsing fever in Panama and found nymphs of *O. talaje* on *Mus rattus.*

O. hermsi. Though both *O. turicata* and *O. hermsi* exist in California, Colorado and neighbouring regions, *O. hermsi* is considered to be the sole transmitting agent in these regions. This tick occurs particularly in the woody and forest regions. The distribution of this tick and that of relapsing fever closely coincides. It is thought that one means of acquiring relapsing fever in timbered regions may be contact with wood where these ticks are found.

O. erraticus. This tick is synonymous with *O. marocanus* and has been found naturally infected in Dakar, Western Africa. It has a limited distribution in Spain, Morocco and Algeria. It is found in rodent and owl burrows and is responsible for local outbreaks of relapsing fever.

O. papillipes (*O. tholozani*). This tick has a very wide distribution over a large area from Persia to Central Asia. In India, they are confined to the northern and western parts of Pakistan and in Kashmir. Wherever it occurs, it is a vector of great importance. This tick has been described under different names in different countries, e.g., *O. papillipes* in Turkestan, *O. tholozani* in Persia, *O. asperus* in Syria and *O. crossi* in India; all belong to the same species.

Cross and Patel (1921) reported successful transmission of the trypanosome of surra by the bites of *O. lahorensis* and *O. crossi* in the Punjab; this has, however, not been confirmed.

This is an extremely common tick and can be readily found at any time of the year in cracks in mud walls of animal shelters, also in the floors of sheds for domestic animals lying hidden in the dust. It is found quite commonly also in human dwellings. The larva is a rapid feeder and is rarely found attached to the host. In Turkestan and Palestine the adults are found in caves inhabited by bats. In Central Asia, *O. papillipes* occurs chiefly in human dwellings.

In the Middle East and Central Asia *O. lahorensis* and *O. papillipes* generally overlap in their distribution and sometimes both are suspected to be carriers of relapsing fever. As *O. palpillipes* have greater opportunities of feeding on human blood it is but natural that greater emphasis on its acting as a carrier in nature has been laid. Further Pavlovskii (1932) found *O. lahorensis* to be an unsuitable host for the development of relapsing fever spirochaete in comparison to *O. papillipes.*

It has been claimed that in Turkey *O. lahorensis* is the carrier of relapsing fever. It is essentially a sheep tick in Turkey. It has, however, a very wide distribution in Central Asia and North Western India. In Central Asia the young stages remain attached to their hosts during the winter months. They leave their hosts in spring. The adults are found in the dust and in fissures in walls in sheep enclosures, around rat holes and in sheds where cattle, camels, etc., take shelter at night. When opportunities occur, they freely attack man.

Thus any species of *Ornithodorus* is as a rule capable of transmitting all strains of relapsing fever normally transmitted by ticks belonging to this genus. The tick should normally feed on the infected animal in the nymphal stage and it is able to transmit the infection to the host, either human or animal, on which it feeds in the adult stage. When the species of tick is not the carrier of the spirochaete in nature, the infection tends to die out sooner than when the tick is the natural carrier.

Dissection of Ticks. A gorged female should be starved till much of the ingested blood has been digested. The tick is held gently between the thumb and index finger of the left hand and with a razor make an incision at the lateral border of the abdomen. This is carried forwards as far as the head, and backwards round the posterior margin of the body. The dorsal cuticle may now be separated from the ventral cuticle by inserting a fine needle at the junction of the two and passing it around the entire margin of the body. The ventral integument is next fixed by small pins to the dissecting paraffin block. Further dissections are performed in 4 per cent formalin solution under a dissecting microscope. The dorsal integument is now gently separated from the body by cutting with a knife-edged needle the tracheal and other attachments. As soon as the skin is reflected forwards, the intestine and the ovary which lies transversely, will come into view. The salivary glands are also easy to find.

Collection and Examination of Ticks. Ticks should be removed with forceps or tweezers. The danger of leaving the head of the tick embedded in the skin after the tick has been removed should be borne in mind. This can be avoided by killing the tick with pyrethrum, and pulling it off with forceps after 10 to 15 minutes. The pyrethrum may be used either in the liquid form as extract or as an ointment prepared with the dry powder in vaseline. Larvae are small objects and they can be dislodged by scraping with the blunt side of the knife blade. In collecting ticks care should be taken that all members from a particular host are collected in separate tubes, preserved in 75 per cent alcohol and are properly labelled showing especially the locality and the host.

When ticks are preserved in alcohol, the tick dies with its legs contracted. It will be almost impossible to straighten the legs subsequently. If living ticks are placed in a mixture of 30 per cent alchool to which 2 to 3 drops of ether have been added, they will die with their legs extended. After 24 hours they should be transferred to 70 per cent alcohol. Formalin should be avoided in preserving ticks as it causes the body to become hard which renders it unfit for examination.

For identification, they should be dried between folds of blotting paper, suitably placed on plasticine and examined under the microscope. In some cases slide preparations will be needed and the carbolic acid method, after the parts have been treated with caustic potash, will prove satisfactory.

Breeding of Ticks. Gorged ticks when placed on lightly moistened sand in a wide mouthed bottle will generally oviposit without any difficulty. A few twigs of moistened grass are also placed inside. After the tick has been introduced, a piece of cloth is tied round the open end of the tube. After the larvae have emerged, a week must elapse before they are placed on a host for feeding. If the larvae are now released on an animal there are great possibilities of loosing them altogether. They have a tendency to wander about after engorgement for seeking shelter for resting and digesting the ingested blood or they may even escape before feeding. Some ticks exhibit a natural aversion for feeding on the common laboratory animals. In such cases, calves should be selected. When the tick is placed on the ear or abdomen of a white rat, the movement of the tick should be carefully watched till it has taken a firm attachment to the skin of the host. The rat should next be placed in a cage inside a big glass jar containing carefully sifted white sand.

Some prefer to raise ticks in specially prepared cloth bags which are tied over the ears of rabbits, around scrotum of rams and goats or on the abdomen of calf. A far as possible, the larvae should not be let loose on the body of an animal. When rabbits are used, the jacket should enclose the body in such a way that the head, neck and legs are kept out of it. To prevent larvae and nymphs from escaping out of the bag after they have completed their feed, a false lining may be fixed by buttons to the inside of the bag. The ticks after feeding and detachment from the host will take shelter between the jacket and the false lining. The inside of the bag should be examined daily and when necessary at longer intervals. The engorged tick, after it has stopped feeding, should be picked up with forceps and left in a glass tube on a bed of moist sand till the blood has been digested and metamorphosis has taken place. It is a point to note that ticks do not feed immediately after moulting and about a week should be allowed to elapse.

When solitary examples have to be reared, they may be left in a battery jar with a white rat or a mouse and at the bottom finely sieved white sand is placed. If a larger jar is available, in place of a mouse a guineapig may be selected. In this way nymphs, larvae and adults may be fed without the fear of loosing them.

Clearing, Staining and Mounting. For clearing, a 10 per cent solution of potash is recommended. It should always be used cold. If necessary the cuticle can be stained with picric acid in xylene, fuchsine, Bismarck Brown, orange G or osmic acid (1% aqueous solution) etc., and then dehydrated by transference through graded alcohols, cleared in clove oil before mounting in Canada balsam.

Control. Control of hard ticks has been dealt with comprehensively in Part II.

Regarding soft-ticks, fumigations by sulphur dioxide or HCN do not cause any inconvenience to *O. moubata* even when the ticks are exposed in open vessels to the gases. Turpentine is lethal to ticks as it is to mites but it should be diluted with kerosene. Addition of alcohol and soap will enhance its killing properties. The composition of the mixture is as follows: Turpetine—30 c.c.; 25% alcohol 50 c.c., Kerosene —5 c.c., Soft soap—q.s. This kills ticks in a few minutes.

Another effective but less expensive solution is as follows : Paradichlorobenzene —1 lb., Kerosene—1 gallon. For 2000 sq. ft. of surface sprayed, about 12 to 14 gallons will be necessary. It has proved very effective against *O. papillipes* in Central Asia and *O. moubata* in Africa.

Ornithodorus ticks appear to be relatively sensitive to DDT but distinctly more to BHC and the dusting of the powder in the places where these ticks are found will appreciably reduce their number. This is a very useful method of checking an epidemic of tick-borne relapsing fever.

[For references, see end of chapter XVI.]

CHAPTER XV

MITES

Family **Trombiculidae** (Harvest or Velvet Mites). These belong to the suborder Prostigmata in which the spiracles lie on the lower part of the basis capituli.

Trombiculid mites are found in greatest abundance and diversity in areas with a hot moist climate and an abundance of vertebrate life. They are reported from all the zoogeographical regions of the world and parasitize in their larval stage, representatives of four of the five classes of vertebrates, the amphibians, the reptiles, the birds and the mammals. In the tropics they are found even on mountains.

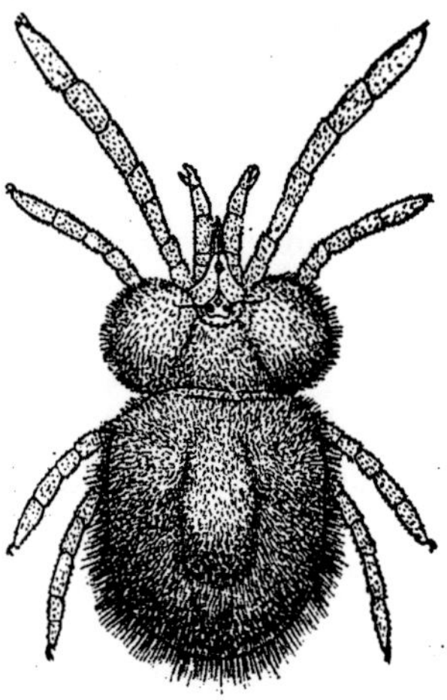

FIG. 167.
Trombicula irritans.
(After Ewing).

The typical members of this family are large bright red mites and from the velvety nature of their integument, they are known as velvet mites. They resemble the common harvest mites of Europe. The cephalothorax is well demarcated from the rest of the body, which is densely covered with hairs. Eyes may be present or absent. The last joint of leg IV is not, or very slightly, shorter than the penultimate, and the last joint of leg I is usually swollen. The body is hairy. The palpi are 5-jointed, the penultimate one being provided with an accessory finger-like process. The legs are 7-jointed. Along the median line of the cephalothorax there is commonly a crista or dorsal groove.

The characteristic feature of the members of this family is that only during the larval stage they are parasitic and at other stages of their extistence they are free-living. Though the range of their hosts is wide, yet rats and bandicoots are the preferred ones.

Larva. The parasitic red mite is the hexapod larva of *Trombicula*. It is of ovoid shape and of reddish or greyish yellow in colour. The mite at this stage is so small that it is hardly visible to the naked eye. The larval mites are found in nature parasitic on many animals, such as dogs, cats, fowls, snakes, rats, shrews, bandicoots etc., being found in the interior of their ears in clusters. The palps are stout and consist of 5 segments of which the last one bears long hairs and the penultimate a claw. The dorsal scutum is rectangular and bears 7 plumose hairs, one pair of the latter is long and tactile and attached to the pseudostigmata which lie on the scutum. One pair of eyes, most conspicuously red in colour and consisting of groups of granules, are placed on the dorsum near the posterior corner of the scutum. The integument is soft and finely folded bearing numerous plumose hairs.

The larva is active in its movements. At the time of feeding it buries the whole length of its chelicerae in the skin. The method of feeding is unique. It feeds through a long tube called stylosome, a term coined by Jourdain in 1899; the mechanism of its formation is not known. It is probably formed by the salivary secretion after the mouth parts have penetrated the skin of the host. An irritant secretion of the saliva is injected which has probably the power of causing lysis of the tissues. Its walls are of a solid nature. This tube, however, remains in situ when the mite has ceased to feed and has already withdrawn its porboscis. Kalra in his "Reports on Histopathological Investigations" has described the internal structure of *T. deliensis* larva and the lesions produced in the skin of the rat by the attached mites (War Office Report on Scrub Typhus Investigations in S. E. Asia, March, 1947; London War Office).

FIG. 168. Diagramatic representation of the stylosome after withdrawal of the mouth parts. (After Ewing).

The food of the larva consists of lymph and tissue juice of the host which is the product of an inflammatory reaction. Neither the nymph nor the adult is parasitic nor predacious. The adult mite is found in a free state on the ground or under fallen leaves and decayed vegetation. It is comparatively large and on account of its scarlet colour and velvety smooth appearance it is very attractive to the sight.

The trapping of rats can always give information as to the presence of mites in the soil and examination of the rats can give information as to the presence of mites.

Important genera. Among a large number of genera of these mites, mention may be made of only two, *Trombicula* and *Schongastia.* Their importance lies in the fact that in the larval stage only they are ectoparasites of rodents, especially rats and bandicoots, on whose "blood" they feed, and which are the animal reservoirs of *Rickettsia orientalis.* Among the two, *Trombicula* is by far of greater importance than *Schongastia.*

In both the genera the dorsal plate of cephalothorax of the larva has a median seta on or near its anterior margin. The character of the pseudostigmatic organs helps to distinguish them. In *Trombicula* (*Trombicula* and *Eutrombicula*) the pseudostigmatic organs are setiform or flagelliform and are provided with barbs.

Genus **Schongastia.** This genus has been created to apply to those *Trombicula* larvae which have peculiar club-shaped pseudostigmatic organs. This genus besides *Schongastia* also includes—*Neoschongastia* and *Euschongastia.* The anterolateral setae of dorsal plate are large, barbed and similar to posterolateral setae; each chelicera has a row of dorsal teeth; palpal claw is usually provided with two prongs.

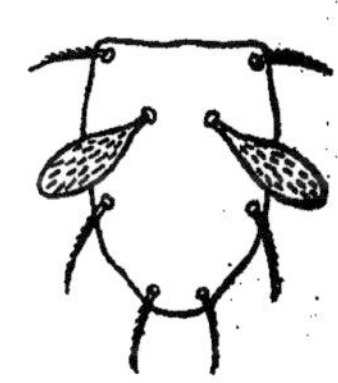

FIG. 169. Scutum of *Schongastiela bengalensis.*

Genus **Trombicula.** The characters of *Trombicula* are the following: Each chelicera has a single dorsal tooth; ventral tooth present; palpal claw is divided into three prongs; dorsal abdominal setae usually more than 30.

T. autumnalis. It is the European harvest mite-bug. The larvae are especially prevalent and noxious in the autumn when they are found in large numbers in gardens. Men are bitten at this season. It also attacks sheep, goats, cows, horses, moles, voles etc; rabbits and hares may be literally covered with the larval mites.

T. irritans Riley. (*Eutrombicula alfreddugesi* Oud.). It is the American harvest mite and is prevalent throughout the United States. It has the same feeding habits as the European harvest bug.

T. akamushi Brumpt. It is distributed in Japan, Formosa, South and Eastern China, Korea, Malaya, Borneo, Philippines, Celebes and New Guinea. In Japan it occurs in silt-covered fields on the banks of rivers and freely attacks man. Amongst its hosts have a wide variety of rodents especially a vole, *Microtus montebelloi* and in Formosa, the common house rat, *Rattus rattus refuscens*. In Malaya it is particularly common on chickens and congregate in clumps under the wing or in the breast feathers. It is also noticed in the ears of rats. The larva is bright red in colour. The adult is found in the soil of infested fields. It is a tiny creature, pale greyish or reddish in colour, with one pair of rudimentary eyes.

T. pseudoakamushi Hatori. It closely resembles *T. akamushi* except that the latter is the "thin-haired" species, while the former has coarse and thick hairs on the integument. It is regarded as identical with *T. hirsti*. The larvae are generally found attached to the loose skin of chicken. They are also fairly common in the ears of rats where they occur in clusters.

FIG. 170.

A. *Trombicula akamushi*. (After Hirst). B. *Trombicula deliensis*. (Atter Fletcher, Lesslar and Lewthwaite). C. Scutum of *T. deliensis* (After Fletcher *et al.*).

T. hirsti Sambon (*T. minor*). It occurs in North Australia, New Guinea and Philippine Islands. Its hosts are bandicoots. It attacks man and it has been suspected to be a vector of typhus but this has been based on incomplete evidence. *T. wickmanni*: It is prevalent in large numbers in certain areas in New Guinea, Philippines and Borneo. In New Guinea it has the gaura pigeon as its principal host. It has been reported to bite man. *Schongastia schuffneri*: It is a small red mite with colourless hairs and has been reported as biting man in Sumatra. It is seldom found on rats. It has not been recorded from Malaya.

T. deliensis Walch. It is a yellowish white or pale orange mite with colourless hairs and is widely distributed in the tropical regions of South Eastern Asia, e.g., British New Guinea, Admiralties Islands, North Queensland, Sumatra, Malaya, Burma especially the Northern part, India, Indo-Pakistan border, Ceylon, and Maldive Island. Its prevalence is directly related to the intensity and length of the monsoon period, i.e., there is a gradual decrease of the mite population from the east to the west of India. It is found even at considerable heights in mountains. They are found on rats, bandicoots and shrews which are the animal reservoirs of the virus of scrub typhus, and which is carried by a member of the subfamily Trombiculinae. In the Australian region in addition to rats, bandicoots and hydromys carry *T. deliensis* on them in large numbers. The bandicoots of the Australian region are marsupials and somewhat resemble rabbits in appearance and habits. They are one of the common animals of the bush and large range of animal ectoparasites can be obtained on them. While the larval mite can be easily obtained in the ears of rats where they are found in clusters of 50 to 60 or more often lying embedded in serum and scarf, they are mostly found on the hind legs of shrews just on or below the hairy margin. They can also be obtained, though in small numbers, from the earth around rat burrows. Sometimes the larvae are found on dogs, cats and even fowls. The larva remains on the host for 3 to 4 days.

Shrew is a term applied to the members of the family Soricidae of the mammalian order Insectivora. *Crocidura cerulea*, the common musk-shrew of India, commonly frequents human habitations. *Hydromys* is an Australasian genus of rodents known as water rats and beaver rats.

People coming out of mite-infested fields or jungle occasionally carry a large number of larvae on them without even knowing it. Larval mite can also be picked up by a person sitting or lying on the ground. The bites as a rule do not excite any irritation but generally within 24 hours a small papule may appear. The parts itch intolerably. A blister may or may not form. It is rare to find larvae after once the blister has formed.

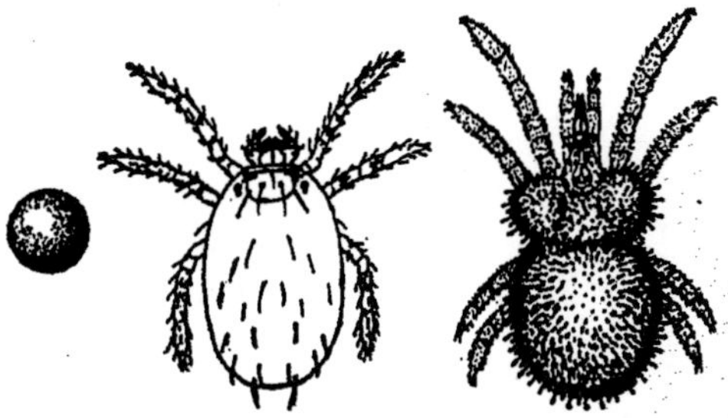

FIG. 171.
Egg, larva and nymph of *T. deliensis.*

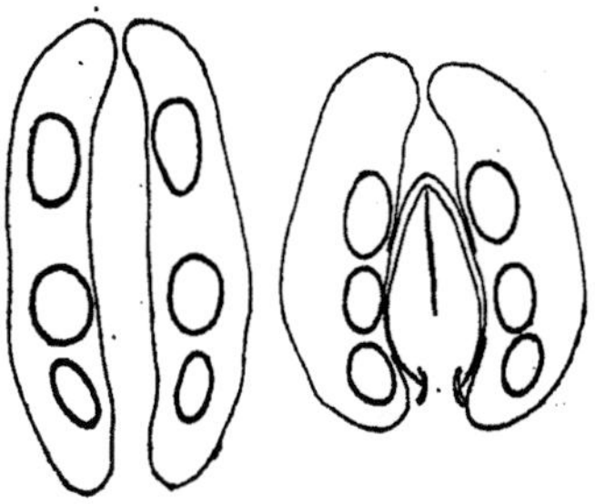

A B

FIG. 172.
External genital opening in (A) female and (B) male *T. deliensis.*

Within the last few years its life history has been worked out by several workers, e.g., Jayewickreme and Niles (1946, 1947), Wharton and Carver (1946), Jenkins (1947, 1948), Cockings (1948) and Krishnan *et. al.* (1949). The important features in its life history are that the mite in its nymphal and adult stages are carnivorous and normally the eggs of Collembola, Isopoda and other soil-inhabiting animals are probably the most important food of the adult and nymphal mite in nature. In the laboratory the nymph and the adult readily feed on moistened mosquito eggs. The different stages in the life of this mite are : egg—7 days; deutovum with the egg shell broken and the larvae remaining quiescent enclosed in the membrane—7 days; larva—remains attached to the host for 3 to 5 days and then drops off to turn into the next stage in 2 to 4 days; nympho-chrysalis, a quiescent stage lasting 7 days in which the nymph develops within the larval integument; nymph—7 to 14 days; imago-chrysalis, the second resting stage in the cycle of development of the mite—7 days; adult, in which the sexes are distinguishable. When the food is insufficient the adults become cannibalistic.

These mites are most active during the rainy season and are found in moist places covered with scrub jungle. The females when fed properly will oviposit for about five months.

The essential requirements for rearing *T. deliensis* in the laboratory are : moisture, proper type of food, e.g., powdered leaves, coarse sand and mosquito eggs, and prevention of the growth of fungus by putting a few earthworms in the culture bed. The larval mites are allowed to feed to repletion on rats. If the rat cage containing the rat is placed over a pan containing water, the mites drop off and float in the water. They can thus be obtained for breeding purposes.

(*a*) **Scrub typhus.** The name scrub typhus has become popular as a synonym for tsutsugamushi disease because the disease is acquired during exposure in terrain covered by secondary growth or scrub vegetation where *T. deliensis*, through which the disease is contracted, are found. A few scattered cases of scrub typhus which apparently were contracted in virgin jungle occurred in the United States Army personnel, stationed in North Burma and New Guinea during World War II.

FIG. 173.
Collection of larvae of *T. deliensis* for rearing to the adult stage and for breeding. The larval mites after feeding drop off from the ears of rats. The bowls contain water.

The disease is transmitted by a mite of the subfamily Trombiculinae from a rat or bandicoot. The mite becomes infected in the larval stage and transmits the infection to man or rat only in the larval stage of the next generation when it bites again. The rickettsiae, *R. orientalis*, are carried through various stages of development and are transmitted through the eggs. The infected rat does not become ill and remains healthy except that its spleen is enlarged. Scrub typhus exists almost in every island from north-west Honshu down to north Australia. In India it has been reported from Bengal, United Provinces, Punjab, Central Province, Bombay Presidency, Madras, Mysore, Assam and Bihar.

Mite infested areas are usually open and particularly cultivated areas in the jungle covered with low scrub and tall grass. It has been noticed that a freshly opened up tract in a jungle is free from mites but as soon as food becomes available, rats, shrews and bandicoots are readily attracted to such areas. This explains the reason why cases of typhus appear generally sometime after the establishment of a camp in jungle.

R. orientalis, the rickettsia concerned, is transmitted by the larval forms of certain mites, the two important ones being *T. akamushi* in Japan, China, Formosa and Philippines and *T. deliensis* in the rest of the areas, including India. In certain areas in New Guinea and possibly in the neighbouring islands mites of the genus *Schongastia* have been incriminated.

Although the virus of tsutsugamushi is normally carried by *Trombicula* mites, *Xenopsylla cheopis* is not only able to retain the virus for 11 days, it is also able to transmit it by biting (Nicolle and Sparrow, 1934).

(*b*) **Rickettsialpox.** It has been so named on account of its clinical resemblance to chickenpox. The causative agent, *Rickettsia acari*, has been isolated from persons ill with this disease and from *Mus musculus*, the house mouse (Sussman, 1946; Shankman, 1946). The rickettsia has also been recovered from the mouse-mite *Allodermanyssus sanguinensis* Hirst, inhabiting the domicile of ill persons in New York city from where the disease had been reported (Huebner, Stamps and Armstrong, 1946; Huebner, Jellison and Pomerantz, 1946).

Collections of Mites in the Field: (1) From the ears of captured rats and bandicoots after they have been killed, and from the hind legs of shrews. (2) On boots after a person has stood in the jungle or on a low seat for about 15 minutes. (3) After clearing a small area of grass and shrubs, a piece of white paper is spread out on the cleared ground. (4) In suspected places clean rats and mice may be left in wire cage on the ground for 12 hours. (5) By taking samples of superficial soil, breaking it up and floating in water. This method is not suitable for recovering a large number of nymphs from the soil as they are readily wetted.

For a list of mites found on rats in India and recorded by Kalra (1951): see in Siphonaptera.

Prevention. (1) Systematic trapping and burning of rats and bandicoots around newly established camp sites. Poison baits should not be laid as there would be a greater chance of dispersal of the mites from dead rats. (2) The shrub and low vegetation should be cut down or preferably burnt. If possible, a bulldozer may be employed. When the top soil is dried out, the area becomes considerably safe. Even the removal of grass and undergrowth from a site goes a long way in reducing the mite population. (3) Sleeping and squatting on the ground should be strictly forbidden until after the jungle has been cleared. No one should be allowed to walk except along paths made safe against infiltration by mites. (4) DDT dust spray is ineffective. (5) DMP—By applying Dimethyl phthalate on garments. The DMP is poured into a pot and the ends of the fingers of one hand are dipped in the fluid which is then spread over the palms of both hands by rubbing. Both the palms are now applied over every part of the garment spread out on a table, particular attention being paid to the seams. Both the inside and the outside are equally treated. As a rule two ounces will be sufficient for the treatment of one pair of socks (6 smears for each), trouser (30: each leg—10, trunk—10), bush shirt (25: each arm—5, body—15), cotton vest (6) and underwear (6). A smear means dipping of the

FIG. 174.
Application of DMP on garments.

fingers in the fluid. Though one application should remain effective for six weeks, in field conditions, however, it should be repeated at the end of a fortnight. Regarding crude Benzene hexachloride and other chemicals for control of ectoparasitic mites see in Part II.

Family **Tarsonemidae.** The members of this family belong to the suborder Heterostigmata. They are so-called on account of their possessing different forms of spiracles. The Tarsonemidae is a family of extremely small mites, which are mostly parasitic on plants and on other insects. In this family the spiracles are situated on the ventral surface just in front of the first pair of legs. Some of the legs are provided with long stalked terminal suckers. The females are recognised by the presence of a club-shaped appendage, which projects between the bases of the first and second legs on either side of the body. In size they are minute but they multiply very rapidly. Amongst other things they infest straw and hay. The straw becomes quickly covered with a mould-like coating which consists of a mass of mites amongst which the large pregnant female can be readily spotted with the naked eye.

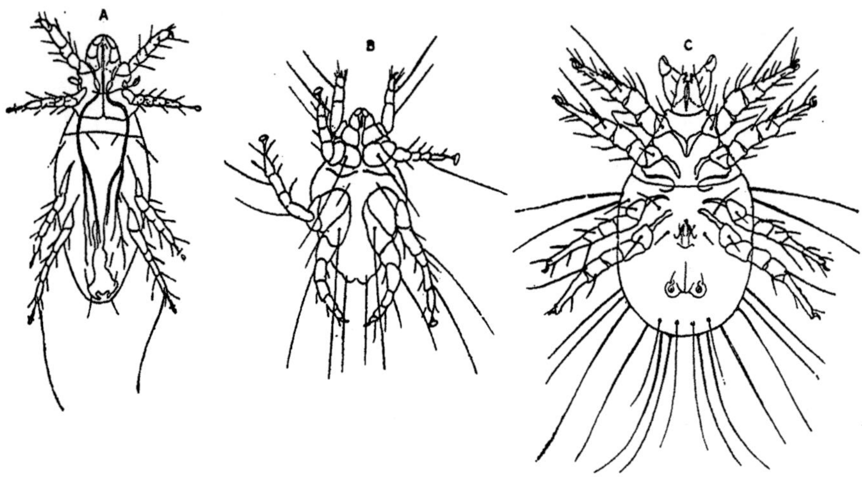

FIG. 175.
A & B. *Pediculoides ventricous*, female and male. C. *Tyroglyphus longior* var. *castellani*. (After Castellani and Hirst.)

Genus **Pediculoides.** In this genus the pedipalps are stout and curved and comprise almost the whole head. The species *P. ventricosus* Newport, is so-called on account of the fact that in the pregnant female the lower part of the abdomen becomes swollen to form a large brood-sac inside which the embryos develop. These on hatching live within the mother till they are mature. They escape through the already thinned abdominal wall. This mite is normally parasitic on insects and their larvae and is found on straw and different kinds of grain in tropical countries. Outbreaks of dermatitis acquired through sleeping on straw mattresses and unloading cotton seed have been traced to this mite.

Family **Tyroglyphidae.** The members of this family and also of Sarcoptidae (itch mites) have no external openings for respiration and are therefore included in the suborder Astigmata. Mites belonging to the family Tyroglyphidae generally attack provisions and groceries and are known as forage mites. They are also found in decaying leaves. They are often found swarming in stores and cause damage to sugar, flour, grain, cheese, cereals, dried fruits etc. In this family there is a distinct cephalothorax and the chelicerae project like a rostrum. The body is elongate.

The life history of some of these mites is remarkable. Its growth from the egg stage takes place either in the normal way as occurs in other mites, i.e., the larva is followed by a nymph which develops into the adult mite, or sometimes the nymphal stage may pass through another stage called *hypopus*. The hypopus is exceedingly small and does not feed. It is mainly adapted for the purpose of emigration by attaching itself to some active creature such as a fly or a mammal. As soon as it reaches the proper environment, it becomes detached and moults again into the nymphal stage and then turns into an adult.

The important genera which may concern the public health worker are :

(1) *Tyroglyphus* : common species are *T. longior* Gerv., *T. siro* L. (2) *Glycyphagus*. *G. domesticus* de G. (3) *Aleurobius*. *A. farinae*. (4) *Carpoglyphus*.

The dorsal integument in *Tyroglyphus* is smooth but is rough in *Glyciphagus*. Some of the hairs of the body of *Glycyphagus* are feathered and the claws of the legs are very inconspicuous, whereas in *Tyroglyphus* the opposite conditions are found. In *Tyroglyphus* the presence of a clavate or thickened hair on base of tarsi I and II and a suture between cephalothorax and abdomen will settle the identity. Both may occasionally give rise to dermatitis (grocer's itch).

Cheese on prolonged storage often becomes infested with *T. longior*. *T. longior* var *castellani* are responsible for copra itch contracted from handling infested copra. Water-itch or tea bush dermatitis is caused by *Rhizoglyphus parasiticus* Dalgetty., belonging to the family Tyroglyphidae. The dermatitis usually occurs on the foot in tea-gardens during the rainy season (Castellani and Hirst, 1912).

Grocer's itch is a troublesome dermatitis for which *G. domesticus* is responsible.

Invasion of the human respiratory, alimentary and urinary tracts by Tarsonemid and Tyroglyphid mites has been mentioned. *A. farinae* infests flour. *Glycyphagus* are often found in sugar and *Carpoglyphus* in dried fruits.

In *Aleurobius* the pedipalps are distinct from the chelicerae. The integument in *Carpoglyphus* is smooth, suture between the cephalothorax and abdomen absent and the tarsal claws are not extension from the claws but arise from a membrane attached to the last tarsal segment.

Family **Sarcoptidae.** These cause itch in man and mange in domestic animals. All are parasitic. They are liable to attack both human beings and domesticated animals causing scabies and intolerable irritation. Mites belonging to Tyroglyphidae and Sarcoptidae have no respiratory openings and are therefore placed in the suborder Astigmata.

Genera of Sarcoptidae. (1) **Sarcoptes.** Anal opening is not dorsal but terminal; the dorsal surface of the body is provided with pointed scales and spines. These include the itch mite of man and other animals. (2) **Notoedres:** Anal opening is dorsal; dorsal surface of the body with only short, sharp setae. These produce similar diseases in cats and rats. It is much easier to find *Notoedres* than *Sarcoptes* from the infested parts. *Sarcoptes* and *Notoedres* are the two burrowing mites. (3) **Psoroptes** : Distinguished from *Sarcoptes* by the tarsal suckers having jointed pedicles and chelicerae non-segmented. They produce scabs and do not burrow. It is responsible for mange in horses, cattle, sheep and rabbit. (4) **Otodectes cynotis** Her., produce similar diseases in dogs and cats. In *Otodectes* the tarsal suckers have unjointed pedicles on the first and second pair of legs in the female and on all four pairs in the male. The fourth pair of legs in the female is small. In the male the copulatory tubercles are not prominent.

Genus **Sarcoptes.** Two genera of burrowing mites are responsible for mammalian scabies, namely *Sarcoptes*, which affects man and larger animals and *Notoedres* which at times may affect man but is principally a parasite of smaller animals, such as cats and rodents. The genus *Sarcoptes* was formerly subdivided into numerous species, named according to the host upon which the parasite was found, but principally owing to the work of Warburton (1920) and of Buxton (1921), these species are now generally accepted as varieties of a single species *S. scabiei*. Many of the species that are normally confined to one host can live on other animals and on man though for a short time.

Mites of the genus *Sarcoptes* are extremely small, globular in shape and of a whitish colour. There is no definite segmentation of the body though it is clearly divided into two regions by a fold of the integument. These two regions are generally called the notothorax and the notogaster.

The integument shows transverse striations and is provided with scales, hairs and spines. The legs are short and stumpy and are arranged into two groups. Each is provided with a powerful claw. Legs one and two terminate in ambulacra, long tubular processes ending in a bell-shaped sucker. In female mites these ambulacra are not found on legs 3 and 4, each of which is produced into a long bristle. In the male the fourth pair of legs terminate like the anterior legs in ambulacra, while the third has a long bristle. The first and second legs are similar in all respects. Both possess 5 joints. The third and fourth pair of legs are smaller and each consists of 4 joints. The anus is terminal.

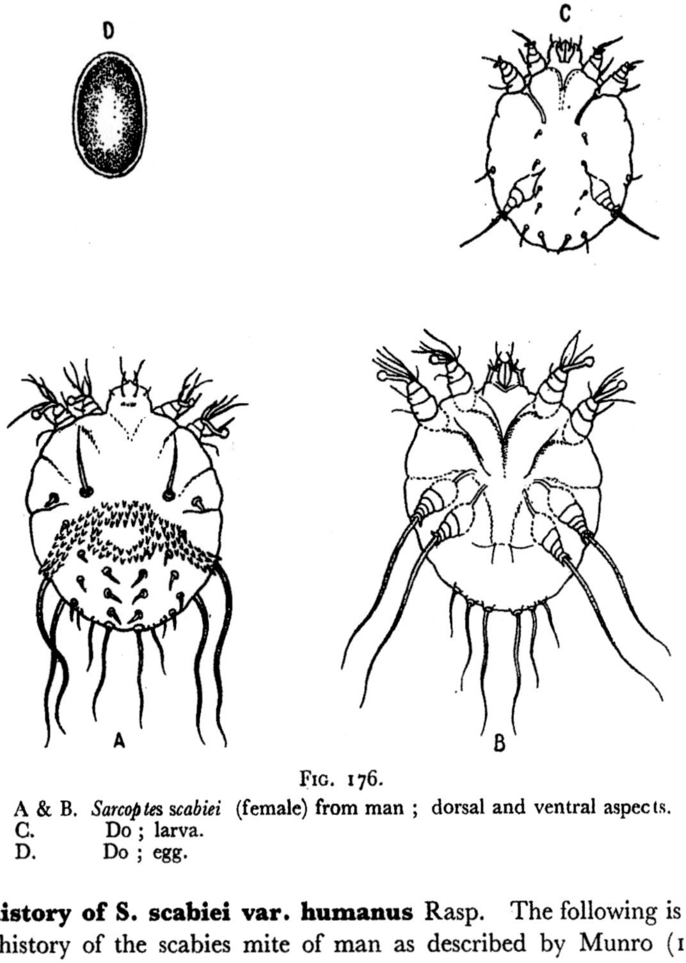

FIG. 176.
A & B. *Sarcoptes scabiei* (female) from man ; dorsal and ventral aspects.
C. Do ; larva.
D. Do ; egg.

Life history of S. scabiei var. humanus Rasp. The following is an account of the life history of the scabies mite of man as described by Munro (1919). The female burrows into the epidermis of the skin. The burrow is a fine winding line and is never deeper than the Malpighian layer. The parts of the body which she selects for her burrowing and oviposition are well defined and are as follows : the interdigital spaces, the wrists and ulnar margins of the wrists, the elbows and anterior folds of the axillae, the penis, scrotum and buttocks, the back of the knee and the ankles and toes. The ovigerous female proceeds to form the egg burrows after the last moult. Under normal conditions the burrowing corresponds more or less to the time spent in bed, which is roughly eight hours. Oviposition commences almost simultaneously with the burrowing. The number of eggs laid by the *Sarcoptes* of man, according to Gerlach (1857), is more than 21 and may be between 40 and 50. The maximum number of eggs in one burrow, as observed by Munro, was 14. The hatching time of the egg varies from 70 to 80 hours. Blanchard (1890) states that development may commence before oviposition and that the eggs are viviparous. The same thing was noticed by Munro in the case of sheep itch mites. Ova have never been obtained from acari removed from the body of the host.

Munro experienced great difficulty in obtaining larvae and nymphs even when numerous scabies cases were available. In searching for these stages he removed complete burrows and surrounding skin. This method yielded larvae and on one occasion a first-stage nymph.

According to Munro larvae are found in egg burrows and also in the scrapings of the skin in the neighbourhood of the burrows. They can also be obtained in the neighbourhood of isolated vesicles. The larvae when released on the skin will bore into the hair follicles where a vesicle makes its appearance and from which larvae can be recovered.

According to the same author there are two nymphal stages though he was unable to obtain any specimen of the first nymphal stage. The second nymphal stage or immature female is found in burrows which are smaller and are unaccompained by vesicles. Pockets or short blind branch burrows may be found opening into them, which are the work of the male. Mating occurs at this stage. After moulting in the burrow and leaving the cast skin in it, she turns into a mature female and proceeds to make the egg burrow proper.

The duration of the different stages as found by Munro is as follows: Egg: 2 to 3 days; Larva : 1 to 3 days; First nymph : 1 to 2 days; Second nymph : 2 to 4 days. The whole life cycle from egg to egg covers a period of from 9 to 15 days. Mellanby (1943) holds the view that both larvae and nymphs enter hair follicles for shelter and food. In our observations on patients and on volunteers artificially infested, we did not come across a single male individual and only one specimen of a larva taken from a burrow which had just escaped from the egg shell, was found. The rest were all nymphs of two sizes, and mature females. In most cases the females were gravid.

The female after depositing the full complement of eggs dies at the blind end of the burrow.

In experimentally transmitted infestations Mellanby (1944) observed that itching is generally experienced at least a month after the introduction of the parasite, and after six weeks the irritation is usually sufficient to cause disturbance of sleep. The appearance of this symptom so late after the entrance of the mite into the skin of the host has been considered by Mellanby as an indication that the symptoms are the result of sensitisation produced by the parasite. On the other hand, we are inclined to hold from our observations that a large number of parasites are necessary to cause disturbance of sleep, or to attract the attention of the patient to the fact that he needs medical attention. If a large number of ovigerous females are introduced, marked symptoms appear within a week. It has been observed that the papules are associated with the most intense itching.

So-called severe cases of scabies showing well-marked clinical symptoms may have large mite populations. In the presence of secondary infection mites cannot thrive, and under such conditions it is difficult to find them.

Unless it is a fertilised female, the transfer of a single individual from a patient to a volunteer will lead to no results and eventually the parasite will die. But if the object is to induce an artificial infestation, either younger stages of both sexes must be transferred or better, a fertilised female.

Mellanby (1941) shares the same views with Hebra (1868) that the possibility of the migration of mites from the body to clothing, bedding etc., is small. In fact Mellanby has found that prolonged contact, e.g., sleeping on the same bed is necessary for the spread of the disease. On the other hand, MacCormac and Small (1917), Munro (1919), Lydon (1941) and the author hold opposite views.

In course of time, as a rule, an immunity against re-infestation develops.

Sarcoptes scabiei var *crustosae* (Furst) : Norwegian scabies or crusted scabies is a well-defined condition for which a mite resembling the ordinary scabies mite of man but presenting minor difference only in the size of the spines, has been thought to be responsible (Buxton, 1921). This variety of scabies is at present considered to be in no way different from the ordinary type.

Sarcopticidal drugs. (1) Sulphur : Though it is not directly sarcopticidal, it is able to kill the mite when applied to the affected parts. (2) Benzyl benzoate : (Benzyl-benzoate—20% with an equal quantity of alcohol and soft soap). It is

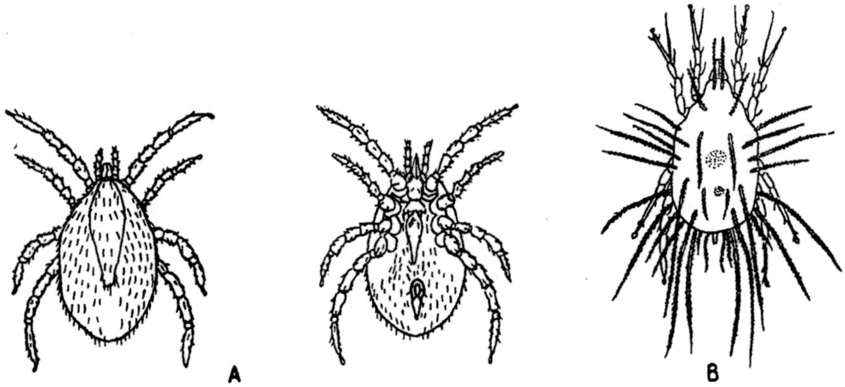

FIG. 177.
A. *Liponyssus bacoti* (after Hirst) ; dorsal and ventral aspects. B. *Glycyphagus domesticus.* (After Hirst.)

directly sarcopticidal. (3) Turpentine emulsion : (Oil of turpentine 20 parts; soft soap or soap shavings, q.s.; 80% alcohol 80 parts). It is the best form of ambulatory treatment of scabies. It is directly sarcopticidal and is as efficacious as Benzyl-benzoate being at the same time considerably cheaper (Roy and Ghosh, 1944). (4) 'Tetmos'—Tetraethyl thiuram monosulphide : It has been claimed to be a cure for scabies (Gordon and Seaton, 1941, 1942). According to later reports it gives considerable relief but does not bring about a cure. (5) Danish treatment with sulphur. (6) Phenyl cellosolve : It is deadly toxic to the itch mite and is not irritating to the skin. It is mixed with vegetable oil. In babies an extensive area of the skin on the body including the face and scalp may be affected with itch and phenyle cellosolve, suitably diluted with any bland vegetable oil, offers the best method of treating such cases. (7) Benzyl hexachloride in 1 per cent solution in liquid paraffin (Taylor,

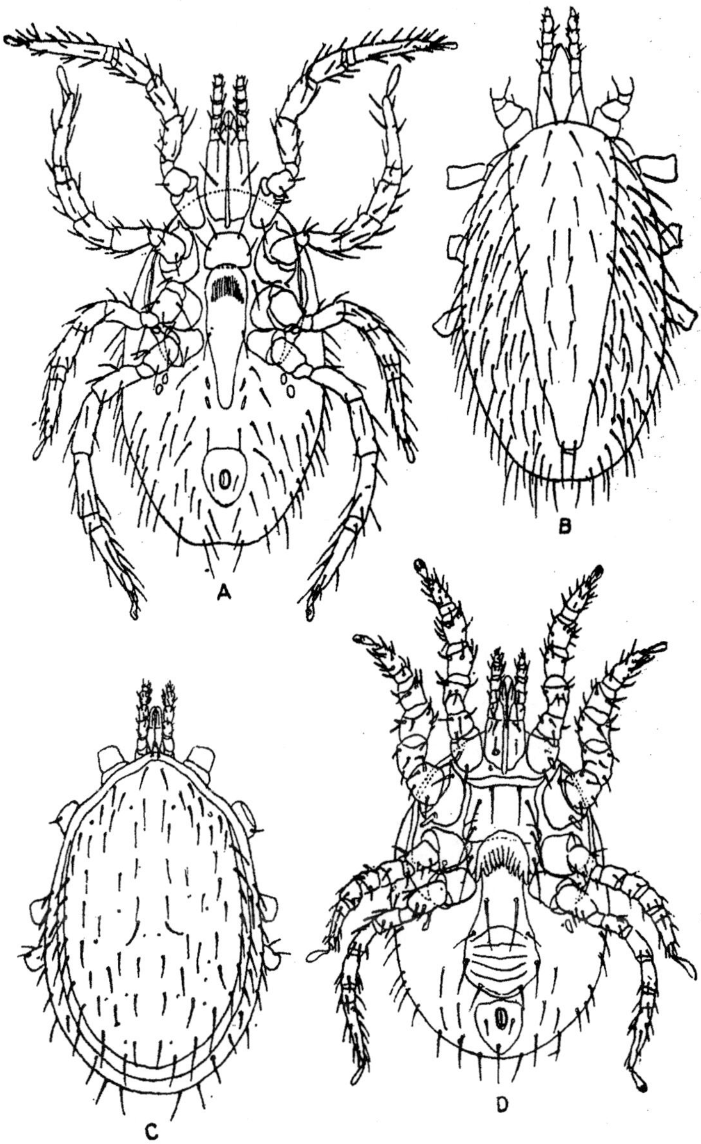

FIG. 178.

A & B. *Dermanyssus muris*, (from Hirst) ; ventral and dorsal aspects.
C & D. *Laelaps nuttalli* (from Hirst) ; Dorsal and ventral apects.

1945; Everdringen, 1948). The powder may also be dusted on the skin. (8) Dimethyl phthalate has been found of great value in infections in rats with *Notoedres muris* (Osborne, 1947). (9) Goldman's modification of benzyl benzoate treatment for scabies consists in the application of a 25 per cent emulsion in 2 per cent lanette wax. The lanette wax is melted on a water bath, benzyl benzoate is added and the mixture is brought to 60° to 70° C. The mixture is then poured into water previously brought to the same temperature, stirred thoroughly and allowed to cool (Goldman, 1939).

Mange in animals quickly responds to sulphur treatment. Recurrence is likely to occur unless the treated animal is separated from the rest. Gammexane dips are as beneficial as sulphur. Sarcoptic mange of pigs can be effectively and quickly controlled by using 0.025% chlordane as an emulsified solution and sprayed on the animals. For a detailed account of the treatment of mange, see in Part II.

Family **Gamasidae.** The mites belonging to this family have the respiratory openings between the coxae of the third and fourth pair of legs and are placed in the suborder Metastigmata. The majority of the Gamasidae are free-living, only a few being parasitic. There has been a tendency to include the parasitic species in one separate family Dermanyssidae, which are found on birds, reptiles, small mammals, bats and small rodents. Among those found on rats, the following genera are important. (1) *Liponyssus* (*Leignathus*), (2) *Dermanyssus* and (3) *Laelaps*.

Genus **Liponyssus.** The chelicerae are slender and pointed. *L. bacoti* Hirst : This mite occasionally attacks man. It can transmit endemic typhus from rat to rat. Smith and Mehta (1937), however, found that this species was unable to take up and develop the virus of typhus of the X 19 type. This species has been incriminated with dermatitis in man. *L. bursa* Bert., the tropical fowl mite, occasionally attacks man. Those which live in close proximity to pigeons are known to be bitten by these mites.

Fig. 179. *Demodex folliculorum*. (After Blanchard).

Genus **Dermanyssus.** It is essentially a chicken mite. In this genus the chelicerae are united to form a very long style. Various cases of *D. avium* attacking human being have been reported. *D. muris* Hirst: This species is found on rats. *D. gallinae* Redi, attack fowls and are found in cracks in walls in the day time. They occasionally attack man.

Genus **Laelaps.** The fingers of the chelicerae of the female are rather short, fairly stout and armed with distinct teeth. Coxal and ventral shield with stout bristles and spines. They are true parasites and suck the blood of their hosts. They parasitise small animals. *L. nuttalli* Hirst, and *L. echidninus* Berl., are found on rats. One genus of Gamasidae, *Pneumonyssus* is found in the lungs of monkeys.

Liponyssus bacoti and other Gamasid mites do not ever fix themselves to their hosts in the way the larva of a Trombiculid mite does. After repletion, they seek cracks and holes generally in buildings in which to mould or oviposit. The development from the egg to the adult stage may be completed in 15 days.

Family **Demodecidae.** (Follicle mite). Genus **Demodex** Owen. They are elongate and worm-like. The posterior extremity is tapering. They have no respiratory openings and the respiration takes place through the cuticle. They are parasitic.

D. folliculorum Simon. While the dog follicle mite, *D. canis* Leydig causes a severe acariasis in the dog known as red mange the follicle mite, *D. folliculorum* is generally considered harmless. It is supposed to be a common parasite of man in all parts of the globe. It attacks the hair follicle and sebacious glands producing no symptoms. In sections the mite can be readily recognised by its striations on the cuticle as these are clealy visible. Nicholas (1943) states that the mite is associated with various types of naevus, epithelioma and a variety of other conditions. They are much more common on the face and particularly the nose than other areas. Indeed they are very unusual on the trunk.

Promising control of demodetic mange in dogs has been obtained by oral and subcutaneous administration of BHC at a dosage rate of 20 to 80 mg/kg body weight of the pure gamma isomer in poppyseed oil (Borchert, 1950).

CHAPTER XVI

ORDERS ARANEIDA & SCORPIONIDEA & OTHER ARACHNIDA

Order **Araneida.** The spider is characterised by the body being markedly constricted between the cephalothorax or prosoma and the abdomen or the opisthosoma. The chelicerae, a two-jointed structure, carries the poison gland at the base. The body is clothed in hairs. The cephalothorax is covered by a carapace or shield on its dorsal surface. The cephalothorax carries the usual mouth appendages which consist of a pair of pedipalps and a pair of chelicerae, and on its ventral surface four pair of legs attached to the sternum. A large number of eyes are placed in two rows on the carapace of the cephalothorax.

The chelicera consists of two joints. At the tip lies the curved claw of the poison fang. The poison is used for killing its prey. The palpi perform diverse functions. They are six-jointed limbs. The distal joints are used by young and female spiders as sensory organs while in the mature male they perform a function quite unique in the animal kingdom. They are used for insemination of spermatozoa by these organs into the genital opening of the female. The semen as it appears externally is picked up by the palpi and transferred during mating. The posterior part of the opisthosoma contains the spinnerets, 6 in number, and the lung books. The alimentary system is a complicated structure. The spider is a fluid feeder and feeds on small insects which are caught in its web. It sucks only the juice of the insect and the carcass is left behind. The male can be recognised by the palp, the extreme tip being bulbed. In the adult the arrangement of the eyes is used in the separation of the families and genera. All spiders are oviparous. After the eggs are laid, a cocoon is spun round them.

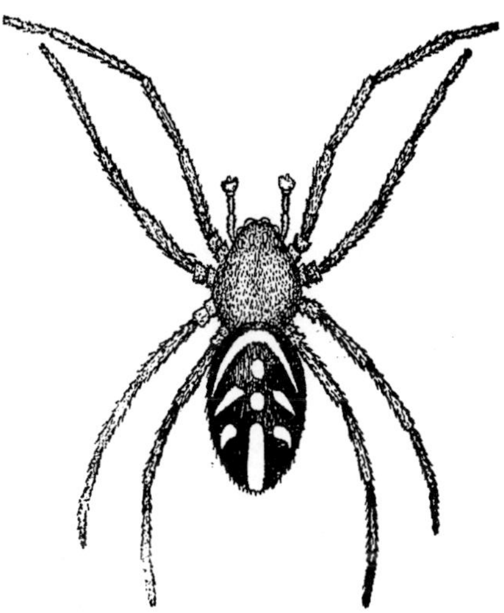

FIG. 180.
Latrodectus mactans, male. (After Chamberlin and Ivie).

Natural enemies of spiders are: (1) birds; (2) lizards; (3) monkeys; (4) spiders themselves; (5) Ichneumonidae: these lay eggs either in the spider's cocoon or in the body of the spider itself; (6) wasps: they catch the spider, sting it to death and carry it to the nest where it is cut into pieces on which the young ones are fed.

Spiders are grouped in two suborders, Mygalomorphae and Arachnomorphae. The latter group contains the important spiders.

Genus **Latrodectus.** This is very widely distributed almost all over the world though the extremely poisonous varieties have been described chiefly from American and from Australian regions.

L. mactans. It is popularly known as the "black widow," "hour glass" or "shoe-button spider." The black widow is so called on account of its colour which is shining black and its habit of killing its mate. The name hour glass has been given to it on account of a conspicuous red spot, shaped like an hour glass, on the ventral surface of the abdomen. There is also a circular reddish spot at the tip of the abdomen surrounding the anus. It occurs in almost all the States in North America, and in southern parts of Canada, but more commonly found in Central and South America as far as Peru, and the West Indies. It is found in nature under logs and stones, and in out-buildings.

Majority of cases of spider bite occur on the genitalia both in male end female individuals at the time of defaecation. Male spiders seldom cause any trouble whereas the females are highly poisonous. The bite is felt as a shaıp sting and though not immediately painful, it is soon followed by a dull numbing pain which extends to other parts of the body such as the chest, back and abdomen. There is profuse perspiration, the abdomen becomes board-like; there is breathing difficulty on account of spasms of the thoracic muscles. There is often nausea and vomiting, a rise in blood pressure and leucocytosis. The symptoms appear within 4 hours and may last for two or three days.

The poison glands of a female spider is characteristically large in comparison to the size of the creature and each gland weighs from 1.5 to 2 mg. The average lethal dose is 0.032 mg. of dired venom, i.e., the amount of venom which will kill 50 per cent of rats.

L. hasseltii Thor. It is the red-back spider. This is found in New Zealand, Australia, in the Southern Islands of the Philippines and in India. It is poisonous and is characterised by its small size, and by the orange or reddish yellow longitudinal stripe on the dorsal aspect of the abdominal segments.

In South Africa *L. geomatricus* Koch, is the commonest spider found in out-houses.

Treatment of spider bite. (*a*) Incision and rubbing in of strong solution of potassium permanganate. (*b*) Intravenous administration in 25 per cent of strength of magnesium sulphate. (*c*) Injection of morphia to relieve pain and counteract shock. (*d*) Injection of antiserum (prepared by Mulford) in 5 c.c. dose and given intramuscularly. The antiserum is prepared by injecting an extract of cephalothorax of the spider in a horse and given daily for at least a fortnight.

Order **Scorpionidea.** Scorpions are prevalent over a large part of the globe and in India they are particularly more common in certain parts, e.g., in Central Provinces and Rajputana, than in others. At times they are found congregated together. They are viviparous and give birth to a large number of young ones at any one time. The young ones resemble the parents. Scorpions are found in houses in dry and dusty places which are seldom disturbed. In hilly places they lie hidden under stones.

The body consists of cephalothorax and abdomen. The cephalothorax bears the two pair of mouth appendages and four pair of walking legs. The pedipalpi are 7 jointed structures, the last two pieces forming the pincers for catching and holding the prey. On the upper surface of the cephalothorax are placed two sets of simple eyes. There are a pair of large eyes close to the median line and a group of smaller ones on either side. The abdomen is composed of 7 segments. On the ventral surface of the first segment lies a pair of rounded plates one on either side of the middle line which covers the opening of the genital ducts and is called operculum. The second abdominal segment bears the comb-like appendages called pectines. From the 3rd to the 6th are the openings of the lung books, one pair on each segment. The tail consists of six segments, the last one bears the pointed sting. Within this lies a pair of poison glands, the ducts of which open below the point of the sting. Scorpions have no external organs that characterise the sex.

Scorpions feed on flies and other insects including cockroach. In the cold months it hibernates and in the summer it is active. Its activity is markedly diminished during the rainy season.

Family **Buthidae.** Sternum long, tibial spurs present. It contains two subfamilies: (1) Buthinae. (2) Centrurinae. Buthinae : The principal genera are *Buthus* and *Archisometrus*. Most of them are found in the Old World. Subfamily Centrurinae: they are New World scorpions. The principal genera are *Tityus*, *Centurus* and *Isometrus*.

Genus **Buthus.** Sternum triangular; 3 distinct dorsal keels on carapace. This is distributed in Europe, Asia and Africa. *B. tamulus* Fab: 2nd and 3rd segments of tail distinctly larger than wide; pectinal teeth 28 or more. Widely distributed in the Ethiopian Region, China, India but absent from Ceylon and Burma. *B. citrinus* Hemp. and Ehrebn: Common in North Africa and South Europe on the Mediterranean. Its bite is extremely painful and distressing symptoms may follow. *B. quinquestriatus* : Occurs in Palestine and Egypt. The black scorpion, *B. occitanus* in Egypt is harmless.

Genus **Tityus.** *T. bahiensis* Perty and *T. serrulatus* Lutz., are extremely common scorpions in Brazil and both are pisonous.

Family **Scorpionidae.** No distal spurs; claws of the legs overlapped by the sides of the tarsus. Genus *Palamnaeus* : *P. fulvipes* Koch. Legs reddish yellow; humerus of pedipalps smooth. *P. swammerdani* Simon. Humerus coarsely granular on the under side at the base. In North Africa *Primurus australis* is the most dangerous species and during the hot season its sting proves fatal to children. The other poisonous scorpion is *Pr. liouvillei.*

Family **Centrurinae.** Sternum small, long and triangular; usually a spur or tooth on ventral margin of fixed finger of mandible. Genus *Centrurus.* Distributed in the Americas. In South America *Centruroides* (*Centrurus*) *sculpturatus* and *C. limpidus* are looked upon as extremely poisonous.

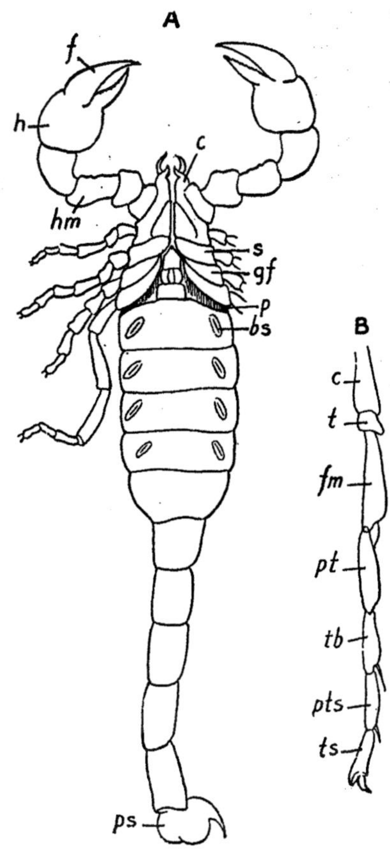

FIG. 181.

A. Ventral view of scorpion.
bs, breathing slits ; c, chelicera ; f, finger ; gf, genital flap ; h, hand ; hm, humerus ; p, pecten ; sp, poison sac ; s, sternum.
B. Leg of scorpion.
c, coxa ; fm, femur ; pt, patella ; pts, protarsus ; t, trochanter ; tb, tibia ; ts, tarsus.

The venom of different species of scorpions varies very greatly in virulence. The toxic principle is in the nature of a toxalbumin, a neurotoxin which resembles cobra venom and acts on the medulla and motor end plates causing death by paralysis of the respiratory system. Although the sting of a scorpion is very painful, the poison as a rule does not produce general symptoms in adults but in children under 5 years of age the sting frequently causes death. The mortality among young children stung by *B. quinquestriatus* is even 50 per cent. In Trinidad the case mortality rate under 5 years is 25 per cent (Waterman, 1938). Baerg (1929) reported that the Mexican scorpion *Centruroides suffusus* was responsible for 1608 deaths in course of 36 years. The high death rate in Egypt led Todd to prepare an antivenene.

Caius and Mushkar (1932) were of the opinion that even the largest species of scropion in this country is unable to cause the death of a child though marked symptoms may be produced. Chaudhuri (1949) has reviewed 160 cases of scorpion sting in children which occurred in Calcutta. The mortality rate has been recorded as 15 per cent. The common site of the sting is in the fingers, hands, toes or legs. Earlier to this Basu (1939) had reported from Calcutta deaths from scorpion stings. The symptoms are excruciating pain, local redness followed by profound constitutional symptoms indicating affection of the cardiovascular system leading to peripheral circulatory failure.

In infants and children the diagnosis is difficult because often no history of scorpion sting is available. The presence of salivation, vomiting and glycosuria, slow pulse and hurried respiration are points in favour of scorpion sting and go against snake bite. Scorpions are active only during the summer when scorpion stings may occur.

Treatment. (1) Morphia should be administered by injection immediately, even to children. (2) Glucose (10 per cent in solution) given rectally or subcutaneously followed by an injection of insulin. (3) Pituitarin and saline administration may be necessary to counteract shock. (4) Anti-venom, if available. Horses and cattle are immunised against scorpion venom. The anti-serum is specific to the particular scorpion. (5) Injection of Emetine hydrochloride gr. 1 locally in the area; pain is relieved in one minute and in five minutes the patient is free from pain (Lakshminarayana, H. S.-*J. Ind. Med. Assoc.*, 1953, 22 : 431).

Order **Pentastomida.** (Tongue worms). They bear considerable resemblance to Arachnids although they possess annulations like worms; but such annulations are only superficial. They are devoid of antennae. All live parasitic lives and suck blood in different stages of their post-embryonic existence.

The body is flat and worm-like. The mouth is distinct. Two pair of retractile hooks encircle the mouth. The sexes are separate. All are oviparous. The larval and adult lives are generally passed in two different types of hosts, one herbivorous and the other carnivorous. The life history is imperfectly known.

There are only two important genera which are pathogenic to man, these are *Linguatula* and *Porocephalus.*

Genus **Linguatula** Froh. The body is attenuated posteriorly. Hooks are simple, equal and disposed archwise. They are parasitic in mammals. The adult is found in domestic animals of cosmopolitan distribution and the nymphal form has been reported from cattle, rodents and ungulates.

L. serrata Froh. It is the most common of the tongue-worms in mammals and is distributed very widely. Leuckart traced its life history and showed that there was a complete metamorphosis and that an alteration of hosts was essential for its development. Both larvae and nymphs have been recorded from man at autopsy and usually lodged in the liver. The adult is also known as a human parasite. Among the domestic animals the dog is most frequently attacked.

Genus **Armillifer** (*Porocephalus*). The identification of the genus *Armillifer* is based on clearly defined anatomical features particularly the type of claws and their relation to the oral opening. They occur in the Ethiopian and Oriental Regions inhabiting in the adult stage the respiratory tracts of pythons. Nymphs have been recorded from man especially from the liver though they are ordinarily found in mammals and birds.

The annuli in the adult are strongly marked by prominent ridges; mouth just above the inner hook line. The posterior half of the body is curved.

A. moniliformis. Occurs in the Oriental Region inhabiting in its adult stage the respiratory tracts of Indian pythons. Nymphs have been recorded from the liver of man.

A. armillatus : These belong to the Ethiopian Region inhabiting in its adult stage the pythons of Africa and other snakes. They are known to attack man in Africa, both adults and larvae being found encysted in the liver.

Order **Xiphosura.** (King crab). On account of their large size and resemblance to crabs they are popularly called King crab. They represent a group of primitive Arachnida which are different in several ways from all other members of the class. They are marine creatures being found in water less than ten fathoms deep in the Atlantic Coast of America and in a few areas near Japan, Malaysia and India.

The body is characterised by a large shield or carapace and posteriorly there is a long unjointed spine. There are two pairs of eyes. The appendages which lie in a hollow are of the usual Arachnida type. The food consists of soft marine animals such as molluscs. They are oviparous.

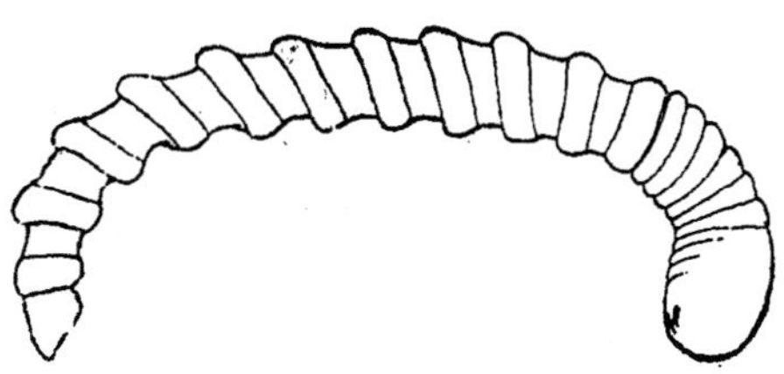

FIG. 182.
Armillifer armillatus, male. (After Sambon).

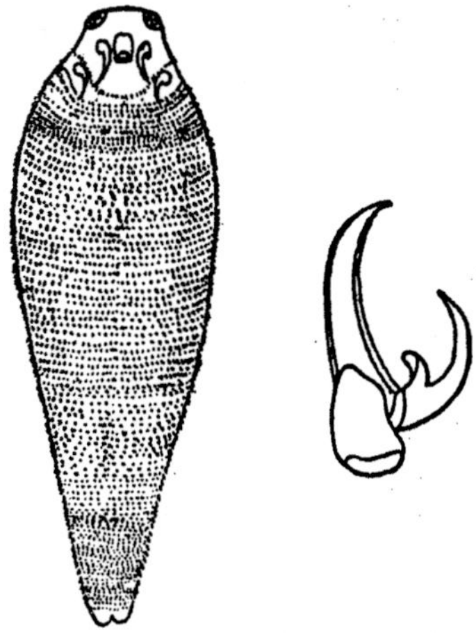

FIG. 183.
Linguatula serrata (nymph) and oral hook.
(After Roy and Ganguli).

Order **Solifugae.** They are confined to hot countries. They abound in Africa, Tropical Asia and Central and South America. The Solifugae are spider-like. It differs from all other orders of the class Arachnida in having the body segmented into head, thorax and abdomen. The whole body and the limbs of the animal are covered with hair-like setae.

The mouth parts consist of pedipalps and chelicerae. The base of the chelicera is swollen and its distal end is provided with pincer-like fangs. Antennae are absent.

The thorax is composed of three independent segments and carries four pair of legs. The legs are long and hairy. Some of the basal segments of the hind legs carry peculiar club-shaped appendages called malleoli. There is only one pair of simple eyes on the suture between the head and the thorax.

They are very active and are difficult to catch. They are nocturnal in habits and prey on small insects which they catch by means of the pincers on the chelicerae.

Genus **Chelifer.** They are widely spread nearly over the whole world. They are carnivorous and prey on small insects. They hide under stones. In general appearance they bear a close resemblance to scorpions but the tail is absent. Like spiders they make nests in which the eggs are laid.

Order **Pedipalpi.** Although they resemble scorpions, they differ in having the pedipalps not chelate, the first pair of legs being antenna-like and the tail very slender and without the sting.

They are predacious.

REFERENCES (Arachnida)

Adler, S., Theodor, O. and **Schieber, H.** (1937). *Ann. Trop. Med. Hyg.*, 31 : 25. **Audy, J. R.** (1947). War Office Rep. on Scrub Typhus Investigations in South East Asia. London War Office. **Badger, L. F.** (1933). *Pub. Hlth. Rep.*, 48 : 507. **Baerg, W. J.** (1929). *Ann. Ent. Soc. Amer.*, 18 : 471. **Basu, B. C.** (1930). *Ind. Med. Gaz.*, 65 : 326. **Basu, U. P.** (1939). *Amer. J. Trop. Med.*, 19 : 385.

Bishopp, F. C. and **Smith, C. N.** (1938). *Circ. u. s. Dept. Agric.*, no. 478, 25. 13. **Blanchard, R.** (1890). *Zoologie Medicale*, Paris. **Bone, G.** (1939). *C. R. Soc. Biol.*, 130: 84. **Borchert, A.** (1950). *Mh. Vet. Med.*, 127. **Browning, H. C.** and **Kalra, S. L.** (1948). *Ind. J. Med. Res.*, 36: 279. **Brumpt, E.** (1933). *C. R. Soc. Biol.*, 113: 1366. **Bushland, R. C.** (1946). *Amer. J. Hyg.*, 43: 219. **Buxton, P. A.** (1921). *Parasitology*, 13: 114, 146. **Caius, J. P.** and **Mhaskar, K. S.** (1932). *Ind. J. Med. Res.*, Mem. No. 24. **Castellani, A.** and **Hirst, S.** (1912). *J. Trop. Med. & Hyg.*, 15: 374. **Chaudhuri, K. C.** (1949). *Ind. J. Pediatrics*, 16: 208. **Christophers, S. R.** (1906). *Scientific Mems., Govt. of India*, No. 23. **Christopers, S. R.** (1907). *Br. Med. Jl.*, *i*: 76. **Cockings, K. L.** (1948). *Bull. Ent. Res.*, 39: 281. **Cornwall, J. W.** and **Patton, W. S.** (1914). *Ind. J. Med. Res.*, 2: 569. **Cowdry, E. V.** and **Ham, A. W.** (1932). *Parasitology*, 24: 1. **Cross, H. E.** and **Patel, P. G.** (1921). *Punjab Dept. Agric. Lahore Vet. Bull.* No. 6. **Da Rocha Lima, H.** (1916). *Arch. f. Schiffs. u. Trop. Hyg.*, 20: 17. **Davis, G. E.** (1941). *Pub. Hlth. Rep.*, 56: 1799. **De Meillon, B.** (1942). *South Afr. Jl. Med. Scie*, 7: 226. **Dick, G. W. A.** and **Lewis, E. A.** (1947). *Trans. Roy. Soc. Trop. Med. & Hyg.*, 41: 295. **Dodd, S.** (1921). *Jl. Comp. Path. & Therap.*, 34: 309. **Dutton, J. E.** and **Todd, J. L.** (1905). *Liver, Sch. Trop. Med. Mem.* No. 17. **Eltringham, H.** (1930). Histological and Illustrative Methods for Entomologists Humphrey Milford, London. **Everdringen, W. A. G.** (1948). *Belg. Tijdsch. Geneesk.*, 4: 2. **Ferguson, E. W.** (1924). *Med. Jl. Australia*, 11th year, *ii*: 346. **Fielding J. W.** (1943). *Ibid*, *i*: 435. **Francis, E.** (1938). *Pub. Hlth. Rep.*, 53: 2220. **Fulton, F.** and **Joyner, L.** (1945). *Lancet*, Dec. 2: 729. **Gerlach, A. C.** (1857). Kratze und Raude. Berlin. **Goldman, L.** (1939). *Arch. Derm. Syph. Chicago*, 39: 878. **Gordon, R. M.** and **Seaton, D. R.** (1941). *Ann. Trop. Med. & Parasit.*, 35: 247. **Gordon, R. M.** (1942). *Brit. Med. Jl.*, *i*: 685. **Gregson, J. D.** (1941). *Proc. Ent. Soc. B. C.* no. 37, 9, Vernon. **Griffiths, J. T.** (1945). *J. Parasit.*, 31: 341. **Guyer, M. F.** (1936). Animal Micrology, Chicago, University of Chicago Press. **Hadwen, S.** and **Nuttall, G. H. F.** (1913). *Parasitology*, 6: 298. **Huebner, R. J., Jellison** and **Pomerantz** (1946). *Pub. Hlth. Rep.*, 61: 1677. **Huebner, R. J, Stamps** and **Armstrong** (1946). *Ibid*, 61: 1605. **Hindle, E.** and **Cunliffe, N.** (1913). *Parasitology*, 6: 353. **Hindle, E.** and **Merriman, G.** (1912). *Ibid*, 5: 203. **Hoeppli, R.** and **Feng, L. C.** (1933). *Chinese Med. Jl.*, 47: 29. **Hunter, W. D.** and **Bishopp, F. C.** (1911). *U. S. Dept. Agric., Bureau of Entomology, Bull.* 105. **Inniss, K. V. A.** (1927). *Trans. Roy. Soc. Trop. Med. & Hyg.*,.3: 607. **Jayewickreme, S. H.** and **Niles, W. J.** (1946). *Nature*, 157: 878. **Jayewickreme, S. H.** (1947). *Ibid*, 160: 578. **Jolly, D. W.** (1952). *Vet. Rec.*, 64: 76. **Jenkins,, D. W.** (1947). *Amer. Antom. Soc.*, 40: 56. **Jenkins, D. W.** (1948). *Amer. J. Hyg.*, 48: 22. **Kalra, S. L.** (1947). *Jl. Ind. Army. Med. Corps.*, 3: 10. **Kalra, S. L.** (1951). *Ibid*, 7: 166. **Kirk, R.** (1938). *Ann. Trop. Med. & Parasit.*, 32: 357. **Knowles, R., Das Gupta** and **Basu** (1932). *Ind. J. Med. Res.* Mem. No. 22. **Koch, R.** (1905). *Deutsche Med. Wochenschr.*, 47: 1866. **Krishnan, K. V., Smith, Bose, Neogy, Ghosh Roy** and **Ghosh** (1949). *Ind. Med. Gaz.*, 84: 39. **Linduska, J. P.** and **Morton, F. A.** (1947). *Amer. J. Trop. Med.*, 27: 771. **Lounsbury, C. P.** (1904). *Jl. Comp. Path. Therap.*, 17: 113. **Lydon, F. L.** (1941). *Jl. Roy. Army Med. Corps.*, 76: 23. **Manteufel, P.** (1908). *Arb. a. d. Kais. Gesundheitsamte*, 29: 337. **Marchoux, E.** and **Couvy, L.** (1913). *Ann. Inst. Past.*, Paris, 27: 453, 620. **Mac Cormac, H.** and **Small, W. D. D.** (1917). *Br. Med. J.*, *ii*: 384. **Mc Culloch, R. N.** (1946). *Med. J. Australia*, *i*: 717. **Megaw, J. W. D.** (1921). *Ind. Med. Gaz.*, 56: 361. **Mellanby, K.** (1941). *Br. Med. Jl.*, *ii*: 405, **Mellanby, K.** (1943). *Scabies. Oxford Univ. Press.* London. **Mellanby, K.** (1944). *Parasitology*, 35: 197. **Monteiro, J. L.** (1933). *Mem. Inst. Butanton*, 8: 9. **Moskwin, I. A.** (1929). *Zeitschr. f. Parasitenk.*, 2: 73. **Mukerji, D.** (1937). *Current Sci.*, 6: 16. **Munro, J. W.** (1919). *Jl. Roy. Army Med. Corps.*, 33: 1. **Neuman, R. O.** (1909). *Munchen Med. Wochens.*, 56: 477. **Nicolle, C.** and **Sparrow, H.** (1934). *C. R. Acad. Sci., Fr.*, 199: 1349. **Nuttall, G. H. F.** and **Strickland, C.** (1908). *Parasitology*, 1: 302. **Nyka, W.** (1945). *J. Path. & Bact.*, 57: 317. **Olsufev, N. G.** (1939). Abstr. in *Rev. Appl. Ent.*, B, 27: 241. **Osborne, H. G.** (1947). *Aust. Vet. Jl.*, 23: 52. **Parker, R. R.** (1923). *Amer. J. Trop. Med.* 3: 39. **Parker, R. R.** (1924). *Pub. Hlth. Rep.*, 39: 1057. **Parker, R. R.** (1935). *J. Infect. Dis.*, 57: 78. **Parker, R. R.** (1937). *5th Pacific Sc. Cong.*, 3367. **Pavlovskii, E. N.** (1932). Abstract in *Rev. Appl. Ent.*, B, 21: 97. **Pavlosky, E. N.** and **Chodukin, N. J.** (1929). *Zeitsch. f. Parasitenk.*, 2: 90. **Pijper, A.** and **Dau, H.** (1933). *Nederl. Tijdschr. Geneesk.*, 77: 2030. **Regendanz, P.** and **Reichnow, E.** (1932). *Zbt. Bakt.* I Orig., 124: 471. **Ray, H. N.** (1945). Presidential address, Section of Zoology and Entomology. 32nd. Indian Science Cong., Nagpur. **Ross, I. C.** (1926). *Ibid*, 18: 410. **Roy ,D. N.** and **Ghosh, S. M.** (1944). *Ind. Med. Gaz.*, 79: 589. **Sapre, S. N.** (1940). *Ind. J. Vet. Sci.*, 10: 346. **Sen, S. K.** (1935). *Parasitology*, 27: 355. **Sen, S. K.** (1941). *Ind. J. Entom.*, 3: 51. **Sergent, E.** (1928).

Arch. Inst. Pasteur d' Algerie, 16 : 257. **Shankman, B.** (1946). *New York State J. Med.*, 46 : 2156. **Sharif, M.** (1924). *Agric. Res. Inst., Pusa.* Bull. 152. **Shortt, H. E.** (1936). *Ind. J. Med. Res.*, 23 : 885. **Smith, D. J. W.** (1940). *Australian J. Exp. Biol. & Med.* Sc. 18 : 103. **Smith, D. L. W.** (1942). *Ibid*, 20 : 213, 295. **Smith, C. N.** (1944). *Ann. Ent. Soc. Amer.*, 37 : 325. **Smith, C. N.** and **Gouck, H. K.** (1947). *J. Econ. Entom.*, 40 : 790. **Snyder, F. M.** and **Morton, F. A.** (1946). *J. Econ. Entom.*, 39 : 385. **Southwell, T.** (1924). *Ann. Trop. Med. & Parasit.*, 18 : 515. **Spencer, R. R.** and **Parker, R. R.** (1925). *Pub. Hlth. Rep.*, 40 : 2159. **Spencer, R. R.** and **Parker, R. R.** (1930). *U. S. Public Hlth. Hyg. Lab.* Bull., 154. **Strickland, C.** and **Roy, D. N.** (1939). *Ind. J. Med. Res.*, 27 : 251. **Sussman, L. N.** (1946). *New York Med.*, 2 : 27. **Taylor, E. L.** (1945). *Vet. Res.*, 57 : 210. **Todd, C.** (1909). *Jl. Hyg.*, 9 : 69. **Topping, N. H., Heilig** and **Naidu** (1943). *Pub. Hlth. Rep.*, 58 : 1208. **Trager, W.** (1939). *Jl. Parasitology*, 25 : 57. **Warburton, C.** (1920). *Parasitology*, 12 : 265. **Waterman, J. A.** (1938). *Trans. Roy. Soc. Trop. Med. & Hyg.*, 31 : 607. **Weigl, R.** (1930). *C. R. Soc. Biol.*, Paris, 103, 823. **Welt, L. G.** (1947). *Amer. J. Trop. Med.*, 27 : 221. **Wharton, G. W.** and **Carver, R. K.** (1946). *Science*, 104 : 76. **Wheeler, C. M.** (1938). *Amer. J. Trop. Med.*, 18 : 641. **Wilson, G. G.** (1948). *Bull. Ent. Res.*, 39 : 269. **Woodland, J. C., Mc Dowell** and **Richards** (1943). *J. Amer. Med. Assoc.*, 122 : 1156.

CHAPTER XVII

Class : CRUSTACEA & OTHER ARTHROPODA

This class includes a very large number of animals of a diverse character and comprises crabs, lobsters, prawns, shrimps, wood-lice etc. With few exceptions, they live in water. They are abundant in all parts of the world and are generally free-living. They breathe by lungs or cutaneously. The head always carries two pair of antennae by which this class can be readily distinguished from others. Other appendages of the head consist of 3 pair of jaws. The chitin is strongly calcified in a majority of crustaceae. The sexes are separate.

Most of these creatures are unimportant to the medical officer who is concerned with only a few, especially the cyclops or water-fleas.

Sub-class **Copepoda.** It includes elongated crustacea usually with distinct segments. There is only one important genus, *Cyclops*. These are minute creatures and are typical examples of free-living copepoda. They are popularly known as waterfleas on account of their characteristic jerky movements while swimming. They are semi-transparent and though they are visible to the naked eye their structure can only be studied under the microscope.

The body consists of an anterior broad mass which is called the cephalothorax and an elongated posterior part, the tail or abdomen. The anterior mass includes the head with which is fused the first thoracic segment. Its appendages are : (*a*) a pair of antennules; (*b*) a pair of antennae; (*c*) a pair of mandibles; (*d*) a pair of maxillae; and (*e*) a pair of second maxillae, the last three comprising the gnathites or mouth organs.

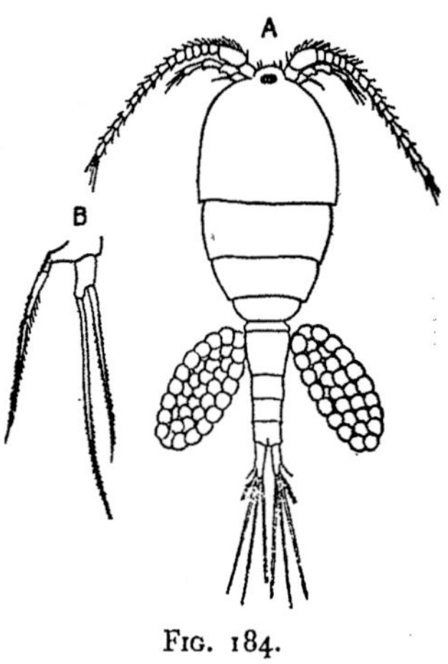

FIG. 184.
A. *Cyclops leuckarti.*
B. Fifth foot of *C. leuckarti.*
(After Ward and Whipple).

Antennules are very large. These are the principal organs of locomotion. The antennae are much smaller. Between the insertions of the antennae in the mid-dorsal line lies the small pigmented eye. The mouth is situated ventrally. Behind the cephalothorax lie 5 thoracic segments. Ventrally the first four thoracic segments bear swimming feet; the 5th segment bears a pair of vestigial limbs.

The abdomen or the tail consists of 5 distinct segments which are devoid of any appendage. The last segment branches into two feathered filaments. The eggs are carried in external ovisacs attached to the first abdominal segment.

The alimentary canal is a narrow tube which runs from the mouth to the anus, the latter being located at the site where the last abdominal segment forks.

Cyclops exist almost all over the globe. They are found not only in clean but also in muddy water. They generally prefer stagnant pools, tanks, and ditches.

In order to discover the existence of cyclops in a ditch or well, a net of the pattern of a butterfly or a fly net (the net being replaced by muslin), is dipped in water and when lifted, the water will quickly drain out through the cloth. The presence of cyclops is noticed as soon as the net is immersed in a bowl of water. A small sample of water collected in a test tube is held against a dark background. The cyclops, if present, can be quickly recognised by the characteristic jerky movement.

Mesocyclops leuckarti and *M. hyalinus*, both of which have wide distributions, are the two important species from the point of view of disease relationship.

Dracontiasis. The spread of guinea-worm disease is directly related to the presence of cyclops which serves as the intermediate host of this disease. Guinea-worm infestation is not confined to man alone. The parasite is able to attain to the mature state in horses, dogs and other domesticated animals. It has been recorded also from leopards. Man becomes infected by ingesting infected cyclops with drinking water.

The geographical distribution of dracontiasis is limited to certain parts of India, Africa, Southern Asia, West Indies and America.

The cyclops actively hunts the larvae and ingests them. Within 5 to 6 hours after ingestion all the larvae have passed into the body cavity of the cyclops by actively boring head first through the stomach wall (Moorthy, 1938; Moorthy and Sweet, 1939).

The adult female worm, which is the only sex known, discharges millions of embryos in a milky fluid through the vaginal opening. The infective stage is reached in the cyclops in about 8 weeks time and in pups the development of the worm up to the production of the ripe, motile larvae, takes 9 to 10 months. The embryos enter the cyclops by way of the alimentary tract (Leiper, 1906). The guinea-worm larvae may live in water for 7 days but in brackish water both the embryo and the cyclops are quickly killed.

Southwell and Kirshner (1938) have shown that even when 15 larvae are present, the cyclops is not affected in any way. In nature the infection is found in the younger forms of both sexes and none in those bearing ovisacs. Hydrochloric acid in 0.2 per cent strength, which approximates the strength of acid in the gastric juice, kills the cyclops and activates the resting guinea-worm larvae within its body cavity. The larvae escape by breaking through the cuticle. Even a lower concentration of acid and also bile will kill the cyclops and activate the worm (Rao, 1936).

For detecting guinea-worm embryos in the bodies of cyclops, captured cyclops are palced in test tubes from which most of the water is pipetted off and kept at 35° to 36°C. for 1 to 2 days after which the contents are examined under a microscope. The crustaceans are macerated and broken up, while the *Dracunculus* larvae retain their normal structure and can easily be recognised (Issajev, 1934).

Diphyllobothrium latum. Besides man, the infection has been reported from domestic animals such as dogs and cats. The eggs are discharged in water where they hatch and the embryos produce infection in the cyclops. Later fish becomes infected by swallowing infected cyclops. When such a fish is eaten by man, dogs or cats in a raw or insufficiently cooked state, the worm develops into the adult stage and is lodged in the small intestine.

Cyclops is also an intermediate host of *Gnathostoma spinigerum* Owen, producing gnathostomiasis in man and animals. Man becomes infected by eating uncooked or improperly cooked fish already infected by eating infected cyclops. In man the infection is a rare condition. It is common in cats, dogs and other carnivora.

Prophylaxis of guinea-worm disease is easy if it is remembered that man becomes infected by drinking water from step-wells and ponds which harbour infected cyclops. These are sources of infection to man in India and on the West Coast of Africa. The following measures are recommended.

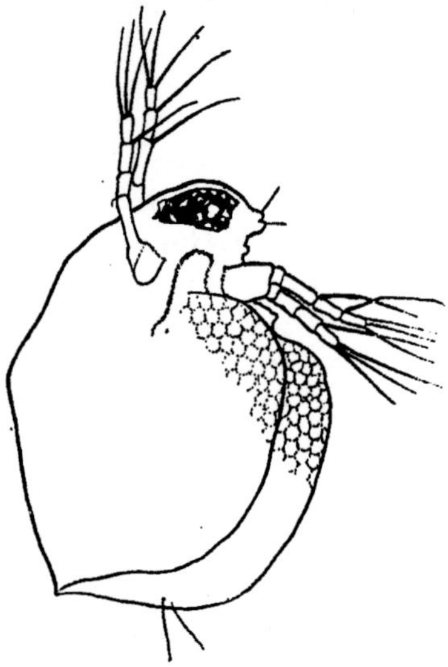

FIG. 185.
Daphnia.

(*a*) Conversion of the step-wells into draw wells. (*b*) Periodical treatment of wells with lime. (*c*) Treatment with potassium permanganate. (*d*) Addition of Perchloron and Copper sulphate, 3 lb. and 1 lb. respectively per 100,000 gallons of water. (*e*) Boiling or straining drinking water through a piece of thick muslin. (*f*) Stocking step-wells with fish (*Saccobranchus fossilis, Anabas scandens*).

Genus **Daphnia.** These live in stagnant pools and may be mistaken for cyclops from which they differ in the body being unsegmented and lying within a bivalve shell from which the beak-like head and the antennae project. Daphnia do not take any part in the spread of guinea-worm disease.

Class **Myriapoda.** (Centipedes and Millepedes). These are tracheate arthropoda. The head bears many-jointed antennae, a pair of eyes, and 2 or 3 pair of jaws. The body consists of a number of similar segments, each bearing either one pair of legs or two pair.

The tracheae open by a series of stigmata usually in considerable numbers on the sides and lower surface of the segments.

This class has been grouped into two sub-classes, Diplopoda and Chilopoda. The former are known as Millepedes and the latter Centipedes. The genital aperture in millepedes is situated far forwards towards the anterior end of the body. The commonest genus is *Julus* which lives on vegetable food.

Chilopoda (Centipede). In Chilopoda the first pair of legs of the trunk is specially modified to act as maxillipedes or poison-jaws by means of which the centipede inflicts its poisonous bite. The body is composed of a large number of closely similar segments. Each segment bears a pair of jointed legs. The antenna is long. At the extremity of the pointed terminal joint of the maxillepede opens the duct of the poison gland. The genital aperture in centipedes is situated at the posterior extremity of the body. It feeds on small insects and is oviparous. The venom glands or salivary glands are difficult to dissect as they lie in a mass of fat body and other tissues.

Cornwall in an account of three Indian species stated from experiments that the venom is little toxic to laboratory animals. The bite is extremely painful and the pain lasts for 4 to 6 hours. In children other symptoms may supervene, such as nausea, vomiting, diarrhoea, glycosuria etc. Death from centipede bite is rare, though it has been reported. In Canada *Scutigaro forceps* Raf., inflicts painful bite to man (Judd, W, W. *Ent. News*, 1952, 63 : 238).

Order **Orthoptera.** This order includes such insects as cockroaches, grasshoppers, locusts, crickets, mantids, earwigs, stick-insects etc. They are mostly destructive of vegetation and except the cockroach, they are hardly of any importance to us. The cockroach is undoubtedly a great nuisance in almost every house though it is not directly concerned with the carriage of the common disease organisms affecting man.

Family **Blattidae.** This family includes the cockroach of which there are many species described. Among them only a few may be regarded as domestic varieties. These include *Blattella germanica*, *B. orientalis*, *Periplaneta australis* and *Periplaneta americana.*

FIG. 186.
Millepede.

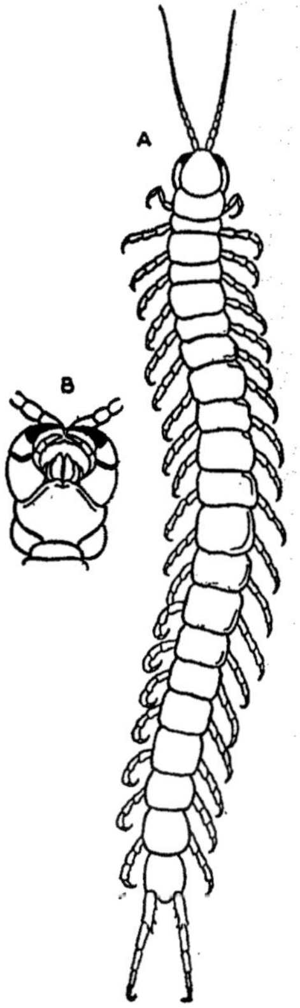

FIG. 187.
A. Chilopoda.
B. Ventral view of head.

Blattella germanica. This is known as the little German cockroach. It is about ½ inch long. There are two longitudinal dark markings on the pronotum. The egg capsule contains 28 to 56 eggs in a pocket.

Blattella orientalis. The colour is dark brown. It is about 1 inch long. In this species the wing cases in male do not cover the abdomen completely and the wings of the female are almost rudimentary. The pronotum is uniformly covered without any dark markings. The egg capsule contains 16 eggs in each.

Periplaneta australis. It is much larger in size being over one inch in length. The Australian cockroach is distinguished by the yellow ring on the first thoracie segment and the yellow streak on the side of the wing case. They have fully developed wings which completely cover the abdomen. This cockroach is seldom found in dwelling houses.

Periplaneta americana L. This is about 1½ inches long and is particularly associated with ships. The pronotum has a faint yellowish border. It has fully developed wings which completely cover the abdomen. Eggs are enclosed in a bi-valve chitinous covering called an egg-pod. The number laid by a single female is usually 10 to 15; sometimes even 25 egg capsules are known to be produced by a female. The average interval between oviposition is 4 to 5 days when the females have ample opportunities for pairing. There are normally 16 eggs in a pod. The metamorphosis is incomplete or gradual. There are one pre-nymphal and six nymphal stages. The period from hatching to the last moult takes about 12 months and the total life cycle from hatching to the death of the adult is about 2 years.

Natural enemies. The chalcid *Spalangia orientalis* Graham, commonly attacks the ootheca of cockroach. Some Hymenoptera, scorpions and lizards prey on cockroaches.

Disease relationship. On account of the habits of cockroaches, there is no reason why they should not act as carriers of cholera, typhoid, dysentery and ova of intestinal worms and protozoal cysts.

Control. See in Part II.

Order **Isoptera.** (Termites). On account of their highly developed social mode of life, they are often called white ants. They live in tunnels caused by hardened clay. They are nocturnal in habits.

The workers are blind, wingless and without any sexual development. The soldiers are larger and have very large heads. In every colony there is a pair of sexually developed individuals, which are called king and queen. There are also a large number of larvae in a colony from which workers, soldiers and sexually mature winged king and queen develop. From time to time these sexually matured insects fly in swarms. These flights always occur in sunset. During their flight a large proportion of them are preyed upon by birds.

Order **Corrodentia.** (Book-lice). They are exceedingly minute insects without any wing and are soft-bodied. They possess a pair of eyes and a pair of long antennae. They are of common occurrence in places where books and papers are stored. Almost any old book will contain them. They have chewing mouth parts and live on paper and bindings.

Fig. 188. Ephemerid.

Order **Thysanura.** (Bristle tails). A common example of this order is the silver-fish insect, *Lepisma,* found in books. They feed on paper.

They are small and wingless insects having long antennae. They all possess two long tail-like filaments projecting from the last abdominal segment.

Order **Collembola.** (Spring tails). They are minute, wingless insects which are characterised by the presence of a pair of jointed structure attached to the ventral surface of the last abdominal segment, by means of which the insect is able to jump. They are provided with 4-jointed antennae. They are found in damp places.

Order **Plecoptera.** (Stone flies). They are known as stone-flies. The larvae are aquatic and are found in flowing streams. The larvae hide under stones. They prey on other insect larvae.

Order **Ephemerida.** (Mayflies). They are so-called because in the adult stage they do not live more than one or two days. The larvae are aquatic and live on other insect larvae. They possess long antennae and well-developed trachael gills on both sides of many of the abdominal segments. There are two anal cerci which originate from the last abdominal segment. The adult insect is delicately built and has the front pair of wings very large and the hind pair is generally suppressed. The mouth parts are vestigial and the insect does not take any nourishment. There are two long filaments attached to the last abdominal segment.

Order **Odonata** (Dragon flies). They are of much interest to us as the fly both in the larval and adult stages preys on other insects and their larvae, some of which are injurious to our health.

Dragon flies have large heads, elongated bodies and stiff wings. The antenna is short, eyes prominent and mouth parts are formed for biting. The wings are peculiar. The hind pair of wings may be slightly larger than the front pair or the hind wings are of the same size.

The larva is aquatic. It preys on insect larvae by means of the curious elongated labium which are extended for catching prey. The last segment is provided with tracheal gills.

Order **Neuroptera.** This is so-called on account of a network of veins (nerves) in the wings. This order includes a large number of insects which are of considerable economic importance to us as both larvae and adults are rapacious. The veins are net-like. The larvae are aquatic.

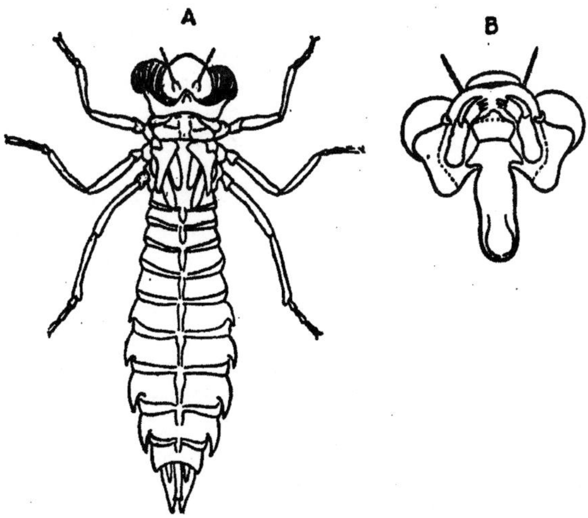

FIG. 189.
A. Larva of *Æschna* (Dragon-fly). (After Miall).
B. Mouth parts of larva of *Æschna* (After Miall).

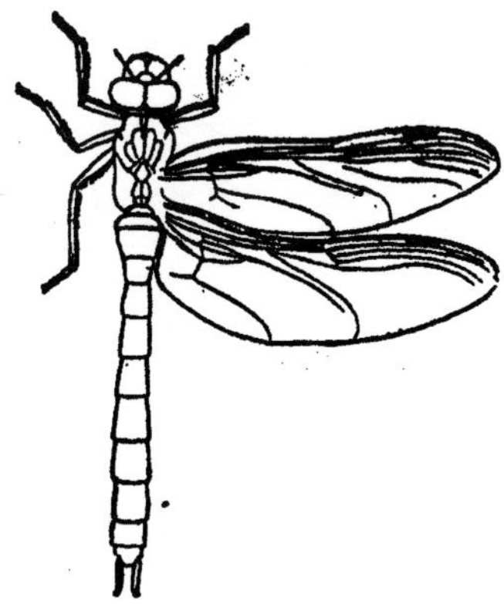

FIG. 190.
Dragon Fly (Odonata).

Order **Mecoptera** (Scorpion-flies). Both adults and larvae are carnivorous. The adult head is produced in the form of a beak. There are few cross veins in the wings.

Order **Trichoptera** (Caddis-flies). The larvae are aquatic and are known as caddis-worms. They live in tubes which are open at both ends. Some larvae are carnivorous.

REFERENCES (Crustacea)

Issajev, L. (1934). *Med. Parasit. Dis.*, Moscow, 3: 231. **Leiper, R. T.** (1906). *Br. Med. Jl.*, *i*: 19. **Moorthy, V. N.** (1938). *Amer. J. Hyg.*, 27: 437. **Moorthy, V. N.** and **Sweet, W. C.** (1939). *Ibid*, 27: 301. **Rao, S.** (1936). *Ind. J. Med. Res.*, 24: 535. **Southwell, T.** and **Kirshner, A.** (1938). *Ann. Trop. Med. Parasit.*, 32: 193.

CHAPTER XVIII

NOTES ON ENTOMOLOGICAL TECHNIQUE

Equipment for Collection of Insects : (a) Butterfly net (large) and a fly net (small). (b) Cone-shaped paper bags for temporarily keeping insects. (c) Corked specimen tubes and test-tubes. (d) Methylated spirit. (e) Cyanide killing bottle. (f) Suction tube. (g) 4 oz. wide-mouthed glass bottles for keeping living aquatic insects after collection. (h) A suitable wooden carrier with compartments for holding bottles and with a handle for carrying. (h) Cotton-wool, muslin, string, grease pencil, lead pencil, paper, note book etc.

Preservation of Entomological Specimens in the Laboratory : They may be mounted on suitable-sized pins and preserved in store boxes or they may be preserved in dry glass tubes with tight-fitting corks. A few fragments of naphthalene are placed at the bottom of the tube covered by cotton wool placed over them. The insects are placed over the cotton wool.

They may also be preserved in fluid media and the solutions that are commonly used for this purpose are : (i) 4 per cent formaldehyde solution. (ii) 70 per cent rectified spirit. (iii) Borax—5 g., Glycerine—2.5 c.c., Formaldehyde (40 per cent)—100 c.c., Distilled water added to make 1 litre. This is strongly recommended in tropical climates. (iv) Freshly prepared Bles's fluid : Alcohol (70 per cent)—90 c.c., Formaldehyde—7 c.c., Glacial acetic acid—3 c.c.

Small objects such as stomach and salivary glands of mosquitoes, insect larvae, cyclops, fleas, ticks etc., may be preserved for an indefinite period in alcohol or in formalin solution in tubes which are drawn out at one end to a fine point at some distance from the object, by means of a Bunsen flame.

How to Kill Adult Insects and Insect Larvae : (i) Adult insects may be allowed to die in test-tubes. However, the longer these insects live, the greater are the chances of the wings and legs being damaged. (ii) In chloroform vapour insects die slowly, the legs and wings remain fully stretched which is the ideal state in which to mount insects on pins. It is therefore desirable that the minutest amount of chloroform vapour to stupefy the insects should be used. The chloroform should never be poured on cotton-wool but should always be introduced inside the test-tube in the form of vapour by means of a pipette. (iii) Potassium and Sodium cyanide which are deadly toxic to insects, are most commonly used. The cyanide remains effective for a long time provided the bottle which contains the cyanide has a tight-fitting cork. To make such a bottle or tube, place some cyanide of potassium and sodium in the bottom; this is fully covered with a layer of dry plaster of Paris, over which is poured liquid plaster of Paris which is allowed to set. This is covered by several layers of filter paper and the bottle or the tube is kept tightly corked. The bottle must have a very wide mouth. (iv) Insects and their larvae are killed and preserved at the same time in 70 to 80 per cent alcohol, 4 per cent formaldehyde or in Bles's fluid. (v) All Dipterous larvae are readily killed when dropped in water heated to 80° C. Mosquito larvae retain all their fine hairs when killed in this way.

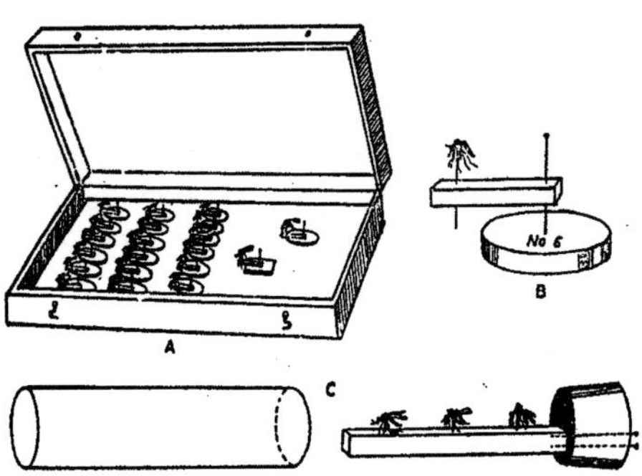

FIG. 191.

A, Box for storing pinned insects. B, A mosquito mounted on a pin. C, Easy method of keeping a large number of mosquitoes in a tube. In this way preserved insects can be despatched by post.

Mounting Mosquitoes and Flies on Pins : Place the mosquito on its back on a piece of cork. For mosquitoes use No. 1, 0, 00 (nickel coated) pins. Hold the head of the pin with a pair of forceps, pass the pointed end through a piece of pith and then through the thorax of the mosquito, taking particular care that the end does not pass through any of the joints of the legs and the sharp end is just visible on the dorsum of the thorax. Great care should be taken to prevent rubbing or injury of any kind. If the specimens are dry, they may have to be placed in a relaxing jar for a day or two. An ordinary office pin is then passed through the other end of the pith, and the pinned insect preserved in the store box.

Preparation of Store Boxes : Store boxes for storing pinned specimens must be light and air tight. They are generally made of teak wood. They should be of a standard size and the lid must be well fitting. A cork carpet is cut to fit the floor. The cork may be fixed to the floor by a few small pins or by means of paraffin wax. A mixture of paraffin (melting point 163° F) 80 parts and flake napthalene 20 parts is melted and a small quantity is run into the box. The cork is at once put in and this is held down by the paraffin. After some time more of the mixture is then run in, sufficient to cover the cork completely. The box is ready for use when cold.

As a preventive against mould and insects like mites and ants, creosote mixed with chloroform is commonly used. This should be poured on some cotton-wool wound round the head of the pin, the latter being fixed to the cork. Too much of the chloroform and creosote mixture is likely to cause damage to specimens due to condensation of creosote on the wings and body of the insects. Ordinary cigar boxes will make good store boxes for field work.

Setting Boards : These are made of long pieces of cork with a central depression running along its whole length to accommodate the body of the insect. The insect is placed astride this hollow and the wings are stretched on the two sides of the depression. They are kept in position by placing thin strips of paper across them, the paper being fixed to the board by means of pins.

Relaxing Box : For mounting old or dry specimens, they may have to be placed in a relaxing jar for a day or two. A relaxing jar or box may be made of any small box with a tight-fitting top. Some moist sand is placed at the bottom and a few drops of carbolic acid is added to prevent the growth of moulds. Over the sand a few layers of blotting paper are placed. The jar is now ready for use. The dried insects are placed carefully on paper to prevent any undue moisture. The jar is covered and allowed to stand in a warm place. In a day or two the mosquitoes or any other dried insect will become softened and they can be removed and dealt with as desired for mounting on pins or for permanently mounting them.

How to Mount Entomological Objects Permanently on Glass Slides : The specimens are killed and left in Bles's fluid for 12 to 24 hours. They are then passed through ascending grades of alcohol starting from 80 per cent to absolute alcohol and then alcohol+xylol. They are next cleared in xylol or clove oil and mounted in Canada balsam.

When carbolic acid is used, the insect is drowned in the liquid acid; the latter is then gently heated for 5 to 10 minutes and thereafter the object is left in the acid for 15 minutes to 1 to 2 hours till it looks transparent which indicates that it has been properly cleared. The carbolic acid is now drained off and a few drops of oil of cloves is put in its place. The object is now ready for mounting in Canada balsam on a slide.

Lactophenol is also a good clearing agent. Its composition is as follows : Chloral hydrate—0.5 g., Distilled water—1 c.c., Glycerine—1 c.c., Lactic acid—2 c.c., Glacial acetic acid—2.4 minims, Formol —0.5 c.c. **Berles's Fluid** is not only a good clearing but also a good mounting medium especially for small objects and is therefore extensively used in entomological work. It takes about 2 to 3 days for the gum to harden. The edge of the cover-glass should always be ringed with wax. It has the following composition : Chloral hydrate—160 g., Gum arabic—15 g., Glucose—10 g., Glacial acetic acid—5 c.c., Water—20 c.c. **Langeron's Fluid** is an excellent clearing and dehydrating medium for both fresh and dry insects. The latter are sufficiently softened for dissections within about 4 to 6 hours' immersion in this fluid. For its composition see page 114.

Bhatia's Medium : Bhatia claims to have obtained extremely satisfactory results with all types of insects, e.g. mosquito larvae, fleas, lice, ticks etc. This is prepared by dissolving at room temperature 50 gm. of rosin (crystallised pine rosin) in 75 c.c. of eucalyptus oil in a stoppered bottle. It takes about 48 hours to dissolve the rosin completely but this period can be reduced to about 2 hours by placing the container in an automatic shaker. The solution should be allowed to stand for a few hours and then decanted after which it will be ready for use. Its consistency can be varied by adding more eucalyptus oil as required.

The medium acts as a clearing as well as a mounting agent which sets hard by itself within 2 to 3 days. The slides with the mounts need not be placed in an incubator. It is not necessary to seal the edges of the mounts with any cement.

For making permanent mounts of mosquito larvae, the following procedure may be adopted. (1) The larvae are killed in hot water and put into 90 to 95 per cent alcohol, or they may be killed in alcohol by putting them directly into it. (2) Change the alcohol or transfer the specimens to fresh alcohol which may be absolute or 95 per cent. Keep specimens in it for about 15 to 20 minutes. It is preferable to prick the larvae in the thoracic region while they are in alcohol. The larvae need not be completely dehydrated before they are remounted. (3) Pick up one larva at a time and place it in the centre of a slide, remove any excess of alcohol with a piece of blotting paper and then place a few drops of the medium on it. After the larva has been properly arranged in the desired position with the help of a needle, place a cover-slip over it.

Transportation of Mosquitoes : For transporting living mosquitoes by post under all conditions Barraud's cage is very suitable.

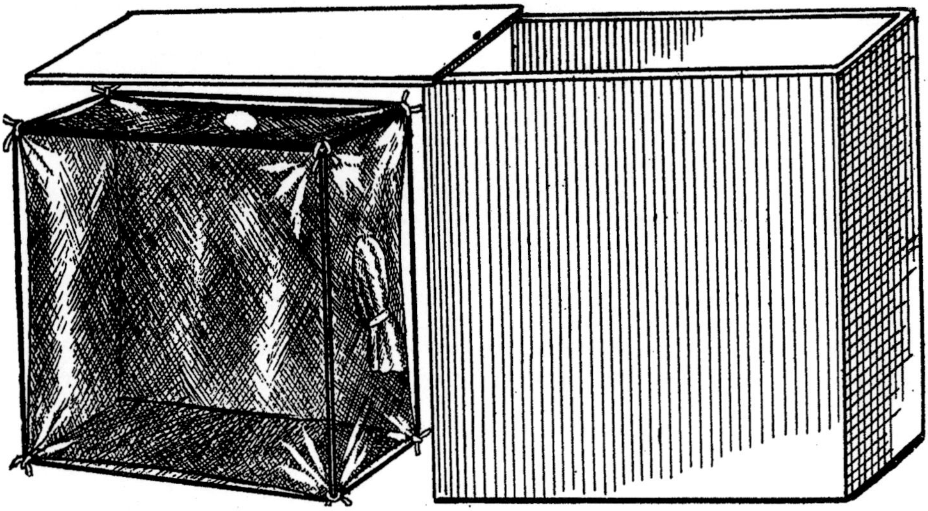

B A

FIG. 192.

Barraud's box (A) with sliding lid in which is placed Barraud's cage (B) containing living mosquitoes for transportation by post or railway.

A framework is constructed by tying firmly together at the corners with tapes six squares of strong wires measuring about 6 inches along each side. On the inside of this frame a cage of small mesh mosquito netting is fastened by means of tape sewn to the corners of the net, to one side of which a small sleeve is attached large enough to admit a test-tube. The cage is placed in a wooden box. The mosquitoes are introduced into the cage through the sleeve, which is then secured with tape. A folded pad of wet lint is placed at the bottom of the box, the pad being sufficiently thick to press against the bottom of mosquito netting. For the nourishment of the mosquitoes some raisins are deposited on the top of the cage and kept in place by a thick pad of cotton wool.

In the case of dead mosquitoes these are put in dry tubes, one specimen in each tube between two layers of wool so that all remain separate. A preservative in the form of naphthalene or a drop of creosote should be used.

Pinned mosquitoes are arranged side by side on a flat piece of cork, the latter fixed by pins to the cork of a glass tube.

Transportation of Larvae by Post : These are packed in a corked tube, half of which is filled with spirit and the rest filled with small strips of paper; this will prevent the objects from being shaken during transit.

Sticking Fluid : For sticking damaged insects ordinary gum is useless. Celluloid chopped up in amyl acetate makes a sticky mixture and this is generally used for sticking legs and wings of damaged insects.

Preparation of Different Strengths of Alcohol from Rectified Spirit : The lower strength of alcohol should always be made from rectified spirit which contains 96 per cent of alcohol. The quantity of spirit and water to be used in order to prepare a certain strength is shewn below :

26 : 74—25%. 31 : 69—30%. 41.5 : 58.5—40%. 52 : 48—50%. 62.5 : 37.5—60%. 73 : 27—70%. 78 : 22—75%. 83 : 17—80%. 93.5 : 6.5—90%.

Section Cutting : Section cutting includes three different operations, e.g., (1) fixation in any of the well-known fixative agents : (2) clearing or dehydration in ascending grades of alcohol and (3) embedding in paraffin (single embedding) or in celloidin first and then in paraffin (double embedding).

Fixation and Dehydration : For general insect work two fixatives are commonly used. These are (1) Bouin's fluid and (2) Bles's fluid.

Bouin's Fluid: Saturated solution of picric acid in water—75 c.c., Formaldehyde, 40 per cent—25 c.c., Acetic acid—5 c.c.

Bles's Fluid: 70 per cent alcohol—90 parts, Formalin (40 per cent Formaldehyde)—7 parts, Glacial acetic acid—3 parts.

Bles's fluid may be used either cold or hot. The object may be left in either Bles's or Bouin's fluid for 24 to 48 hours. As far as possible living insects are dropped in the fixative and after 6 to 12 hours one or two nicks are made in its body to allow the fixative to penetrate into every part of the insect. The object is generally transferred from Bouin's fluid to 30 per cent and from Bles's fluid to 70 per cent alcohol. Thereafter dehydration is effected in ascending grades of alcohol, e.g., 30, 50, 70, 80, 90, 96 per cent and last of all in absolute alcohol. Up to 80 per cent alcohol a mosquito or any other small object does not generally require treatment for more than 6 hours, but in each of the higher strengths the object should be kept for 12 hours. Two changes of absolute alcohol are desired. From absolute alcohol pass the object through absolute alcohol and xylol in equal parts for 1 hour, xylol for ½ hour when it is ready for embedding.

Henig's Solution (particularly recommended for mosquitoes and sand-flies) : Nitric acid—16 parts; Chromic acid—0.5 per cent, 15 parts; Saturated solution of mercuric chloride in 60 per cent alcohol—24 parts; Saturated solution of picric acid in water—12 parts.

Fix for one day; then put in a mixture of Lugol's solution and 70 per cent alcohol for 2 hours. Pass through two changes of 70 per cent alcohol till no trace of iodine remains.

If necessary the whole object may be stained in Ehrlich's haematoxylin for 3 days after fixation in Henig's solution.

The subsequent process of dehydration is carried out in the usual way.

Embedding : Double embedding in celloidin and paraffin : On account of the chitinous exoskeleton of insects, double embedding with celloidin and paraffin will yield more satisfactory results than single embedding with paraffin alone. For double embedding, a thick and a thin solution of celloidin dissolved in clove oil are required. The thick solution is first prepared. It has the consistency of thick treacle. This is diluted with clove oil to make a very thin solution. It must be remembered that celloidin dissolves in clove oil slowly and it takes several days to prepare the stock solutions.

The object is transferred from xylol to thin solution of celloidin and left for two days and thereafter transferred to the thick solution for 2 days.

Now take a cover-glass and dip it in melted paraffin. Take a large drop of thick celloidin solution on the paraffin-coated cover-glass and wait till all air bubbles have disappeared. Quickly transfer the object to the thick celloidin on the cover-glass taking particular care that in doing this no new air bubbles are introduced. The object will gradually sink in the celloidin solution and if necessary another drop of celloidin may be poured on the object in order to cover it completely. Lift the cover-slip with forceps and turn it over so that the drop hangs down with the object in it. Quickly lower it into chloroform in a corked tube. The celloidin will soon harden into a jelly and as soon as the paraffin is dissolved, the object is detached from the cover-glass.

The object is now trimmed down so as to leave a moderate thickness all round it. It is now put in a tube containing chloroform into which shreds of paraffin wax with a melting point 58°F., are gradually dropped till the mixture is saturated with paraffin wax. Leave the object in the paraffin-chloroform mixture for 24 hours when the object will sink at the bottom. It is now ready for paraffin embedding.

Paraffin embedding should be done in an incubator at a constant temperature of 58 to 60° F. The object is transferred from the chloroform mixture to paraffin and thereafter to two subsequent changes of paraffin. After this the block is made in a paper capsule or in watch glass smeared with glycerine. As soon as the paraffin begins to harden, the watch glass is gradually plunged in water beginning from its edge.

Section Cutting : Sections are cut generally 7 to 10 m.mm. thick with a microtome.

Section Staining : The sections are arranged on a glass slide in water, the slides having been just previous to this carefully smeared with a minute drop of egg albumen. Gentle heat applied to the water will cause the sections to stretch. The slide is now put away for one or two days till they are completely dry.

For staining, pass the sections through (1) Xyol. (2) Absolute alcohol. (3) Descending grades of alcohol. (4) Distilled water, (5) Delafield's or Ehrlich's haematoxylin for 2 to 5 minutes. (6) Rinse in water; blueing in tap water for a few hours is essential. (7) If overstained, dip in a jar of weak solution of HCI (0.5 per cent) for 5 to 30 seconds. (8) Rinse quickly in water. (9) Blue again in tap water. (10) 4 per cent aqueous solution of eosin for 1 minute. (11) Rinse rapidly in water. (12) Dehydrate in ascending grades of alcohol till absolute alcohol is reached. (13) Mount in a drop of Canada balsam and cover with a cover-glass.

After treatment with haematoxylin and blueing in water, the object may be passed through ascending grades of alcohol till 70 per cent spirit is reached. At this stage any overstaining with haematoxylin is corrected by 1 per cent solution of picric acid in alcohol but the object must not be left for more than half to one minute. Then treat with 70 per cent and 90 per cent alcohol. It is now stained with 2 per cent eosin solution prepared in 90 per cent alcohol, for 1 minute. Dehydration is now effected with 95 per cent alcohol, absolute alcohol, xylol, and clove oil. After this sections are mounted in a drop of Canada balsam in the usual way.

Staining Whole Objects : The entire object when small, may be stained in Ehrlich's acid haematoxylin for 3 days or in Mayer's haemalum or acid haemalum for one to three days. The object may either be mounted whole or may be used for section cutting. Mayer's acid haemalum is extremely useful for demonstrating embryos of guinea-worm undergoing development in cyclops. Mosquitoes, sand-flies etc., may be first stained either with haematoxylin or haemalum and then mounted on slides.

Ehrlich's Acid Haematoxylin : Haematoxylin—2 g., Absolute alcohol—100 c.c., Glycerine —100 c.c., Distilled water—100 c.c., Glacial acetic acid—10 c.c. Potash alum in excess. Dissolve haematoxylin in alcohol and then add the rest. Ripen it by exposing it to light and oxygen. Oxygen can freely enter through the cotton-wool plug.

After treatment with acid haematoxylin leave the object in tap water for a few hours for blueing.

Mayer's Haemalum : Haematin—1 g., Alcohol, 90 per cent—50 c.c., Pot. alum—50 g., Aqua dist.—100 c.c.

Mayer's Acid Haemalum: Add 20 c,c. of glacial acetic acid to the above. When objects are overstained, differentiate in 0.5 per cent HCI. in distilled water. Immerse in ordinary tap water for 2 hours (add a few drops of strong ammonia). Dehydrate, clear in ascending grades of alcohol and mount in Canada balsam.

Eosin in Carbolic Acid : Entire objects may also be stained in eosin dissolved in pure carbolic acid; this is simple and there is little possibility of overstaining. Imms recommends the following : *Solution I*—Basic fuchsin (90 per cent dye content)—0.3 gm., Ethyl alcohol (95 per cent strength)—10 c.c. *Solution II*—Carbolic acid—5 gm.; Distilled water—95 c.c. Solutions I and II are then mixed in proportion for use. Slow staining by leaving the material in the solution overnight is better than rapid staining. The latter would require warming the stain containing the object in watch glass which will accelerate the process. In case of overstaining the appropriate degree may be achieved by transference to 95 per cent alcohol containing 3 per cent by volume of concentrated hydrochloric acid. Carbol-xylol is used for clearing and the mounting is done in Canada-balsam prepared in xylol.

Delafield's Haematoxylin : Prepare 400 c.c. of a saturated solution of ammonia alum (1 part of alum to 11 parts of distilled water). Dissolve 4 gm of haematoxylin in 25 c.c. of absolute alcohol in a flask with cotton-wool plug and expose it to light for about a week. Filter. Add to the filtrate 100 c.c. of galcial acetic acid and 100 c.c. methylic alcohol. Leave in a warm place for 6 weeks.

Frozen Sections : Frozen sections of 7 to 10 m.mm. thickness are sometimes necessary for the demonstration of fat cells. Such sections can not be mounted permanently. These are temporarily mounted in a drop of glycerine.

Room for Insect Breeding : Temperature 78—82° F., and Humidity 79 to 80 per cent.

Dissecting Fluids : (1) Normal salt solution : 8.5 g., Water—100 c.c. (2) Ringer's Solution : Sodium chloride—0.8 g., Calcium chloride-0.02 g., Pot. chloride-0.04 g., Sodium bicarbonate—0.02 g., Distilled water 100 c.c. (3) Pickets' Liquid : 1 to 3 per cent solution of manganese chloride in distilled water.

Part II

INSECTICIDES and INSECTS & RAT CONTROL

by

A.W.A. Brown

PREFACE

This section on the control of insects by chemical insecticides was contributed in response to a request made by Dr. D. N. Roy in 1948, when he was preparing the second edition of his book. At the time Dr. Roy felt that his hands were full enough as it was with the biological aspects of medical and veterinary entomology, without entering the rapidly changing field of chemical control. He, therefore, forwarded his request to Canada, looking for a contribution from a sister country in the Commonwealth of Nations.

This request was gladly met, for the undersigned has felt that the subcontinent of India was the place where insect control by chemicals could do the most good to the most people. It may seem odd to many readers that the contributor chosen should be a Canadian, but there are good reasons. Canada has an interest in the welfare of the world, and is an active participant in the Colombo Plan. She is represented in the deliberations and in the actual administration of the World Health Organization. She is in close touch with the United States of America and its Point Four Programme. And she is acquainted with those departments of the United Kingdom government which are engaged in improving the hygiene of tropical Africa.

A Canadian entomologist is also fortunately placed with regard to access to scientific developments emanating both from the United States and also from the United Kingdom and western Europe. Entomological organizations are consequently strong in his own country. He becomes acquainted with new insecticides and new techniques for their application more rapidly than most. And so it is with pleasure that he offers this section in the spirit in which it was requested by Dr. Roy, as a contribution to distant fellow-citizens.

The reader will note that this section is not fully supplied with references. Only the most recent citations are given, or those not already listed in my "Insect Control by Chemicals," published by John Wiley and Sons, New York, in 1951.

A. W. A. BROWN
The University of Western Ontario, London, Canada

PART II

CONTENTS

Chapter I. The chemistry of insecticides

Brief history of insecticides 297
The chlorinated hydrocarbons 300
Other synthetic organic compounds 305
Organic compounds of botanical origin 307
Inorganic compounds 311
Fumigants 312
Oils, solvents and detergents 314
Formulation of insecticides 317

Chapter II. The toxicology of insecticides

Mode of entry of insecticides 320
Mode of action of insecticides 328
Methods of testing isecticides 334

Chapter III. The application of insecticides

Toxic hazards of insecticides 341
Apparatus for spraying 347
Apparatus for producing aerosols 351
Apparatus for dusting 355
Aerial application 356
Residual spraying 357

Chapter IV. Insecticidal control of noxious Diptera

Recent history of malaria control 360
Mosquito control 364
Mosquito repellents 373
Control of Phlebotomus, Simulium and Culicoides 375
House-fly control 378
Control of biting flies on livestock 384
Control of flies parasitic on livestock 387

Chapter V. Insecticidal control of other species of medical importance

Control of fleas 390
Control of lice on man and animals 391
Control of bed-bugs, including fumigation 394
Cockroach control 396
Control of ectoparasitic mites 398
Control of ticks and other arthropods 399
Rodent control 402

CHAPTER I

THE CHEMISTRY OF INSECTICIDES

BRIEF HISTORY OF INSECTICIDES

Ancient Remedies. The earliest insecticide reported in Western literature is the sulphur dioxide produced by burning the "pest-averting sulphur" mentioned in the Homeric poems a millennium before the birth of Christ. Ointment containing elemental sulphur is described as a treatment for scabies in the "Materia Medica" of Dioscorides, compiled in the 1st century A.D. Admixture of lime with sulphur was found to be highly toxic in the 18th century, and its use was described by Erasmus Darwin in 1800. The formula for lime-sulphur, discovered by Grison at Versailles in 1851 by boiling equal parts of lime and sulphur together in water, was adopted as a sheep-dip against sarcoptic mange in Australia around 1860. Its use was later extended to control biting lice and sarcoptic mange on horses and swine.

The insecticidal and repellent value of bituminous tars was known to the Romans and is described in Pliny's "Historia naturalis" of the 1st century A.D. Mineral oil was reported by Marco Polo at the close of the 13th century to be a cure for mange of camels in Asia. The use of spirit of tar for ant control, and oil of turpentine for control of scabies, evidently dates from 1787. Raw linseed oil is a time-honoured remedy to control biting lice on livestock, and scaly leg mite on poultry. Whale oil appeared as an insecticidal wash in 1800, followed by fish oil in 1820. The insecticidal value of soaps has been recognized since the end of the 18th century, and they are a constituent of most live-stock dips and washes.

False hellebore was used in classical Roman times to control body lice and the breeding of house-flies, as well as a poison for rats and mice. The various insecticidal plants used by the Chinese to control bed-bugs, fleas and mosquitoes were described by the pharmacologist Li in 1596. Pyrethrum originated as the Persian flea powder obtained from the flower-heads of the flea-grass or Persian camomile (*Pyrethrum roseum*, now called *Chrysanthemum coccineum*) growing on the southern side of the Caucasus mountains. The powder was exported to Russia by an Armenian called Jumtikoff between 1818 and 1828; it reached Vienna in 1846 and Paris in 1850. By this time it had been discovered (circa 1840) that the buhach (*C. cinerariaefolium*) of Dalmatia also yielded an excellent pyrethrum powder. This powder was employed to control fleas, bed-bugs and lice on man, and later poultry lice. The insecticidal qualities of tobacco, known by the late 17th century, were first utilized in 1854 in sulphur-tobacco dips for the control of sheep scab. Such dips were used until quite recently to control psoroptic mange and sucking lice on cattle, and nicotine has been applied to roosts for the control of mites and lice of poultry. The rotenone poisons had already been used in south-east Asia for some time when their insecticidal value was first reported in 1848. Quassia, first used in Europe as a flavoring for drinks, was described as an insecticide in 1825, namely as the toxic ingredient in sugary baits for house-flies.

The use of arsenicals for insect control is very ancient. The Chinese used powdered arsenic sulphides to control garden insects in the 9th century. White arsenic is quoted as an ingredient of ant baits in 1669. Arsenical dips were first introduced for sheep scab in 1810, and sodium arsenite has been a standard ingredient for dips to control sucking lice for over a century. Mercury has been used in India for centuries as a preservative for stored foods against insect attack; a small amount of the liquid is put in a shell alongside the food, and the vapour acts as a fumigant. In 1822 corrosive sublimate (mercuric chloride) was reported as an insecticide for destruction of bed-bugs.

Chemicals as Insecticides. The age of applied chemistry dates from the mid-19th century, when the fractionation of coal tar offered a host of new organic compounds. Phenol and the cresols were discovered to be insecticidal in 1867. The phenol of commercial carbolic acid was used in ointments for control of scaly-leg mite, and the cresols in cresylic acid were adopted as ingredients for dips to control mange on sheep and fleas on domestic animals. Commercial creosote has been, until quite recently, the favoured constituent of dips for the control of ticks on cattle, scabies on pigs, and biting lice on horses. It also proved useful as a spray for the breeding places of fleas, bed-bugs and house-flies. The heaviest coal-tar fraction, anthracene oil or carbolineum, first used about 1890 on dormant fruit-trees in Germany, has been particularly valuable for the treatment of hen houses against the poultry mite.

During this period, mineral oil or petroleum was obtained from wells and fractionated by distillation. The kerosene or coal-oil fraction was introduced as an insecticide in 1865. It was heavily applied on to the breeding sites of fleas, house-flies and bed-bugs. Various formulae for kerosene emulsions were developed between 1868 and 1880, one of which has been employed to control sucking lice on cattle. The use of special petroleum fractions as larvicidal oils for mosquito control dates from 1892.

New gaseous compounds were developed as insect fumigants. Carbon disulphide appeared for this purpose in 1854, hydrogen cyanide in 1877, chloropicrin in 1918, ethylene oxide in 1928, and methyl bromide in 1932. All of these fumigants have been used to eliminate bed-bugs and cockroaches from houses, and to disinfest clothing of lice. The slow fumigant power of naphthalene was discovered in 1882; besides being used in moth balls for the protection of clothing, it was also developed for control of fleas and bed-bugs. Admixed with a little creosote and iodoform, naphthalene was used in the NCI louse powder at the close of World War I. It was gradually superseded as a store preservative by PDB (paradichlorobenzene), whose fumigant properties were discovered in 1912.

Inorganic compounds were also developed as insecticides, starting with the appearance of phosphorus paste for cockroach control in Germany. An arsenical pigment used since 1814 as a wallpaper dye, called Paris green, was discovered in 1867 to be an excellent insecticide for the Colorado potato beetle. Acid lead arsenate appeared in the U.S.A. in 1892. After the first great war, Paris green was widely developed as a larvicide for anopheline mosquitoes, and played an important part in the development of aerial dusting in the period between the world wars.

Sodium fluoride was first patented as a cockroach powder in 1896. This powder has also been generally adopted to control Mallophaga on poultry and biting lice on livestock. Sodium fluosilicate came to be used as a substitute for sodium fluoride in cockroach powders, this being first reported in 1921. Borax and boric acid have been used in cockroach and house-fly maggot control since the early years of the present century.

The botanical insecticides were further developed, and the chemistry of the active components was elucidated. The active materials in pyrethrum flowers appeared in kerosene extracts in 1916; the 1920's witnessed a rise in the use of these extracts, sold under the name of "Flit," as space sprays against house-flies, stable-flies, sand-flies and mosquitoes. In 1924 the chemical structure of the pyrethrin esters was established. Derris powder was patented as an insecticide in Britain in 1911, and came into general use around 1920. Cube was first tested in the U.S.A. in 1924. These rotenone-containing materials afforded a safe method of controlling head lice, fleas, biting lice, mites and ticks. The chemical studies culminated in the establishment of the structure of rotenone in 1932. Even in the middle of the 20th century, rotenone remains the best insecticide available for control of the warble fly and sheep ked.

More complex synthetic organic insecticides were developed. The potassium salt of DNOC (dinitro-orthocresol) appeared as an insecticide in 1892. Butyl carbitol thiocyanate was produced commercially in 1929, and by 1934 a group of organic thiocyanates were available as contact insecticides. In that year phenothiazine was found to be a powerful mosquito larvicide, and subsequently it was widely developed as an anthelmintic.

Modern Synthetic Insecticides. The age of modern insecticides dates from 1939, when Paul Muller in Basle discovered the extraordinary contact action against house-flies of DDT, a compound first synthesized by his fellow-Swiss, Zeisler, in 1872. This was formulated by his employer, the E. R. Geigy Company, into Neocid dust and Gesarol liquid concentrate, which became known to the United Kingdom and the U.S.A. in late 1942. The Americans made enough DDT in 1943 to supply it for louse control to the U.S. army, and in January, 1944, they performed a mass treatment in Naples which stopped a typhus outbreak and completely protected their own forces there. The extraordinary effectiveness of DDT against malaria mosquitoes was abundantly proven in Italy and the southwest Pacific theatres of war during the succeeding year. The paramount suitability of DDT for control of other biting flies and indeed of most insects of medical entomological importance earned a Nobel prize for Muller, with the citation that this insecticide promised to revolutionize public health in the tropics.

Meanwhile, other chlorinated hydrocarbons were discovered to be highly insecticidal. The potency of BHC was discovered in France in 1941, and independently in England in 1942. Chlordane appeared in 1945 in the United States, simultaneously with toxaphene, methoxychlor and DDD. Aldrin and dieldrin were put out by Julius Hyman of Denver, Colorado, in 1948. The purified gamma isomer of BHC came to be produced commercially at the end of 1940 under the name of lindane. In 1951 isodrin and endrin appeared as isomers of aldrin and dieldrin.

In Germany, the insecticidal activity of certain organic phosphates was discovered by Schrader of I. G. Farbenindustrie, and HETP was put out for use in 1944. Certain of the new compounds that he synthesized were developed by the allied victors immediately after the war, particularly parathion in the U.S.A. and schradan in Britain. The first organic phosphate safe enough for use in medical entomology, namely malathon, appeared in the U.S. in 1951.

During this modern period, American workers have made notable advances in the use of pyrethroid materials. The first synergist for pyrethrins, sesame oil, was discovered in 1940. The use of compressible gases such as Freon for propellents of aerosols was developed in 1942. The piperonyl compounds appeared as pyrethrin synergists in 1946. The synthetic substitute for pyrethrins discovered by Schechter and his American colleagues, called allethrin, was produced commercially in 1950.

REFERENCES

Essig, E. O. (1931). A History of Entomology. MacMillan Co., New York, pp., 1029 **Shepard, H. H.** (1939). The Chemistry and Toxicology of Insecticides. Burgess Publishing Co., Minneapolis, pp. 383 mimeographed. **Shepard, H. H.** (1951). The Chemistry and Action of Insecticides. McGraw-Hill Book Co., New York, pp. 504

THE CHLORINATED HYDROCARBONS

DDT. P,P′-Dichloro-diphenyl-trichloroethane, or more correctly 2,2-bis (p-chlorophenyl)-1,1,1-trichloroethane, is a white crystalline solid, but is produced technically as a white amorphous powder. Its specific gravity is 1.556 at 25°C., and its vapour pressure is 1.5×10^{-7} mm. Hg at 20°C. This exceptional lack of volatility is responsible for its outstanding properties as a residual insecticide.

DDT is prepared by condensing 1 mole of chloral with 2 moles of chlorobenzene in 98 per cent H_2SO_4 at 15°C. The average commercial product contains 70 per cent of the P,P′ isomer, the remainder consisting mainly of o, P′-DDT. The melting point of purified DDT is 109°C., and its setting point is between 103° and 105°C. Technical DDT shows a melting point of 90°C., and it is considered a saisfactory product if the setting point is not below 88°C.

DDT is almost completely insoluble in water, and is the least watersoluble organic compound known. Its true solubility is 0.0002 parts per million of water, although it can form colloidal solutions up to 0.2 p.p.m. concentration. While DDT is quite freely soluble in most organic solvents, its solubility in aliphatic oils is limited (refined kerosene 4 per cent, no. 2 fuel oil 7 to 10 per cent); however, certain aromatic oils are superior solvents (*APS*-202 45 per cent, *Velsicol AR*-50 55 per cent).

Table 1. Solubility of Certain Chlorinated Hydrocarbons in Organic Solvents. grams per 100 cc. at 25°—30°C.

	DDT	Lindane	Aldrin	Dieldrin	Toxaphene
Acetone	58	44	159	54	450
Ethyl alcohol	2	6.4	9.0*	4.9*	12
Carbon tetrachloride	45	6.7	..	..	450
Cyclohexanone	116	37	..	..	..
Benzene	78	29	350	75	450
Xylene	53	25	..	..	450

* in methyl alcohol.

DDT is very stable at normal temperatures; thermal decomposition develops at 195°C. It is labile in alkali, and is decomposed by oxides or chlorides of Fe or Al;

thus it is incompatible with kaolin, fuller's earth and dolomite, and with nicotine alkaloid. DDT is not normally decomposed by sunlight. It is almost completely stable to ultra-violet light when in solid form, but in oil solution it is slightly decomposed. Its water-insolubility completely prevents its leaching by rain. However, tropical temperatures and humidities decompose DDT appreciably faster than temperature climates.

DDT is formulated in 5 per cent solutions for household space sprays, odourless kerosene being the solvent along with a small amount of xylene or methyl-naphthalene as co-solvent. It is also put up as 25 per cent emulsion concentrates, in which the solvent is methyl-naphthalene, solvent naphtha or turpentine mixed with 15 per cent of an oil-soluble emulsifier (*e.g. Triton X*-100, *Teepol X*). With the addition of casein as a stabilizer, the concentrate may be shipped as a mayonnaise cream containing 27 per cent water.

Since DDT becomes waxy when warm, it is mixed with an equal amount of pyrophyllite or talc before being ground to a fine dust. Addition of wetting agent results in a 50 per cent wettable powder, 1 lb. of which is diluted with 1 gallon of water for residual spraying in houses, and 1 lb. per 100 gallons for agricultural spraying. Wettable powders with 75 per cent DDT can be made, but they do not suspend well in water. Simple dilution with more talc or pyrophyllite results in a DDT dust; concentrations of 1, 2, 3 or 5 per cent are used for agricultural purposes, 10 per cent dusts for roach powders, and 50 per cent dusts for rat-borne plague control.

DDT may be included in paints and washes for the treatment of walls to make them residually insecticidal. This stratagem is successful with casein paints, calcimines and chalk whitewashes. In limewashes the DDT is decomposed by the alkali, but if enough insecticide is added it may retain its toxicity for at least 2 months. A concentration of 3 to 5 per cent DDT in flat oil paints is very effective. With enamel paints the insecticide is sealed off by the glossy outer film, but they may nevertheless be rendered effective if as much as 20 per cent DDT is added. Very effective coatings may be obtained with this concentration of DDT incorporated in a urea-formaldehyde resin.

DDT may be chemically estimated by dehydrochlorinating it in alcoholic KOH and titrating the chloride by Volhard's method; this is applicable to the assay of formulations. The greater sensitivity of a colorimetric method is required in the assessment of residues in foods and animal tissues. The Schechter-Haller method, and the simplified modification of Alessandrini, is the most satisfactory; the DDT is converted to its tetranitro derivative which gives a blue colour with sodium methylate. The Stiff-Castillo method, in which a red colour is obtained with xanthydrol, may also be used.

Methoxychlor and DDD. The methoxy analogue of DDT, 2,2-bis (p-methoxyphenyl)-1,1,1-trichloroethane, is a white solid which melts at 88°C. Produced by the condensation of chloral with anisole, the technical material is pale buff in colour and is 88 per cent pure. It may be determined quantitatively by the Schechter-Haller method, in which case a pink colour is obtained.

DDD. P,P′-Dichloro-diphenyl-dichloroethane or tetrachloro-diphenyl-ethane (TDE) is an analogue of DDT which has one less chlorine atom, being 2,2-bis (p-chlorophenyl)-1,1-dichloroethane. It is a white crystalline solid with a melting point of 109°C., and is produced by condensing dichloroacetal with chlorobenzene. It is marketed as a 50 per cent wettable powder (*e.g. Rhothane WP*-50) and 25 per cent solution in an aromatic oil (*e.g. Rhothane S*-216). The solubilities of methoxychlor and DDD resemble those of DDT.

DFDT, Lucex and Dilan. The difluoro analogue of DDT, namely 2, 2-bis (p-fluorophenyl)-1, 1, 1-trichloroethane, is a solid melting at 45°C. It is appreciably volatile, and thus has less residual persistence than DDT. It was produced commercially in Germany under the name of *Gix*. 2-(p-Chlorophenyl)-1, 1, 2, 2-tetrachloroethane has also been used as an insecticide in Germany under the name of *Lucex*.

Dilan (compound CS-708) is the trade name applied to a mixture of 1 part of 2-nitro-1, 1-bis (p-chlorophenyl) propane (Prolan or compound CS 645A) with 2 parts of 2-nitro-1, 1-bis (p-chlorophenyl) butane (Bulan or compound CS-674A). These compounds are solids with a melting point of 79° and 59°C. respectively, stable to acid and weak alkali, and the mixture has an odour of almonds. The materials are soluble in organic solvents, and an emulsion concentrate is made of 25 per cent *Dilan* in pine oil.

BHC. Benzene hexachloride, or more correctly 1, 2, 3, 4, 5, 6-hexachlorocyclohexane, has the empirical formula $C_6H_6Cl_6$, from which is derived its original code name of 666. Commercial BHC is prepared by the chlorination of benzene in the presence of ultra-violet light; it consists of buff-coloured crystals which have a pronounced musty odour, due to impurities and to breakdown products which subsequently develop. The material is a mixture of optical isomers, 55 to 70 per cent consisting of the beta isomer, and 10 to 13 per cent being the gamma isomer. It is the gamma isomer that is the active insecticidal principle, and the toxicity of BHC is proportional to its gamma content. Commercial preparations are now available which contain as much as 92 per cent of the gamma isomer (*e.g. Hi-gam*). Purified preparations which contain more than 95 per cent of the gamma isomer are given the name of lindane*.

The isomers of BHC are appreciably volatile; the vapour pressure value for the gamma isomer is b 1×10^{-5} mm. Hg.at 20°C. and it is practically volatized at 120° C,; the beta isomer is slightly less volatile. This volatility is sufficient to prevent BHC from showing adequate residual properties, but endows it with a definite fumigant action on insects. The purified crystals of lindane are colourless, and comparatively odourless when first produced.

The isomers of BHC are moderately soluble in organic solvents, the beta isomer being the least soluble. The solubility of the gamma isomer is 3 per cent in paraffin oil, 18 per cent in heavy naphtha, and even higher in the *Velsicols* and other aromatic solvents. The BHC isomers are relatively insoluble in water, that of the beta isomer being 5 parts per million, and the gamma isomer 10 p.p.m..

The isomers of BHC are stable to the effects of light, and are approximately as thermostable as DDT. However, they are labile in alkali, the beta isomer being the most resistant to alkaline dehydrochlorination. Gamma-BHC decomposes to a marked extent in containers which are not completely non-corrosive; this decomposition may be inhibited by the addition of sodium thiosulphate.

*On the other hand, gammexane is a certain commercial product analogous to BHC.

BHC may be quantitatively assessed by the total chlorine method developed for DDT. The gamma isomer content may be assessed by the hydrolysable chlorine method by treating the BHC with N alcoholic KOH at 0°C. There are also spectroscopic, cryoscopic and chronatographic methods, but lindane residues are still most conveniently determined by bioassay.

Chlordane. Technical chlordane, or chlordan, is the name given to a product, formerly called compound 1068, containing not less than 60 per cent of 1, 2, 4, 5, 6, 7, 8 8-octachloro-4, 7-methano, 3a, 4, 7, 7a-tetrahydroindane, with the remainder being related dicyclopentadiene derivatives. It is a highly viscous liquid, dark amber to brown in colour, with a terpene-like odour. The specific gravity is 1.56 at 60°F., the viscosity 6900 cp. at 25°C., and the refractive index 1.56 at 25°C. The volatility of chlordane is intermediate between that of DDT and BHC, the vapour pressure of the refined product being approximately 1×10^{-5}mm. Hg at 25°C. Chlordane consists of both cis and trans forms, the latter being 10 times as insecticidal as the former and the product is contaminated with inseparable materials such as trichloro-237, hexachlor, and heptachlor, the last-named compound being slightly more insecticidal than trans-chlordane.

Chlordane is insoluble in water, but it is completely miscible with organic solvents, including both aliphatic and aromatic oils. Technical chlordane begins to lose its toxicity at temperatures above 60°C. It is readily detoxified by alkaline dehydrochlorination, and should not be stored in galvanized iron or steel containers.

Emulsion concentrates containing 45 or 75 per cent chlordane are made by mixing the technical product with kerosene and adding 10 per cent of a wetting agent. Wettable powders are made by spraying chlordane heated to 130°F. on to diatomaceous earth in a ribbon blender, to give a 50 per cent mixture; the wetting agent is then added. Alternatively the 50 per cent mixture may be diluted with pyrophyllite to give 5 or 10 per cent dusts.

Chlordane may be quantitatively determined by the total chlorine methods employed for DDT. It may also be assessed by the hydrolysable chlorine method, 2 atoms of chlorine being removed per molecule.

Aldrin and Dieldrin. *Aldrin* (formerly called compound 119 or octalene) is 1, 2, 3, 4, 10, 10-hexachloro-1 : 4, 5 : 8-diendomethane-1, 4, 4a, 5, 8, 8a-hexahydronaphthalene; dieldrin (compound 497 or octalox) is the 6, 7-epoxy derivative of aldrin.

Aldrin is a white crystalline solid, with a melting point at 100-103°C. It is much less volatile than lindane, and is intermediate between chlordane and DDT in residual effectiveness. It is insoluble in water, but is 89 per cent soluble in refined kerosene. It is stable to alkali and metallic chlorides. Residues on plants may be determined by the Danish-Lidov colorimetric method.

Isodrin (compound 711), a highly insecticidal stereoisomer, is a white crystalline solid that is slowly decomposed on being heated above 100°C. It is highly soluble in aromatic solvents, less so in paraffin oils, and insoluble in water.

Dieldrin is an odourless white crystalline solid, with a melting point of 173°C. It is almost non-volatile, and shows a residual toxicity whose persistence is comparable to that of DDT. It is less soluble in organic solvents than aldrin, being only 4 per cent soluble in base oil. It is completely stable to alkali. Dieldrin is the most toxic and residually effective of the present insecticides.

Endrin (compound 711), a highly insecticidal stereoisomer, is a white crystalline solid that melts with gradual rearrangement on being heated above 200°C. Its solubility characteristics resemble those of isodrin.

Toxaphene. Toxaphene (formerly called compound 3956) is a mixture of 16 to 18 insecticidal isomers to which the name *chlorinated camphene* has been applied. It is produced by the chlorination of catalytically rearranged turpentine, and its insecticidal activity is maximal when the chlorine content is between 65 and 69 per cent by weight, corresponding to an empirical formula of $C_{10}Cl_{10}H_8$. The material is a cream-coloured waxy solid with a mild terpene odour, melting between 65° and 95°C., and its specific gravity is 1.66. It is scarcely volatile and is insoluble in water. Toxaphene is highly soluble in organic solvents and oils, the solubility being 75 per cent in lubricating oil, 260 per cent in fuel oil and 280 per cent in kerosene. It is stable at room temperatures and oil solutions may be stored in clear glass bottles without appreciable deterioration. Although toxaphene is dehydrochlorinated by alkali, iron containers are permissible at normal temperatures. Toxaphene is chemically determined by the total chlorine method, using the nitrobenzene modification in the Volhard titration.

Chlorinated Biphenyls. *Octochlor* (2, 3, 5, 6, 2', 3', 5', 6 -octachlorobiphenyl) has been used successfully in 10 per cent solution as a substitute for DDT in residual sprays against *Anopheles* in Madagascar. *Arochlor* 1242 is a commercial product that consists of a group of chlorinated biphenyls, whose distillation range lies between 325 and 360°C.

Chlorinated Acaricides. A great number of chlorinated compounds, many of them analogues of DDT and some containing sulphur, have been developed for the control of Tetranychid mites on orchard trees and field and greenhouse crops. They are as follows :

DMC	1, 1-Bis (p-chlorophenyl)-ethanol
DCPM	Bis (p-chlorophenoxy)-methane
K-6451	p-Chlorophenyl p-chiorobenzenesulphonate
Genitol 923	2, 4-Dichlorophenyl benzenesulphonate
R-242	p-Chlorophenyl phenyl sulphone
Aramite	Chloroethyl-p-t-butylphenoxy-methyl ethyl sulphite

Of these compounds, DMC is the only one to have application in medical entomology, since it has been found to act as a synergist for DDT against resistant house-flies. It is di (p-chlorophenyl) methylcarbinol, or more correctly, 1, 1-bis (p-chlorophenyl)-ethanol. A crystalline solid, it melts at 70°C., and is stable to alkali although decomposed by acid. It is highly soluble in benzene and ethanol, and soluble to the extent of 7 per cent in naphtha. At present it is sold as a 25 per cent wettable powder under the trade name of *Dimite.*

Other DDT Synergists. The following chlorinated hydrocarbons have been found effective in restoring the toxicity of DDT against resistant house-flies :

MR-60 Di-p-chlorophenyl-chloromethane
MR-30 Di-p-chlorophenyl-ethynylcarbinol (Compound 876)
K-3926 Di-p-chlorophenyl-ethane

Chlorinated Chalcone : p-chlorophenyl 1, 2-dichloro-2-(chlorophenyl)-ethyl ketone.

REFERENCES

Brown, A. W. A. (1951). Insect Control by Chemicals. John Wiley and Sons, New York. pp. 817. **Department of the Army** (1947). Insect and Rodent Control. Technical Manual TM-5632. Superintendent of Documents, Washington. pp. 156. **Director-General, Army Medical Services** (1946). The use of DDT in the Field. Manager of Publications Delhi, pp. 47. **Julius Hyman and Company** (1951). Entomological Progress Reports. Compound 711 : 32 pp. Compound 269 : pp. 57. **March, R. B.** (1952). The resolution and chemical and biological characteristics of some constituents of technical chlordane. *J. Econ. Ent.* 45, 452-456. **Metcalf, C. L., Flint, W. P.** and **Metcalf, R. L.** (1951). Destructive and Useful Insects. McGraw-Hill Book Co., New Tork. 3rd Ed. pp. 1071. **World Health Organization** (1951). Malaria Conference in Equatorial Africa. Tech. Rep. Series No. 38, pp. 32 (reference to octochlor).

OTHER SYNTHETIC ORGANIC COMPOUNDS

Organic Phosphates. The most important organic phosphate products used as insecticides are as follows :

HETP	Hexaethyl tetraphosphate
TEPP	Tetraethyl pyrophosphate
Schradan	Octamethyl pyrophosphoramide
Parathion	Diethyl p-nitrophenyl thionophosphate
EPN	Ethyl p-nitrophenyl thionobenzenephosphonate
Malathon	Dimethyl dithiophosphate of diethyl mercaptosuccinate

Of these, HETP and TEPP (the active principle in HETP) are higly toxic to warm-blooded animals, highly labile, and are used for quick contact kill of aphids. *Schradan* (also called OMPA) is a systemic insecticide, being taken into the sap of plants and thus protecting them against sucking insects. Parathion is a powerful and persistent insecticide, highly toxic to higher animals; but this compound, along with EPN and malathon, has been used in mosquito larvicides and residual sprays against bed-bugs and DDT-resistant house-flies.

Parathion (also called E-605 and available commercially in 95 per cent pure form as *Thiophos* 3422) is an amber liquid with a boiling point of 375°C. and a vapour pressure of 0.006 mm. of mercury at 24°C. It is slightly soluble in water, to the extent of 20 parts per million. It is completely soluble in fat solvents and aromatic oils, but almost insoluble in paraffinic oils. It is stable to acid, but is hydrolysed by alkali; at temperatures exceeding 130°C., it isomerizes.

EPN is an amber liquid of specific gravity 1.27, much less volatile than parathion. It is slightly water-soluble but dissolves in many organic solvents. At present it is produced as a 27 per cent wettable powder.

Malathon is an amber liquid of specific gravity 1.23, with a melting point of 3°C. and a vapour pressure of 0.00004 mm. Hg at 30°C. It is soluble in water to the extent of 145 p.p.m.; like parathion it is insoluble in paraffinic but soluble in aromatic oils. Malathon hydrolyses at pH's above 7 and below 5.

Methyl-parathion, in which the ethyl groups of parathion are replaced by methyl, is considerably more volatile and may be less hazardous than parathion. It now replaces parathion in the formula E-605, and is mixed 2 to 1 with parathion in Metacide. It has been used as wall sprays against bed-bugs, and is claimed to be synergistic with DDT against resistant house-flies. Another organic phosphate tested in residual sprays against house-flies is Compound 3456, which is diisopropyl p-nitrophenyl thiophosphate.

Dinitro Compounds. The following derivatives of dinitrophenol have been used in agricultural entomology in dormant sprays against aphid eggs, summer sprays against orchard mites, or locust poisons :

DNOG	4, 6-Dinitro-ortho-cresol
DNCHP	2, 4-Dinitro-6-cyclohexyl-phenol
DNBP	4, 6-Dinitro-2-sec-butyl-phenol
Arathane	Dinitro-capryl-phenyl crotonate

They show too great an affinity to stain the skin, and have too high a mammalian toxicity, to be generally adopted in medical or veterinary entomology. However, DNOC and DNBP may be employed for area control of harvest mites or chiggers.

Organic Thiocyanates. These insecticides were developed for their knock-down properties in fly sprays and the majority are commercially known as *Lethanes.* Lauryl thiocyanate, C_{12} H_{25} SCN, is the main constituent of "lorol" thiocyanate, produced by treatment of the alkyl chlorides of the mixed fatty acids from cocoanut oil with sodium thiocyanate. *B*-Thiocyanoethyl laurate, $C_{12}H_{25}O_2C_2H_4SCN$, is the principal constituent of a mixture of similar esters up to the palmitate; a 50 per cent solution of this material in petroleum distillate is marketed as Lethane 60. *B*, *B'*-Dithiocyanodiethyl ether, $C_4H_8O(SCN)_2$, also called n-butyl-carbitol thiocyanate, is marketed in 90 per cent solution as Lethane A-70. It is also produced as a 13.5 per cent dust called Lethane B-71, and a 13.5 per cent wettable powder called Lethane B-72. *B*-Butoxy-*B'*-thiocyano-diethyl ether, $C_8H_{17}O_2SCN$, is produced as a 50 per cent solution, with Lethane 384 as the commercial name. Lethane 384 Special is a mixture of 12.5 per cent *B*-butoxy-B'-thiocyano-diethyl ether with 37.5 per cent *B*-thiocyanoethyl laurate in an equal volume of petroleum distillate.

The insecticide *Thanite* is a dark aromatic liquid produced by treatment of the camphene from pine oil or turpentine with chloroacetic acid and then with an alkyl thiocyanate. The product contains 80 per cent of isobornyl thiocyanoacetate, along with fenchyl and bornyl thiocyanoacetate, and the thiocyanoacetates of secondary terpene alcohols and tertiary terpene alcohols such as terpineol.

Phenothiazine. The compound produced by the cyclization of diphenylamine with sulphur is known as thiodiphenylamine or phenothiazine, $C_{12}H_9NS$. It is a light yellow crystalline solid with a melting point of 185°C.; the commercial product is however dark green. Phenothiazine is insoluble in water; it is slightly soluble in acetone or ethanol but insoluble in chloroform. It is oxidized on exposure to air and sunlight to phenothiazone and/or thionol; this process is hastened by alkali. It may be quantitatively determined colorimetrically by treating with bromine water and reading at 520 millimicrons wavelength.

Other Sulphur Compounds. *Lauseto-neu*, a preparation containing chloromethyl-p-chlorophenyl-sulphone, was developed in Germany as an outstanding louse ovicide. *Tetraethyl thiuram monosulphide* (TETMOS) has been used for control of scabies mites on man, mixed in the form of Tetmosol soap.

Miscellaneous Organic Insecticides and Acaricides. *Pyrolan* (compound G-22008) is the name given to 1-phenyl-3-methylpyrazolyl-(5)-dimethylcarbamate, a colourless crystalline solid, which has given good residual control of DDT-resistant house-flies in Switzerland and Sardinia. It is soluble in water to the extent of 0.1 per cent, and it is slightly volatile.

Valone is the name given to 2-isovaleryl-1, 3-indandione or 2-pivalyl-indandione, which along with the tert-butyl-indandione (*tertiary butyl valone*), has been found to be a lousicide which is effective on oral ingestion by rabbits.

Pentachlorophenol, a flaky crystalline solid with a pronounced creosote odour, is marketed as a wood preservative, to give protection from fungous decay and attack by termites or lyctus beetles. *B-Naphthol*, a pleasant smelling solid with a melting point of 122°C., the standard material for impregnating orchard tree bands, has been used to control the head louse. *Benzyl benzoate*, a water-insoluble liquid freezing at 18.5°C., is used as a clothing impregnant for protection of men against trombiculid mites.

ORGANIC COMPOUNDS OF BOTANICAL ORIGIN

Nicotine. This alkaloid as obtained from natural sources is laevorotatory beta-nicotine, and has the shortened name of 1-*B*-pyridyl-a-N-methyl-pyrrolidine. The correct nomenclature is 1-3-(1-methyl-2-pyrrolidyl)-pyridine, or 1-1-methyl-2-(3-pyridyl)-pyrrolidine. It is a liquid which boils at 247°C. and whose specific gravity is 1.009. Nicotine is an alkaloid which readily forms salts with acids. It is appreciably volatile, its vapour pressure being approximately 0.1 mm. Hg. at 25°C., and it exhibits fumigant action.

Nicotine is commercially prepared by its extraction from the coarse parts of the tobacco plant, *Nicotiana tabacum* or *N. rustica*, by treatment with alkali and steam distillation. The yield ranges from 2 to 5 per cent of the dry weight of the plant, and may reach as much as 14 per cent in the latter species.

Nicotine alkaloid is miscible in water in all proportions at atmospheric temperatures and it is miscible with the organic solvents. It is usually marketed as the dibasic salt nicotine sulphate in aqueous solution containing 40 per cent of the alkaloid (*e.g. Blackleaf* 40), because this form is more stable and less volatile. When it is added to alkaline water or to soap solutions, the free alkaloid is liberated and shows a higher insecticidal activity than the undissociated sulphate.

Nicotine is compatible with all insecticides and fungicides except definitely acid materials. It is commonly formulated in 3.6 per cent dusts with alkaline carriers such as hydrated lime, limestone and other alkaline carbonates, which are termed active carriers because they release the alkaloid. Certain acidic dust diluents act as sorptive carriers, such as pyrophyllite, talc, kaolin and bentonite. Nicotine bentonite (*e. g. Blackleaf* 155) is so stable that the preparation is known as fixed nicotine. Nicotine is quantitatively assessed by precipitation with silicotungstate and gravimetric determination of the crystalline silicotungstate.

Anabrsine is B-pyridyl-a-piperidine, or more correctly levo-2-(3'-pyridyl)-piperidine or 3-(2-piperidyl)-pyridine. It is a liquid boiling at 281°C. and has physico-chemical characteristics similar to nicotine. It is obtained from the new twigs of *Anabasis aphylla*, a woody shrub ranging from Russian Turkestan to North Africa, in which it constitutes 1 to 2.5 per cent of the dry weight.

The Pyrethrins. The pyrethrin compounds as found in pyrethrum flowers consist of 4 esters, representing the possible combinations of 2 different alcohols with 2 different acids, as follows.

Pyrethrin I : ester of pyrethrolone with chrysanthemum monocarboxylic acid.
Pyrethrin II : ester of pyrethrolone with chrysanthemum dicarboxylic acid-monomethyl ester.
Pyrethrin I : ester of cinerolone with chrysanthemum monocarboxylic acid.
Pyrethrum II ester of cinerolone with chrysanthemum dicarboxylic acid-monomethyl ester.

These compounds are obtained from the dried flower heads of *Chrysanthemnm cinerariaefolium*, a native of Jugoslavia but now cultivated in Kenya and Japan. The content in most samples ranges between 0.5 and 3.0 per cent; they may be extracted with alcohol, acetone or ethylene dichloride. After evaporation of the solvent, the pyrethrins appear as a viscous oleoresin. The pyrethrin esters are very slightly volatile, and are soluble in oils and organic solvents. They are very slightly soluble in water, in which they are hydrolysed, and the process is accelerated by acids or alkalis (such as lime or soap). They are also oxidized on exposure to air, but may be protected by antioxidants such as hydroquinone, tannic acid, o-aminophenol or p-aminophenol. Stored pyrethrum powders lose about 20 per cent of their insecticidal activity in a year; whole flowers are less susceptible to decomposition than impregnated dusts, and oily concentrates less than aqueous suspensions.

The ground pyrethrum flowers may be used directly as insect powder, but there is a great waste of the insecticide due to its being locked in the plant cells. The pyrethrins may be extracted in kerosene and then compounded to a content of 3 per cent and sold as concentrates (*e.g. Pyrefume Super* 30). Or they may be diluted in odourless kerosene to a 0.1 per cent content and sold as household fly-sprays. The pyrethrin esters may be extracted in ethylene dichloride and impregnated on to talc, gypsum or diatomaceous earth; the pyrethrum marc left behind after extraction may also be used as the dust diluent.

Allethrin. Allethrin, first synthesized in 1949, is the dl-2-allyl-4-hydroxy-3-methyl-2-cyclopenten-1-one ester of dl-cistrans-chrysanthemum monocarboxylic acid. It is the product of esterification of the allyl analogue of cinerolone with the synthetic cistrans chrysanthemum acid. It is thus an analogue of the natural cinerins, and is more stable and the more easily synthesized. It is now produced commercially in 75 to 92 per cent purity (*Pyresyn*), a clear brownish resin of sp. g. 1.01 and refr. ind. 1.504. It is marketed as a 20 per cent concentrate and in 2 per cent solution. The commercial product is thus cistrans-allethrin; it is one to three times as toxic as the pyrethrins to house-flies, slightly less toxic than them to mosquitoes, and much less toxic than pyrethrins to cockroaches. Trans-allethrin on the other hand has been found to be considerably more effective than pyrethrins to *Anopheles quadrimaculatus*, although there is thought to be little difference between cis and trans allethrin in toxicity.

The toxicity of allethrin is not as susceptible to an increase by synergists as that of the pyrethrins. Piperonyl butoxide and propyl isomer make it over twice as toxic, and octacide increases the toxicity slightly, against house-flies. Sulfoxide, octacide and N-isobutyl-hendecamide have been used as synergists for allethrin against body lice.

The pyrethrins, and allethrin also, may be quantitatively determined by the mercury reduction method using Deniges reagent, by a colorimetric method involving hydroxylamine, by hydrogenolysis, or by the steam-distillation and titration method of Seil.

Pyrethrin Synergists. Certain compounds, themselves slightly or not insecticidal, have the property of extending the toxicity of pyrethrins, so that very much less pyrethrum is needed in the formulation for equal results. Usually the formulation contains 8 to 10 times as much synergist as pyrethrins. The synergists fall in three groups on the basis of their chemical nature.

(*a*) *Methylenedioxyphenyl compounds* : Piperonylcyclonene (piperonyl cyclohexenone), a condensation product of alkyl-3, 4-methylenedioxystyryl ketone with ethyl acetoacetate,
Piperonyl butoxide, or (3, 4-methylenedioxy-6-propylbenzyl) (butyl) diethylene glycol ether, a yellow oily liquid of sp. g. 1.06 and boiling at 180°C. at 4 mm. Hg.,
Sesamin, constituting 0.25 per cent of the sesame oil obtained from the plant *Sesame indicum*,
Piperine, a crystalline solid melting at 129°C. obtained from black pepper,
Propyl isome, a condensation product of isosafrole with n-propyl maleate,
Sulfoxide, the n-octyl sulphoxide of isosafrole, a viscous brown liquid.

(*b*) *Terpene derivatives*: DHS Activator, the ethylene glycol ether of pinene, a liquid of sp. g. 0.985 and boiling between 257 and 273°C.,
Pine oil, and its products terpin diacetate and Thanite.

(*c*) *N-substituted Amides and Imides* : IN-930, or N-isobutylundecylenamide, used in louse powders, N-Isobutylhendecamide, and also methylvaleramide.
Octacide, or N-octyl bicycloheptene dicarboximide, produced commercially as MGK-264 (formerly Van Dyke 264).
Diisopropylacetamide, amylphthalimide and dipropylcinnamamide have been found to be synergistic with DDT against house-flies.

Of these synergists, the most important at the present time is piperonyl butoxide, widely used in the protection of livestock against flies and capable of increasing the toxicity of pyrethrins 14 times. Propyl isome can increase it 7 times and octacide 3 times. Piperonylcyclonene is used mainly against insects feeding on plants. Sulfoxide is as effective as piperonyl butoxide against house-flies and shares its property of stabilizing pyrethrins against decomposition.

Rotenone. The fish-poison rotenone is a white solid, whose melting point is 163°C. It is almost completely insoluble in water (0.17 p.p.m. at 25°C.) Rotenone is soluble in the organic solvents, particularly in chloroform, but is relatively insoluble in carbon tetrachloride. Only 0.05 per cent may be dissolved in kerosene, which may be increased to 0.2 per cent with co-solvents.

Rotenone is oxidized by exposure to air, the process being catalysed by light and alkali, but it is stable in kerosene solution. This compound is incompatible with alkaline materials such as nicotine and lime-sulphur; clays and talcs should be avoided as dust diluents.

Rotenone is obtained from the roots of *Derris elliptica* and other congeners which grow in southeast Asia and Indonesia; the dried product called tuba or derris has an average content of 4 to 5 per cent ranging up to 13 per cent. It is also obtained from the roots of *Lonchocarpus* spp. native to Central and South America; the dried product is called timbo or cube and contains an average of 8 to 10 per cent rotenone, with a maximum of 20 per cent. It is also found in various parts of plants belonging to the genera *Tephrosia*, *Millettia*, *Mundulea* and *Pachyrhizus*. The roots of derris or cube are dried in the sun and ground in hammer mills; the powders are then blended to give a rotenone content between 4 and 5 per cent. Insecticidal dusts may be made by diluting with pyrophyllite or diatomaceous earth to a 1 per cent rotenone content, or by impregnating the inert carrier with the chloroform extract of the ground root.

In addition to rotenone, derris resins contain a number of related compounds of the fish-poison group, of which deguelin is the most important. The rotenone content of a derris preparation may be gravimetrically assessed by crystallization from carbon tetrachloride and recrystallization from ethyl alcohol. A colorimetric method is available; when the acetone extract is treated with KOH and then acidified with HNO_3 in presence of sodium nitrite, a red colour is produced.

Veratrine Alkaloids. These are a group of alkaloids found in the roots and seeds of liliaceous plants. Of these, probably the most insecticidal is cevadine, $C_{32}H_{49}NO_9$, and protoveratrine $C_{32}H_{51}NO_{11}$. *Hellebore* root from *Veratrum album* of Eurasian mountain regions contains these two alkaloids, as well as jervine and pseudojervine. Green hellebore root (*V. viride*) contains cevadine, jervine and pseudojervine. Hellebore is adapted for application to garden fruits shortly before picking, since it detoxifies rapidly. *Sabadilla* is obtained from the seeds of certain species of *Schoenocaulon* which grow in Mexico; it contains cevadine, veratridine, sabadilline and sabadine.

Other Botanical Principles. The insecticidal product *quassia* includes a group of compounds which resemble the rotenoids in possessing two methoxy groups, in having similar empirical formulae ($C_{22}H_{30}O_6$ in each case), and in being slow poisons; however they are water-soluble. They include quassiin (m.p. 205° C.) and its isomer neoquassiin (m.p. 225°C.), obtained from chips of the wood of *Quassia amara* of Dutch Guiana. A third compound called picrasmin (m.p. 218°C.) is found in addition to these two in *Picrasma excelsa* of Jamaica.

Ryanodine, the active principle of *Ryania* dust, is an alkaloid whose empirical formula is $C_{25}H_{35}NO_9$ or $C_{26}H_{37}NO_9$. This material is obtained from the root and stems of *Ryania speciosa* of Trinidad.

The alkaloid *ricinin*, $C_8H_{10}N_2O_2$, is an insecticidal principle in the castor bean *Ricinus communis*. The protein ricin, which also occurs in the bean, has no insecticidal properties but is extremely toxic to man.

Affinin is the name given to N-isobutyl-2, 6, 8-decatrienoamide, which is as highly insecticidal as the pyrethrins. It is obtained from the roots of *Heliopsis longipes* of Mexico, a plant which had first been misidentified as *Erigeron affinis*. *Scabrin* is the N-isobutyl amide of a straight-chain C_{18} fatty acid, obtained from the roots of *H. scabra*. It is more toxic than the pyrethrins, and its toxicity is extended by pyrethrin synergists, especially sulfoxide.

Other botanical principles which have been used in China or which have proved very promising in American tests include the following :

Seeds of the croton tree, *Crolon tiglium.*
Fruit of the cork tree, *Phellodendron amurense.*
Roots of the thundergod vine, *Tripterygium wilfordii.*
Flowers of *Rhododendron* and *Andromeda* spp. of China.
Seed kernels of the mamey tree, *Mammea americana.*
Extracts of the male fern, *Aspidium filix-mas.*
Bulbs of the cyclamen, *Cyclamen elegans* (miticidal).
Bark of the prickly ash, *Xanthoxylum herculis.*

Antibiotics of Animal Origin. Iridomyrmecin, a compound isolated from the ant *Iridomyrmex humilis*, has proved to be highly insecticidal, especially against DDT-resistant house-flies. It is a lactone whose empirical formula is $C_{10}H_{16}O_2$.

REFERENCES

Buchland, R. C. *et al.* (1945). *Soap San. Chem.* 21 (11): 119. (Hydroquinone as a pyrethrin antioxidant). **Connolly, P. J.** (1952). *Pest Control*, May 1952 : 24-25. (Sulfoxide as a pyrethrin synergist. **Dove, W. E.** (1947). *Amer. J. Trop.* Med. 27: 339-345. (Properties of piperonyl butoxide). **Eddy, G. W.** (1951). *J. Econ. Ent.* 44': 109-111. (Aminophenol as a pyrethrin antioxidant). **Fales, J. H.** *et al.* (1951). *J. Econ. Ent.* 44 : 23-28 and 250-252. (Cis-trans allethrin and trans-allethrin). **Gersdorff, W. A.** and **Mitlin, N.** (1952). *J. Econ Ent.* 45 : 519-523. (Scabrin). **Gersdorff, W. A., Nelson, R. H.** and **Mitlin, N.** (1951). *J. Econ. Ent.* 44 : 921-927. (Synergists for allethrin). **Pavan, M.** (1952). *Boll. Istit.* Sieroterap. Milanese Serafino Belfanti, 31 : 195-208 and 232-245. (Iridomyrmecin). **Roark, R. C.** (1952). A digest of information on allethrin U.S. Dept. Agr. Bulletin *E*-846 :58 pp.

INORGANIC COMPOUNDS

Arsenical Compounds. The salts of arsenious acid are derived from arsenious oxide, As_2O_3, called white arsenic. This oxide is a by-product of the roasting of mineral ores which collects in the flues since it sublimes at 120-150°C.; it is the raw material from which all the arsenicals are manufactured. It is soluble in alkaline water, and has been used in insect baits. By dissolving white arsenic in aqueous NaOH, *sodium arsenite* may be prepared, which contains a mixture of the orthoarsenite Na_3AsO_3 and the metarsenite $NaAsO_2$. This material, which contains 82 per cent As_2O_3, has been widely used in cattle dips.

The arsenites of calcium, zinc or copper have been tested on field crops but have not gained general acceptance. A complex of copper metarsenite and copper acetate, of the probable composition $(CH_3.COO)_2\ Cu.3Cu(AsO_2)_2$ has been widely used under the name of *Paris green* for the control of *Anopheles* larvae. It is also known as *Mitis green* or *Schweinfurtergrun*, and contains 57 per cent As_2O_3.

The arsenates include acid lead arsenate, basic lead arsenate, and calcium arsenate. The first two compounds are widely used in orchard spraying, the last compound in crop dusting. Magnesium, aluminum, ferric, barium and zinc arsenates have also been used. None of these arsenicals have had application in medical or veterinary entomology.

The percentage of arsenic in the usual arsenical insecticides may be assessed by treatment with cuprous chloride, distillation of the arsenious chloride, and titration with iodine. The determination of arsenic in residues is accomplished on a micro scale by the Gutzeit method or bromate method.

Fluorine Compounds. *Sodium fluoride*, NaF, is a white crystalline solid, whose melting point is 98°C. and specific gravity 2.79. Approximately 5 per cent soluble in water, it has been used in insect baits and traps. It is commonly employed in roach and ant powders; usually blended with pyrethrum, borax and inert diluents, it is ground to a particle size of 5 to 10 microns.

Sodium fluosilicate, more correctly sodium fluorosilicate or silicofluoride, has the empirical formula Na_2SiF_6 or 2NaF. SiF_4. It is made by combining sodium carbonate with the product of hydrofluoric acid and silica. The white crystalline solid has a specific gravity of 2.75; it decomposes before melting. The commercial product consists of powdered particles whose median diameter by mass is about 25 microns, and which has a high bulk density of 30 cu. ins/lb. Sodim fluosilicate is 0.75 per cent soluble in water at 25°C.; it is used in roach powders and moth-proofing fluids.

Other fluorine compounds, hitherto used only as agricultural insecticides, are *barium fluosilicate* ($BaSiF_6$) and *cryolite* (Na_3AlF_6). Fluorine compounds may be quantitatively assessed by ashing and conversion to lead chlorofluoride. The amount of Ag required to combine with the Cl in the PbClF is measured by titration of the remaining $AgNO_3$ with potassium thiocyanate.

Miscellaneous Inorganic Insecticides. *Tartar emetic* or potassium antimonyl tartrate, $K(SbO)\ C_4H_4O_6.1/2H_2O$, is prepared as a white powder by heating antimony oxide in a solution of acid potassium tartrate. It has a solubility of 5.3 per cent in water, and is often used in poison baits for ants. *Borax*, $Na_2B_4O_7$, and *boric acid*, H_3BO_3, have been used for the control of maggots, and in ant traps and cockroach powders. The latter compound has been included in anti-blowfly dressings for sheep.

Sulphur and Lime-Sulphur. Elemental sulphur (in the rhombic form) is a yellow solid melting at 113°C. and with a specific gravity of 2.07; it is insoluble in water. It originally appeared as an acaricide in the form of "*flowers of sulphur,*" crystals recondensed after sublimation. This material is nowadays more finely ground, the extreme being "micronized sulphur," produced in an air-grinding apparatus. More finely divided particles are obtained in "colloidal sulphur," precipitated from several dissolved sulphur compounds in various ways. Another source is "*flotation sulphur,*" produced by oxidation of the H_2S recovered in aqueous alkali from coal gas fumes.

When 3 parts of sulphur are heated with 2 parts of hydrated lime, $Ca(OH)_2$, in aqueous suspension, an orange-coloured solution is obtained. In the resulting lime-sulphur solution, between 60 and 94 per cent of the sulphur is in the form of calcium polysulphides, ranging from Ca sulphide up to Ca pentasulphide or higher. These compounds have not been isolated in pure form, but they are considered to be the active insecticidal principles in lime-sulphur.

FUMIGANTS

Methyl bromide is the most generally useful of insect fumigants at the time of writing. It has a slight sweetish odour, and is non-inflammable except in extremely high concentration. It is slightly soluble in water, and considerably soluble in fat solvents; it is not adsorbed on fruits to any extent, but is adsorbed into nuts and fatty foods. It is a favoured fumigant for mills, warehouses, ships and freight-cars, in which it is used at 1 to 3 lbs. per 100 cubic feet. The gas is released from canisters or cylinders in which it is kept liquid under its own vapour pressure.

Ethylene oxide is a relatively odourless gas with good penetrating properties; moreover it is non-corrosive and leaves no residue. The dosage employed in mills and warehouses is 2 lbs. per 1000 cubic feet. Since ethylene oxide is inflammable, it is mixed with 9 times its volume of CO_2 (*e.g. Carboxide*) which enhances the toxicity by keeping the insects' spiracles open.

Sulphur dioxide, a gas with a suffocating odour, may be generated by burning sulphur or by emitting it from 150-lb. cylinders. It is effective against insects at a concentration of 5 lbs. per 1000 cubic feet, and is employed in rodent extermination. Sulphur dioxide is 10 per cent water-soluble, and the resulting sulphurous acid is highly corrosive.

Table 2. Properties of Insect Fumigants

Compound	Boiling Point C.	Spec. Grav. of Liquid	Spec. Grav. of Gas	Lower limit of Inflammability %	Formula
Sulphur dioxide	−10	1.50	2.3	Non-inflamm.	SO_2
Methyl bromide	5	1.732	3.2	13.5	CH_3Br
Ethylene oxide	11	0.887	1.5	3.0	$(CH_2)_2O$
Hydrogen cyanide	26	0.697	0.94	5.6	HCN
Carbon tetrachloride	76	1.595	1.65	Non-inflamm.	CCl_4
Acrylonitrile	77	0.801	..	Inflamm.	$CH_2{=}CH{-}CN$
Ethylene dichloride	84	1.257	3.5	6.2	$CH_2Cl.CH_2Cl$
Trichloroacetonitrile	85	1.440	4.95	Non-inflamm.	$CCl_3{-}CN$
Chloropicrin	112	1.692	..	Non-inflamm.	CCl_3NO_2

Hydrogen cyanide has been widely employed as a fumigant for buildings. Although it has a penetrating odour of bitter almonds, it is toxic in such low concentrations that it requires the admixture of 5 per cent chloropicrin or cyanogen chloride as a warning gas. It is inflammable only at very high vapour concentrations. HCN is highly adsorptive; in addition, it is readily soluble in water and fatty materials, and hence is highly absorptive. It may be generated by atmospheric humidity from calcium cyanide and calcium acid cyanide; the commercial powders yield 55 per cent of their weight as HCN. It also may be released from cellulosic discoids on which it has been held by adsorption. The traditional method is to drop NaCN into dilute H_2SO_4; this may be weighed into paper or zinc-foil bags, or compressed into eggs, 1 lb. being dropped into 5 lbs. of 50 per cent H_2SO_4.

Carbon tetrachloride is not inflammable, which makes it valuable as a fumigant in dwellings in spite of a low insecticidal toxicity. It is often used as a diluent for ethylene dichloride or ethyl acetate. When CCl_4 is employed alone for grain fumigation, a dosage of 30 lbs. per 1000 cubic feet is required. *Acrylonitrile* is a useful low-vapour-pressure fumigant for dead spots in flour mills and for bed-bug control. Being inflammable it may be mixed with an equal amount of CCl_4, and the mixture proves to be a more effective insecticide than acrylonitrile alone. *Trichloroacetonitrile* has been used as a louse fumigant in Germany and for stored products treatments in Italy. It is not inflammable but is highly lachrymatory. *Ethylene dichloride* or 1, 2-dichloroethane has a pleasant chloroform-like odour, but is nevertheless a self-warning gas. Since it is inflammable, 3 parts of ethylene dichloride are diluted with 1 part of CCl_4. It is non-corrosive, and is used as a mill or domestic fumigant at 6 lbs per 1000 cubic feet and may be applied by spraying. It can also be used as a soil fumigant in 15 per cent emulsion. *Chloropicrin* is a highly insecticidal, self-warning sweet-smelling lachrymatory gas that has little fire-hazard and is highly penetrating. Since it is not water-soluble it is not absorbed by foods, and it is not reactive with metals and drapes. It is used for fumigation of mills, warehouses and grain at a concentration of 2 lbs. per 1000 cubic feet; its effectiveness may be enhanced by admixture of CO_2. *PDB* (p-dichlorobenzene) is useful as a soil fumigant and is very widely employed in the form of moth-balls to fumigate clothes in the home. *Naphthalene* has also been employed for the same purpose,

OILS, SOLVENTS AND DETERGENTS

Mineral Oils. These liquids are composed of a mixture of hydrocarbon oils, which may fall into any one of 4 classes : paraffins, straight-chain saturated compounds of empirical formula C_nH_{2n+2}; olefines, unsaturated straight-chain compounds with at least one C=C double bond; aromatic compounds, unsaturated cyclic compounds, containing benzene and similar rings; naphthenes, saturated cyclic compounds containing cyclohexane or cyclopentane rings. As obtained from natural strata, oils vary according to their content; for example, Pennsylvania oils are rich in saturated paraffins and are classed as paraffinic; Mexican Gulf and Californian oils contain a large proportion of aromatic and naphthenic compounds and are classed as asphaltic.

Within these classes, oils may be classified according to their molecular weight. This may be assessed by measurements of their boiling range and their viscosity. The viscosity is measured in the Saybolt universal viscosimeter by the time in seconds that is required for 60 ml. to flow through an orifice at 100°F.; it is expressed as seconds Saybolt or s.s.u. British determinations are in Redwood units, and European work is based on the Engler viscosimeter. It is therefore best to express it as the absolute viscosity in poises or centipoises at 25°C. or other specified temperature. Another criterion of oils is their specific gravity, which may be measured and expressed as degrees Baum or degrees A.P.I. (American Petroleum Institute).

Oils are fractionated from crude petroleum oil by their distillation and recondensation at different levels in distillation columns. At present the liquid fractions in order of their volatility, apart from the volatile natural gas, are gasoline, kerosene, gas oils and lube oils; the solid residue is petrolatum or paraffin wax (from paraffin crudes) and petroleum pitch or asphalt (from asphaltic crudes). The cyclized hydrocarbons may be separated as naphthas. The unsaturated hydrocarbons may be separated from the saturated hydrocarbons by refinement. The contents of unsaturated hydrocarbons in an oil may be assessed by sulphonation of the oil with 4 times its volume of 38N H_2SO_4 at 60°C. The purity of the oil as far as saturated hydrocarbons are concerned may be gauged by the proportion of unsulphonated residue and expressed as per cent U.R.

Kerosenes are employed as solvents for pyrethrins in fly sprays and for DDT and other chlorinated hydrocarbons in sprays applied from aircraft at low volume expenditures. The lighter kerosenes have a boiling range of 360-480° F., and the heavier of 425-510° F. Their viscosity is approximately 1 to 2 centipoises at 25°C., as determined on a special Saybolt thermoviscosimeter. The kerosenes may be refined to give odourless kerosene and other refined grades (*e.g. Deo-Base*), or they may be deodorized by the addition of deodorants (*e.g. Neutroleum alpha*). The fuel oils have similar boiling and viscosity characteristics as the kerosenes. Distillate fuel oil is a recommended mosquito larvicide with a boiling range of 430-630°F., a viscosity of 35-40 s.s.u., and a gravity of 27-33 A.P.I.

Lubricating oils are not employed as solvents for agricultural insecticides, but they may be used in mosquito control. The lighter lube oils have a viscosity of 70-110 s.s.u., and the heavier grades extend up to 330-360 s.s.u.

Naphthenes are generally too inflammable for insecticidal use, since no solvent should be sprayed which has a flash-point below 125°F. But the highest fractions have a flash-point of 140°F., which makes them safe for the treatment of upholstery where a rapidly evaporating solvent is required (*e.g. Stoddard's Solvents, Varsol*).

Methyl-naphthalenes are components of asphaltic petroleum oils as well as of the coal-tar oils. They are themselves insecticidal, and have proved themselves particularly valuable as initial solvents for DDT (*e.g. Arosol* 151B). The physical characteristics of various commercial solvents in this class are shown in Table 3. Highly refined fractions are now available (*e.g. Velsicols*); these non-volatile solvents may also be used to stabilize the droplets in liquefied gas aerosols.

Table 3. Properties of heavy aromatic solvents

	Flash Point °F.	Initial Boiling Point °F.	Final Boiling Point °F.	API Gravity at 60°F.	Color Reading Union Colorimeter	DDT Solubility at 30°C.
Amsco Solv H-La	231	400	612	14.2	4.0	40.8
Bronoco Hi-Sol No. 4	212	380	550	16.0	1.5	44.5
Bronoco Hi-Sol No. 100	195	382	673	20.0	4.5	40.8
Hi-Solv Oil No. 473	195	306	512	21.5	2.5	34.6
Indocene 90	203	392	554	18.0	1.0	47.0
Saf-T-Sol No. 1	175	312	638	22.5	4.0	37.9
Sovacide 544 B	270	424	642	13.0	7.0	42.2
Sun No. 3	210	340	594	20.5	5.0	34.6
Umex 4060	215	392	536	21.0	5.0	40.8
Velsicol AR-50G	190	398	556	20.1	1.0	39.0
Velsicol NR 70	270	372	680	12.5	3.5	40.8

Tar Distillate. This material is the condensed distillate from the destructive distillation of coal to produce illuminating gas and coke. It consists of a mixture of aromatic compounds (benzene, naphthalene, anthracene and other ring compounds with their methyl derivatives), tar bases (pyridine, quinolines, *etc.*) and tar acids (phenol, cresols, etc.). Satisfactory tar distillates start to boil at 410°F. (210°C.) and not more than 65 per cent have boiled at 671°F. (355°C.). The fraction boiling below 210° C. is the light oil containing benzene, toluene and xylene; that boiling between 210 and 240°C. is the middle or carbolic oil containing phenol (carbolic acid) and naphthalene; boiling between 240 and 270°C. is the heavy or creosote oil, and above 270°C. the anthracene oil. Cresylic acid is a commercial product derived from the creosote fraction and containing a mixture of the ortho, meta and para cresols. The residue from which the acids and bases have been removed by refinement is called neutral tar oil. The heavier fractions have been marketed for dormant orchard sprays in Europe, under the name of *carbolineum*.

Natural Oils and Soaps. The natural oils and fats are mixtures of the triglycerides of fatty acids, notably lauric (C_{12}), myristic (C_{14}), palmitic (C_{16}) and stearic (C_8) acid. Raw linseed oil has proved effective in the control of biting lice on mammals. Cottonseed, rape, castor, cocoanut, soya bean, palm and corn oils, whale, cod, herring, menhaden and sardine oils, neatsfoot oil, wool grease and lard, have occasionally been used in insect control products, but mineral oils are usually much cheaper.

The soaps are manufactured by alkaline treatment of these natural fats and oils. Saponification with potash produces the potassium salts of the fatty acids (soft soaps), and with soda the yield is the sodium salts (hard soaps). These soaps are water-soluble, but may be precipitated by the calcium or magnesium ions in hard water. The principal value of soap as a detergent or emulsifier, wetter or spreader, lies in its capacity to reduce the surface tension of water. For the maximum reduction of surface tension, the most effective fatty acid group is in the neighbourhood of C_{12} or lauric acid. The richest source of these fatty acids is cocoanut oil; the corresponding alcohols produced by reduction are also valuable as raw materials for contact insecticides, being produced under the name of *Lorol.*

Synthetic Detergents. The problem of insolubility in hard waters may be overcome by making a soap with a sulphuric acid group substituting for the carboxylic group, viz. R-OSO_3H instead of R-COOH. The sodium salts of these alkyl-sulphuric acids are employed, and they are not rendered insoluble by cations which normally make for hardness. As in the true soaps, the bulk of the detergent molecule lies in the anionic component. Similar anionic detergents may be made from alkyl-sulphonic acid, from aryl sulphonic acids, from alkyl-aryl sulphonic acids and from sulphonic acid attached to the alkyl groups of esters and N-substituted amides (*e.g. Dreft, Nacconol, Penetrol, Teepol X, Turkey Red Oil*). Non-ionic detergents may be made from the mono-esters of fatty acids with poly-alcohols, and from alcohol polyethers (*e.g. Lissapol N, Triton X*-100, *Tween* 80). Cationic detergents are rare, and are known as "invert soaps." The synthetic detergents may be dissolved in concentrated solutions of insecticides in oil, and these emulsion concentrates will emulsify themselves on dilution with water. *Triton X*-100 and *Tween* 80 have been found to be the best emulsifiers for concentrates of chlorinated hydrocarbons. Or they may be milled with insecticidal dusts to constitute wettable powders which will readily disperse in water suspension.

Miscellaneous Materials. Saponin, a glucoside from the soapwort *Saponaria officinalis*, has been occasionally used because it greatly reduces the surface tension. Casein reduces the surface tension of water, and has been widely used as a spreader, besides being effective as a sticker. It is dispersed in water with the aid of 3 parts of lime (*e.g.* lime-casein, "calcium caseinate"). Gelatin and flour have also been used as wetting agents and adhesives.

Dust Diluents. In order to reduce their cost, and to prevent waste of superfluous material, insecticidal compounds may be diluted with finely ground minerals. The most important are the silicates of aluminium or magnesium (*e.g.* talcs, pyrophyllites, bentonites and clays). Another important group are the diatomaceous earths, consisting mainly of silica (SiO_3). A general classification is shown in Table 4.

Diatomite is a highly spicular form of silica that is the principal constituent of the diatomaceous earths and keselguhrs. *Celite* is a commercial grade of diatomite ground to an average particle size of 5 microns, whereas *Supercel* and *Santocel C* ("silica aerogel") are ground to below 1 micron. Silica may also be used in the form of the mineral tripolite found in tripoli clays.

Calcium oxides may be produced by roasting calcite or dolomite, but when used as dust diluents they may be expected to be in the slaked form known as hydrated lime, Ca $(OH)_2$. *Calcite* or mineral calcium carbonate may be used in insecticidal dusts as ground limestone or whiting. *Gypsum* or hydrated calcium sulphate has been used as a dust diluent, but is not favoured because it has a high bulk density.

Table 4.—Classification and Properties of Dust Diluents

			Bulk Density lbs/cu. ft.	Specific Gravity	Reaction pH
Oxides	Silicon	Diatomite	9-11	2.0-2.3	5-8
	Calcium	Hydrated Lime	28-32	2.1-2.2	12-13
Sulphates		Gypsum	49-57	2.3	7-8
Carbonates		Calcite	48-67	2.7	8-9
Silicates	1. Talc		30-52	2.7-2.8	6-10
	2. Pyrophyllite		28	2.7-2.9	6-7
	3. Clays	*a.* Montmorillonite	38-44	..	9-10
		b. Kaolinite	30-35	2.6	5-6
		c. Attapulgite	27-31	2.6	7

Talc is essentially a hydrated magnesium metasilicate. The ground dusts appear in a variety of forms : foliated or flat; spicular or fibrous; and granular or massive, when ground from soapstone or steatite. They show a great range in the pH reaction of various samples, many being considerably alkaline. *Pyrophyllite* is a hydrated aluminium silicate that is rather closely related to the clays. Being non-alkaline, it is especially favoured as a dust diluent for rotenone, pyrethrum and the chlorinated hydrocarbons. The usual varieties of pyrophyllite have flat or plate-like particles, and the laminae may be separated by heat : hence the name.

Bentonite is a type of clay mineral whose principal constituent is montmorillonite. It exhibits the property of swelling with water, and it is able to hold nicotine and pyrethrin very tenaciously. *Kaolinite* is the principal mineral in china-clays and kaolin clays. The particles may be flat to granular, but never spicular. Another type of clay mineral is *attapulgite.* The dusts (*e.g. Attaclay*) are often described as fuller's earths, and have fibrous or spicular particles.

Elemental sulphur has also been used as a dust diluent for insecticides. Finely ground botanicals, including the flours of wheat, soyabean, tobacco, walnut-shell and wood, have been used as diluents for pyrethrins and rotenone.

FORMULATION OF INSECTICIDES

Dusts. Since it is wasteful and often hazardous to use an insecticidal compound in its native state, it is customary to dilute it with a solid or liquid carrier. The simplest method is to mix it with an inert powder, usually a mineral such as talc or pyrophyllite, to form an insecticidal dust. Equal amounts of the solid insecticide and the mineral are ground together in a Raymond kiln mill or micronizer mill. This is called a straight-mix dust, and it is usually further diluted with carrier to give a concentration of 2, 5 or 10 per cent of the insecticide, in which form it is marketed.

Liquid insecticides such as chlordane or parathion may be sprayed directly on to a moving layer of the dust diluent in a ribbon blender, to produce an impregnated dust. Solid insecticides may be dissolved in a volatile organic solvent and similarly sprayed on to the dust diluent, to give an impregnated dust. Since it offers more insecticidal surface, the insecticidal activity of a 3 per cent impregnated dust is roughly equivalent to that of a 5 per cent straight-mix dust.

The average particle diameters of most high grade insecticidal dusts, and wettable powders, range between 2 and 6 microns. In order to increase their flowability, they may be admixed with a "conditioner" such as diatomaceous earth. Since the finest particles tend to hang in the air and drift away from the target, approximately 1 per cent of an oil such as *Mentor* 29 may be added to the dust.

Wettable Powders. The inorganic insecticides such as lead arsenate, which have been used in agriculture for many years, were finely ground and suspended in water for spraying. The modern chlorinated hydrocarbons are ground down with an equal amount of inert mineral diluent, and to the mixture is added a small quantity of detergent or emulsifier, thus producing a 50 per cent wettable powder (50 WP) of the insecticide. With the very powerful insecticides, lower proportions of insecticide to diluent are used, *e.g.* dieldrin 25 per cent, malathon 20 per cent, parathion 15 per cent. The effect of the detergent is to allow water to enter the microscopic spaces between the dust particles, and thus to throw it readily into suspension. Since it does not prevent the suspension settling out again, spray tanks for wettable powders should have agitation devices, unless they are small enough to be frequently shaken. The use of 5 per cent suspensions of DDT, in which 1 lb. of a 50 per cent wettable powder are mixed with 1 gallon of water, demands that the formulation be readily suspendable. A non-ionic synthetic detergent such as *Triton* X-100, a phthalic glyceryl alkyl resin, has proved excellent in this regard.

Spray Concentrates. The modern insecticides are soluble in highly aromatic oils. Therefore they are marketed as spray materials in the form of concentrated solutions, the carrier being one of the heavy, comparatively non-volatile oils listed in Table 3. The nature and characteristics of these proprietary solvents are seldom known to the purchaser. Isophorone is used as an auxiliary solvent for lindane, and toxaphene is often diluted in the naphthene called *Varsol.*

These primary solvents put the insecticide into solution. For oil sprays, as in residual spraying or application from aircraft, these concentrates are further diluted with an oil diluent such as fuel oil or kerosene. In order that they may be diluted with water, a small amount of an emulsifying detergent such as *Triton X*-100 or *Tween* 80 is added to the primary solution, thereby making an emulsifiable concentrate. Nearly all concentrates now marketed are emulsifiable.

DDT is usually produced in 25 or 30 per cent concentrates in polymethylnaphthalene solvents such as the Velsicols. Chlordane and toxaphene may be distributed in 80 per cent concentrates, since they are miscible with oil in all proportions. Malathon now appears as a 50 per cent concentrate, and lindane and dieldrin are sold in concentrates containing approximately 20 per cent of the insecticide. Pyrethrins are marketed as 3 per cent extracts, and are diluted to 0.1 per cent in refined oil for space spraying; for livestock spraying, a 1 per cent emulsifiable concentration containing 10 per cent synergist is a common formulation.

The figure quoted for concentration is on a weight/weight basis. For example, a 10 per cent concentrate contains 10 grams of the insecticide in each 100 grams of the concentrate. If quoted on a weight/volume basis, it would imply that there is

1 lb. of the insecticide for each 1 gallon (10 lbs. if the specific gravity of the solvent were 1) of the liquid solution. The Imperial gallon contains 4.546 litres, but the U.S. gallon contains only 3.785 litres and is equal to 0.8327 Imperial gallons; 1 Imperial gallon is equivalent to 1.2 U.S. gallons.

REFERENCE

Chisholm, R. D. and **Koblitsky, L.** (1951). U. S. Department of Agriculture Bulletin *ET*-296. pp. 2 (Rules for expressing concentration of formulations).

CHAPTER II

THE TOXICOLOGY OF INSECTICIDES

MODE OF ENTRY OF INSECTICIDES

A small organism such as an insect presents a very large surface area relative to its volume, and thus its cuticle is of extraordinarily great importance. Being waterproof and hydrophobic, this cuticle is at the same time lipophilic and thus particularly receptive to organic liposoluble compounds as insecticides.

Moreover, the tracheal system of the insect offers many points of entry at the spiracles disposed along the length of the body. Consequently the insect is susceptible not only to fumigant vapours, but also to the inhalation of poisonous liquids of low surface tension, as for example the space sprays which achieve knockdown of houseflies and mosquitoes.

Although insecticides are generally poisonous to all multicellular animals, their apparent specificity for insects is largely due to their ability to penetrate the cuticle. When applied by contact to the cuticle, rotenone, lindane and DDT are just as poisonous as if they had been directly injected into the haemolymph, in cases where the insects are moderately sclerotized. The less hardened and impermeable its cuticle, the more susceptible an insect will be to contact poisoning by rotenone. Indeed, the penetrability of a chemical may be of more importance than its intrinsic toxicity in deciding whether it is a contact insecticide.

Certain nerve poisons may reach a site of action without entering the haemolyph at all, since they exert an effect at the cuticular level on the sensory nerve endings at the hair sensilla, campaniform sensilla, the pulvilli or the tarsal sense organs. For this reason it may occasionally be found, as with DDT or nicotine on some insects, that the contact toxicity exceeds the toxicity by injection.

Classification of the Mode of Entry. Insecticides have been classified according to the manner in which they are administered to insects and their mode of entry into the body, into the following 4 groups :

Stomcach. Application to the food and entry through the midgut; *e.g.* arsenicals and fluorides.

Contact. Application to the body surface and entry through the cuticle and tracheae; *e.g.* pyrethrins, rotenone, nicotine.

Fumigant. Application as a vapour and entry through the tracheae; *e.g.* HCN, chloropicrin, methyl bromide.

Residual : Application to surfaces and subsequent uptake through the cuticle of legs and mouthparts; *e.g.* DDT, chlordane, dieldrin.

This is not a true classification, since any given insecticide will be found to partake of a measure of several of the types of toxicity. Nicotine, which is considered primarily as a contact insecticide, is also a stomach poison and a fumigant. The vapour of nicotine may not only enter through the tracheae, but may also penetrate directly through the cuticle. Arsenicals and fluorides exhibit a degree of contact toxicity, yet the insect's cleaning movements may bring them to the mouth where they also enter as stomach poisons. Chlordane, lindane and aldrin are capable of showing all four types of toxicity, whereas DDT and dieldrin lacking fumigant activity show only three.

The position of an insecticide in this classification is decided by its physical characteristics. Stomach insecticides are those compounds which have not sufficient liposolubility to be applied by contact or volatile enough to act as fumigants. Contact insecticides are those that are solid or liquid at working temperatures, but they also show fumigant toxicity if their vapour pressure is appreciable, *e.g.* nicotine, lindane or chlordane. If the vapour pressure is negligible they exhibit residual toxicity, e.g. DDT or dieldrin. Chlordane has a vapour pressure at such a level that both types of toxicity are apparent. Pyrethrins are not volatile enough to exert an appreciable fumigant toxicity and are not sufficiently stable to offer residual toxicity for a long period. Contact insecticides usually show stomach toxicity if the opportunity arises; they fail to do so when they are destroyed in digestion, *e.g.* pyrethrins, or fail to be absorbed, *e.g.* rotenone or phenothiazine.

Permeability of the Cuticle. It has frequently been demonstrated that insecticides can penetrate directly through the cuticle of unsclerotized or newly-moulted insects, by applying wax or glass cells to limited areas of cuticle, by exposing only the appendages, or by plugging the spiracles and excluding the mouth and anus by collars. The penetration of nicotine vapour through the abdominal cuticle of *Heliothis*, and of derris toxins through the body-wall of *Bombyx*, has been clearly proved when the spiracles have been sealed to preclude tracheal entry. Although the most penetrating contact poisons are liposoluble compounds, even oil-insoluble materials such as arsenicals and fluorides are capable of entering through the cuticle to some degree.

The cuticle may be regarded as consisting of a series of barriers. The outermost barriers include the hairs of caterpillars, the elytra of beetles, the waxy threads of certain aphids, the cuticular cement of assassin bugs, or the armour of certain scale insects.

The first continuous barrier to be met is the epicuticle, whose importance is considerably greater than its meagre thickness would indicate. This consists of fatty molecules oriented on a framework of tanned lipoprotein called cuticulin. If the lipoid is dissolved by saponification in alkali or by treatment with fat solvents, the permeability of the entire cuticle to contact poisons is greatly increased. Immersion of the isolated cuticle in boiling water increases its permeability but does not destroy its selective action.

The next barrier is provided by the exocuticle, the outer layer of the cuticle proper in which the protein has become tanned, and the combination of chitin with the tanned protein results in a hard, dark and rigid sclerotization. It is a general observation that heavily sclerotized insects such as adult beetles or orthopterans show more resistance to contact insecticides than unsclerotized aphids or caterpillars. Cuticular penetration in these types of insects takes place mainly at the intersegmental membranes; only immediately after a moult is the cuticle vulnerable as a whole.

The endocuticle itself, consisting of alternating lamellae of chitin and protein, interposes a thick barrier to protect the underlying hypodermal cells. Although it is moist and in direct connnection with the body fluids, it does nevertheless impede the entry of contact poisons.

Once the cuticle is penetrated, the contact poison may exert a direct effect on the hypodermal cells, as has been shown for DNOC and pyrethrins in certain insects. It may destroy the haemocytes in the body cavity, as contact dusts of sodium arsenite do to locusts. Or it may be carried by the haemolymph to be deposited in all the tissues. Another mode of transport is along the lipoid sheaths of the nervous system, the initial entry being made at the nerve endings at the cuticular trichogen cells.

Vulnerable Points of Entry. It is usually found that the most vulnerable regions for the contact application of insecticides are the head and thorax, since here the insecticides can most quickly reach the vital centres. The mouthparts, the antennae, the ventral cervical region, and the thoracic sternites between the legs constitute regions of rapid insecticidal effect. The labella on the mouthparts of the house-fly offer a ready route for the uptake of DDT from deposits.

The cuticular grease of cockroaches may serve to spread insecticides which it can dissolve. Borax and fluoride dusts have been found to accumulate in the ventral region between the coxae of the legs. It is on the ventrum that the insecticide is closest to the nerve cord, being separated from it only by the epidermis and basement membrane. The wings may also serve as a route of entry in species that have a circulation of haemolymph through them.

The legs are the most vulnerable region for contact poisoning of locusts, where it is twice as effective as on the head or abdomen. The trochanters on the legs, which bear the campaniform sensilla, have been deduced to be the most vulnerable regions of the cockroach to DDT poisoning.

The tarsi, on which the insect walks, form a very vulnerable point in flies and mosquitoes for the entrance of DDT. Here only a thin cuticle covers the cells of the tarsal chemoreceptors. Moreover, the pulvilli on which the house-fly walks contain a gland cell, opening to the outside by hollow tenent hairs, which produces a secretion capable of dissolving DDT, and which connects with a nerve bundle passing up the leg. Contact of the pulvilli of *Glossina* with pyrethrum causes almost instantaneous paralysis; with DDT, contact for 2 seconds is enough to ensure death at some time later. When powdered DDT crystals are applied to its feet, the house-fly spreads them all over its body, and into its mouth and digestive tract, by cleaning movements. A similar movement of powdered poisons has been observed on the surface of the cockroach.

Permeability is a function of the entire cuticular area in unsclerotized larvae, particularly those of Lepidoptera, Diptera and certain Coleoptera. But in the hardened cuticles of adult Coleoptera, Hymenoptera and Orthoptera, penetration is restricted to the unsclerotized intersegmental membranes.

The cuticle has been found to be considerably more permeable in the spots which bear the hair sensilla. Here the insecticide may penetrate the areolae or sockets, the setal membranes or plaques which lead to the trichogen cells and sensory nerve endings. The ducts of the dermal glands may provide channels of entry in some cases. The pore-canals may also increase cuticular permeability; if the pore-canals of *Rhodnius*, which are normally tortuous, are straightened, shortened and widened by stretching the cuticle, the rate of penetration of pyrethrins shows a threefold increase.

Penetration by Contact Poisons. The cuticle of insects is impermeable to strongly dissociating compounds, a property contributed by the lipoid epicuticle.

Whether the poison is lipophilic or not, the rate of entry of the undissociated molecule is much faster than that of the dissociated ionic component. Mosquito larvae are killed much faster by sodium arsenite in solution at pH 5, where the arsenic is in the form of undissociated arsenious acid, than at pH 11 where it is contained in dissociated arsenite ions. Similar results are obtained with the insecticide DNOC which is an acid highly dissociated at neutrality and completely undissociated at a pH of 2. It is found that dosage levels which killed 100 per cent of *Ephestia* eggs at pH 2, gave no mortality at pH 5.

Nicotine as a free alkaloid is a weakly dissociating base; its watersoluble salts, such as the sulphate, are highly dissociated. Basic solutions containing nicotine alkaloid kill mosquito larvae more rapidly than acidic solutions where it is in ionic form. A more striking circumstance is that nicotine is absorbed faster into *Periplaneta* from vapours of the undissociated alkaloid than from immersion in more concentrated aqueous solutions of nicotine sulphate. On the other hand, both molecular nicotine and ionized nicotine paralyse and kill *Periplaneta* at equal rates when they are injected into the body fluid, showing the difference to have been due to the rate of cuticle permeability.

The non-dissociating and entirely apolar insecticide known as DDT penetrates cuticle with particular facility, the lethal dose necessary for contact application seldom exceeding that needed to kill by injection. It has been found that DDT has a peculiar affinity for chitin, which has the ability to adsorb it out of aqueous suspension. The observation that low concentrations of DDT kill more *Aedes* and *Chaoborus* larvae at low temperatures than at higher ones is taken as evidence of the importance of adsorption, of which such a negative temperature coefficient is characteristic. This aid to cuticular penetration is additive to the liposoluble and non-dissociating characteristics which also enhance the contact activity of DDT.

The Action of Inert Dusts. An ancient but effective means of controlling insects by contact was the application of soot, ashes or road-dust. Further refinement has led to the use of activated charcoal, the oxides and carbonates of magnesium and calcium, the siliceous minerals, and finely powdered alumina. These applications are effective at a relative humidity of 100 per cent, but their efficacy decreases with increasing dryness of the atmosphere. Their mode of action involves withdrawal of body water from the insect. In the case of a hygroscopic material such as soot or activated charcoal, this is effected by direct adsorption of water on the dust. In the case of non-hygroscopic but abrasive materials such as pyrophyllite or alumina, the action involves a laceration of the waterproof layer of the cuticle, allowing moisture to be lost to the surrounding air.

Furthermore, these inert dusts can absorb and remove the outer layers of wax or grease overlying the epicuticle. This disruption is effected by the insect's own movements where they result in rubbing the abrasive dust over the epicuticle. In response to this abrasion, the hypodermal cells migrate to the affected areas, and there secrete a soft waxy bloom. Since this resecreted wax is not closely packed or oriented, any dust still remaining on the body surface will readily mop it up also. If the insect is in dry air it may die of desiccation before sufficient new wax has been secreted to waterproof it.

This abrasion will also facilitate the cuticular entry of contact insecticides. A preliminary rubbing with powdered alumina has been found to increase the speed of action of rotenone or nicotine on the assassin-bug *Rhodnius* by approximately 50 times. As a practical application it has been discovered that dilution of rotenone with an abrasive dust, such as pyrophyllite or diatomite, increases its insecticidal effectiveness.

Penetration of Materials into the Egg. The chorion enveloping the egg consists of two layers; the thick exochorion is made of lipoprotein, while the thin many-layered endochorion is protein in nature. It would appear that insecticidal oils must penetrate through the minute pores present in the chorion of some eggs, or through the thin region at one end known as the micropyle. In eggs of *Aedes aegypti*, no penetration of petroleum oils can be detected at any point through the tough chorion.

Oil-resistant eggs such as those of aphids are attacked by means of phenols, cresols or tar oils, which soften and eventually disintegrate the chorion. On the other hand, lime-sulphur hardens the chorion, the gaseous exchange is blocked, and the embryo dries up inside it.

There is evidence however that petroleum oils may exert their lethal effect, which is more properly regarded as an inhibition of hatching, without penetrating the chorion at all. The petroleum oil that inhibits the hatching of *Grapholitha* eggs may be removed by rapid rinsing, which would indicate that it cannot have penetrated below the surface of the egg. If this highly-refined oil is removed within 5 hours of application, the eggs are not affected; but if it remains for 24 hours or more, the eggs are all killed. The effect of the oil is to depress the respiratory rate, and the embryo fails to hatch. With *Aedes* eggs, this is due to the failure of sufficient oxygen to penetrate the chorion. In the leaf-roller *Grapholitha*, it is a case of failure of toxic metabolites to be eliminated to the outside.

Wetting and Spreading of Liquids. The cuticle of most terrestrial insects is covered with a layer of fat, grease or wax, which has little affinity for water and is described as hydrophobic. When aqueous solutions make contact with such a surface they gather themselves into droplets that roll off when the insect moves. This resistance to wetting is enhanced by hairs, spines and microtrichia, which trap pockets or air and prevent the aqueous liquid from contacting the cuticle. Once contact has been established, the extent to which the liquid applied spreads depends essentially on the degree of adhesion between the liquid and the cuticle. The adhesion tension may be determined from $T_L \cos\theta$, where T_L is the surface tension of the liquid, and θ is the contact angle, *i.e.* the angle which the advancing edge of the liquid droplet makes with the cuticle under it.

Where the contact angle approximates O°, the liquid has as much attraction for the solid as for itself, and it will spread as far as the amount of molecules in the liquid droplet allows. If the contact angle approximates 180°, there is no adhesion between the liquid and the solid, and the liquid will gather itself into a spherical droplet as dictated by its own surface tension. A contact angle of 90° indicates that the liquid has half as much attraction of the solid as for itself. The liquid will remain on the solid, and will wet it, but the droplets spread only as far as a right-angled contact angle allows. Any reduction that is effected in T_L will increase the ability of the liquid to spread and reduce its contact angle with the solid. The surface tension of water, namely 76 dynes per centimetre, is too high for spreading to occur on the wax of the insect cuticle since the contact angle is 180°; but if it is reduced by the addition of soap, saponin or a sodium alkyl sulphonate, the contact angle will be sufficiently reduced to allow wetting and spreading to occur. Addition of 0.5 per cent sodium oleate to distilled water reduces the surface tension to 25 dynes per centimetre and the contact angle to 22° on house-fly or cockroach cuticle.

The higher the melting point of the epicuticular wax, the greater the contact angle made by aqueous droplets on it. The hard wax of *Tenebrio* larvae is thus highly hydrophobic, and the greasy waxes of cockroaches are considerably less hydrophobic. The larval cuticles of the house-fly maggot and of the caterpillars *Polia* and *Mamestra*, are both hydrophilic and lipophilic. The cuticle of *Psylla* nymphs allows water to spread to a low contact angle because it is encrusted with hydrophilic carbohydrate material.

Tracheal Penetration. Aqueous solutions and suspensions of insecticides, having high surface tension, are unable to enter the tracheae. But mineral oils and aqueous solutions of wetting agents, with the surface tension reduced to about one-half that of water, are able to do so. Emulsions may penetrate tracheae either as the oil fraction after breaking, or as the intact emulsion. The liquid must be able to spread into the tracheae by advancing along the tracheal walls, and its movement is decided by the physical laws that governed its spread on the cuticle.

Energy is available for tracheal penetration, as for cuticular spreading, only if the angle of contact is less than 90° and $T_L \cos\theta$ is a positive value. Thus only those liquids which show contact angles of less than 90° can penetrate, and their penetration will continue provided the properties of the tracheal wall further inside the insect do not change to increase the angle of contact. It has been found that for entry of the propodial spiracles in *Apis*, the surface tension of the spray liquid should be below 40 dynes per centimetre, and for tracheal penetration in *Tenebrio* it should not exceed 35 dynes per centimetre.

Kerosene rapidly penetrates the tracheal system, but it is often pushed out again by respiratory movements and lost by evaporation. More viscous materials, such as refined lube oil, linseed or cottonseed oil, penetrate completely and remain in the tracheal system. As the oil encounters tracheae of ever-diminishing calibre the speed of penetration decreases in direct proportion to the diameter of the trachea, until the advance becomes no longer perceptible. *Vice versa*, the further the oil penetrates into the finer branches, the more difficult its expulsion becomes.

It is considered that the entry of oils is usually opposed by the air pressure in the trachea, backed up by the hydrostatic pressure of the fluid in the tracheoles. Indeed oils which had stopped penetrating have been observed suddenly to advance, as if the trachea had broken and the pressure had been relieved. However, the reverse may occur and penetration be aided by respiratory currents. The penetration of soap solutions into the tracheae of sphingid larvae fails to occur in specimens which have just previously been killed with cyanide. Observations on the penetration of oils into mosquito larvae suggested that tracheal penetration might be aided by reduced pressure in the sections yet unfilled. On the other hand, mosquito larvae may sometimes be protected against entry of oils by the outer tracheae collapsing inwards; the toxic aromatic oils are most liable to cause this effect and thus may kill more slowly than the less toxic aliphatic oils.

Once well within the tracheal system, the oils may diffuse into the haemolymph through the walls of the tracheae or tracheoles. The fate of solutions of pyrethrins in oil dyed with gentian violet, applied to *Periplaneta*, suggests that diffusion takes place through the walls of the tracheae into the haemolymph and thence to the ganglia. In *Culex* larvae, similar dyed solutions penetrate directly to the ganglia by way of the tracheoles which ramify within them.

It is surprising to find that insecticidal dusts can enter the tracheae of insects in certain instances. There is evidence that sodium arsenite particles may enter the outer sections of the tracheae of flying locusts. Derris dusts have been observed to enter the tracheae of *Melophagus* and to be carried in the postero-anteral air circulation, a mode of entry more effective than cuticular penetration.

Susceptibility to Contact Poisons. The most important single factor determining the susceptibility of a particular insect species to an insecticide is the extent to which the poison is prevented from reaching its site of action in the vital organs and tissues. The habitat and habits of the insect may set up the first barrier.

A mobile species such as the cockroach is hard to control, for it will escape from the area treated. On the other hand aquatic larvae in a pool are not only defenceless, but can no more escape the dissolved insecticide than insects exposed to a vapour in a fumigation chamber. However, mosquito larvae may be repelled by aromatic oils and fail to rise to contact the toxic surface films.

Active insects that live an exposed aerial life may show less susceptibility to contact insecticides because of a thorough cuticular sclerotization. But the tracheae may be vulnerable points in some species, especially those with well-ventilated respiratory systems such as house-flies. The gross thickness of the cuticle is an important factor in determining resistance to contact insecticides. The decreasing susceptibility of larvae and nymphs as they proceed into the older stages may be mainly ascribed to this factor. It has been found that the susceptibility to contact poisoning of caterpillars with many pore-canals is much greater than those with few pore-canals. Yet mosquito larvae are highly susceptible to contact poisoning despite an entire absence of pore-canals in the cuticle.

Many species of bees, wasps and flies are susceptible to contact insecticides when in the adult stage, in spite of possessing thick hardened cuticles. This is due to the abundance of cuticular sense organs, which serve as points of exposure outside the main defences of the insect. Conversely many larvae, and particularly maggots, are comparatively resistant to contact poisoning despite their unsclerotized cuticles, due to the scarcity or lack of skin sense organs.

The flies *Musca* and *Glossina*, and to a lesser extent the bee *Apis*, are highly susceptible to residual deposits of DDT, since they bear tarsal chemoreceptors with cuticular sensilla on the plantar surface of their tarsi. On the other hand, the hemipteran *Oncopeltus*, which lacks tarsal chemoreceptors, is comparatively immune to deposits of DDT over surfaces on which it walks. When a number of species were compared, *Musca* and *Aedes* proved to be exceptionally susceptible, the threshold lethal deposits of DDT being 1000 times lower than those for *Pediculus*. The body-louse was followed in susceptibility by *Cimex*, *Blatta*, *Tribolium* and *Sitophilus*; finally *Rhodnius*, *Ornithodorus*, and *Tenebrio* larvae proved to be the most resistant.

In conclusion, certain generalizations may be made about the relationship of cuticular structure with susceptibility to contact insecticides. On the one hand resistance is conferred by cuticular thickness, the presence of hairs and spines, cuticular sclerotization, and a repellent epicuticle. On the other hand, susceptibility is increased by the presence of hair sensilla or microtrichia, an abundance of pore canals, and a receptive epicuticle.

Susceptibility to Fumigants. The susceptibility of insects to fumigants is most closely correlated with their ability to withstand desiccation. The more effective the control of tracheal ventilation, the greater their ability to exclude the poisonous vapour. It is probable that the degree of tracheal ventilation of an insect, and consequently its susceptibility to a particular fumigant, depends on whether it is stimulated or stupefied by the initial exposure to the vapour. Different compounds evoke different responses, carbon disulphide being typically stimulant and hydrogen cyanide typically stupefying or narcotic.

Carbon dioxide in low concentration causes the spiracles to open, and so does a moderate vacuum due to the concomitant shortage of oxygen. Addition of small amounts of carbon dioxide enhances the toxicity of fumigants to *Tribolium* or *Sitophilus*, but not to some other species. The effectiveness of hydrogen cyanide against *Tribolium* rises steadily as the air pressure drops in the fumigation chamber.

In holometabolous insects it is generally found that the larva is the most susceptible stage to the toxic effect of fumigants. This occurs despite the fact that in most insects, such as *Tribolium*, the larval respiratory rate is lower than in the adult. The pupa with its low respiratory rate is the least susceptible stage, except in certain cases where its resistance is exceeded by the egg.

The Effect of Temperature upon Mortality. Consideration of this topic is clarified by bearing in mind that two factors are involved in the toxic action of an insecticide. The first factor represents the speed and degree to which the poison is taken up by the insect, and acts upon its tissues; and the second involves the speed and degree to which the poison is eliminated by excretion, breakdown or detoxification. With decisive doses of quick-acting poisons, it may therefore be expected that mortality is quicker and greater at a high temperature than at a low temperature of application. With marginal doses of slow-acting poisons, the expectation is that recovery will be less at the low temperature than at the high, and that the mortality will be greater though it develops at a slower rate. The temperature of application decides the rate of uptake or the immediate toxicity, and the post-treatment temperature governs the rate of recovery.

REFERENCES

Beament, J. W. L. (1945). *J. Exp. Biol.* 21 : 115-131. (Chemistry and physics of cuticle lipoids). **Busvine, J. R.** (1938). *Ann. Appl. Biol.* 25 : 605-630. (Metabolic rate and fumigant toxicity). **Hoskins, W. M.** (1940). *Hilgardia* 13 : 307-386. (Insect physiology and toxicology : review). **Klinger, H.** (1936). *Arb. physiol. angew. Ent.* 3 : 49-69 and 115-151. (Cuticle structure and insecticide entry). **Pal, R.** (1950). *Bull. Ent. Res.*, 41 : 121-139. (Hydrophilism in certain larval cuticles). **Richards, A. G.** (1950). The Integument of Arthropods. University of Minnesota Press, St. Paul. pp. 400 **Roy, D. N., Ghosh, S. M.** and **Chopra, R. N.** (1943). *Ann. Appl. Biol.*, 30 : 42-47. (Entry of pyrethrins into roach). **Roy, D. N.** and **Ghosh, S. M.** (1944). *Bull. Ent. Res.*, 35 : 161. (Mechanism of action of a contact insecticide). **Sarkaria, D. S.** and **Patton, R. L.** (1949). *Trans. Amer. Ent. Soc.* 75 : 71-82. (Pulvilli as entrances for DDT). **Shafer, G. D.** (1915). *Michigan Agr. Exp. Sta. Tech. Bull.*, 21 :

67 (Mode of action of roach powders). **Sun, Y. P.** (1947). *Minnesota Agr. Exp. Sta. Tech. Bull.* 177 104 (Action of fumigants on insects). **Wigglesworth, V. B.** (1945). *J. Exp. Biol.*, 21 : 97-114. (Effect of oils and dusts on cuticle).

MODE OF ACTION OF INSECTICIDES

Physical Poisons. These materials characteristically exert a physical rather than a biochemical effect. However, these inert oils and dusts are logically reinforced by insecticidal compounds when they are applied.

(*a*). Heavy mineral oils and tar oils. Highly refined mineral oils may exert a purely asphyxiant effect and kill insects such as scales by excluding the air. Dormant oils smother insect eggs for a sufficient length of time to asphyxiate them, but their action is hastened by the toxic impurities which are almost always present in such materials.

(*b*) Inert dusts. These materials withdraw the body moisture from the insect by two types of action. Abrasive dusts, *e.g.* aluminium oxide, effect the water loss by lacerating the epicuticle. Water-adsorbent materials, *e.g.* charcoal, remove water as a consequence of their hygroscopic properties. Certain dusts may show a combination of both characteristics.

Protoplasmic Poisons. The action of these poisons appears to be associated primarily with the precipitation of protein.

(*a*) Heavy metals, *e.g.* compounds containing Hg and Cu.

(*b*) Alkaloidal reagents, *e.g.* nitrophenols and nitrocresols.

(*c*) Acids. Acid radicals are toxic, but only the comparatively apolar fatty acids are contact insecticides. Hydrochloric acid may be produced in the tissues from the fumigant chloropicrin, hydrobromic acid from methyl bromide, and sulphurous acid from sulphur dioxide.

(*d*) Formaldehyde and ethylene oxide.

(*e*) Fluorides, fluosilicate and fluoaluminates, borates, arsenates, and arsenites. These inorganic stomach poisons appear primarily to destroy the cellular protoplasm of mid-gut epithelium.

Respiratory Poisons. These include HCN, H_2S and CO, which are respiratory poisons not only in the sense of being fumigants, but also in that they block cellular respiration. They combine with cytochrome oxidase and other compounds which contain Fe, and thus inhibit their catalytic action.

Nerve Poisons. The action of these poisons is associated primarily with their solubility in the tissue lipoids.

(*a*) Chlorinated hydrocarbons : CCl_3, ethylene dichloride, PDB, DDT, lindaae, etc.

(*b*) Aromatic and olefinic hydrocarbons : napthalene, kerosene, gasoline, etc.

(*c*) Botanical insecticides : pyrethrins, nicotine, sabadilla, hellebore, etc.

(*d*) Organic phosphates : HETP, TEPP, parathion, etc.

(*e*) Miscellaneous : CS_2, valone, piperine, aniline, etc.

Poisons of a more general nature. (*a*) Chlordane, toxaphene, aldrin, dieldrin; these chlorinated terpenes do not induce neurotoxic symptoms until a latent period has passed. (*b*) The Lethanes, Thanite; these organic thiocyanates have an immediate depressant effect. (*c*) Rotenone, ryania, phenothiazine; these are muscular depressants, of which the first two may induce slight neurotoxic symptoms.

Toxicology of the Nervous System. Nerve poisons characteristically induce the appearance of symptoms in 4 stages; excitation, convulsions, paralysis and death. The narcotic fumigants show typically only the 3 stages of excitation, paralysis and death, the convulsions or spasms being omitted. With the irritant vapours the stage of paralysis also is lacking. The paralytic stage induced by the contact nerve poisons may be either (*i*) a flaccid paralysis where the muscles are relaxed, as with rotenone poisoning, which is analogous to narcosis, or (*ii*) a tetanic paralysis where the muscles are immobilized in a contracted position, as in DDT poisoning.

Evidence is accumulating to show that the central nervous system of the insect is physiologically different from that of the mammal. Instead it closely resembles the parasympathetic autonomic system of vertebrates, whose synaptic transmission depends on mediation by acetylcholine which must then be rapidly removed by cholinesterase.

The concentration of the enzyme cholinesterase in the central nervous system of the insect is extremely high. It is poisoned by eserine, but is unaffected by nicotine or curare. It is markedly inhibited by the organic phosphate insecticides, as also by sodium fluoride and the derivatives of phenothiazine. Anticholinesterase activity is not shown by DDT *in vitro*, but there is evidence that it may have this effect in the living insect.

Hydrocarbon Oils. It has been a matter of controversy whether larvicidal oils kill by asphyxiating or by poisoning the insects. In truth both effects make their contribution to the death of the mosquito larvae. When kerosene is applied to the surface of water infested with *Culex*, the larvae become lethargic and sink to the bottom in 10 to 20 minutes. Since a similar effect is produced by medicinal paraffins and highly-refined lube oil, neither of which are poisonous, it is reasonable to conclude that this immobilization of the insects is due to asphyxiation or anoxia. While these larvae are still under narcosis, and before death has set in, their nerve cells are found to show a characteristic clumping of the nuclear chromatin. The condensed nucleus is surrounded by a halo of clear fluid, and the Nissl granules take on a reticulate arrangement. Such a condition has been induced by anoxia alone, by depriving larvae of *Sciara*, *Chironomus* and *Drosophila* of oxygen. It has also been established that chromatin clumping is accompanied by increased nuclear acidity, due possibly to the accumulation of carbon dioxide.

In practice most insecticidal oils, such as kerosene and fuel oil, contain olefines and impurities which add a truly toxic effect. For unrefined kerosene or fuel oil induce mortality in *Culex* larvae even if the water under the oil is continuously oxygenated. In this case death is preceded by convulsions and twitches; and pre-mortem examination of the central nervous system reveals a degeneration of the fibres and separation of the inner fibre tracts from the outer cell layer, without the chromatin clumping characteristic of asphyxiation. Nevertheless, kerosene and fuel oil kill *Aedes* larvae much quicker if the underlying water is oxygen-free.

The hydrocarbon oils produce a narcotic effect by reason of their solubility in tissue lipoids, Xylene injected into the tracheae of *Culex* is preferentially taken up by the nerve cord, where the fibre sheaths are rich in phosphatides and cholesterol.

As a consequence probably of the reduced gaseous exchange of the fibres themselves, the action currents of the nerve are depressed, and the dehydrogenase activity of this tissue is decreased. More lengthy exposure to lipoid solvents, xylene or hydrocarbon oils results in cytolysis and dissolution of fatty tissue. The first stage in this process is the destruction of the ultra-structure of the lipoid sheath of the nerve fibre. An increase in the phenoloxidase and deyhdrogenase activity also occurs, and may result in the accumulation of toxic quinones.

The narcotic effect of hydrocarbon oils is particularly noticeable in an active subject such as the adult house-fly. If it is sprayed with a large quantity of refined kerosene, after a struggle it is "knocked down" by a progressive paralysis which passes from the hind legs forward. After a period of 5 to 15 hours under narcosis, probably resulting from anoxia of the nervous tissue, the flies completely recover. The narcosis of knockdown, like surgical anaesthesia, is thus completely reversible. However, if the spray consists of an unrefined kerosene which contains the usual toxic impurities, a certain proportion of the house-flies will die under narcosis.

Pyrethrins. The pyrethrins are poisonous to all animals which possess a nervous system. Insects are much more susceptible than the higher animals, because of their small size and thin cuticular defences, and because their coldbloodedness prevents enzymic detoxification of the pyrethrin esters from proceeding fast enough.

Pharmacologically the pyrethrins very much resemble veratrin in their effect on the nervous system. When applied to the insect nerve cord, they first excite a massive discharge of many types of nerve impulse, followed by repeated synchronized discharge of continuous trains; eventually the response to stimulation falls and finally fails altogether. It is considered that the pyrethrins enter the nerve sheath as a consequence of their lipoid affinity, and that they spread through the body by way of the nerves and more particularly their outer sheaths.

When nerves are poisoned by pyrethrins, the first effect is a decay of the ultrastructure of the protein micelles in the axon. Later the ultrastructure of the lipoid of the myelin sheath degenerates, and vacuoles appear in the neurone cells, while the nuclear chromatin dissolves. Finally when the nerve ultra-structure has entirely disintegrated and the insect is in the early stages of paralysis, the typical pyrethrin lesions of the ventral nerve cord and brain appear.

These lesions appear mainly in the segmental ganglia in the brain and to a less degree in the connectives. Owing to the dissolving of the lipoid sheaths, many nerve fibres show lysis, and clear empty spaces appear in the fibre tracts. In the nerve cells, the cytoplasm becomes vacuolated, the chromatin clumped, and the tigroid of the Nissl granules dissolves away. A general indication of histological damage is afforded by the cells of the nerve cord staining violet with toluidine blue.

Synergists. The addition of certain compounds to pyrethrins so increases the toxicity of sublethal concentrations that the mixture is insecticidal where each material separately was without effect, the toxicity of the ensemble being greater than the sum of its parts. It has been found that if the adjuvant is sprayed before the pyrethrum, the synergistic effect is just as strong; sesamin applied 24 hours before the pyrethrins makes house-flies just as susceptible to knockdown as when the materials were sprayed together. But there is no synergism if the adjuvant is applied after the pyrethrum; although there is some effect if it is applied within an interval of less than 30 seconds.

If the insect is pretreated by topical application of the adjuvant to one part of the body, the synergistic effect on knockdown is shown even when the pyrethrins are later applied to another part of the insect's body.

This pretreatment effect has been shown by the piperonyl synergists, sesamin, IN930 and propyl isome, and is evidently a general phenomenon. Evidently the synergist effects an alteration in the insect so as to predispose it to pyrethrum poisoning and render it incapable of recovering from the narcotic effect of the pyrethrins. It has been established that DHS activator causes lysis of the non-fibrous cellular components of the house-fly's brain, IN930 causes chromatolysis of the cell nuclei, and that sesamin induces vacuolation of the large nerve cells. Incidentally, both synergists effect changes in muscle also, to accentuate the appearance of the nodes and Krause's membrane. It has therefore been suggested that the synergists operate by so damaging the nerve cells that they can no longer repair the injury which the pyrethrins inflict on the nerve fibres. It has also been found that piperonyl butoxide partially inhibits the hydrolytic detoxification of pyrethrins by lipase taken from the house-fly; but the lipase level in the living cockroach is not reduced by treatment with this synergist.

Nicotine. This alkaloidal poison has a particular action on the synapses between nerve fibres which are located in the ganglia. Nicotine has been found to block synaptic transmission in the isolated nerve preparation of the crayfish. Application of nicotine to a thoracic ganglion of *Periplaneta* results in violent tremors of the leg, which disappear when the limb is severed from its connection with the ganglion. Injection of nicotine into the isolated leg does not induce tremors, since this insecticide has no effect on nerve fibres, but on the ganglion only. When nicotine is applied to the brain of a roach whose haemolymph has been immobilized by cauterization of the heart, tremors develop through the entire body, which disappear if the brain is removed by decapitation. Thus the effects of nicotine, which acts on the telephone exchange, as it were, can mask the symptoms of DDT, which act somewhere along the lines or at the peripheral ends. However, nicotine does exert a transitory effect on the nerve axon of arthropods, causing repetitive discharge like DDT or the pyrethrins do. Low doses increase the action currents in the isolated insect nerve cord, and really high concentrations are required to block the currents, an effect which is reversible on removal of the poison.

Rotenone. Insects poisoned by rotenone, such as caterpillars and cockroaches, show a drop of some 50 per cent in the rate of oxygen consumption. This may be a result of depressed mechanical respiration, since derris dusts have been found to stop the movements of the tracheal opercula of *Bombyx* and *Heliothis* larvae. The depression of air intake in turn may decrease the rate of heartbeat. When derris is applied as a dust to *Bombyx*, there is first a latent period of 40 minutes in which no outward sign of poisoning is evident, but towards the end of the period the heart-beat drops slightly from the normal rate of 70 per minute. Then there follows a 10-minute period when the caterpillars are highly active and the heart rate shows irregularity, short bursts of stimulation alternating with occasional stoppage; and a further 10-minute stage of ataxia during which the heart rate falls to 20 beats per minute, a level which is maintained during the subsequent long paralytic period in which the larvae lie quiescent on their sides.

Preparations containing rotenone have been employed as contact insecticides as well as stomach poisons, but they have no fumigant effect. The contact action is well-marked in soft-bodied insects such as aphids, mosquito larvae and the sheep-ked *Melophagus*. It is considered probable that derris dust is dissolved in the body exudations and carried across the intersegmental membranes. However, rotenone appears to encounter more difficulty in penetrating the insect cuticle than nicotine or the pyrethrins. It is unable to poison by contact those species of sclerotized beetles which are highly susceptible to applications of pyrethrins.

DDT. This insecticide is a nerve poison, and when applied in relatively high concentrations it may act directly on the motor nerves. In low concentrations it exerts no effect on the motor nerves, but it does set off a series of impulse trains in the sensory nerves. If the sensory nerves running from the campaniform sensilla on the trochanters of the cockroach are poisoned with DDT, the motor fibres of the crural nerve will be affected and they will cause tremors to appear in the leg. In order to translate the effect on the sensory fibres into symptoms in muscular movement excited by the motor fibres, an intact reflex arc is necessary. Blocking the ganglion, or physically removing it, completely inhibits the production of symptoms by low doses of DDT, but high concentrations still exert a direct effect. However, the fast twitches, characteristic of the poisoned cockroach, appear only if the reflex arc is intact. This has led to a minority view that DDT has its effect by first facilitating and then blocking the synapses in the ganglia.

The effect of DDT on isolated arthropod nerve is to multiply the impulses passing through the treated region, each single impulse becoming converted into a train or volley of impulses, repeating at the rate of 250 to 500 per second, each train lasting from one-tenth to one second. This repetitive discharge may even occur in the absence of any detectable stimulus. It may therefore be concluded that DDT in low concentration acts on the sensory nerves, whose patternless impulse trains act on the motor nerves via the reflex arc, to produce the characteristic twitches and spasms of the muscles. This instability of peripheral nerve may be related to the sheath becoming impermeable to Ca ions as a consequence of the affinity of DDT for the cholesterol in it. The highly sensitive ultrastructure of insect nerve is not impaired by DDT, nor has this insecticide produced any histopathological change in the nerve cord of the insect.

The characteristic symptoms of DDT poisoning are prevented when the nervous activity of the insect has been previously depressed by application of atropine, or barbiturates such as phenobarbital or nembutal. Moreover, these drugs have the power to abolish the symptoms of DDT poisoning even after they have already started to show in the insect. It may be observed that the symptoms of DDT in the cockroach closely resemble those caused by eserine (physostigmine), a drug whose pharmacological action is derived from its inhibition of cholinesterase in nerve. In fact DDT, like eserine, greatly increases the amount of free acetylcholine in nerve; and the effect of DDT is additive to that of eserine.

As the insect enters the condition of tremors or DDT jitters, its rate of respiration rises steadily to a peak that is two to five times the normal level. DDT-poisoned insects, such as *Popillia* beetles, were found to lose weight rapidly, all their body glycogen being consumed and their fat being partially oxidized by the time of their death.

They completely lost their resistance to desiccation, and the adult beetles were found to die of water-loss sooner than the larvae which live in moist soil. Poisoned *Phormia* adults also lost water faster than the normal, and their calorific metabolites disappeared even more rapidly; the respiratory quotient rises from 0.90 up to 0.96, indicating that carbohydrates were being consumed. Nevertheless they died before these reserves decreased to starvation level. Roaches poisoned by DDT were found to lose nearly all their glycogen and much of their fat, while they accumulated the ketone bodies, acetone and acetoacetic acid. However, there was no evidence that acidosis due to ketone bodies developed in the DDT-poisoned *Phormia*. It has therefore been suggested that DDT-poisoned insects die from the exhaustion that results from the muscular tetany during the final stage of paralysis, and from intoxication by the metabolites which accumulate in the muscles in their continuous state of contraction. The effect of depressants such as atropine or barbiturates would be to relieve the nervous and muscular symptoms long enough for the DDT to be removed from the system by detoxification or excretion; these depressants cannot protect against the action of pyrethrins, which are neurotoxins that also destroy nerve ultrastructure. If caterpillars can be induced to feed in spite of their symptoms, they will recover from DDT poisoning. The milkweed bug is able to metabolize almost all the DDT injected at moderate sublethal dosage levels. DDT-resistant strains of *Musca* have been found capable of rapidly metabolizing the insecticide to its dichloroethylene derivative, which the normally susceptible strains are unable to do.

Other Chlorinated Hydrocarbons. Lindane or gamma-BHC is an extremely potent and fast nerve insecticide, the lethal dose being rapidly taken into the insect in both solid and vapour form. The characteristic symptoms of excitation, convulsions and paralysis follow in rapid succession. The respiratory rate rises almost immediately to a high peak, and thereafter gradually falls. The amount of free acetylcholine in nerve increases. Repetitive discharge trains appear in the afferent nerves upon the slightest stimulus.

On the other hand chlordane appears to be initially a depressant, since muscle tonus and movements become weak. But later while in this state the insect shows exaggerated responses to stimuli. This situation is reflected in the respiratory picture, which does not rise until a latent period of several hours has passed. Similarly this material depresses the action potentials of nerve, but after some hours repetitive discharge trains make their appearance. Similar increases of oxygen consumption and action potentials have been obtained after a passive latent period in the case of toxaphene, aldrin and dieldrin.

Organic Phosphates. There is good reason to consider that the organic phosphates are lethal to insects because they are anticholinesterases, although there is a residuum of controversy on this point. Parathion inhibits the cholinesterase of the cockroach both *in vivo* and *in vitro*, and symptoms of parathion poisoning in the honey-bee are directly correlated to the increasing inhibition of cholinesterase in its brain. In experiments on bees and houseflies, the insecticidal activity of the other organic phosphates related to parathion have been found to be directly correlated to their ability to poison the cholinesterase of the test insects. But the anticholinesterase effect of parathion *in vitro* may be due to impurities, and *in vivo* to derivatives of which it is a precursor.

REFERENCES

Chadwick, L. E. (1952). *Amer. J. Med. Hyg.*, 1 : 404-411. (Mechamism of DDT intoxication). **Chamberlain, R. W.** (1950). *Amer. J. Hyg.*, 52 : 153-183. (Piperonyl butoxide and lipase inhibition). **Dresden, D.** (1949). Physiological investigations into the action of DDT. Drukkerjj Van der Wiel, Arnhem, Holland. pp. 114 **Hartzell, A.** (1945). Contrib. Boyce Thompson Inst., 13 : 443-435. (Histopathology of pyrethrum). **Knipling, E. F.** (1951). Mosquito News 11 : 197-210. (Oxygen-free water hastens larvicidal action of oils). **Lalonde, D. I. V.** and **Brown, A. W. A.** (1954). *Can. J. Zool.* 31, (in press). (Effect of insecticides on action potentials of nerve). **Metcalf, R. L.** and **March, R. B.** (1949). *J. Econ. Ent.*, 42 : 721-728. (Organic phosphates as anticholinesterases for insects). **Richards, A. G.** and **Cutkomp. L. K.** (1945). *J. New York Ent. Scc.*, 53 : 313-355. (Effect of insecticides on nerve ultrastructure). **Richards, A. G.** and **Cutkomp, L. K.** (1945). *J. Cell. Comp. Physiol.* 26 : 57-61. (Cholinesterase of insect nerve). **Roeder, K. D.** (1939). *Biol. Bull.* 76 : 183-188 and J. *Cell. Comp. Physiol.* 14 : 1-9. (Action of drugs on insect nervous system). **Shafer, G. D.** (1911). *Michigan Agr. Exp. Sta. Bull.*, 11 : 56. (Studies on action of insecticides). **Tischler, N.** (1935). *J. Econ. Ent.*, 28 : 215-220. (Physiological action of derris). **Trappmann, W.** (1938). *Zeit. Pflanzenkrankh. und Pflanzenschutz.* 48 : 514-538. (Classification of insecticides). **Wigglesworth, V. B.** (1950). The Principles of Insect Physiology. 4th Ed. Methuen, London. pp. 544

METHODS OF TESTING INSECTICIDES

There are numerous methods of testing insecticides, each adapted to the particular type of insect and the nature of its susceptibility to attack. There are laboratory methods designed for screening new compounds for insecticidal activity and for assessment of the biological activity of products and formulations, and there are field methods designed to evaluate the relative effectiveness of the most promising commercial insecticides for the particular insect problem. Laboratory techniques of testing have been well covered by publications of the American Association for the Advancement of Science (1943) and the Chemical-Biological Coordination Centre of the U.S. National Research Council (1954).

Laboratory Methods. These are designed to evaluate the contact, fumigant and stomach toxicities of the chemical. Contact toxicity is the most important quality for most modern insecticides, and it is assessed by spraying solutions of the material and establishing their biological effectiveness. Three types of application are of interest : space sprays, settling sprays and residual deposits.

For space sprays the standard method of testing has involved the Peet-Grady chamber, a 6-foot cube of capacity 216 cubic feet into which 12.5 c.c. of the spray is emitted from a special fine atomizer*. A standard strain of house-flies (*Musca domestica* L.) is exposed in a central swinging cage. The procedure has recently been simplified by the Pest Infestation Laboratory, Slough, which substitutes a smaller chamber of 1 cubic metre capacity and obtains a very fine spray from an artist's spray brush (Aerograph Spray Co., Sydenham).

Settling sprays are tested by means of a spray tower at the top of which the spray solution is atomized. The droplets fall freely and evenly to the base of the tower, where the insects are exposed in petri-dishes or crystallizing dishes. Precautions are taken to insulate the tower against temperature differences that may cause air currents, and against static electricity; the Potter tower is designed with these safeguards. In

* DeVilbiss atomizer, No. 5004 ; The Devilbiss Company, Toledo 1, Ohio, U.S.A.

another type designed by Hoskins and Caldwell, the spray is emitted along a horizontal cylinder which leads into a vertical one. The glass cylinders used in petrol pumps, measuring 29 inches high and 12 inches in diameter, make satisfactory spray towers. The spray nozzle is of a precision air atomizer type; and the air is supplied, usually at a pressure of 15 p.s.i., by a supply of compressed air provided with a reducing valve and pressure gauge. A measured amount of the insecticide solution is dispensed from a burette into the air-stream at the nozzle.

The chemical compounds being tested are dissolved in a refined kerosene, such as *Deo-Base*, or in benzene thickened with kerosene or medicinal paraffin. Tests are made with a given amount of solvent, such as 4 c.c., containing different concentrations of the compound under test. Starting with 8 per cent, for example, the tests proceed to lower concentrations; control sprays are made with the solvent alone. The dosage actually falling on the petri-dish may be estimated by dyeing the spray with 0.5 per cent Williams red dye, and taking up the deposit in 5 c.c. of a xylene-kerosene mixture; the intensity of colour of this 5-c.c. sample is compared with the colour intensity of a standard 1 in 100 dilution of the dyed spray. The density of deposit is expressed as cubic millimetres of spray per square centimetre. Thus for each solution tested, the dosage of the compound may be calculated in micrograms per square centimetre.

Insects suitable for spray-tower technique include the cockroach *Blattella germanica* and the louse *Pediculus corporis*. After exposure to the spray the insects are quickly removed from the dishes to fresh jars, and are observed for mortality 24 hours later. The technique of topical application has recently been developed, in which a single droplet of solution may be applied to an insect such as the house-fly by a platinum loop or micropipette.

Settling-tower methods may be employed for the assessment of dusts. An amount such as 3 grams is placed in a dust cup, and is blown by a sudden short blast of air through a tube which enters the top of the tower. The concentrations of the insecticidal compound in the dust may be varied by making individual impregnated dusts, or by progressively diluting a concentrated dust with an inert diluent. The density of dust deposit may be estimated by weighing the material recovered on tared large microscope coverslips.

Residual deposits may be tested by spraying glass plates, approximately 6 inches square, with a paint sprayer or by exposure in the spray tower. Alternatively, glass containers may be dipped in an acetone solution of the compound, and then inverted to allow excess solution to drain off before drying. Other surfaces such as baked mud and unpainted wood may be employed. The test insects, usually house-flies or mosquitoes, are confined to the surface deposits in a cage. Lousicides may be tested by dipping squares of cloth in an acetone solution of the compound.

The testing of fumigants on a laboratory scale is best performed by the Lindgren method. Here the fumigant is volatilized in a glass bottle of not less than 1-litre and preferably of 20-litre capacity. The bottle is first partially evacuated, and then a given weight or volume of the liquid fumigant is drawn in by opening the tap of a small burette attached to the stopper. After volatilization is complete, the test insects, which are usually the flour beetle *Tribolium confusum* or one of the two species of *Sitophilus* weevils, are introduced in small cylindrical screened cages. They are left in the vapour for 5 hours at 25° C., at the end of which time they are removed to jars of flour or grain and left for 3 days before the percentage mortality is assessed. The vapour concentrations are adjusted and expressed in milligrams per litre of air. In the case of aquatic insects, such as mosquito larvae, the usual method is to add to a litre of water a small amount of an acetone solution of the insecticidal compound. The concentrations are adjusted and expressed as parts of the insecticide per million parts of water (p.p.m.).

The testing for stomach toxicity is difficult to carry out and is of academic interest only; in the case of caterpillars it may be accomplished by making leaf sandwiches of the insecticide. However, the general toxicity of the compound, including stomach, contact and fumigant actions, may be simply carried out in the case of stored products insects like *Tribolium*, *Sitophilus* and *Ephestia*. The compound is intimately mixed with the flour or grain in which the insect lives and feeds, and the concentrations tested are progressively reduced.

As a measure of the essential toxicity of a compound, methods of injecting it directly into the insect have been developed, using a 1/4-c.c. tuberculin syringe fitted with a fine needle* (*e.g.* B. and D. No. 29). The plunger of the syringe is actuated by a micrometer screw-head**. The large American roach *Periplaneta* is a suitable test insect for injection. It is thus possible to obtain the true lethal dose in micrograms per gram body-weight.

In all the above types of tests, the percentage mortalities obtained at the various dosages and concentrations are employed to calculate the median lethal dosage or LD50 of the compound for the test insect. This is done by a graphical method, by plotting the logarithm of the dosage (as abscissa) against the percentage mortality (as ordinate) on a probability scale. If logarithm-probability paper is not used, the same result may be obtained on semi-logarithm paper by plotting the probits (probability units) of the percentage mortality as the ordinate on an arithmetic scale***. Tables of the probits corresponding to the percentage values are published in a paper by Bliss, or in text-books on insecticides. The dosage-mortality figures when so plotted will fall on a straight line; from this line it is possible to read the dosage at which 50 per cent mortality may be expected, which is the m.l.d. or LD50. Since the concentration required for more complete mortality, say 90 per cent, is frequently of greater interest, it is equally possible to read off the LD90 from the graph. Thus compounds may be compared with one another on the basis of their LD50 or LD90 values. It will be found that the dosage-mortality lines for DDT and most chlorinated hydrocarbons are shallow and gradual, while those for fumigants are steep.

Field Methods. In testing insecticides in the field, methods must be adjusted to the particular insect problem, but certain principles illustrated in the laboratory methods may well be kept in mind. In assessing area sprays or pools or tracts of land, it is the lbs. of active insecticide per acre which is the important feature, and to which the quantities should be adjusted. In the spraying of buildings and structural surfaces, it is the milligrams of insecticide per square foot which is the important

* Gauge No. 29, Becton and Dickinson, retailed by Ingram and Bell.

** Micrometer No. 263; The L. S. Starrett Company, Athol, Mass., U.S.A.

*** Semi-logarithm probit paper, copyright Winthrop-Stearns, Inc., 33 Riverside Ave., Rensselaer, N.Y., U.S.A.

consideration. The total amount of spray solvent should be kept constant, and only the concentration of insecticide in it is varied by adjustment. This is based on the view that the amount of active insecticide deposited, regardless of the quantity of solvent or carrier in which it is contained, is more important than the degree of concentration of the insecticide in the solvent. This is borne out by data from sprays but not from dusts. In the case of dusts, as applied for example for louse control in clothing or flea control along rat runways, it is the concentration of insecticide in the dusts which should be adjusted and tested, rather than the dosage per unit area.

The evaluation of the resultant percentage mortality in the field is seldom possible. Instead, it is deduced from the decrease in population density, and from this the percentage control is derived. Population assessments vary with the species concerned. For example, with adult *Anopheles* mosquitoes, the index is taken from the numbers found in houses. With *Aedes* mosquitoes, and certain species of *Anopheles*, assessments may be made from captures taken in standard light traps. The New Jersey light trap consists of a 25-watt bulb above an electric fan*. Alternatively, a landing rate of *Aedes* mosquitoes may be taken from the number of mosquito landings on the front part of the trousers, during one minute; or a biting rate may be taken from wrist to elbow in one minute. Similar methods may be applied to assess populations of adult blackflies. Landing rates of tabanid flies may be taken by exposing 3-foot cubical tents of black cloth and counting the landings of these insects on them during 3 minutes. Populations of adult house-flies may be obtained by exposing the Scudder grill, a small portable framework of wooden slats, on which the landing rates may be counted.

Populations of adult *Anopheles* in woodland areas may be sampled by erecting pyramidal shelters of wood, 6 feet high, with the sides measuring 4 feet ascending to 1 foot. Nail kegs or red-painted boxes have been employed. Model huts of sectional construction with fibre board walls measuring 7 by 4 by 7 feet high, with a 4-foot low wall, have been used in Africa. These huts may be baited with animals or men, and are provided with window traps of calico cloth. Criteria of mortality after spraying may be taken with the aid of sheets exposed on the floor to assess those mosquitoes dying in the hut, and window traps to assess those which made their exit.

Larval populations of mosquitoes in pools may be assessed by dipping water samples with a long-handled dipper of standard dimensions, a diameter of 5 inches being most usual. The dipper may have a screen bottom, the catch being transferred to a clean white pan of water. A more complicated method consists of a plastic cylindrical sleeve sampler into which the water is drawn by means of a small hand-operated suction pump.

Sampling Methods. Qualitative methods of ascertaining the extent of coverage of a spray on the ground, whether emitted from machines or aircraft, include the simple expedient of stretching a roll of paper (e.g. adding-machine tape, toilet paper) across the pattern of deposit. This paper may be treated with a chemical (e.g. haematoxylin or tannic acid), on which a characterizer compound added to the spray liquid (e.g. ferric chloride) develops a coloured spot to mark each droplet. Or a surface may be treated with polyvinyl acetate and p-nitrobenzeneazo-B-naphthylamine in acetone; droplets falling on this golden coating appear as black

* Available from The Concession Supply Co., Toledo, Ohio, U.S.A.

circles*. Quantitative determination of the density of deposit on the ground may be effected by using dyed spray (*e.g.* Williams red, *i.e.* Dupont oil red, Calco blue Na, Calco oil red N-1700 or waxoline red for oils; Rhodamine B, croceine scarlet, potassium permanganate or coloured inks for water-base sprays) and placing glass plates or petri-dishes at appropriate intervals across the track. The deposit on the plate or in the dish is colorimetrically assessed by washing it off with an aliquot of solvent and comparing it for colour density with the original spray liquid diluted by a known amount. A deposit of 7 cu. mm. of spray in a 75 cm petri-dish is equivalent to 1 gallon per acre. This deposit, washed ofl with 5 c.c. of solvent and compared with a 1 :1000 dilution of the spray liquid, will show a relative colour value of 7 over 5.

The droplet size may be determined by exposing glass plates coated with material that prevents their spreading on the surface, so that they remain as a partially flattened droplet standing up sufficiently to be sized. For aqueous droplets, coatings are made of resin in castor oil, Gelva resin, and polyisobutylene. At present aluminium stearate, mannitan monolaureate (NNO) and Drifilm 9987** are used as oleophobic coatings for the study of oil sprays; the last material, a silicone compound, is the most satisfactory. The droplet that has fallen on the coating spreads to form a lens whose diameter is determined by the final contact angle that its margins make with the coating. For most spray liquids on the coatings commonly used, the ratio of the droplet diameter D to the lens diameter A, *i.e.* the spread factor D/A, is between 0.25 and 0.60. Determination of the focal length f of the lens by reflecting parallel beams of light through it on the stage of a microscope allows the derivation of the value of the contact angle θ of the lens margin, and hence the value of the spread factor D/A. For example, the spread factor of fuel oil droplets on the Drifilm 9987 silicone is 0.5; thus the diameter of the droplets is one-half the diameter of the lenses they produce.

The droplet diameter may be measured directly by using glass plates smoked with a fluffy coating of carbon (*e.g.* lamp black) or magnesium oxide. Here the droplet punches a hole in the coating before spreading out on the glass underneath. With magnesium oxide, the hole is slightly larger than the drop, so that a correction factor of 0.85 has to be applied to the measured value (this is for droplets $>20\mu$; for $<15\ \mu$ the factor is 0.75). Or depression microscope slides may be used containing an aqueous solution of glycerin, oleic acid and triethanolamine, into which the oil droplets fall and maintain their spherical shape. For trapping aqueous droplets, a mixture of mineral oil and petrolatum jelly is used.

Measurements of droplet size are made under the microscope, using an ocular micrometer checked by a stage micrometer; if one or both is unavailable at the time, a rough standard of comparison is provided by a human red blood corpuscle, which measures 7 microns in diameter. The most suitable ocular micrometer is Fairs' modification of the Patterson-Cawood graticule.

On unsized papers of close texture, and on filter paper, droplets will fail to give a roughly circular stain. If these are measured before the edges become ragged, a rough idea of the droplet size may be obtained on the assumption that the spread factor is approximately 0.16, the stain being 6 times as wide as the droplet. Its value

* Check papers made to show droplets in colour are patented by and available from Independent Crop Dusting, Inc., Campbell, California.

** Available from the General Electric Company, Schenectady, New York, U.S.A.

may be determined by blowing off small uniform droplets from the tip of a capillary microburette. The number of droplets falling on the paper is related to the micro-volume delivered, and hence their true diameter can be computed from their individual volume. This is then compared to the diameter of the stain which they produce. The stains may be measured by illuminating magnifiers fitted with a micrometer*. From the sizes of the droplets and the number deposited per unit area, it is possible to calculate the area dosage. This replacement of the colorimetric method is useful when a colorimeter is not available.

Aerosols, which kill insects by air concentrations rather than area deposits, are more accurately determined by air sampling methods. The most useful piece of apparatus is the cascade impactor, in which air is sucked at 17.5 litres per minute through an orifice measuring 19 × 7 mm. to impact the droplets on a glass microscope slide; those that are too small to impact at that speed are carried by a cascade system of 3 jets of successfully smaller orifices to impact at successively higher speeds on 3 more glass slides, which catch droplets of 20, 7 and 3 microns diameter respectively. The slides may be used to determine the product of the concentration of the aerosol and the time over which it operated; this value Ct is expressed in mg.-min./m^3, and is a more useful figure than the concentration alone in mg./l., because it represents the amount of insecticide available to poison the insect. By coating the slides inserted in the cascade impactor, the particle sizes may be determined by the methods outlined above.

Measurements of droplet size should be made as quickly as possible after the sample has been taken, especially where the droplets are taken on coated slides. The slide may be photographed, and the measurements subsequently taken from a projection of the negative. The droplets are measured as falling into a number of size categories. The final sample consists of the total number of droplets in each size category. From this is computed the amount of spray volume in each size category. These figures are then treated cumulatively, to give the volume included in droplets up to each size, and the percentages of the total volume up to each size are computed. These percentages when plotted on a probability scale against the droplet size give approximately a straight line; the droplet size when the cumulative percentage is 50 is the median diameter by mass (or volume). This mass median diameter (m.m.d.) is the point which divides the total volume of the spray into two equal parts, one-half being in droplets smaller, and the other half larger, than the m.m.d.

REFERENCES

Bliss, C. I. (1935). *Ann. Appl. Biol.*, 22 : 134-167. (Derivation of probits of per cent mortality). **Brown, A. W.** *et al.* (1948). *Canad. J. Res., Sect. D.*, 26 : 177-187 and 188-196. (Testing insecticides with spray tower and other methods). **Campbell, F. L.** and **Moulton, F. R.** (1943). Laboratory Procedures in the Chemical Control of Insects. Amer. Assoc. Advanc. Science Publ., 20 : 206. **Carpenter, S. J.** (1945). *J. Econ. Ent.*, 38 : 721-722. (Shelters for sampling *Anopheles* adults). **Davis, J. M.** (1949). U. S. Dept. Agr. *Bull. ET*-272, 7 pp. (Photographic method for sizing spray droplets). **Federal Security Agency** (1947). Malaria Control in Impounded Water. Supt. of Documents, Washington. pp. 279-289. (Field sampling techniques). **Herms. W. B.** and **Gray, H. F.** (1944).

* *e.g.* Luminex Illuminator, available from R. and J. Beck, London, England.
** Available from C. F. Casella and Co., Fitzroy Square, London, W. 1, England.

Mosquito Control. Commonwealth Fund, New York. 2nd Ed. pp. 105, 106, 117. (References to New Jersey light traps). **Hoskins, W. M.** and **Caldwell, A. H.** (1947). *Soap San. Chem.*, 23 (4); 7 (Horizontal spray tower). **Magoon, E. H.** (1935). *Bull. Ent. Res.*, 26 : 363-371. (Portable stable trap). **Potter, C.** (1949). *Proc. 2nd. Int. Congr.* Crop Protection. 12 pp. (Methods for assessing insecticidal activity). **Scudder, H. I.** (1947). U. S. Pub. Health Reports 62, 681-686. (Grill for estimating housefly populations). **Wilkinson, P. R.** (1951). *Bull. Ent. Res.*, 42 : 45-54. (Types of trap huts). **Yamagushi, T.** (1952). *Ann. Meeting Amer. Mosq. Control Assoc.*, Salt Lake City, Utah, March 24-27, 1952. (Sleeve sampler for mosquito larvae). **Yeomans, A. H.** (1949). *U. S. Dept. Agr. Bull. ET-267* : 7 pp. (Determination of mass median diameter of droplet samples).

CHAPTER III

THE APPLICATION OF INSECTICIDES

TOXIC HAZARDS OF INSECTICIDES

The hazard of accidental poisoning of humans by insecticides may occur in five situations : in the manufacturing plant, in the home, in formulation and mixing, in application for control measures, and in contamination of food. The first, third and fourth situations involve the breathing, swallowing or skin absorption of sufficient amounts of the insecticide to be dangerous. Such accidents may be prevented by the wearing of a respirator or dust mask, and of protective coveralls and gloves which are frequently changed. In the second place, accidents in the home may be minimized by proper labelling of the materials and by indicating their poisonous nature with cautionary signs or by dyeing them a warning colour. Lastly the production of harmful residues on foods may be avoided by proper care during the control application, or by using insecticides which offer no residue hazards.

The toxicity of economic poisons is determined by the pharmacologist using standard procedures. The best single criterion of the toxicity of an insecticide is the LD_{50} by oral administration, the median lethal dosage level that kills approximately 50 per cent of the experimental animals, which are usually white rats. It is expressed in milligrams of the compound per kilogram of body-weight of the animals, or mg./kg. The chronic toxicity is usually determined by feeding a given concentration of the compound in the entire diet, and determining the maximum level in parts per million (p.p.m.) which fails to induce symptoms or death.

DDT. The toxic hazard of this chemical is very much less than that of the insecticides it has replaced. When orally ingested, the median lethal dose of DDT is 250 mg./kg., a level which indicates that its toxicity is low. The symptoms of DDT poisoning, as observed in experimental animals and occasionally in man, include nervousness, lack of appetite, and ataxia, followed by muscular tremors in the advanced stages; death may occur from respiratory failure due to ventricular fibrillation. There may be liver damage, and an accumulation of DDT in the body fat. There are cases of human deaths from ingestion of DDT in oily solution; but when DDT is undissolved, in non-fatty foods, it is surprising how large a dose may be taken with impunity. Similarly, DDT in oil solution may be absorbed through the skin and engender nervous symptoms; but DDT-powdered and DDT-impregnated clothing have been worn without ill effects by millions. It is possible to take in a harmful dose from fine sprays and aerosols, and even more so from fine dusts, by remaining in a high concentration for a long period of time. Therefore it is recommended that a dust-mask be worn over nose and mouth when spraying operations are conducted in buildings. Handlers of DDT solutions should wear rubber or plastic-covered gloves, as well as rubber aprons or canvas coveralls that can be discarded and cleaned. Antidotes that may relieve DDT symptoms include the barbiturates, calcium gluconate and magnesium sulphate.

In spraying DDT it is a wise precaution to cover food against contamination. Small quantities of this chemical, when ingested daily over a long period, will accumulate in the fatty tissues of the body, since very little is broken down. This may be enough to give neurotoxic symptoms or to impair the oxidative enzymes of the body. Contamination of their feed with DDT will cause dairy cattle to accumulate sufficient insecticide in their fatty tissues to yield a significant level (from 0.1 to 10 p.p.m.) of DDT in their milk. When fed to babies, such milk could give toxic symptoms, as it has to other young mammals in the laboratory. Butter would contain 20 times as much DDT content as the milk for which it is made. Contamination of milk also results if dairy cattle are treated with DDT sprays or dips. Therefore in the United States the use of DDT on dairy cattle or their feed is prohibited. A possible substitute for cattle spraying and dipping is methoxychlor, which is virtually non-toxic, having a median lethal dose of 7000 mg./kg.

The use of DDT in large operations in the field, particularly in aircraft spraying, is not without hazard to certain other forms of animal life. None of the mammals are affected by sprays, even when applied at 25 lbs. per acre, a hundred times the dosage used in mosquito control. Wild birds are not generally affected; but when they are tied to a particular site in the nesting season, and particularly when they are insectivorous, they may suffer (and the nestlings more so) when the dosage reaches 5 lbs. per acre. Frogs and other amphibians may be killed when the DDT expenditure is 2 lbs. per acre. DDT sprays have no effect on earthworms, slugs or snails.

Aquatic life is still more susceptible to DDT spraying, particularly animals that inhabit shallow pools, such as tadpoles and crayfish. Again, fish which inhabit still waters, such as sunfish, carp or goldfish, are more susceptible than those in streams, such as trout. The young fry are more susceptible than adult fish, and they generally occur in shallower water. At a dosage of 1 lb. per acre, many of the fish in shallow water may be killed, after showing a period of locomotory ataxia. There is little hazard to fish at 0.25 lbs. of DDT per acre. However, if a safe insecticide is required for mosquito spraying, DDD may be substituted. It is fully as effective as DDT against *Aedes* larvae; but is only one-tenth as toxic as DDT to goldfish. DDD is almost non-toxic to higher animals, having a median lethal dosage of 2500 mg./kg.

Other Chlorinated Hydrocarbons. BHC is less poisonous than DDT, its acute oral toxicity being represented by a median lethal dosage of 1000 ng./kg. Also it is not so serious a contaminant of foods, since it can be broken down in the body. It has been orally administered as a cure for pinworms in Germany, although not without toxic hazard. Only very high skin doses of BHC penetrate into the body fat or reach the milk secretion. But emulsion sprays containing 0.05 per cent of the gamma isomer will kill calves. Its toxicity is almost entirely due to its content of the gamma isomer; lindane, in fact, shows a median lethal dosage of 125 mg/.kg. Like DDT, lindane is a stimulant of the central nervous system.

Chlordane is about as toxic as DDT, since the acute oral m.l.d. figures range down to 250 mg./kg.; the chronic toxicity of chlordane as a food contaminant is also the same as with DDT. The symptoms of chlordane poisoning include anorexia, hyperexcitability and tremors, resembling those caused by DDT. It is possible to be poisoned by skin contamination with chlordane, since it is a heavy liquid. Protective clothing should be worn while handling it, and contaminated clothing must be discarded. It is also possible to be chronically poisoned by continuous exposure to a slight vapour concentration of chlordane accumulated in a closed room. One of the effects of chlordane is at least 4 times that of DDT. The toxicity of heptachlor is about twice that of chlordane, of which it is a constitutent.

Dieldrin is one of the most toxic members of the chlorinated hydrocarbon insecticides, with a median lethal dose of 50 mg./kg. It is quite poisonous when applied to the skin, so that rubber gloves and protective clothing are essential when it is handled, and full facial gas-mask when it is sprayed on buildings. The symptoms of poisoning, which may appear some days after the harmful exposure, are loss of appetite, loss of weight, accumulation of fat in the liver, and finally convulsions. The neurotoxic effects may be relieved by phenobarbital. Aldrin is also readily absorbed through the skin, and it is slightly more toxic than dieldrin. However, persons engaged in the manufacture of chlordane, aldrin and dieldrin have not exhibited any deleterious effects. Toxaphene is about 4 times as toxic as DDT by oral ingestion, with a median lethal dose of 60 mg./kg. It is rather readily absorbed through the skin. It engenders the nervous symptoms characteristic of this class of chemicals. Its chronic toxicity is the lowest of the chlorinated hydrocarbons, since animals may withstand considerable contamination of the diet with toxaphene, although it may accumulate in the fatty tissues.

DMC is less toxic than DDT, having a median lethal dose of 500 mg./kg. It also shows a low level of chronic toxicity as a contaminant of the diet. The compounds making up the insecticide *Dilan* offer about half as much toxic hazard as DDT.

Organic Phosphates. These compounds as a class are very poisonous, the median lethal doses for TEPP and parathion being less than 5 mg./kg. They are very toxic on absorption through the skin, and the vapour of parathion is highly hazardous. Their effect derives from the inhibition of the cholinesterase of nerve; the symptoms include salivation, sweating, diarrhoea, micturition, and muscular fibrillation. The antidote atropine relieves the nervous symptoms. In the United States, assays of blood cholinesterase levels are made to check against gradual poisoning of humans exposed to organic phosphates.

Despite these hazards, parathion has been used as a mosquito larvicide, and as a residual insecticide against house-flies outdoors. Substitutes are : EPN, which is one-fifth as toxic as parathion; the methyl analogue of parathion, which is not as toxic by skin absorption; and malathon, which is one-hundredth as toxic as parathion. Needless to say, all these materials should be handled with rubber gloves, rubber or plastic clothing, and respirators with approved canisters (*e.g.* Mine Safety Appliances MSC-1 or CR-45779).

Other Synthetic Organic Insecticides. The *Lethane* thiocyanates used in space for house-fly control show no hazard at the dosages normally employed. When used for louse control only certain of them are irritant to the skin, and then only to certain Caucasians. But all the thiocyanates, when applied in large doses, are skin irritants and are absorbed in toxic amounts. The median lethal doses by oral ingestion are approximately 500 mg./kg. for *Lethane* 60, *Lethane* 384 and *Lethane A*-70, and approximately 1000 mg./kg. for *lauryl thiocyanate* and *Thanite*. The thiocyanates are paralytic poisons, and the insecticides mentioned above damage the liver and other vital organs.

Phenothiazine is so non-toxic that it is administered orally to eliminate stomach bots and parasitic worms. For example, a dog can take a 3-gram dose without show ing any deleterious symptoms. Pyrolan is an anticholinesterase nerve poison, with a median lethal dose of 60 and 90 mg./kg. for rats and mice respectively.

The dinitro compounds such as DNOC, DNCHP and DNBP, extensively used in orchard protection and locust control, have median lethal dosages of approximately 50 mg./kg., and are readily absorbed through the skin; they have virtually no application in medical and veterinary entomology.

Insecticides of Botanical Origin. The pyrethrins show the least toxic hazard of all insecticides commonly used. With a median lethal dose of 1000 mg./kg. or higher, they have caused no cases of serious poisoning. However, persons subject to pollen allergies may show the 'pyrethrum idiosyncrasy' to the powder. And pyrethrins injected intravenously are very toxic, doses of 6 mg./kg. causing convulsions and paralysis. Nevertheless pyrethrins are the safest materials to use as aerosols in houses and sprays on livestock, and they involve no residue hazards on foodstuffs. Allethrin has a toxic level no higher than the pyrethrins. The synergists piperonyl butoxide and n-propyl isome are also virtually non-toxic, with median lethal doses exceeding 10,000 mg./kg.

Rotenone has a toxicity level that varies greatly according to the form in which it is administered and the kind of animal which ingests it. On the one hand, it has been used as an anthelminthic for livestock at 50 mg./kg. and for man even at 200 mg./kg., yet on the other hand pigs may be killed by drinking the run-off from warble-fly sprays applied to cattle. Human suicide is possible from fresh derris root, yet there are no fatalities attributable to the use of rotenone as an insecticide. It is not absorbed through the skin, but it may cause dermatitis and affect the eyes and the mucous membranes of the nose and mouth. In advanced stages of poisoning, rotenone depresses the respiratory rate.

The veratrin alkaloids present in hellebore and sabadilla dusts are highly toxic when injected intravenously. Yet the dusts themselves involve virtually no hazard. Ryania dust has also proved to be less toxic than rotenone.

Nicotine is one of the most powerful poisons employed as insecticides. The minimum lethal dose for man is considered to be 60 mgms. (*i.e.* 1 mg./kg.). It is readily absorbed through the skin. Symptoms of poisoning include vertigo, hallucinations, rapid respiration and prostration. Anabasine is no less toxic than nicotine. Application even of dilute solutions requires protective clothing and face masks. Neither of these compounds now have any place in the control of insects affecting man and animals.

Inorganic Insecticides. The toxicity of arsenious oxide (white arsenic) is represented by its median lethal dose of 2 mg./kg. for man and 20 mg./kg. to the rabbit. Sodium arsenite is about one-tenth as toxic as white arsenic. The As_2O_3 is not a cumulative poison, and it is claimed that the body may develop a tolerance for the poison. However, repeated dipping of livestock in sodium arsenite solutions containing approximately 0.2 per cent As_2O_3 involves the absorption of arsenic through the skin, thickening and desquamation of the hide, an increase in wound infection, occasional skin cancer, and the risk of oral poisoning by swallowing the dip, or of bronchopneumonia by contamination of the lungs. The symptoms of acute poisoning are a general depression and a progressive paralysis spreading from the hind legs forward. The antidote for accidental ingestion of arsenious oxide or sodium arsenite is a freshly prepared solution of ferric hydroxide and magnesium oxide, administered orally immediately after the accident.

Sodium fluoride is not as toxic as the arsenites, but its use as roach powder has caused numerous deaths by its being mistaken for common salt. The median lethal dose of NaF is about 300 mg./kg. for laboratory animals, and the toxic dose for man is about 5 grams. The symptoms are severe gastro-intestinal disturbances followed by epileptiform convulsions and pallor. Antidotes are milk or dilute calcium chloride, and calcium gluconate injected intramuscularly. Sodium fluosilicate is slightly more toxic than sodium fluoride, with a median lethal dose of 200 mg./kg., but in its use as a mothproofer of clothing it involves no hazard to the wearer.

Solvents. The petroleum oils used as solvents for DDT and other chlorinated hydrocarbons present a slight but definite hazard from skin contamination, inhalation, and accidental ingestion. Tests made on men handling various DDT solvents showed that those of lower boiling point were the most toxic. Calves have been killed by the application of kerosene sprays to the skin; the oil not only interrupts the dermal transpiration of water and the control of body temperature, but the toxic impurities penetrate to poison the tissues. The drinking of kerosene has been fatal to children. Skin contact with kerosene has caused dermatitis in some people, which disappears later on habituation. Presumably the hazards from fuel oil are similar.

The heavy aromatic solvents, such as polyalkylnaphthalenes (*Velsicols*) have about the same toxicity as the kerosenes. When continually applied to the skin they have caused a hardening and irritation. When added to the food of dogs, they have caused weight loss and slight damage to liver or kidney.

Xylene, often used as an initial solvent in livestock emulsion dips, is safe when in a final concentration of 2.5 per cent. However, emulsions containing 3.6 per cent xylene have caused skin desquamation in horses, skin irritation in cattle, and a transient paralysis in sheep. Concentrations of 10 per cent cause tremors, convulsions and death. Vapour concentrations of xylene in excess of 100 parts per million in a room may be dangerous, although not as toxic as benzene vapour. Naphthalene vapour may cause headache, nausea, visual disturbances, and the appearance of blood in the urine.

Two solvents employed for domestic use with DDT have proved entirely safe; these are cyclohexanone and isophorone, which are not truly hydrocarbons, and they are expensive. Dichlorodifluoromethane ("Freon"), used as the solvent in DDT aerosol bombs, is non-toxic even when in 20 per cent concentrations in the air. A cheaper substitute, methyl chloride, is definitely toxic; it is used only in greenhouses, where respirators are worn when it is applied.

Fumigants. The hazards of accidental exposure to toxic fumigants are considerably modified where the gas gives warning of its presence by irritation or offensive odour. For example, sulphur dioxide, ethylene oxide and chloropicrin are self-warning, but hydrogen cyanide, carbon disulphide and (to some extent) methyl bromide are not. The presence of organic halides such as methyl bromide may be tested by the halide leak detector, and of hydrogen cyanide by methyl orange papers. The choice of canisters for respirators used in fumigation is important. The standard canister for organic chlorides and esters is the Type GMA, while hydrogen cyanide, acid gases and chloropicrin each require their special canister. Inflammability hazards have been mentioned in Chapter I.

Hydrogen cyanide is one of the most toxic and hazardous gases. Concentrations above 0.3 per cent in air are rapidly fatal, and concentrations above 20 p.p.m. are dangerous if inhaled for more than an hour. Moreover, HCN is rapidly absorbed by the skin, and can penetrate cloth garments to do so. The initial symptoms are salivation, throat irritation, and vertigo, followed by headache and high pulse rate. The antidote to be applied by a physician is the gradual intravenous injection of sodium nitrate, followed by 25 grams of sodium thiosulphate. Before the arrival of the doctor, the patient may use a small ampoule of amyl nitrite, which he holds under his nose at intervals. The action of cyanide is to inhibit the cytochrome oxidative system of the tissues.

Methyl bromide is not highly toxic, the dangerous dosage level for one hour being 2000 p.p.m. Continuous exposure to 35 p.p.m. has caused mild intoxication in some men, and skin lesions in others. It has a delayed action; the victim of a serious exposure may later suddenly collapse, enter coma after convulsions, and die. Methyl bromide is a nerve poison, and can cause profound injury to the central nervous system. It inhibits the important enzyme succinic dehydrogenase. Methyl bromide has had a very good record as a fumigant when used by qualified personnel.

Sulphur dioxide is a safe fumigant, being non-toxic at doses (200 p.p.m.) in which a man finds it intolerable to remain. At this concentration it causes inflammation of nose and throat, increases fatigue, and may promote an oedema of the lungs. Chloropicrin is highly poisonous, but the dangerous level of 2 p.p.m. (for one hour) is detectable by the nose, and the eyes begin to water. Higher exposures may result in irritation and oedema of the lungs, which often prove to be fatal.

Carbon bisulphide is a liposoluble nerve poison, and may be absorbed through the skin. Concentrations above 500 p.p.m. are dangerous for exposures of more than one hour. The initial symptoms include irritability and faulty vision, with severe doses leading to coma and convulsions. Prolonged exposure to sublethal concentrations results in anaemia and general debility. Despite its putrid smell, carbon bisulphide is not self-warning.

Ethylene oxide is relatively safe; although the dangerous concentration for one hour is 2000 p.p.m., it irritates the eyes and nose to the point that it is intolerable. Severe exposure leaves the subject gasping for breath, and leads to inflammation of the lungs. Ethylene dichloride is less toxic, concentrations up to 1000 p.p.m. being relatively safe, whereas the odour can be detected at much lower concentrations. Symptoms of poisoning commence with vertigo and nausea, followed by muscular weakness and tremors.

Carbon tetrachloride is so safe that it is used in household solvents and fire extinguishers. The dangerous concentration for one hour's exposure is 5000 p.p.m., while it can be detected by smell at only 70 p.p.m. Severe exposure is marked by irritation of mucous membranes, headache and nausea, followed by stupor and convulsions. Prolonged exposure to lower concentrations lead to cirrhosis of the liver, kidney damage and jaudice. Trichloroacetonitrile is about as toxic as ethylene oxide, and is self-warning. Paradichlorobenzene (PDB) offers no toxic hazard under ordinary conditions of use; but men confined to buildings saturated with its vapour are in danger of liver damage.

REFERENCES

Committee on Pesticides. (1950-1952). *J. Amer. Med. Assoc.*, 144: 104; 145: 728, 147: 571, 149; 367 and 1135. (Reviews of toxic hazards of organic phosphates, DDT, lindane, toxaphene and electric vaporizers). **Communicable Disease Centre** (1952). Operational and Clinical Memoranda on Economic Poisons. TDB 5-1-52. Mimeographed Publications. **Frings, H.** and **O·Tousa, J. E.** (1950). *Science* 111: 658-660. (Toxicology of chlordane). **Hayes, W. J.** and **Ferguson, F. F.** and **Cass, J. S.** (1951). *J. Trop. Med.*, 31: 519-522. (Toxicology of dieldrin). **Lehman, A. J.** (1948). *Quart. Bull. Assoc.* Food and Drug Office. U. S. 12: 82-89. (Toxicity of important insecticides to mammals). **Princi, F.** and **Spurbeck, G. H.** (1951). *Arch. Indust. Hyg. Occup. Med.*, 3: 64-72. (Manufacturing hazards of chlordane, aldrin, dieldrin). **Radeleff, R. D.** and **Bushland, R. C.** (1950). *J. Econ. Ent.* 43: 358-364. (Hazards of chlorinated hydrocarbons to cattle). **Stammers, F. M.** and **Whitfield, F. G. S.** (1947). *Bull. Ent. Res.*, 38: 1-73. (Toxicology of DDT : review).

APPARATUS FOR SPRAYING

Compression Sprayer. The most important piece of spraying apparatus in medical entomology is the compression or pneumatic sprayer (Plate I Fig. 1). In this type of equipment the spray liquid is pushed to the nozzle by a body of compressed air. The specifications for suitable models for residual spraying have recently been drawn up by the World Health Organization. The usual size has a capacity of 4 U.S. gallons and consists of a cylinder approximately 20 inches high and 75 inches in diameter. When empty it should not exceed 12 lbs. in weight, complete with fittings. It may be slung over the shoulder with one strap, or carried on the back with two straps. The cylinder should be resistant to chemicals, of stainless steel or high-quality galvanized iron, and should be seamless or seam-welded, instead of rivetted as in the old-fashioned "Four Oaks" type. It is preferable that it be graduated with embossed capacity markings. The filler cap, formerly threaded, is now of the inner-seal type that is held closed by the pressure; there should be a Schrader type pressure release valve.

The cylinder is filled three-quarters full with spray liquid, and air is pumped by hand into the remaining space by means of a vertical pump measuring 15 inches long by 1 3/4 inches in diameter, strong enough to withstand pressures of 130 pounds per square inch. The pressures developed by hand pumping range up to 100 p.s.i., but the recommended pressures for residual spraying range from 25 to 40 p.s.i.; several pumpings are necessary to compensate for the increasing air space as the liquid is discharged. A pressure-regulating reducing valve and a pressure gauge are useful refinements. Compression may be provided by a small air-compressor powered by a truck motor, or by a 10-ounce cylinder of carbon dioxide. The former system of compression effects a slight economy in time and cost, but is feasible only under special circumstances.

The sprayer is equipped with 30 to 36 inches of rubber hose with 3/4 inch inside diameter. It should be of synthetic rubber to provide resistance to chemicals, and not more than 35 per cent of natural rubber is permissible. The hose is attached to the extension lance, which preferably should have a gooseneck attachment for spraying difficult corners. At the base of the lance is a trigger-type, quick shut-off, spray control valve; a squeeze valve on the hose has been used in some models. The nozzle should be of the flat fan type, with 60 to 65 degrees included angle; it should deliver between 0.2 and 0.4 gallons per minute at 40 p.s.i. pressure. The strainer screen protecting it should have mesh openings one-half its shorter diameter, and the tank outlet should be protected by similar screening but with 20 times more area. Examples of American compression sprayers are the Hudson and Dobbins, and British sprayers are the Ross and Kent.

Knapsack Sprayer. This name is conventionally reserved for hydraulic-type sprayers in which the liquid itself is pumped to the nozzle (the pressure being measured off an air by-pass). The container is constructed to fit the back of the operator, is fitted with two straps, and usually has a capacity of 5 imperial gallons. The knapsack sprayer is widely used in the spraying of gardens or allotments, and the pump is activated by a lever at the side of the tank. Recently, American workers have found that the 5-gallon knapsack sprayer (Hudson Ozark) with a lock cap and lip, fitted with a trombone pump which forms part of the spray lance, and carrying nozzle no. 123, is the most satisfactory equipment for residual spraying (Plate II, Fig 5).

Stirrup Pump. The simplest hydraulic sprayer is the stirrup or bucket pump, which is a plunger pump clamped in a pail, delivering by a hose to a rod and nozzle. The pumps may be single or double-acting; in the latter case they discharge on both the downward and upward strokes. It is possible to generate 150 p.s.i. pressure on a good stirrup pump. The length of the barrel in the single-barrel pump should not exceed 22 inches, and that in a double-barrel pump should not exceed 8 inches. The pump should be supplied with 3 feet of hose, and an extension rod, spray control valve and nozzle of the same design as the compression sprayer. The plunger cup should be of chrome leather impregnated with graphite. Stirrup pumps are useful substitutes in malaria control programmes if compression sprayers cannot be obtained.

Hand Sprayer. This name is applied to sprayers consisting of a cylindrical plunger-type air pump carrying a can of liquid at the distal end. The capacities are 1 pint to 1 quart (U.S.) and the weight is 1 1/2 lbs. empty. They are of two types.

The intermittent type is the familiar Flit-gun, which is an air atomizer in which each compression stroke of the air piston aspirates the liquid up a tube, 0.1 inches in diameter, and atomizes it at the top where the jet of air and the liquid are opposed at right angles (Plate. I Fig. 2). In some models the mixture of air and atomized droplets is carried forward to a nozzle, whose orifice may be adjusted to give fine or coarse spray.

The continuous type is really a compression sprayer, since the pump compresses a reservoir of air in the can and pushes it to a nozzle. There is a trigger just behind the nozzle, which may carry a device to regulate the volume of discharge of air. It is recommended by W.H.O. that hand sprayers have the correct nozzle for the purpose in hand, namely coarse atomization for residual spraying, fine for larviciding, and very fine for space spraying with aerosols. For larviciding it is recommended that the nozzles be extended into a 26-inch lance to prevent the operator from stooping, and to allow him to poke it through the underbrush. The largest hand sprayers (there is a 3-quart model) are provided with a handle behind the can and a strap for carrying. Some hand sprayers are supplied with a pipe-fitting on the pump handle which may be connected to a supply of compressed air.

Oil Spray Gun. Air atomizers related to paint sprayers and oil spray guns have been specially developed for insect control (*e.g.* the DeVilbiss model). To economize on the requirements for air compressors, the air consumption has been reduced to only 1 cubic foot per minute, which is much more economical than the

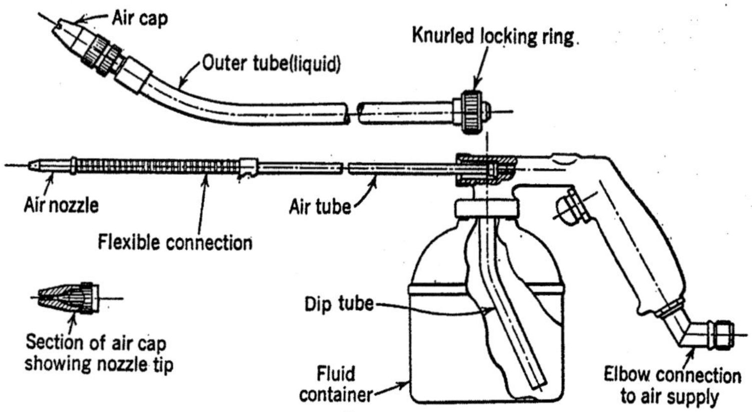

FIG. 1.
Oil spray gun.

conventional paint-sprayers. They have a long nozzle and angled tip, and the atomizer carries the container of spray liquid and is fitted with a pistol-grip trigger. The air is emitted along an inner male tube running within a larger female tube which carries the liquid (Fig. 1). The air issuing from the nozzle aspirates and atomizes the spray liquid presented in an annulus around it. The atomization is very thorough and spitting does not occur. This unit is attached by 20 feet of hose to a spraying power plant consisting of a 2.5-cubic foot compressor driven by a gasoline engine of at least one horse power.

Fine-atomizing sprayer. In this equipment air is used both to push the liquid to the nozzle, and also to atomize it at the nozzle. The Husman model, developed by the U.S. Bureau of Entomology, consists of a rotary-vane compressor supplying 8.3 cubic feet per minute at 8 to 15 p.s.i. pressure to a group of 6 air-mix-nozzles. Complete with a 1.25-h.p. engine, the unit weighs only 125 lbs., and can emit spray at 20 gallons per hour. The atomization is such that the spray may be applied as residual deposits to buildings, or be drifted downwind to control adults for a distance of 100 yards or more.

Electric Sprayer. In order to eliminate the need for bulky pressure tanks, high-speed rotary compressors have been developed that can keep up to the loss of pressure that occurs during spraying. These are built in to an electric motor, a 1/10 h.p. model developing and holding a pressure at the atomizer nozzle of 25 p.s.i. The spray liquid is contained in metal cylinders or glass jars of 1 pint to 3 quarts capacity, and may be carried with the motor in a frame or attached to it by flexible air-tubing. Its use being confined to buildings supplied with electric power, it is known as the electric sprayer (*e.g.* Hudson, Piezo). A modification of this type of equipment is known as the mistorizer (*e.g.* West, Sprayer Corp. of America). Here the spray liquid is atomized directly into the rotary compressor where the blades beat it up, and then after being warmed a few degrees it is emitted under pressure through a fine nozzle; in this way a fine mist is produced. Both the conventional electric sprayers and the mistorizers may be fitted with a time switch for 1 to 30 minutes emission, and can be installed as permanent automatic aerosol machines in food storage establishments. A further modification is the fan type sprayer, in which the air is moved by a blower impeller mounted on the electric motor. The smallest model, with an 1/10 h.p. motor, is able to shoot the spray for a distance of 8 to 10 feet; the largest model with a 1 1/3 h.p. motor can attain 60 feet (*e.g.* Mistmaster). In this latter heavy-duty sprayer, which is nevertheless still portable, the mixture of air and spray droplets emerges at a speed of 320 m.p.h. Thus inaccessible places may be reached with a directed fine spray, which is a characteristic advantage of electric sprayers as a class.

Mist Blower. In this type of equipment the hydraulic atomization of the spray liquid at nozzles or orifices is aided by impelling a body of air past the source of the spray. The air blast serves also to carry the spray to the target, and in this way fine sprays can be aimed. In some spray-blowers it is the air-blast which is mainly responsible for atomization, as in the turbine sprayer where the liquid is first emitted from simple orifices. In other types the greater part of the atomization originates at the nozzles and the air blast serves mainly as a carrier for the spray, as in the speed sprayer with its banks of nozzles. None of these machines require more than 70 p.s.i. of hydraulic pressure, since the atomization and carry are aided by the blower. This allows the heavy plunger pumps to be replaced by light centrifugal pumps, and the machines are thus portable on wheel-barrows or truck bodies.

The smallest model consists of 4 capillary jets emitting into a nozzle 1.25 inches in diameter on the end of 3 feet of hose (*e.g.* Portamist); the high velocity blower is driven by a 1 1/2 horse-power gasoline engine. A larger unit has spray jets emitting into a muzzle 8 inches in diameter (*e.g.* Aeromist), and the centrifugal fan is driven by a 13 h.p. engine. The largest practical model has a 10 1/2-muzzle through which air is driven by a turbine driven by an 18 h.p. engine (*e.g.* Buffalo Turbine). These mist-blowers are suitable for larviciding where pools are frequent, often in woods and bushes, and are accessible by road. The two sizes of Lawrence mist-blower are commonly used by mosquito abatement organizations in the U.S.A.

Power Sprayer. Machines of this type are hydraulic sprayers in which the spray liquid is impelled to nozzles by a positive-displacement plunger pump. These are widely used for the control of insects affecting livestock. They are designed for emitting aqueous suspensions and emulsions at a high volume output; the emission rate ranges from 5 to 85 gallons per minute, the pressure from 200 to 1000 p.s.i., and the power plants from 1 to 45 horse power. The units are mounted on trailers or skids, and rubber hoses lead from them to spray guns or spray brooms. Hollow-cone nozzles are employed, consisting of a disc affixed on top of an eddy chamber, with a central hole usually 4 sixty-fourths of an inch in diameter. The treatment of herds of cattle for control of warble fly, for example, is done with power sprayers emitting rotenone suspensions at 400 p.s.i. pressure; the animals are in a corral or paddock. Alternatively the herd may be driven through a chute under an arched boom containing approximately 6 nozzles supplied by a power sprayer.

Rotating-disc Mist-blower. Machines of this type consist of a sheaf of thin metal discs mounted together on a shaft, so that they are almost contiguous at the

PLATE I

FIG. 1
Back-portable compression sprayer.
Courtesy John Wiley & Sons.

FIG. 2
Hand sprayer.
Courtesy H. D. Hudson Mfg. Co.

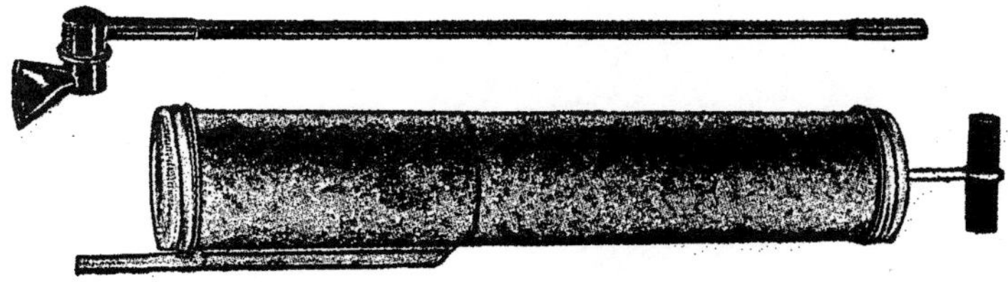

FIG. 3
Hand duster, with extension tube.
Courtesy H. D. Hudson Mfg. Co.

PLATE II

Fig. 4
Aerosol application from a helicopter.
Courtesy John Wiley & Sons.

Fig. 5
Knapsack sprayer, with trombone pump.
Courtesy H. D. Hudson Mfg. Co.

periphery (*e.g.* the Hession Microsol). The insecticide solution is brought by way of the hollow shaft into the narrow spaces between the discs. When the sheaf is rotated at high velocity by an electric motor the solution is forced outwards through the very narrow peripheral spaces by centrifugal force, and is atomized as it is thrown off into the air. A similar principle of atomization is utilized in the spinner-disc employed in aerial spraying. A large field model is available which consists of 21 discs each inches in diameter; it is rotated at 6500 r.p.m. by a 6 1/2 h.p. motor. The spray is directed to the target by an air blower operating at 4500 cubic feet per minute. Rates of emission vary up to a maximum of 250 gallons an hour; at 20 g.p.h. the droplet size is 50-100 microns. This model has a place in larviciding operations, just like other mist-blowers. The atomization increases with the speed of rotation and varies with the square root of the disc diameter and inversely with the delivery rate. Smaller models are available for the treatment of buildings, and can thus produce fine sprays approaching aerosol size without involving the use of heat which decomposes labile insecticides such as pyrethrins.

APPARATUS FOR PRODUCING AEROSOLS

Exhaust Generator. In machines of this type, oil solutions of insecticides are finely atomized by the blast of hot exhaust gases from a gasoline engine. The shatter afforded by the high-velocity blast is enhanced by the heat decreasing the viscosity of the oil. The droplet size is further reduced by vaporization which occurs before the aerosol is committed to the outside air.

The oil may be injected into the air-blast from the sides of the exhaust pipe; this arrangement has been used in aerosol generators installed on the exhaust stacks of aircraft. However, it is preferable to use a coaxial tube for injection, so that the oil pipe is conducted inside the exhaust pipe for some distance. When the oil is emitted fairly close to the exhaust outlet, it has already been heated to a highly fluid state. Since the emission orifice is generally quite large, gravity feed affords sufficient pressure for injection. Such installations have been made on portable 1 1/2 h.p. gasoline motors, and have been used with success in barns and warehouses. Here the exhaust is emitted through a 1/4-inch hole in a pipe-cap screwed on to the end of the stack; the 3/16-inch oil injection-tube is carried up to the hole in the cap, leaving enough room (a little less than 1/32 inch) for the exhaust gases to escape around it and atomize the oil (Fig. 2). When kerosene is emitted at one gallon per hour, this installation atomizes it to an aerosol whose average droplet diameter is 15 microns. If the machine is running at too high a temperature, the normal white aerosol takes on a bluish tinge since the droplets have become too small.

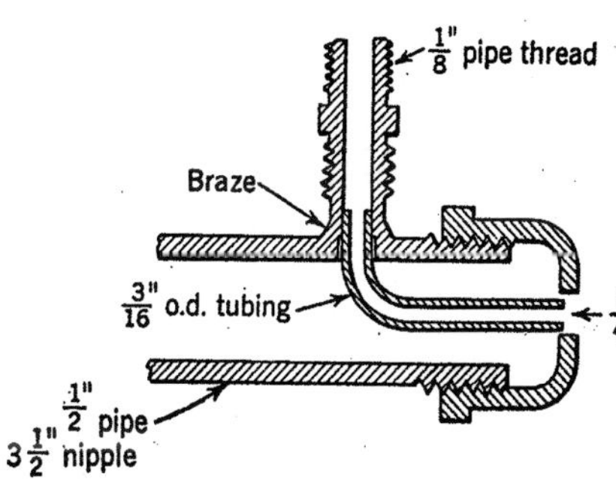

Fig. 2.
Exhaust aerosol generator.

A similar atomizing unit has been made to slide over the end of the exhaust pipe of a truck or tractor. Here the end of the injection tube is closed by a 1/2-inch cap which is pierced with slots in the sides to allow emission. It is centred within a 3/4-inch pipe nipple, which serves as the orifice and which is attached to the end of the exhaust stack by a pipe-adapter. The insecticide solution may be delivered either by a gear-pump, a pressurized tank, or by gravity feed. Droplet size may be decreased by restricting the insecticide supply, increasing the speed of the engine, or by moving the installation closer to the engine. Exhaust aerosol generators have been installed in the military 5-cwt. truck known as the "jeep." They were originally constructed with a narrow throat (on the venturi principle) similar to that employed on aircraft, but with a coaxial oil feed. There is now a considerable body of information published on the performance of the various "plumber's nightmares" devised for attachment to the exhaust stacks of various vehicles. They are useful in controlling *Aedes* mosquitoes in open marshes of moderate size.

Thermal Generator. Special equipment has been developed in which air is heated and pumped past hydraulics prayer nozzles. An example is the TIFA or Todd insecticidal fog applicator (Plate III, Fig. 6). In this heavy machine, air is pumped at 150 cubic feet per minute through a combustion chamber heated by burning gasoline to a temperature regulated at a level usually between 800° and 1200° F.; this is done by adjusting a fuel pressure regulator to give the desired gasoline supply. The oil solution of insecticide is sprayed through nozzles into a steel cup, where it is given a whirling motion with a little hot air before it is carried through the combustion chamber and out of the muzzle or foghead by the main blast of hot air. It is delivered by a centrifugal pump which has a capacity of 50 g.p.h.; the maximum pressure of delivery is 25 p.s.i., but it may be adjusted to the desired level by an insecticide pressure regulator. The rate of delivery, which is normally 20 g.p.h., may be governed by a metering valve; since the effect of changing the rate of flow is to change the degree of atomization, this valve is called a "particle size selector." For example, if the injection rate is increased from 10 g.p.h. up to 50 g.p.h., the mass median diameter of the resulting aerosol is increased from 5 microns up to 50 microns. In the TIFA, a 6 1/2 h.p. motor is sufficient to run the air pump, fuel pump and insecticide pump which constitute the moving parts of the equipment.

At a combustion chamber temperature of approximately 1000° F., much of the oil solvent is volatilized, so that the concentration of insecticide in the droplets of aerosol is higher than in the original solution. There may be some decomposition of the insecticide at this temperature, and even DDT shows 35 per cent destruction in paraffin oil solution and 50 per cent in oil emulsion. But this loss may be minimized by employing less volatile oils such as lube oil. With light oils there may be a danger of explosion, if the flash-point is below that of kerosene; high-flash oils such as Sovacide and Tificide have been especially developed for this machine. When the combustion chamber is held at 850° F., and the insecticide solution is injected at 15 g.p.h., the droplets in the cloud average 20 microns when no. 2 fuel oil is the solvent, and 15 microns when kerosene is the solvent.

This type of fog is well adapted to eliminate flying mosquitoes and blackflies as it is drifted by the wind across the infested area. In application it is recommended that the area be traversed at not more than 5 m.p.h. at intervals of 100 to 300 yards,

PLATE III

Fig. 6
Thermal aerosol generator.
Courtesy John Wiley & Sons.

Fig. 7
Rotary-type knapsack duster.
Courtesy John Wiley & Sons.

PLATE IV

Fig. 8
Residual spraying with flat fan nozzle.
Courtesy World Health Organisation.

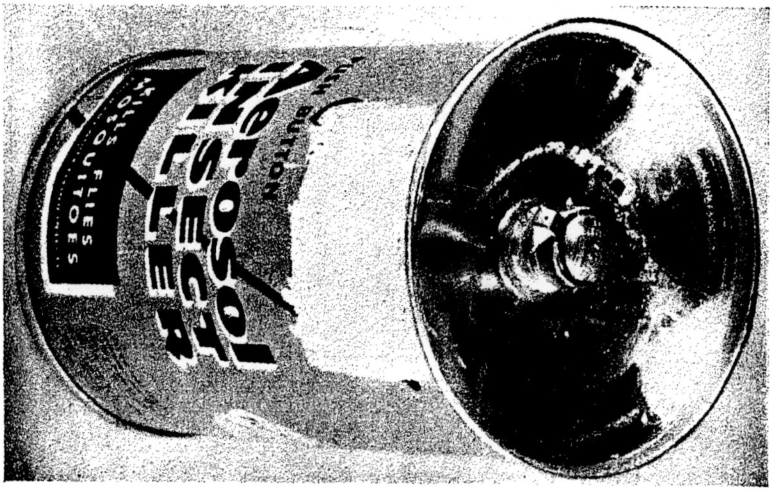

Fig. 9
Low-pressure aerosol bomb.
Courtesy John Wiley & Sons.

and that 5 per cent solution of DDT be emitted at approximately 20 gallons per hour. At these dosages, which approximate 0.05 lbs. of DDT per acre, complete control of mosquitoes may be expected for a distance of 400 yards in the open and 200 yards in woodland; blackflies may be controlled at a dosage of 0.2 lbs. per acre. It is considered that the optimum droplet diameter is 20 to 30 microns for application in the open, and 10 microns for woodland. Treeless open areas should be treated in the evening, whereas woodland may be treated during the day provided the wind speed in the open is not less than 10 m.p.h.

Jet Generator. A recent generator, the Dyna-Fog, utilizes a pulse-jet engine combusting gasoline at the rate of 60 explosions per second. The insecticide solution is introduced into the exhaust pipe, and is atomized and partially vaporized, producing an aerosol with a spectacularly high content of screening smoke. At emission rates approximating 20 g.p.h. the mass median diameters are 5 to 10 microns, and the aerosol gives good control of mosquitoes in woodland up to 400 yards downwind. The machine is light, but the insecticide is held in a separate 20-gallon tank. The Swingfog pest control unit is lighter and portable by hand ; at 80 pulsations per second its jet emits insecticide solutions and emulsions at 2 g.p.p.

Steam Generator. In this aerosol machine the insecticide solution or emulsion is atomized by the escape of steam from superheated water. In the Hochberg-LaMer modification of the Besler smoke generator (U.S. Army C.W.S. Disperser E-12), equal quantities of insecticide solution and water are separately pumped at 20 g.p.h. through coils of tubing in a combustion chamber at 300° to 600° F. The steam, escaping through 1/9-inch orifices at 120 p.s.i. pressure, shears the heated oil into droplets which average 10 microns in diameter at the upper temperature, and the resulting aerosol can control adult mosquitoes for a distance of one mile in the open and 200 yards in dense forest. In the C.W.S. Disperser E-15, less water is added to the oil and the coils may be heated to 1800° F., and thus fine aerosols may be obtained with heavy oils. In the most recent modification, the Bes-Kil generator, the insecticide solution does not pass through the heated coils, but is atomized into the steam as it leaves the coils at 500° to 700° F. In this way any thermal destruction of DDT and other insecticides may be avoided. The delivery rates are 40 to 80 gallons per hour, and coarse aerosols with mass median diameters of 30 to 50 microns are produced. The machine is more suitable for orchard work than for the control of adult mosquitoes.

Aerosol Bomb. Aerosols produced by this device result from the extremely rapid volatilization of a compressed and liquefied gas, to which has been added 10 to 20 per cent of a non-volatile oil solution of an insecticide. As the liquefied gas volatilizes, it leaves the oil solution in the form of an aerosol of very fine droplets. The compressible gas employed is dichlorodifluoromethane ("Freon"), which offers no toxic hazard on being inhaled by man or animals. Thick-walled bombs were originally employed and they contained the propellent under a pressure of 80 p.s.i. The smaller bombs were later replaced by thin-walled cylindrical canisters (Plate IV Fig. 9), which contain Freon at a pressure of 25 p.s.i. The usual sizes of containers are 1 lb. for use in tents and houses, and 5 lbs. for business establishments and aircraft. For single application to rooms, soda-siphon cartridges ("sparklets") are available, which contain 7 gms. of DDT-pyrethrum in kerosene-acetone to which 2 gms. of CO_2 has been added as a propellent.

In the domestic aerosol bombs and canisters the nozzle is a simple orifice of 0.01 to 0.02 inches diameter which allows a delivery rate of about 1 gm. per second. It is located either at the end of 4-inch length of capillary tubing, or in the side-wall of a vertical screw-valve. In either case the propellent, in its climb up the tube to the nozzle, decompresses sufficiently to cause bubbles of vapour to form. Atomization is caused by these rapidly expanding bubbles encountering the resistance of the outside air. The degree of initial atomization is a function of the pressure of the dispenser and the type of nozzle; the final size of the droplets at the time they remain airborne is a function of the percentage of non-volatile solvent that remains. The optimum concentration of non-volatile material in the high-pressure aerosols is between 15 and 20 per cent, because they give droplet diameters between 3 and 10 aerosols, the highest kills of house-flies. For the low-pressure aerosols, the optimum concentration of oil lies between 10 and 15 per cent. The solvent usually employed is a heavy aromatic polyalkylnapthalene oil such as *Velsicol.* DDT or lindane, together with pyrethrins or allethrin with synergist, are the insecticides employed.

Insecticidal Smokes. In this category, the insecticide is mixed with some slow-burning material, which is ignited and burned. A simple example is the joss-stick, one end of which has been dipped in plant gum containing camphor or pyrethrum; the incense-like smoke acts mainly as a repellent rather than an insecticide. The particles in insecticidal smokes of pyrethrum, rotenone or arsenicals range between 0.3 and 2 microns in diameter. Slow-burning pellets may be made by compressing a mixture of granular cordite and insecticide with kieselguhr as a coolant. Powders may be compounded of 58 to 60 per cent of an insecticide such as DDT or BHC, 30 to 40 per cent of a burning mixture containing sucrose and potassium chlorate in equal proportions, and 2 to 10 per cent of a coolant such as kieselguhr, diatomaceous earth, MgO or $MgCO_3$. If 1 lb. of this material is placed in a canister and lit by a fuse, it will produce smoke for 3 minutes and is sufficient to treat 15,000 cubic feet of enclosed space. Two-ounce canisters have been developed for greenhouse and domestic use, consisting of a fuse, a starting mixture, and the burning mixture; cooling baffles prevent the entire filling from simultaneously igniting (*e.g.* Murfume). Another type consists of lengths of 1/8-inch fabric cord, impregnated with the insecticidal burning mixture, which are put up in canisters.

When DDT or BHC is burned in canisters, the thermal decomposition of the insecticide amounts to approximately 30 per cent. In the case of DDT, this is attributed to dehydrochlorination by heat; the loss is increased by ferric iron and may be inhibited by the addition of certain chemicals such as picolinic acid. The recovery of DDT in the smoke may be increased by using more highly purified DDT (*i.e.* of higher melting point) or by increasing the amount of sucrose in the burning mixture.

DDT smokes are deposited as supercooled droplets, which readily crystallize upon being lightly brushed. Deposits of lindane rapidly become crystalline and appear as dendritic growths. DDT smokes achieve adequate control of household insects, but they are capable of only slight penetration into the crevices. For house-fly control, DDT smokes give up to 80 per cent reduction in population; to give results equivalent to those obtained with DDT in wall spray, about 8 times the dosage must be applied in the smokes. At present, the odour of BHC smokes renders them unsuitable for domestic application. The results obtained with smokes in the field have at present been unsatisfactory in comparison with thermally generated aerosols.

APPARATUS FOR DUSTING

Hand Duster. In this device a hand-operated plunger pump emits a blast of air in which the dust particles are rendered airborne. The dust container may be located below the plunger tube, or may be an enlargement at its end, or may be the plunger tube itself. The compression stroke of the piston expels a blast of dust-laden air. In some hand dusters a double plunger tube provides a continuous supply of air on both the forward and return strokes. The barrels of hand dusters used in disease vector control should be between 6 and 12 inches in length and 2 to 4 inches in diameter. The dust reservoirs are 60 to 120 cubic inches in volume, and carry 1 to 2 quarts of dust which should be suitably agitated. The discharge tube should be between 1/4 and 1/2 inch in internal diameter and 2 inches long, provided with alternate extension tubes, 6, 12 and 18 inches in length. A bullet-nosed circular discharge nozzle is favoured, with an aperture one-half the area of the discharge tube; a flat fan or fish-tail nozzle is available to fit over it. Hand dusters are particularly useful in introducing insecticide into crevices (Plate I. Fig. 3).

A useful type of hand duster for indoor work consists of a swirl-type dust cup with a horizontal extension nozzle provided with a pistol-grip and trigger; it is operated by a supply of compressed air from the same type of engine-compressor unit that is used with the oil spray gun.

Knapsack Duster. The usual knapsack dusters are actuated by a rotary fan. Hand cranking revolves the fan and positively feeds the dust in the container to the outgoing air; it may also provide agitation. The dust is either fed directly into the fan case from the side, so as to reach the fan blades at a point between the centre and the circumference; or it is fed first into an air chamber which is a "vestibule" of the fan, before passing the fan itself to reach the delivery tube. The rotary type in which the dust is fed on to the fan blades provides the best performance. There is a lever to regulate the rate of dust feed. The tube terminates in a fish-tail nozzle, and it is preferably long enough to reach the ground at a comfortable distance ahead of the operator (Plate III, Fig. 7). Rotary-power knapsack dusters should weigh no more than 15 lbs. empty, and carry from 8 to 15 lbs. of dust. The discharge tube should be not less than 1 1/2 inches in diameter, and provided with 3 sections of extension tube each 14 to 18 inches long. The dust reservoir should be not less than 114 cubic inches in capacity. The duster should be capable of delivering at least 12 ounces per minute at a cranking rate of no more than 35 turns per minute.

The bellows duster is a modification in which the air is supplied by bellows on the top of the cylindrical dust container. A hand lever operates the bellows and agitates screens to allow the dust to flow into a hose and rod tipped with a sieve-type nozzle. Bellows dusters weigh 20 to 25 lbs. when empty.

Knapsack dusters were originally used for mosquito larviciding with Paris green. The advent of DDT has rendered larvicide dusting somewhat obsolete, but it still has a place where the insecticide has to be carried long distances on foot.

AERIAL APPLICATION

Aerial Sprayer. In most contemporary aircraft installations the spray liquid is pumped to a boom fitted with nozzles. Normally the boom extends from wing-tip to wing-tip, and is divided into two parts by the pump which is carried below the fuselage (Plate V, Fig. 10). The boom consists of 3/4 or 1-inch steel tubing, carrying a series of flat-fan or hollow-cone nozzles spaced every 4 to 10 inches. The boom is suspended 9 inches below the lower wing of biplanes, preferably at the trailing edge. In high-wing monoplanes it may be carried obliquely from the fuselage to the wing-tips, or it may be shorter than the wings and carried horizontally.

Rotary-vane or centrifugal pumps are used, light in weight, and generating pressures of 40 to 100 p.s.i. They are driven by 4-bladed outboard propellors turned by the slip-stream at about 2000 revolutions per minute. The pumps emit from 5 to 10 gallons per minute. A quick-acting spray control valve is inserted between pump and booms, with a control lever in the cockpit. There is also a relief valve adjustable for the desired pressure and returning excess discharge to the spray tank. The capacity of the tank is dictated by the load capacity of the aircraft, which is approximately 25 gallons for the Cub high-wing monoplane, 100 gallons for the 450 h.p. Stearman biplane, 150 gallons for the single-engined Beaver and the twin-engined Anson aircraft, and 600 gallons for the Dakota DC-3 (C-47) transport. Swath widths obtained with these aerial sprayers range from 35 yards with light Cub-type planes to 300 yards with the Dakota aircraft.

There are simpler substitutes for the pump, boom and nozzle system. The use of a breaker-bar behind the boom allows simple orifices to be used instead of nozzles. The substitution of venturi-type air-scoops has allowed the pump to be emitted. In fast aircraft such as the Dakota the slip-stream alone is sufficient to shatter the spray, and the solution may be emitted from a straight emission pipe to which it is fed by gravity. The use of rotary brushes has enabled simple gravity feed systems to be used in the Dakota aircraft as well as in light planes. These units consist of an outboard propellor and a sheaf of steel-bristled disc-shaped brushes rotating on a shaft at 2500 r.p.m.; they may be suspended below the wings or the fuselage. A single unit emitting at one gallon per second and mounted on the Dakota aircraft has achieved great success in controlling *Aedes* mosquitoes in Canada. Aerial spraying against malaria mosquitoes has been carried out in the southern U.S.A., Venezuela, Greece and Malaya, and in the southwest Pacific area, Burma, Assam and Italy during World War II; aerial spraying for tsetse fly control has been in progress since the war in various parts of Africa.

Aerial Duster. This consists of a hopper inserted in the front or rear cockpit, a wind-driven agitator, a feed-control gate, and a venturi spreader suspended below the fuselage. The unit is mounted in small aircraft such as Cubs and Stearmans. The hopper is built of aluminium (24ST) or plywood, and the sides slope at not less than 45 to the horizontal down to the throat. There are one or two agitators, preferably of piano-wire, which revolve at 50 to 300 r.p.m., actuated by an outboard propellor at a reduction ratio of 50 to 1. The feed control is usually a horizontally sliding gate made of aluminium or bakelite. The venturi-type airscoop is 30 inches wide, and consists of an apron 8 inches below the fuselage skin and rising to form a throat 3 inches from the skin at the point where the dust leaves the feed-control gate; it also may be constricted laterally. In some aircraft such as the Piper Cub there are two gates, one on each side of the fuselage bracing; the dust hopper is thus of breeches-type design. The rear part of the air-scoop is broadened into a fan with vertical baffles, or divided into two separate outless, to spread the dust horizontally. Light planes can carry from 250 to 500 lbs. of dust, although the Stearman has been reported to carry up to 1000 lbs. Emission rates range from 1.6 to 5 lbs. per second. The swath width obtainable with the Cub J3 aircraft lies between 10 and 15 yards. Aerial dusters were used for malaria control with Paris green larvicides from 1925 until 1944, when the advent of DDT rendered this method obsolete.

Aerosol Generators for Aircraft. Venturi tubes, into whose throat the insecticide is sprayed from two nozzles, have been fitted to the exhaust stacks of aircraft ranging in size up to the Dakota. With Stearman aircraft, both larvae and adults of *Anopheles quadrimaculatus* have been controlled in Alabama and Tennessee. With Anson aircraft, tsetse flies have been controlled in south-central Africa. Todd fog-heads from the TIFA aerosol generator have been fitted to the Bell 47 helicopter and used to control adult black-flies (*Simulium*) in New York state (Plate II. Fig. 4). In general, aerosols are not as efficient as sprays applied from the air in controlling flying insects or their larvae. Moreover, the exhaust generators may exert a back pressure on the engine of the aircraft.

REFERENCES

American Mosquito Control Association (1948). The Use of Aircraft in the Control of Mosquitoes. AMCA Bulletin No. 1, Albany, N.Y. 46 pp. **Brown, A. W. A.** (1951). Insect Control by Chemicals. John Wiley and Sons, New York. pp. 337-466. **Brown, A. W. A.** and **Watson, D. L.** (1953). Studies on Fine Spray and Aerosol machines for control of Adult Mosquitoes. *Mosquito News*, 13: 81-95. **Bureau of Entomology and Plant Quarantine** (1941). Insects in Relation to National Defence. Circular 20, U. S. Department of Agriculture. 56 pp. **Director-General Army Medical Services** (1946). The use of DDT in the Field. Manager of Publications, Delhi. **Federal Security Agency** (1947). Malaria Cntrol on Impounded Water. Superintendent of Documents, Washington, U.S.A. pp. 149-186 and 391-410. **Higgins, A. E. H.** (1945). *British Medical Pulletin*, 3: 229-232. (Design of air-atomizer sprayers). **McClintock, J. A.** and **Fisher, W. B.** (1945). Spray Chemicals and Application Equipment. Horticultural Press, La Grange, Indiana. 319 pp. **World Health Organization** (1951). Technical Report Series, No. 34, pp. 27—36. (Specifications for spraying apparatus). **World Health Organization** (1952). Technical Report Series, No. 46, 35 pp. (Specifications for spraying apparatus).

RESIDUAL SPRAYING

The significance of DDT and certain other modern insecticides lies in their residual contact toxicity. Deposits of these chemicals upon surfaces are lethal to mosquitoes, house-flies, sand-flies, bed-bugs and cockroaches which alight or walk upon them and contaminate themselves with a toxic dose. When applied as a spray in sufficient concentration, these insecticides may remain lethal for many weeks. The formulations employed for this purpose are 5 per cent solutions, emulsions or suspensions mixed on a weight-volume basis to contain 0.5 lbs. of DDT per imperial gallon.

The spray is usually applied with a 3-gallon back-portable sprayer of the pneumatic type, in which air is compressed by hand-pumping. A coarse atomization is preferable, and the spraying pressure should be approximately 40 lbs. per square inch. Flat nozzles are favoured, and the fan of spray is swept across the surface from a distance of 8 to 15 inches (Plate IV. Fig. 8). The surface is systematically covered, but the spray is not allowed to accumulate to the point where it runs off. With 5 per cent solution, 1 quart can be applied to 300 square feet without run-off, and if 2 quarts are applied to every 1000 square feet the dosage of DDT will be 113 mgms. per square foot or approximately 1 gram per square metre.

When 5 per cent DDT in unrefined kerosene is applied to a smooth surface such as glass, metal or sized cement, the insecticide remains in supersaturated solution even although the bulk of the solvent has evaporated. At this time the deposits are extremely toxic to insects which contact them, since they readily adhere to and penetrate through the pulvilli and tarsal segments of their legs. But eventually some stimulus, for example the brushing effect of insect activities, sets the DDT to crystallizing. Bunches or rosettes of large long needles are formed, which remain closely attached to the surface. These are not so readily detached and taken up by the feet of the insects, and so it is found that the change from a "wet" to a "dry" deposit is marked by a loss in residual toxicity to mosquitoes, house-flies and cockroaches. The addition of a nonvolatile solvent such as lanolin would delay this change, but this measure is not suitable for spraying in houses.

When a 5 per cent solution of DDT is applied to an absorbent material such as wall-board, it largely disappears below the surface. As the solvent evaporates and crystals form, they creep out to the surface and produce a bloom or efflorescence; thus the residual toxicity, after having initially fallen, rises again. This efflorescence also occurs on unpainted wood, cardboard, paper and some fabrics. It has been utilized by the incorporation of DDT into urea-formaldehyde resin, which produces an insecticidal coating from which repeated crops of crystals may be obtained by efflorescence.

When a DDT solution is applied to mud, adobe, plaster or unsized cement surfaces, the insecticide is absorbed, but fails to effloresce. Between 80 and 90 per cent of the DDT disappears below the superficial layer, 0.1 mm. in thickness, and does not appear again. And so it is found that the application of 5 per cent DDT solutions to mud walls, for example in East Africa, Morocco and Mexico, produces comparatively poor control of mosquitoes. These muds also contain Fe and Al, which may catalytically destroy the DDT. Foliage and plant materials also absorb DDT solutions; in the bamboo, palm and thatch huts of Amazonia, solutions achieved only 79 per cent kills while suspensions gave 93 per cent mortality of mosquitoes.

Suspensions are therefore recommended for anti-malarial spraying in mud or thatch dwellings, and for house-fly control in wooden barns and barracks. The best material is a 50 per cent DDT wettable powder, suspended at the rate of 1 lb. per gallon. Lower concentrations of DDT in the wettable powder are less successful, because the insecticide is masked by the inert diluent powder; 75 per cent DDT wettable powders are now available which offer a saving in shipping costs, but they are inferior in performance. The insecticide and diluent are ground down to a particle size which can be readily taken up by mosquitoes and house-flies. The optimum diameter for wettable powder particles is about 10 to 20 microns, and they readily

PLATE V

Fig. 10

Boom-and-nozzle assembly on a Stearman aircraft.

Fig. 11

Mixing emulsions Lofstrand sprayers in Thailand.
Courtesy World Health Organisation.

PLATE VI

The two common plague-rats of India.

FIG. 12.
Rattus norvegicus.
(Brown or wharf rat)

FIG. 13.
Rattus rattus.
(Climbing rat: Alexandrine rat: and includes the subspecies, black rat)

lodge in the hairs of the tarsi, and thence are transferred by cleaning movements all over the insect. Whereas the optimum size for uptake is 20 microns, the optimum particle for retention and maximal toxicity is considerably smaller. Such finely-ground powders will not settle out in the spray tank or clog the nozzle.

Emulsions appear to behave in the same way as solutions on many surfaces. However, they are often favoured since they may be made by simple dilution with water of a 25 per cent DDT concentrate in xylene, naphtha or alkylnaphthalene with an added emulsifier (Plate V. Fig. 11). Should the emulsion fail to be absorbed into the material and remain on the surface, the crystals may be coated and stuck down by the emulsifier so that they cannot be taken up by the insect. However, emulsion concentrates are now available in which the DDT particles are preformed in a little water included in the oil; when applied after dilution, the crystals remain on the surface surrounded with oil and thus produce the most toxic residual formulations known (*e.g. Supona-D*).

There are a number of subsidiary considerations in a choice of formulation for residual spraying. Oil solutions are best for treatment of window screens, and should be painted on with a wire brush. In frame buildings, oil solutions introduce a fire hazard while suspensions and emulsions do not. When solutions are applied to fresh paint, varnish or linoleum, the DDT is lost by absorption into it. In high-class homes, suspensions are undesirable because they leave unsightly white residues; on the other hand, solutions may stain certain finishes and wallpapers. On outdoor surfaces, the residue from suspensions is more short-lived and is readily washed off by rain. In this case the addition of 2 per cent pine gum resin to a DDT xylene emulsion has prevented the deposit being washed away. Certain emulsion materials may soften fresh paint and varnish and cause rayon artificial silk to shrink. In spraying DDT or BHC on to walls with limewash or whitewash, the decomposition of the insecticide by the alkali may be minimized if solutions are avoided and application is delayed until the limewash is at least 7 days old.

REFERENCES

Barlow, F. and **Hadaway, A. B.** (1925). *Bull. Ent. Res.* 42: 769-777. (Particle size and uptake by insects). **DaCosta, A. M.** and **Basseres, M. S.** (1947). *Anais VI Cong. Brasileiro de Hig., Rio de Janeiro.* pp. 325-328. (Solutions inferior to suspensions). **Department of the Army** (1947). Insect and Rodent Control. Tech. Manual TM5-632, Superintendent of Documents, Washington, D.C. **Downs, W. G., Bordas, E.,** and **Navarro, L.** (1951). Science 114: 259. (Decomposition of DDT by Fe in adobe). **Federal Security Agency** (1947). Malaria Control on Impounded Water. Superintendent of Documents, Washington, D.C. pp. 203-222. **Hadaway, A. B.** and **Barlow, F.** (1952). *Trans. Roy. Soc. Trop. Med. Hyg.*, 46: 236-244. (Physical factors affecting residual deposits). **Jensen, J. A., Sumerford, W. T.** and **Fay, R. W.** (1950). *Soap San. Chem.* 26 (11), 121-123. (Pine-gum as sticker for DDT emulsions). **Symes, C. B.** (1951). *Brit. Med. Bull.*, 6: 64-73. (Recent progress in malaria control : summary).

CHAPTER IV

INSECTICIDAL CONTROL OF NOXIOUS DIPTERA

RECENT HISTORY OF MALARIA CONTROL

Following the release of DDT for civilian use at the close of the second World War, it has become possible to control and even eradicate malaria over large sections of the globe. In this enterprise WHO, UNRRA, FAO, UNICEF, the Rockefeller Foundation, the Colonial Office and the governments of many States have shown that organized campaigns over large areas can produce splendid results. The method employed is to spray the interiors of house and dwellings with DDT (or BHC and occasionally other chlorinated hydrocarbons) and thus to control *Anopheles*, *Aedes*, *Culex*, *Phlebotomus*, *Musca* and *Cimex*, and the diseases which they transmit.

The routine was first developed in Italy in 1944, and by 1947 an extensive campaign was under way to control *Anopheles maculipennis labranchiae* by spraying the houses and cattle-sheds, larvicidal measures being discontinued. As a result the number of new cases of malaria in the Agro Romano dropped from the annual figure of 435 down to 0 between 1947 and 1950, and the mortality rate in Italy has been drastically cut from 24 per million down to 0 for the entire country in 1949. In Sardinia a programme of house spraying was started in 1946 in combination with larviciding; it succeeded in almost eliminating *A. m. labranchiae*, so that by 1949 there were no cases of primary malaria on the island. It is considered that these measures have probably eradicated this species from the neighbouring island of Elba. In Spain, a campaign of DDT larviciding incepted in 1945 has been extended to house spraying; in Portugal, a campaign was commenced in 1948 using BHC in comparison with DDT on a large scale; *A. m. labranchiae* is the main vector on the Iberian peninsula.

On the island of Cyprus, a programme was started against *A. superpictus* in 1946, and its success was attested by the official statement in January 1950 that malaria had been eradicated. In Greece, a campaign of house and barn spraying supplemented by larviciding from aircraft, which was commenced in 1946 against *A. sacharovi* and *superpictus*, resulted in the elimination of malaria by 1950, accompanied by the disappearance of sand-fly fever and a great reduction in leishmaniasis. In Yugoslavia, measures undertaken since 1947 have given protection to 5 million people, and resulted in a great decrease in the morbidity rate of malaria. Turkey has undertaken a similar programme against these two species of *Anopheles*.

In Asia, a thorough and successful programme was undertaken by Ceylon in 1946 against *A. culicifacies*. By dint of repeated sprayings, as many as 7 per annum, the Dominion government in 3 years has reduced the infant mortality rate by 62 per cent, and the spleen rate in the malarious interior of the island from 77 down to 2.7 per cent.

In India, operations were commenced in 1946 in Bombay State against *A. fluviatilis* and *A. culicifacies*, where 3 sprayings a year were applied. Starting with Kanara and Dharwar districts, the programme later extended to 7 districts with a total population of 5 million people. As a result the annual malaria incidence was decreased by 300,000, the spleen rate fell from 70 down to 7 per cent, and the parasite rate from

20 down to 2 per cent. The incidence of plague disappeared, due to the control of fleas, and dysentery was greatly diminished due to the control of houseflies resulting from the house spraying.

Operations in Ernad and Malnad, Mysore and in the Jeypore Hills in 1949 succeeded in eradicating *A. fluviatilis* from villages which were sprayed. Work done in Orissa in 1948 materially reduced *A. annularis* in numbers. Spraying in Terai in 1949 cut the malaria endemicity, normally as high as 60 per cent, down to zero. The treatment of 193 square miles in East Bengal in 1950 resulted in a greater productivity of labour which increased the rice crop by 15 per cent.

Measures were undertaken against *A. superpictus* in Afghanistan, commencing in 1949 with a small-scale programme in Laghman district, followed in 1950 by a campaign in the Khanabad area; they resulted in reducing the spleen rates by 50 per cent. In 1950 a programme was started in Iran, which is now giving good protection against *A. m. maculipennis*. A model example of malaria eradication was carried out on the island of Bahrein, in the Persian Gulf, in 1948.

In Malaya and the Philippine Islands, however, much less success has been ecnountered in house spraying, since the principal vector *A. minimus* does not settle on walls. Against this species in Malaya, BHC appears to be generally superior to DDT, provided it is applied frequently. Nevertheless, house spraying with DDT in Assam against *A. minimus* has succeeded in reducing the incidence of malaria by 50 per cent. Larvicidal measures are also employed to control *A. minimus* and *A. maculatus* breeding in slow streams. The control of *A. letifer* is possible in Malaya provided the huts are sprayed at least once a month.

However, in northern Thailand the spraying of houses with DDT in 1949-51 has resulted in almost complete disappearance of *A. minimus*, the cessation of malaria transmission, and the termination of an endemic plague of bed-bugs. Consquently, the malaria morbidity figure of 50.5 per thousand in 1949 dropped to 2.1 in 1951. Initial control operations were started in 1951 in Indo-China and Burma. Preliminary work in Java showed that *A. sundaicus* could be controlled for 5 months by a single spraying, and the infant parasite rate reduced from 18 down to 0; DDT sprays are also effective against *A. aconitus*.

In the Pacific Islands, control operations were undertaken by the Allies against *A. punctulatus* in New Guinea and *A. farauti* in the western Pacific, by area spraying with military and civilian aircraft in 1944 and 1945. Despite the excellent results then obtained, this work has now been discontinued.

On the continent of Africa, results obtained by house spraying have not been as successful as elsewhere. In Morocco, the spraying of DDT and BHC on mud walls in that exeptionally hot dry climate achieved mediocre control; therefore larvicidal measures have been substituted. In Egypt and northern Sudan, eradication of *A. gambiae* had already been almost obtained by larviciding with Paris green in 1944-5. In Uganda, house spraying in 1948 did not achieve much reduction in the mosquito population, the poor result being largely attributed to the nomadism of the inhabitants.

The effectiveness of house spraying with DDT against *A. gambiae* in Africa is largely reduced by the readiness of this species to leave houses and "take to the bush," and better results are obtained with BHC. A programme of house spraying with BHC at Ilaro, Nigeria succeeded in 1950 in getting 90 per cent reduction of *A*, *gambiae* and a greater elimination of *A. funestus.* House spraying with BHC in the Belgian Congo has been successful against *A. moucheti* and *A. gambiae.*

In the northern Transvaal, operations with DDT between 1946 and 1950 have succeeded in eliminating malaria and blackwater fever from an area of 42,000 square miles. A combination of larviciding with house spraying has completely controlled *A. gambiae* as well as *A. funestus,* and reduced the infant parasite rate from 62 down to 12 per cent. A parallel programme in Southern Rhodesia, where both BHC sprays and DDT sprays were used, gave similar results. Programmes with emulsions and suspensions of DDT were also incepted in Natal in 1949, and in Swaziland in 1950.

Two islands in the Indian Ocean, off the east coast of Africa, heavily infested with malaria carried by *A. gambiae* and *A. funestus,* have provided most striking malaria control programmes. On the smaller island of Mauritius a programme of house spraying with DDT between January and June 1949 achieved 99.9 per cent control of *A. funestus* and *Aedes aegypti*; there was however 2.9 per cent survival of *A. gambiae,* which was later covered by larviciding. The malaria parasite rate dropped from 9.5 down to 0.36, and the spleen rate from 34.8 down to 3.2 per cent. On the very large island of Madagascar, house spraying with DDT, BHC and octochlor in 1949-50 protected 1 million people by achieving a marked decrease in the numbers of *A. funestus* and *A. gambiae* and in the parasite rate.

On the American continents, the campaign conducted in British Guiana against *A. darlingi,* a species that remains in houses for considerable periods, has given outstanding results. Experimental spraying commenced in 1945, and large-scale operations started in 1947. The spraying of houses with DDT solution every 8 months, without any larviciding, succeeded in eradicating both *A. darlingi* and *Aedes aegypti* from the coastal areas, thus protecting 90 per cent of the colony's population against malaria. The general spleen rate fell to 1.5 per cent, and the parasite rate fell from 55 per cent down to zero. In addition, 80 per cent control of *Culex quinquefasciatus,* the vector of *Filaria bancrofti,* was obtained. A similar campaign now being undertaken in French Guiana has eliminated *A. darlingi* and *A. aegypti* with residual sprays of DDT and BHC suspensions, and is combatting *C. quinquefasciatus* with BHC larvicides.

On the Virgin Islands, DDT treatments have almost eliminated *Aedes aegypti,* and greatly reduced the numbers of *Culex fatigans.* The same results are being obtained on the islands of Trinidad and Tobago, where house spraying is supplemented by larviciding with oil, as well as eradication of the bromeliad plants in which certain of the anophelines of the sub-genus *Kerteszia* breed. Due mainly to the resultant control of *A. aquasalis,* the malaria parasite rate on the islands dropped to 1.5 per cent by 1951.

A large-scale campaign was started in Venezuela in 1947; the villages of the highly malarial Llanos region, an area of 233 thousand square miles, were sprayed with DDT solution. A decisive reduction of *A. darlingi* was obtained; as a result the malarial spleen rates, which had been between 80 and 100 per cent, were reduced in 1951 to a level between 3 and 16 per cent. *Anopheles albimanus* has not been significantly controlled by house spraying, because it is not a domestic species. Colombia and Bolivia have extensive control projects in hand by 1951.

In Brazil, the invasion of *A. gambiae* from Africa had already been eradicated in 1940 by larviciding with Paris green. Oil solutions of DDT are now applied as larvicides as a supplement to house spraying. Solutions of DDT, later replaced by suspensions, were applied to over 3 million dwellings between 1947 and 1950. The control of *A. darlingi* has been so successful that the parasite rate dropped to 0 in most treated areas. Experimental spraying of jungle with DDT from helicopters, however, obtained only temporary reduction of *A. bellator* and *A. cruzii*, species that breed in bromeliad plants.

In Peru, the application of DDT solutions in the coastal valleys to control *A. pseudopunctipennis* has brought the parasite rates, which ranged from 5 to 25 per cent, down to 0 to 0.6 per cent; the rate in Lima has fallen from 13.4 down to 0.06 per cent. In Ecuador, some 800 thousand inhabitants were similarly protected in 1949-50. In Chile, malaria had already been practically eliminated in 1945, but *A. pseudopunctipennis* continues to be attacked by DDT larvicides, recently supplemented by house spraying. In Argentina, the spraying of house with DDT between 1947 and 1949 reduced the annual malaria incidence from 300.000 down to 2,800, so that by 1951 endemic malaria was no longer considered to be a serious health problem.

On the North American continent, house spraying has been practised in many parts of Mexico since 1928. It has resulted in a 99 per cent reduction of *A. pseudopunctipennis*, with complete eradication being obtained in some areas. Control work in the Xochimilco region against *A. aztecus* has reduced the parasite rate to 0. In the Southern United States, a campaign incepted in 1935 using Paris green larvicide had already materially reduced the populations of *A. quadrimaculatus* and *A. m. freeborni*. With the commencement of a programme of annual spraying with DDT at 200 mgms. per square foot, malaria has virtually been eliminated as a public health problem, and control measures may now be restricted to areas surrounding the incidence of new cases of the disease.

REFERENCES

Buxton, P. A. (1952). *Trans. Roy. Soc. Trop. Med. Hyg.*, 46 : 213-226. **Clavero, G.** (1950). *Rev. Sanidad e Hig. Publico.*, 24 : 149-170. **Department of Information** (1950). The Health of a Nation. Kandy, Ceylon. **Dowling, M. A. C.** (1951). *Nature* 167 : 67-68; and *Brit. Med. Bull.*, 48 : 311-314. **Downs, W. G. Celis, S. H.** and **Gahan, J. B.** (1950). *Am. J. Hyg.*, 52 : 348-352. **Giglioli, G.** (1948). Malaria, filariasis and yellow fever in British Guiana. Mosquito Control Service, Medical Department, British Guiana. **Kar, P. K.** (1950). *Indian J. Malariol.* 4 : 305-307. **Mackerras, I. M.** *et al.* (1950). *Coun. Sci. Ind. Res., Australia, Bull.*, 257 : 64 pp. **Mastbaum, O.** (1951). *E. African Med. J.*, 28 : 67-74. **Missiroli, A.** (1950). *Am. J. Trop. Med.*, 30 : 773-783. **Nair, C. P.** (1951). *Nature* 167 : 74-75. **Pampana, E. J.** (1951). *Bull. World Health Org.*, 3 : 557. **Pinotti, M.** (1951). *Trans. Roy. Soc. Trop. Med. Hyg.*, 44 : 663-682. **Rao, T. R.** (1951). *Bull. World Health Org.*, 3 : 639. **Russell, P. F.** (1951). *J. Nat. Mal. Soc.*, 10 : 257-265. **Simmons, S. W.** and **Upholt, W. M.** (1951). *Bull. World Health Org.*, 3, 535-556. **Summary on Malaria.** (1950). *Trop. Med. Bull.*, 48 : 311-314. **Symes, C. B.** (1951). *Brit. Med. Bull.* 8 : 64-73. **Symposium** (1951). *J. Nat. Mal. Soc.*, 10 : 97-196. **Van Tiel, P. H.** and **Winoto, R. M. P.** (1951). *Documenta Neerlandica et Indonerica de Morbis Tropicis.* 3 : 295-319. **Vedemanikkam, J. C.** (1949). *Indian J. Malariol.* 3 : 331-338. **Viswanathan, D. K.** and **Rao, T. R.** (1949). *Indian J. Malariol.*, 3 : 269-312. **Wharton, R. H.** (1951). *Bull. Ent. Res.*, 42 : 1-20. **Williams, L. L.** and **Stowman, K.** (1951). *Mosquito News*, 11 : 1-7. **World Health Organization** (1951). Malaria Conference in Equatorial Africa. Tech. Report Series 38, 72 pp.

MOSQUITO CONTROL

Larvicides : Surface-spreading Oils. The application of oils to the surface of breeding pools has proved a highly effective means of control before the advent of DDT. The mosquito larvae would become contaminated with a toxic dose of the aromatic, unsaturated and phenolic compounds present in the unrefined oils. This necessitated a surface film at least 1 micron thick, so that the oil would penetrate into the tracheae; such a film also prevents fresh oviposition by adults. The successful achievement of a continuous film of oil was found to require a dosage of 15 to 20 gallons per acre on the average, certain situations requiring more. The expenditure usually recommended was 1 pint per 250 square feet, equivalent to 0.5 ounce per square yard, or 25 gallons per acre.

Suitable oils for larviciding should have an adequate content of toxic impurities and should be able to spread over the surface of natural waters. Certain unrefined oils such as *Malariol* (Shell or Burmah Oil Company) have proved especially good; the spreading pressure of the Assam formula is 26 dynes per centimetre. Kerosene, No. 2 fuel oil or No. 2 diesel oil, with spreading pressures of only 8 to 10 dynes per cm., has proved adequate for cleaner North American waters. For use in the tropics it is mixed with crude oil, naval bunker oil being especially good for spreading properties. Larvicidal oils suitable for North American waters should be fluid but not too volatile, and should have viscosities between 31 and 43 S.S.U., specific gravities between 0.83 and 0.86, initial boiling points between 297° and 414° F., final boiling points below 800° F., and spreading coefficients no less than 17 dynes per cm. But it must be remembered that the surfaces of tropical pools may be so covered with bacteria, algae, protozoa, pollen and dust, that they oppose the spreading of oils which have spreading pressures as high as 30 to 35 dynes per cm.

The spreading properties of oils are due to small amounts of impurities, not to the oil itself; the higher paraffins have no spreading power, and the aromatic compounds have only limited ability to spread. Irradiation with sunlight greatly increases the spreading pressure of an oil. The addition of sulphonated oil, or of the products of combined polymerization and oxidation of olefines, greatly increases the spreading pressure. Other spread-aiders include cresylic acid, oleic acid, crude castor oil, or commercial resin, and these are often added in practice. Unfortunately the cresylic acid is too water-soluble and causes contraction of the film, whereas oleic acid tends to come out onto the surface of the water and force the oil film apart. However, with appropriate blending and addition, it has been possible to produce a formula, called *Malariol HS* (high spreading), that is adequate for all tropical waters.

It has been found recently that pine oil and lemon-grass oil will control the breeding of anophelines in small bodies of water at one-quarter the dosage required with *Malariol*. The addition of special detergents to fuel oil, *e.g.* *Triton B*-1956 (a phthalic glyceryl alkyl resin), *Nopco* 1216 (sulphonated sperm oil) or *Amine* 230*X* (a complex with a C_{18} aliphatic chain) can raise its larvicidal activity by 6 to 8 times. Kerosene and isophorone are the two oily materials most susceptible of having their spreading pressure raised by these additives. The best spread-aiders for kerosene are *Triton X*-100, *Triton B*-1956, and *Ethofat* 142/20.

The application of heavy doses of larvicidal oils has the additional property of destroying the vegetation around the water margins, which may be advantageous in some cases to reduce the susceptibility of pools to mosquito breeding, and also to indicate clearly which pools have been treated. Although they are now usually replaced by light doses of oil containing BHC or DDT, heavy doses of malariol oils often have a more persistent effect.

Larvicides : Chemical Insecticides. Paris green has proved to be a powerful larvicide for *Anopheles*, but not for Culicines. It was quite widely applied in the form of 1 to 3 per cent dusts or in suspension in oil and water at dosages approximating 1 lb. of Paris green per acre. Small bodies of water have been disinfested by emulsions of pyrethrum in oil. Phenothiazine applied in dust form at 0.5 to 2 lb. per acre has given good control of *Anopheles* larvae. PDB and tetrahydronaphthalene, copper arsenite and calcium arsenite have proved fairly effective substitutes, although the latter 3 compounds render the water poisonous to livestock (*e.g.* calcium arsenite requires a dosage of 7 lb. per acre).

Larvae of almost all *Aedes*, *Culex* and *Anopheles* mosquitoes are completely controlled by application of 0.1 to 0.2 lb. of DDT per acre to the surface of breeding pools. *Anopheles* may even be eliminated with 0.05 lb. per acre of DDT. The usual formulation is a 5 per cent solution of DDT in kerosene or fuel oil, which expends only 0.2 gallons of oil for the 0.1 lb. of DDT required per acre. Malariol HS itself has been favoured as a DDT solvent in operations against *A. minimus* and *A. maculatus* in Malaya. Alternatively, a 5 per cent emulsion may be made by adding water to a 25 per cent concentrate (containing 25 lb. DDT, 71 lb. xylene or Velsicol, 4 lb. Teepol or Triton detergent). The larvicide is usually applied by a 3-gallon compressed-air sprayer or similar knapsack sprayer, although a large hand sprayer with an air reservoir and atomizing nozzle is quite suitable. When large pools are treated, the atomization should be as fine as possible, so that the spray can drift with the wind for a distance of 25 feet or so. This method of larviciding may be more profitably practised with a mist-blower mounted on a vehicle or towed behind it. For the treatment of small pools, a 1-gallon can is preferable if designed to eject a coarse jet of 0.5 c.c. of liquid at a time. For complete control of larvae, it is recommended that the pools be treated every 1 to 3 weeks; if they do not contain fish, a single application of 1 lb. of DDT per acre is permissible and effective.

Dusts containing 1 to 5 per cent DDT may also be applied at the rate of 0.2 to 0.5 lb. of DDT per acre. By means of a hand-operated rotary duster, the material may be made to drift with the wind over the area. But dusts, like suspensions, are inferior to solutions and emulsions for larvicidal work; solutions are slightly superior to emulsions. The droplet size of the spray is of no importance provided the water surface can be completely covered. Solutions in oil are entirely satisfactory, since the deposited droplets immediately spread to form a film. The spreading pressure of kerosene is definitely increased by addition of a wetting agent; *Triton X*-100 is superior in this respect. The larvae may be poisoned on rising to contact this film, or enough DDT may reach the aqueous phase as the oil dissolves in the water.

Oil solutions of DDT applied at 0.25 to 0.5 lb. per acre from aircraft have given excellent control of *Anopheles*, *Aedes* and *Culex* larvae, whether applied from Cub planes with 15-yard swaths, or from Dakota aircraft with 200-yard swaths. Higher concentrations of DDT in the solution are equally effective, and make a great saving in planeload and flying times. DDT aerosols obtained by introducing 20 per cent concentrates into the exhaust pipes of aircraft have given satisfactory results, despite the tremendous loss by drift. For example, an emission rate of 0.1 per acre results in 98 per cent control of *Anopheles* larvae, despite an actual deposit of only 0.01 lb. per acre. For penetration through forest, the bulk of the droplets should be approximately 50 microns in diameter. Aerosols produced by ground equipment, (such as the TIFA and Besler generators) also control mosquito larvae; the machines are towed across the wind at quarter-mile intervals and emit DDT at the rate of 0.2 to 0.3 lb. per acre.

DDT may be applied to marshes when the ground is desiccated during the dry season, or to rice-fields along with the fertilizer before they are flooded. Such pre-hatching treatments, applied at about 1 to 2 lb. per acre, have given excellent results against larvae of the rice-field mosquito *Psorophora confinnis* and certain salt-marsh *Aedes* mosquitoes. In cold regions similar results may be obtained by applying as little as 0.1 lb. per acre over the snow; solutions are used for aerial spraying, and dusts (which are lighter to carry) are preferred when men on foot make the application.

Larvae of the subgenus *Kerteszia*, which breed in water accumulating at the base of bromeliad plants in Brazil and the Caribbean area, required special treatment. Aerial spraying with DDT has an effect, but the reduction persists for only a few weeks. The present method of attack is to spray 1 per cent copper sulphate in water; copper acetoarsenate is also very effective. The clearing of shore-lines from aquatic growth may be accomplished by spraying with 2,4-D or 2, 4, 5-T at 1 lb. per acre, in suspension or oil solution. Sodium arsenite also is effective and is relatively cheap.

An ingenious method of larviciding is to use an emulsifiable concentrate of DDT or BHC enclosed in gelatin capsules the size of golf balls. Called *Tossits* by the manufacturer, they may be thrown on to pools that are difficult of access, in numbers and spacings arranged to give the correct dosage; their main disadvantage is that the contents cannot spread into the weedy margins of pools. For aerial application through dense vegetation, DDT in xylene solution may be impregnated into small granules of bentonite to give a very coarse dust. For control of mosquitoes in rice fields, the seed may be treated with DDT. Briquettes composed of plaster of Paris, and sawdust, soaked in an oil solution of DDT, have been used to protect open wells in East Africa from mosquito breeding. The latter method has been employed against larvae of *A. minimus flavirostris* breeding in Malayan streams. Drinking-water in jars may be protected against infestation with *Aedes aegypti* by treating them with 1 drop of 2 per cent DDT in ethyl alcohol for every 6 litres of capacity; this protection lasts for 5 weeks even when the pots are emptied and refilled.

Only if resistance to DDT is encountered in the larvae is it advisable to consider using any other insecticide. DDD is not as toxic as DDT to fish, but it is not quite as good a larvicide; methoxychlor also is slightly inferior. However, DDD applied at the rate of 0.014 p.p.m. has eliminated the Clear Lake gnat, *Chaoborus astictopus*, from 65 square miles of lake water in California. Chlordane and toxaphene are definitely inferior to DDT. BHC has proved erratic in performance, although the refined gamma isomer, lindane, is more larvicidal than DDT. BHC applied at 2 gms. gamma isomer per 100 square yards, equivalent to 0.1 lb. lindane per acre, almost completely controlled *Anopheles maculatus* in Malaya. Aldrin and dieldrin are decidedly more effective, and the latter compound may be used at one quarter the dosage of

DDT. For example, dieldrin at 0.05 lb. per acre gave 99 per cent control of the larvae of *Aedes sollicitans* in Florida that are resistant to DDT treatments. In Alaska, lindane has proved even better than dieldrin at this dosage. In Korea, both dieldrin and lindane were more effective than DDT in controlling mosquitoes developing in rice paddies. Against *Anopheles quadrimaculatus*, dieldrin was superior to DDT, heptachlor and parathion, which in turn were much better than lindane, toxaphene, chlordane or Dilan. The most effective larviciding material is a 0.3 per cent emulsion of dieldrin in 1.5 per cent oil, applied at 2 to 4 gallons per acre to give 0.05 to 0.1 lb. dieldrin per acre. Parathion has been used against DDT-resistant salt-marsh *Aedes* in Florida, and EPN against DDT-resistant *A. nigromaculis* in California. Since these organic phosphates are very hazardous to fish and to humans, malathon is recommended as a substitute.

Adulticides : Space Sprays. Closed rooms and tents may be treated with space sprays to kill the mosquitoes inhabiting them at the time. The well-known hand atomizer or flit-gun is the favourite instrument, although telescoping pocket sprayers are now available built on the same principle. Shortly before the advent of DDT, houses were sprayed with pyrethrum emulsions from back-portable compressed-air sprayers with good success. Aerosol bombs or canisters, operated on the compressed-gas principle, are highly efficient and labour-saving. They contain pyrethrins, allethrin, or *Lethane* thiocyanates to obtain knockdown, and DDT to kill the mosquitoes. Aerosol bombs, which contain 0.5 per cent DDT plus 0.02 per cent pyrethrins in *Freon* propellent, are completely effective at an expenditure of 0.33 cc. per cubic metre, equivalent to 1 oz. per 3000 cu. ft. Allethrin has proved to be less effective than pyrethrins against the adults of *A. aegypti*. Lindane or gamma-BHC is 2 to 5 times as toxic as DDT in space sprays, but chlordane and toxaphene are inferior.

Aerosols are used for the disinsectization of aircraft before take-off. The U.K. Ministry of Health, and the W. H. O., recommend a low-pressure aerosol bomb formula containing 3 per cent DDT and 0.4 per cent pyrethrins, using *Freon* or other suitable propellent. The oil base for the insecticide must be non-inflammable and must not deteriorate the Perspex or Plexiglas windows. The aerosol is dispersed at the rate of 1 gm. per second to give a dosage of 10 gms for every 1,000 cubic feet of fuselage space. The aircraft is then kept closed for at least 5 minutes before the passengers come aboard. The formula G-651 approved by the U. S. Public Health Service contains 2 per cent DDT, per cent of a 20 per cent pyrethrum extract, and 8 per cent of methylated naphthalene (*Velsicol AR*-60 or *Sovacide* 544 *G*). The aerosol droplets so produced have a mass median diameter of approximately 25 microns. New developments for disinsecting aircraft include lindane vaporized from Fiberglas screens inserted in the air-conditioning system, and urea-formaldehyde resins containing DDT applied as wall coatings.

Adulticides : Residual Spraying. The standard adulticidal method for the control of *Anopheles* and other disease carrying house mosquitoes is the spraying of DDT upon walls and ceilings. The story of its recent success against malaria has been told in the previous section. The technique of residual spraying, and research into factors determining its success, have been discussed in Chapter III since they apply equally well to control of house-flies and many other domestic insects.

The desposit densities employed in various parts of the world range from 50 to 200 mgms per square foot. The concentration of DDT in the spray is 5 per cent in nearly all cases. Solutions, emulsions and suspensions have each been employed, having been developed in this succession.

A deposit of 200 mgms per square foot (approximately 2 gms per square metre) remains completely lethal to adult *A. quadrimaculatus* for a period of 15 to 36 weeks. After 1 year it may be expected to give 90 per cent control of this species; therefore one annual application at this dosage is the practice in the southern United States. In Venezuela and Brazil, applications at this dosage are made every 7 months against *A. darlingi*. The initial effect of spraying at 2 gms per square metre was found to be 75 to 90 per cent reduction within 5 days of the *A. maculipennis* population in houses in Afghanistan.

A deposit of 100 mgms per square foot applied every 10 months has achieved 99 per cent reduction of *A. darlingi* in British Guiana. Against *A. quadrimaculatus* in the U.S.A. this density remains effective for 2 to 3 months, but in Malaya it keeps the sprayed huts free of *A. letifer* for only 2 to 4 weeks. In Venezuela an application of 100 mg. per square foot every 3 months is a second choice to one of 200 mg. per square foot every 6 months. In Spain, either 2 gm. per square metre annually or 1 gm. per square metre every 6 months, are considered to be equal choices.

A deposit of 50 mgms per square foot has been widely favoured in India and Ceylon. In Ceylon, up to 7 sprayings per year at this dosage have given excellent results against *A. culicifacies*. In South India, applications of 50 mg. per square foot are made every 6 weeks against *A. fluviatilis*, and thorough control is obtained. In the Bombay area, it has been determined that 3 sprayings of 56 mg. per square foot every 6 weeks are superior to a single spraying of 168 mg. per square foot every 18 weeks for control of *A. culicifacies*.

As an overall evaluation of the various deposits, it was concluded from experience with *A. maculipennis* in Greece that the protective period in days was roughly equivalent to the number of centigrams of DDT applied per square metre of surface. Thus 2 gm. per square metre (200 mg. per square foot) could be expected to be effective for 200 days. The recommendation of W.H.O. for residual spraying in Africa, where *A. gambiae* and *A. funestus* are the most important species, is that a deposit of 200 mg. per square foot be applied every 6 months.

The original residual spray formula consisted of a solution of DDT in kerosene. It succeeded in achieving the practical eradication of malaria in British Guiana, where the houses are largely of wood. But in regions where the huts have mud walls, such as Mexico, Morocco and East Africa, poor results were obtained with solutions. The next step was the use of emulsions; these were adopted in the U.S.A., Brazil and the Transvaal, and now constitute the preferred formulation in the Mediterranean countries and India. Finally the advantages of suspensions of wettable powders became apparent, especially for use on the surfaces of mud walls, and they are now employed in Venezuela, British Guiana, Mexico, the U.S.A., the African territories, Mauritius, Ceylon and Malaya. The substitution of suspensions for emulsions for application to the rough clay walls and thatch roofs of many Indian villages has been specially urged, but the convenience of employing local materials (*e.g. Aromex* solvent, soap emulsifier) leaves the balance in favour of emulsions in that country. In the

U.S.A., preliminary trials have shown the effectiveness of residual deposits obtained by spraying hot liquefied DDT from an electrically heated sprayer.

There are evidently considerable differences in the susceptibility of the various species to residual spraying with DDT, and they are attributable mainly to the resting habits of the species concerned. In Africa, *A. gambiae* is controlled less than *A. funestus* and *Aedes aegypti*, because it leaves huts quickly. In Central and South America, *A. darlingi* is far more susceptible to house spraying than *A. albimanus*, *A. punctulatus* and *A. aquasalis*, which do not remain long in buildings. Here *Aedes aegypti* is readily controlled, but *Culex quinquefasciatus* is somewhat more resistant. In Malaya, *A. maculatus* is more susceptible than *A. sundaicus*, while *A. letifer* is more resistant. *Culex fatigans* was scarcely affected by DDT deposits of 100 mg. per square foot, while a level of 200 mg. per square foot only drove them out without killing them; however, a partial kill could be obtained with BHC.

For the residual treatment of walls, DDT is superior to the other chlorinated hydrocarbons. Based on initial kill combined with persistence of toxicity, the order of effectiveness is : DDT > methoxychlor > DDD > DFDT. For *Anopheles quadrimaculatus* and *Aedes aegypti*, the order of residual effectiveness was found to be : DDT > BHC and chlordane > toxaphene. At a deposit of 200 mg. per square foot, BHC and chlordane give complete kills for 4 weeks only, as against 25 weeks with DDT. When heavily applied in poorly ventilated rooms, BHC may remain effective for a year, as against a maximum of 3 months with chlordane. On clay walls, BHC may sometimes prove superior to DDT, but chlordane does not. Dieldrin has proved far superior to DDT both in initial kills and in lasting effectiveness, especially on plaster surfaces. It is now widely used in residual sprays for house-fly control. Pyrethrins applied to surfaces have given excellent kills, but are effective for only a week; a programme carried out against *A. fluviatilis* before the appearance of DDT required spraying to be performed daily. Pyrethrins synergized with piperonyl butoxide have given deposits which may remain effective for 3 months.

The effectiveness of DDT residual deposits has been handicapped by their repellency to some mosquitoes, notably to *Anopheles gambiae* in West Africa. This repellency derives initially from the vapour of the kerosene solvent and later from the stimulus imported by this nerve insecticide for the mosquito to fly off the treated surface. Thus Muirhead-Thomsom concluded that the resulting decline in hut populations during the day does not represent the true situation existing at night. Careful examination of this problem has shown that DDT stimulates the resting mosquitoes to take flight, and if they have failed to take up a sufficient dose of crystals on the tarsi they survive. With DDT suspensions deposited at 200 mg. per square foot in Uganda, 38 per cent of the *A. gambiae* that entered treated mud huts did in fact survive, although in wooden huts only 3 per cent survived. With similar deposits on adobe in Mexico, 35 per cent of the *A. pseudopunctipennis* that entered the hut escaped, but one-half of them died in the next 2 days. When wooden houses were treated in the United States for control of *A. quadrimaculatus*, it was found that 95 per cent of the mosquitoes that made their exit died during the next day. With deposits of 215 mg. DDT per square foot in Malaya, 23 per cent of the *A. minimus* leaving the hut and caught in window-traps were found to survive.

When BHC is used as a wall spray, the repellent effect is not encountered. The resting mosquito usually takes up a lethal dose of BHC, because the material is more toxic, the action is faster, and a fumigant effect is also involved. In the Uganda experiments, only 2 to 3 per cent of the *A. gambiae* entering mud or wooden huts, treated with 28 mgms. gamma-BHC per square foot, succeeded in surviving. In the Mexican experiments, only 20 per cent of the *A. pseudopunctipennis* escaped the hut treated at 20 mg. gamma-BHC per square foot, and of those 85 per cent died. In the Malayan experiment, none of the *A. minimus* reaching the window-trap survived. Where BHC was used most of the mosquitoes were killed in the hut and few died outside, whereas with DDT most of the mosquitoes reached the windows, as was found with *A. gambiae* in Kenya.

In Malaya it was noted that with *A. maculatus* also BHC gave better results than DDT. At present BHC is favoured over DDT for use in regions infested with *A. gambiae* in central Africa, and it is being used in Southern Rhodesia on a large scale. It is used as an alternative to DDT in Madagascar, Mauritius, French Guiana and Portugal. The recommendation of W.H.O. is that it be appled at a deposit of 10 mgms. of the gamma isomer per square foot every 3 months. It has been claimed that the admixture of 1 part gamma-BHC to 5 parts DDT produces a deposit that is more effective against Indian mosquitoes than twice the amount of either insecticide separately. Indian workers have also discovered that admixtures of β-methylanthraquinone, p-benzoquinone or benzil are synergistic with DDT against *C. fatigans*. In Italy, 1 part of DDT is now admixed with 3 parts of chlordane, to combat DDT resistance in *Culex* and *Musca*.

However, it has been the experience of certain workers that DDT is not inferior to BHC against *A. minimus* in Malaya. Moreover it has been found that DDT treatments are superior to BHC for the control of *A. fluviatilis* and *A. culicifacies* in Bombay State.

In many instances, for example, the control of *A. darlingi* in coastal British Guiana, the residual spraying of houses has proved capable of stopping the transmission of malaria. In Italy, larviciding has been abandoned in favour of house spraying alone. But in many parts of Brazil human habitations are not sufficiently abundant, and it is necessary to apply larvicides over the surrounding area. In the Transvaal, it is more economical to treat the breeding pools in urban areas where dwellings are especially numerous and built close together. In Morocco, a return was made to larvicidal measures because of the poor results obtained with house spraying. Treatment of breeding pools is the more reliable method of control for *A. minimus* in Malaya. Occasionally larviciding is used as a supplement for house spraying to complete the control of more resistant species, as for example *A. gambiae* on Mauritius and *C. quinquefasciatus* in French Guiana.

Adulticides : Area Control. Adult mosquitoes in sparsely populated country may be controlled by applying DDT from aircraft to large areas around the settlements. With aerial sprays, virtually complete control of *Anopheles* or *Aedes* may be obtained with 0.25 lb. of DDT per acre. With aerosols from exhaust generators approximatly 0·03 lb. per are is required to prevent the activities of adult mosquitoes. The application of 0.1 lb. per acre is considered adequate for aircraft spraying, since it gives 90 to 95 per cent control, and re-infiltration is so large a factor that re-treatments are necessary every 7 to 10 days. This schedule of weekly treatments with 0.1 lb.

per acre proved effective both for control of *Aedes* spp. in Alaska and also of Anopheles *farauti* in the New Hebrides. If it is desired to put down a single large application to control by residual action mosquitoes subsequently infiltrating the area, a dosage of at least 1 lb. of DDT per acre is required. In Florida a dosage of 10 to 15 lb. per acre gave complete kills of *Aedes taeniorhynchus* and *sollicitans* which settled in the area 15 days after the application, and was found to be still highly effective after 46 days. In Alabama, a 50-foot barrier sprayed at 140 lb. per acre gave almost complete kill of adult *A. quadrimaculatus* that were released within the area it enclosed. In Malaya, the spraying of a 60-foot barrier gave protection against *A. letifer* for 12 days. For residual depsits on vegetation, suspensions are better than emulsions; oil solutions are inferior, presumably because much of the surface deposit is lost by absorption into the foliage. On the other hand, for direct control of flying adult mosquitoes by DDT, suspensions are the least effective and oil solutions are the most suitable.

Aerosol fogs containing DDT and applied by ground equipment are very effective in subduing populations of adult mosquitoes. The thermal aerosol generator called TIFA (Todd Insecticidal Fog Applicator) has given particularly good results. Usually a 5 per cent solution in fuel oil is used by this machine, and it is emitted at a rate of 20 gallons per hour atomized to an average droplet size of 20 microns. If the machine is towed at 2 m.p.h. at right angles to the wind direction, complete kills may be obtained in open country for 400 yards downwind; the rate of expenditure here is 0.3 lb. per 100 yards of frontage, equivalent to 0.03 lb. of DDT per acre. If 25 per cent concentrates are used, approximately the same results are obtained if the vehicle now travels at 10 m.p.h. For treatment of woodland, higher dosages are recommended and a considerable wind-speed aloft (*e.g.* 8 m.p.h.) is required to produce the requisite movement of air within the woods. With the steam aerosol generator (Besler), good reductions of salt-marsh *Aedes* in Florida have been obtained for 300 yd. downwind in bush and one mile in the open when 10 per cent DDT in oil was emitted at 15 gallons per 1000 feet of frontage. Recent work with aerosol formulations for control of *Aedes taeniorhynchus* has shown 5 per cent dieldrin to be superior to 10 per cent DDT, and 2.5 per cent lindane to surpass both.

Resistance to Insecticides. The first instance of the development of resistance to DDT by mosquitoes appeared in central Italy in 1947 with *Culex pipiens autogenicus*. By 1950 this species was resistant to residual deposits of chlordane also. Subsequently it has appeared to be resistant to DDT in Greece. In the new world, *C. quinquefasciatus* has become highly resistant to DDT deposits in Puerto Rico, and there is evidence that it is resistant in a certain region of Venezuela. Larvae of *C. tarsalis* in California also have developed a marked resistance to DDT, as well as to toxaphene, lindane, aldrin and heptachlor.

Resistance of the salt-marsh mosquitoes (*Aedes sollicitans* and *A. taeniorhynchus* to DDT larvicides, which first appeared in Brevard Country, Fla. in 1948, is now widespread throughout Florida. These species are not at present resistant to other chlorinated hydrocarbons. In California resistance first appeared in 1948, involving *A. nigromaculis* and DDT larvicides; in 1950 *A. dorsalis* larvae showed a similar tolerance to DDT.

With the genus *Anopheles*, resistance has not yet interfered with the efficacy of house spraying with DDT in malaria control programmes. Attempts to discover resistance to this insecticide in *A. pseudopunctipennis*, *A. aztecus* and *A. albimanus* in Central America have been unsuccessful. The apparent resistance of *A. claviger*, noted during the malaria control programme in Sardinia, was found to be its normal species characteristic. With *A. quadrimaculatus* in the southern United States, only slight resistance to DDT could be induced by the sixth generation of laboratory selection; while field investigations showed that adults and larvae from sprayed areas were only slightly more resistant than those from unsprayed regions. However, this species may be becoming more "house-shy" in sprayed settlements. Only in Greece does DDT-resistance appear to be serious, where adults of *A. sacharovi* can now rest with impunity on treated surfaces, and *A. maculipennis* adults are becoming more tolerant of DDT than formerly. From field observations it has been suspected that adults of *A. superpictus* are becoming resistant to DDT in Turkey.

REFERENCES

Bertram, D. M. (1950). *Ann. Trop. Med. Parasitol.*, 44 : 255-259. (DDT vs. BHC for *A. minimus*). **Bradley, G. H.** and **Lyman, F. E.** (1950). *J. Nat. Mal. Soc.*, 9 : 113-118. (Annual spraying with 200 mg. DDT per sq. ft.). **Burgess, B. F.** (1949). *Soap San. Chem.*, 25 : 127 *et seq.* (Emulsifiers to spread larvicidal oils). **Busvine, J. R.** (1949). *Med. J. Malaya* 4 : 106-110. (Malariol to prevent oviposition of *Anopheles*). **Busvine, J. R.** (1951). Insects and Hygiene. Methuen, London. pp. 200-201. **Crowell, R. L.** and **Fay, R. W.** (1951). *J. Nat. Mal. Soc.*, 10 : 8-16. (Hot DDT). **Davidson, G.** (1951). *Bull. W. H. O.*, 4 : 329-332. (DDT vs. BHC for *A. gambiae* in Kenya). **Department of the Army** (1947). *Tech. Manual TM* 5-632, Superintendent of Documents, Washington, p. 26. **Downs, W. G.** and **Bordas, E.** (1951). *Am. J. Hyg.*, 54 : 150-156. (DDT vs. BHC for *A. pseudopunctipennis*). **Expert Committee on Insecticides** (1951). *W. H. O. Tech.* Report Series, No. 34, pp. 4-8. **False, J. H.** *et al.* (1951). *J. Econ. Ent.*, 44 : 621-622. (Formula for aircraft disinsectization). **Garrett-Jones, C.** (1951). *Bull. W. H. O.*, 4 : 547-562. (Immediate effect of 200 mg. DDT per sq. ft.). **Ginsberg, J. M.** (1930). *Proc. New Jersey Mosq. Exterm Aisoc.*, p. 57. (Pyrethrum as a larvicide). **Ginsberg, J. M.** (1951). *Mosquito News* 11 : 99-102. (Allenthrin compared with pyrethrins). **Goldacre, R. J.** (1949). *J. Animal Ecol.*, 18 : 36. (Spreading pressures required for field water surfaces). **Hess, A. D.** (1952). *Am. J. Trop. Med. Hyg.*, 1 : 371-388. (Insecticide resistance in mosquitoes : review). **Keller, J. C., Davis, A. N.** and **Mooney, C. I.** (1951). *Mosquito News* 11 : 171-174. (Larvicides for *A. quadrimaculatus*). **Keys, D. N.** (1951). *Med. J. Malaya* 5 : 220-229. (BHC for *A. maculatus* larvae). **Knipling, E. P.** (1951). *Mosquito News* 11 : 197-201. (Emulsifiers to raise toxicity of fuel oil). **Kruse, C. W.** and **Konchady, D.** (1950). *Indian J. Malariol.* 4 : 267-279. (Recommendations for residual spraying in India). **Kruse, C. W., Philen, E. A.** and **Ludvik, G. F.** (1951). *J. Nat. Mal. Soc.* 10 : 23-34. (Larviciding with aerosols from aircraft). **Lindquist, A. W., Roth, A. R.** and **Wzljer, J. R.** (1951). *J. Econ. Ent.*, 44 : 572-577. (Eliminatin of Clear Lake gnat with DDD). **Macdonald, G.** (1950). *Brit. Med. Jour.* 1950, p. 414. (DDT vs. BHC for *A. minimus*). **Matheson, R.** (1950). Medical Entomology. Comstock Publishing Co., Ithaca, N.Y. 2nd. Ed. pp. 384-387. **Missiroli, A.** (1950). *Am. J. Trop. Med.*, 30 : 772-783. (Mixture of chlordane and DDT). **Muirhead-Thomson, R. C.** (1947). *Bull. Ent. Res.* 38 : 449 and *Trans. Roy. Soc. Trop. Med. Hyg.*, 40 : 511. (Insecticides for *A. gambiae*). **Murray, D. R. P.** (1939). *Bull. Ent. Res.*, 30, 211-236 and 29 : 11-35. (Factors determining the spread of oils). **Nair, C. P.** (1951). *Nature* 167 : 74-75. (Spraying with 100 mg. DDT per sq. ft. against *A. letifer*). **Pal, R.** (1951). *Trans. Roy. Soc. Trop. Med.*, 45 : 125-126. (Mixture of BHC and DDT). **Rachou, R. G., Ferreira, M. O.** and **Lima, M. M.** (1950). *Rev. Brasileira Malar.* 2 : 306-317. (Copper sulphate to control bromeliads). **Ranganathan, S. K., Koshi, T.** and **Sitamaran, N. L.** (1949). *Nature* 164 : 1095. (Synergists for DDT against *Culex*). **Reid, J. A.** (1951). *Nature* 168 : 863-865. (Interspecific differences in susceptibility to DDT deposits). **Reid, J. A.** and **Ganapathipillai.** (1951). *Bull. Inst. Med. Res.* Malaya. No. 4. (Larvicidal oils). **Sasse, B. E.** (1950). *Mosquito News* 10 : 144-149. (DDT in ethanol to control *Aedes aegypti*). **Senior-White, R., Ghosh, A. R.** and **Rao, V. V.** (1945). *J. Malaria Inst.* India 6 : 129-215. (Pyrethrins to control *A. fluviatilis*).

Sulivan, W. N. (1952). *J. Econ. Ent.*, 45 : 544-545. (Vaporized lindane for aircraft disinsectization). **Symes, C. B.** (1951). *Brit. Med. Bull.*, 8 : 67. (Urea-formaldehyde resins as surface carriers of DDT). **Symes, C. B.** (1952). *J. Roy. Sanit. Inst.*, 72 : 490-512. (Recent progress in vector control by insecticides : review). **Tew, R. P., David, W. A. L.** and **Busvine, J. R.** (1951). *Monthly Bull. Min. Health* U. K., 10 : (Feb.), 30-38. (Aerosols for disinsectization of aircraft). **Toms, B. A.** (1950). *Bull. Ent. Res.*, 40 : 503-510. (Spreading pressures required for field water surfaces). **Upholt, W. M.** (1947). *U.S. Public Health Reports* 62 : 302. (Suspended mud reduces effectiveness of DDT). **Vedamanikkam, J. C.** (1949). *Indian J. Malariol.* 3 : 331-338. (DDT at 50 mg. per sq. ft. against A. *culicifacies*). **Venkataramanan, K.** (1950). *Indian J. Malariol.* 4 : 449-454. Pine oil and lemon-grass oil as larvicides). **Viswanathan, D. K.** and **Gadre, S. B.** (1950). *Indian J. Malariol.* 4 : 487-507. (Three sprayings vs. single spraying). **Viswanathan, D. K., Rao, T. R.** and **Juneja, M. R.** (1950). *Indian J, Malariol.* 4 : 505-531. (DDT superior to BHC in Bombay). **Wallace, R. B.** (1950). *Med. J. Malaya* 5 : 115-139. (BHC vs. DDT for *A. maculatus*). **Wharton, R. H.** (1951). *Nature* 167 : 854-855 and *Bull. Ent. Res.*, 42 : 1-20. (DDT and BHC for *Culex fatigans*). **Wilkinson, P. R.** (1951). *Bull. Ent. Res.*, 42 : 45-54. (DDT vs. BHC for *A. gambiae* in Uganda). **World Health Organization** (1951). Malaria Conference in Equatorial Africa. Tech. Report Series No. 38. p. 28.

MOSQUITO REPELLENTS

The materials first used for application to the skin to repel mosquitoes include oil of tar, pine oils and certain terpene fractions, pyrethrum extract, and oil of citronella. Pyrethrum resin has also been used in joss-sticks, which are slowly burned to an aromatic smoke which repels mosquitoes without killing them. Oil of citronella, a pleasant-smelling yellow viscous liquid obtained from citronella grass (*Cymbopogon nardus*) in Java and Ceylon, was the most effective of these natural products. There is no appreciable difference in effectiveness between the Ceylon oil and the Java oil. Citronella oil contains 17 to 34 per cent citronellal (which is considered the active agent), 16 to 45 per cent geraniol, some citronellol, and some methylheptenone (which also has been considered the active agent). Lemon-grass oil, a similar but inferior substitute obtained from *C. citratus*, contains 63 to 85 per cent of citral, and some citronellal. The vapour pressure of citronella oil is 0.66 mm. Hg. and that of citronellal is 0.3 mm. Hg. Although very powerfully repellent, oil of citronella does not afford protection in warm climates for more than 45 minutes. The period of protection may be prolonged by incorporating it in a comsmetic cream or diluting it with cocoanut oil. Only in cold climates does citronella compare favourably with the newer highly persistent organic chemicals.

The first synthetic repellent to appear was *Sta-Way Lotion*, a commercial product which contained a mixture of diethylene glycol monobutyl ether acetate and diethylene glycol monoethyl ether dissolved in ethyl alcohol and corn oil; but this material is slightly toxic when absorbed into the body through the skin. It was followed by a number of liquid compounds which are non-toxic to man, and afford protection (against *Aedes aegypti*) for 4 hours after application. Since they have high boiling points, that of DMP being 284° C., they are relatively non-volatile and exhibit low vapour pressures. The most important are :

Dimethyl phthalate (DMP)	v.p. $1{,}2 \times 10^{-3}$ mm. Hg.
2-Ethylhexanediol-1, 3 (Rutgers 612)	v.p. 11×10^{-3} mm. Hg.
Butylmesityloxide oxalate (Indalone)	v.p. 0.9×10^{-3} mm. Hg.

These compounds were mixed in the proportion 6 parts DMP, 2 parts Rutgers 612, and 2 parts Indalone, to give the highly repellent formulation 6-2-2. Also during

World War II the U.S. Navy developed the formulation NMRI-448, containing 7 parts of 2-phenylcyclohexanol and 3 parts of 2-cyclohexylcyclohexanol; it has a vapour pressure of 35×10^{-3} mm. Hg. In the formulation NMRI-201 the second ingredient was replaced by 3 parts of β-tetralol. Whereas Rutgers 612 and DMP act as vapour repellents, Indalone is mainly a contact repellent. The duration of repellency of DMP and 6-2-2 against *Anopheles* mosquitoes are 5 hours for *A. quadrimaculatus*, 7 for *A. punctipennis*, 2.5 for *A. aquasalis*, and less than 1 hour for *A. gambiae*. The period of protection afforded by DMP, but not by 6-2-2, may be doubled by the addition of 8-quinolinol.

Post-war research and development has revealed that propyl N, N-diethylsuccinamate and the dimethyl ester of cis-bicyclo (2, 2, 1)-5-heptene-2, 3-dicarboxylic acid (called dimethyl carbate) are superior to any of the previous compounds against *Aedes* mosquitoes. They are used to repalce Indalone and to combine with DMP and Rutgers 612 in the following formulations :

M-2020 3 parts Dimethyl carbate, 4 parts DMP, 3 parts Rutgers 612
M-1043 3 parts Succinamate, 4 parts DMP, 3 parts Rutgers 612

Whereas the succinamate, the carbate and Rutgers 612 gave the longest protection against *Aedes*, DMP remained the most lasting compound against *A. quadrimaculatus*. These materials, as also 6-2-2, may be used to saturate head-nets or the widemesh fish-net, head coverings employed by the army. The Anti-Malaria Wallet, for example, developed by the Army School of Health (U.K.) in 1944, contains leaves of flannel soaked in DMP between which a 4-mesh head net, a pair of gloves, and a pair of socks are stored.

Indalone is sufficiently persistent to be used as a clothing impregnant which remains repellent to mosquitoes. The compounds 2-butyl-2-ethyl-1,3-propanediol is even better, as also is N-butylacetanilide, a powerful tick repellent. These two materials have been formulated with the acaricide benzyl benzoate and *Tween* 80 emulsifier to give the emulsion concentrate M-1960, which on dilution with water may be used to impregnate cotton drill suits and coveralls. However, clothing impregnated with M-1960 has induced skin rash in a small proportion of Caucasian wearers. The mixture M-2040 has proved to be the more effective of the two as a repellent for *Aedes* mosquitoes. The two formulations have the following ingredients :

M-1960 2-butyl-2-ethyl-1, 3-propanediol ; N-butylacetanilide ; benzyl benzoate
M-2040 N-butyl-4-cyclohexene-1, 2-dicarboximide ; N-butylacetanilide ; hendecenoic acid.

The three constituents are mixed in equal parts, and to the mixture is added 10 per cent of *Tween* 80 emulsifier.

The majority of the mosquito repellents mentioned above are also repellent to mites and ticks, blackflies (*Simulium*) and sand-flies (*Culicoides*).

REFERENCES

Applewhite, K. H. and **H. F. Cross** (1951). Further studies of repellents in Alaska. *J. Econ. Ent.*, 44 : 19-22. **Christophers, Sir Rickard** (1947). Report of the mosquito repellent inquiry. *J. Hygiene* 45 : 176-231. **Granett, P.** (1940). Studies of mosquito repellents. Part II. *J. Econ. Ent.* 33 : 566-572. **MacNay, C. G.** (1939). Sudies on repellents for biting flies. *Canadian Entomologist* 71 : 38-44. **McCulloch, R. N.** and **Waterhouse, D. F.** (1947). Laboratory and field tests of mosquito repellents. Coun. Sci. Ind. Australia Bull. 213 : 28 **Pijoan, M.** (1947). New insect repellents,

Soap San. Chem. 23(3) : 124. **Sarkaria, D.S.** and **Brown A. W. A.** (1951). Studies on the responses of the female *Aedes* mosquito. Part II. *Bull. Ent. Res.*, 42 : 115-122. **Smith, C. N.** and **Cole, M. M.** (1951). Mosquito repellents for application to clothing. *J. Nat. Mal. Soc.*, 10: 206-212. **Smith, C. N.** *et al.* (1952). Mosquito repellent mixtures. *J. Econ. Ent.*, 45 (6), (in press). **Travis, B. V.** (1951). Effectiveness of repellents against several species of *Anopheles* mosquitoes. *J. Nat. Mal. Soc.*, 10 : 197-205 **Travis, B. V.** and **Smith, C. N.** (1951). Mosquito repellents selected for use on man. *J. Econ. Ent.* 44 : 428-429. **Travis, B. V.** *et al.* (1949). The more effective mosquito repellents tested at the Orlando laboratory, 1942-47. *J. Econ. Ent.* 42 : 686-694.

CONTROL OF PHLEBOTOMUS, SIMULIUM AND CULICOIDES

Phlebotomus Sand-flies. The advent of DDT has made the control of *Phlebotomus* sand-flies possible. Residual sprays are very effective against the adults, and the larvae may be destroyed in their breeding grounds. In India, kerosene-water emulsions of DDT, sprayed around the bases of the walls of cattle sheds, stopped larval breeding for 8 to 10 days. In Bengal, residual deposits of 800 mg. per square foot controlled the adults of *P. argentipes*, *papatasii* and *minutus* for 4 months; when 1600 mg. of DDT per square foot was applied the sand-flies did not reappear for 6 months. The recommended treatment in the southern United States is the residual spraying, with 5 per cent DDT in kerosene, of the interiors of quarters occupied in the evening and at night. The exteriors of the doors, windows and screens are also treated, along with the adjacent walls for a distance of two feet around them. A single treatment may be expected to give almost complete reduction for at least 2 months. The spray may also be applied to outdoor resting places, such as tree-trunks, masonry walls, caves, and even vegetation. Excellent and long-lasting control have thus been obtained against *P. verrucarum* in Peru. In Greece and the Levant, the residual application of DDT at 100 mgms. per square foot has given adequate protection against *P. papatasii* for the entire season, and controlled the incidence of sand-fly fever. As a result, this disease has been eliminated from many posts of the Middle East ; and leishmaniasis has been markedly reduced in the Mediterranean islands.

Filter-flies (*Psychoda* spp.) are related to the sand-flies, and their larvae may infest sewage filter-beds in great numbers. They may be controlled by applying DDT solution to the adult resting-places on the rim of the filter-bed, or by periodically adding DDT emulsion to the filter-bed influent so as to maintain 1 part per million of DDT. However, this treatment in Illinois has resulted in the appearance of a DDT-resistant strain of *Psychoda alternata*. It is probable that fine aerial sprays and thermally generated fogs containing DDT will be found to be effective against *Phlebotomus*, since adults of the sand-fly *Styoconops albiventris* in the southwest Pacific area have been controlled by 10 per cent oil solutions emitted from fog machines at an emission rate of 1.7 lb. per 100 yards of frontage.

Simulium Black-flies. DDT is an extremely potent insecticide for the larvae of black-flies (*Simulium* spp.) breeding in running water, since minuscule concentrations cause them to detach from their hold-fast positions and be washed downstream in the DDT-treated water. It was first found that a concentration of 0.1 p.p.m. of DDT was sufficient to clear streams in Guatemala of *S. metallicum*, *ochraceum* and other species for a distance of 1000 yards downstream from the point of application. The same effect was observed on *S. venustum* at Churchill on Hudson Bay, the minimum effective concentration being 0.1 p.p.m. for an exposure of 15 minutes, shorter exposures requiring higher concentrations and longer exposures allowing lower concentrations to be effective. A large-scale application on the South Saskatchewan River, adjusted to give slightly over 0.1 p.p.m. for a 30-minute exposure period, achieved virtually complete elimination of larvae of *S. arcticum* for a distance of approximately 100 miles downstream; no fish were killed at this concentration. Treatment of rivers at dosages averaging 2 to 10 p.p.m. of DDT completely eliminated larvae of *S. neavei*, the carrier of onchocerciasis, over an area of 65 square miles in Kenya; naturally there was high mortality of fish also. In Alaska, larvae of *S. venustum* and *vittatum* were effectively controlled by DDT at 0.3 to 0.7 p.p.m. for 15 minutes. Streams running for a distance of 800 yards through areas treated by DDT at 0.1 lb. per acre for mosquito control were freed of black-fly larvae for 2.5 miles downstream. *Prosimulium hirtipes*, overwintering as a larva, may be controlled in April, whereas *S. venustum*, overwintering in the egg, may not be controlled until it has hatched in May. In Alaska, the large overwintering larvae were found to be more resistant to DDT than those which hatched from overwintering eggs.

There is no evidence of differences in susceptibility between species. But larvicidal doses of DDT are quite ineffective against pupae. Methoxychlor and DDD are equitoxic with DDT, and are superior to chlordane, toxaphene, BHC and even lindane. However, DDD has proved more toxic than DDT to *S. venustum* in New Hampshire, and it is much less hazardous to fish. Lindane at a dosage of 0.2 p.p.m. for 15 minutes is highly larvicidal, and it is toxic to pupae, which DDT is not; toxaphene is considerably inferior. Parathion, aldrin and dieldrin do not compare favourably with DDT as black-fly larvicides. Larvae of *S. albivirgulatum* end *S. fainei* have been eliminated from streams in the Belgian Congo for 8 months by treatment with BHC at 1.5 to 3 p.p.m. of the gamma isomer.

Black-fly larvicides are usually applied to the surface of streams by spraying from ground equipment or aircraft; pouring oil solutions directly from drums has proved hazardous to fish because the oil is not finely dispersed. DDT concentrates with a specific gravity exceeding 1.0 have killed large numbers of suckers (*Catostomus*) and other bottom-feeding fish. Dilute solutions sprayed on the surface are completely effective, the DDT diffusing into the water as the oil droplets gradually dissolve; the addition of a detergent may be expected to increase the oil-water interfaces. Automatic dispensers have been designed which deliver DDT at rates adjusted to the flow of the stream. Insecticidal cakes have been prepared in which DDT is incorporated with PDB, water-soluble wax, plaster or soap, so that it is gradually released into the water.

Adult black-flies may be successfully attacked by DDT in oil solution applied as from ground thermal generators or from helicopters, or as sprays from fixed-wing aircraft or from vertical boom-and-nozzle spray masts erected on a vehicle. Excellent kills of *S. venustum* and other North American species may be obtained in all cases, but migration of black-flies from adjoining areas soon restores the populations to a more normal level. Complete local elimination may be obtained with DDT aerosols applied at 0.2 lb. per acre, but applications of 0.01 lb. per acre used in mosquito control fails to kill all the black-flies. When DDT is sprayed from aircraft at 0.2 lb. per acre, the population reduction on the day following the application averages 70 per cent when the atomization is coarse, and 80 per cent with fine sprays. The uncertainty of adult control has led American workers to rely more on larviciding from the air; excellent control of *S. venustum* and *P. hirtipes* has been obtained by laying down, at quarter-mile intervals, swaths 100 yards wide where the deposit is 0.1 lb. per acre. However, the daily spraying of 0.18 lb. of DDT per acre from aircraft over the Congo River has successfully controlled adults of *S. damnosum* and provided protection against the transmission of onchocerciasis.

The liquid repellents used against mosquitoes are also effective against black-flies. Rutgers 612 has proved superior to DMP, which is better than Indalone. The periods of protection are approximately 4 hours against *S. venustum* and *P. hirtipes* in eastern North America, 2 hours against black-flies in Columbia, and 7 hours against *E. pecuarum* in the southern United States. Recently, propyl N, N-diethylsuccinamate has proved to last longer than these three materials against *S. venustum* and associated species.

Culicoides Sand-flies. The minute larvae of *Culicoides* breeding in the damp organic matter of swamps have proved exceedingly difficult to control. But recent American work has shown that *C. furens* and *C. melleus* may be controlled by spraying their breeding places with dieldrin, lindane, chlordane or heptachlor at a dosage of 1 lb. per acre. With dieldrin, the most powerful of them, each acre receives 15 gallons of an emulsion made by diluting 4 gallons of a 20 per cent concentrate in 100 gallons of water. With lindane, a concentration of 0.2 parts per million in the medium will kill in a day; and BHC containing 12 per cent gamma isomer will control at a dosage of 4 lb. per acre.

DDT and toxaphene are inferior for larval control of *C. furens* and *melleus*, although some control is obtained at 2 lb. per acre. But the Palau gnat (*C. peleliouensis*) has been controlled by dusting the brackish breeding ground with 10 per cent DDT dust applied at 15 lb. per acre. With *C. furens*, the treatment of window screens with 5 per cent DDT solution reduces the biting rate indoors by 95 per cent. But area sprays with DDT do not reduce the adult population, unless a dosage of 0.2 lb. DDT per acre is applied to the vegetation every day.

The biting midge of Scotland (*C. impunctatus*) is controlled in the larval stage by application of gamma-BHC at 100 mgms. per square foot (equivalent to 10 lb. lindane per acre). Attempts to reduce the adult population with 2.5 lb. DDT per acre applied as a suspension to the vegetation were unsuccessful. With *Culicoides* in northern New York State, thermally generated aerosols achieve immediate reduction with DDT at 0.2 lb. per acre, and control lasting for 2 weeks with 2 lb. per acre. Immediate but evanescent control of *C. tristriatulus* and *C. unicolor* in Alaska was also obtained with thermal aerosols of DDT.

DMP is the most effective repellent against the Alaskan *Culicoides*, followed by the mixture 6-2-2, both giving between 40 and 60 minutes of protection. Propyl-N, N-diethylsuccinamate, Rutgers 612 and dimethyl carbate are inferior and afford protection for 10 to 20 minutes only.

REFERENCES

Applewhite, K. H. and **Cross, H. F.** (1951). Further studies of repellents in Alaska. *J. Econ. Ent.*, 44: 19-22. **Brown, A. W. A.** (1952). Rotary brush units for aerial spraying against mosquitoes and black flies. *J. Econ. Ent.*, 45 : 620-625. **Brown, A. W. A.** *et al.* (1951). Control of adult mosquitoes and black flies by DDT sprays applied from aircraft. *Mosqaito News* 11 : 75-83. **Collins, D. L., Travis, B. V.** and **Jamnback, H.** (1952). The application of larvicide by airplane for control of blackflies. *Mosquito News* 12 : 75-77. **De Foliart, G. R.** (1951). A comparison of several repellents against blackflies. *J. Econ. Ent.*, 44 : 265-266. **Department of Health for Scotland** (1948). The second report on control of midges. H. M. Stationery Office, Edingurgh, 11 pp. **Dorsey, C. K.** (1947). Populalation and control studies of the Palau gnat on Peleliu, Western Caroline Islands. *J. Econ. Ent.*, 40 : 805-813. **Ghosh, S. M.** (1950). The control of *Phlebotomus* (sand-flies) with DDT and BHC (gammexane). *Indian J. Malariol.* 4 : 175-184. **Gjullin, C. M., Sleeper, D. A.** and **Husman, C. N.** (1949). Control of black fly larvae in Alaskan streams by aerial applications of DDT. *J. Econ. Ent.*, 42 : 392. **Gjullin, C. M., Cross, H. F.** and **Applewhite, K. H.** (1950). Tests with DDT to control black fly larvae in Alaskan streams. *J. Econ. Ent.*, 43 : 696-697. **Glasgow, R. D.** and **Collins, D. K.** (1946). The thermal aerosol fog generator for large sacle applications of DDT and other insecticides. *J. Econ. Ent.*, 39 : 227-235. **Goulding, R. L.** and **Curran, R. F.** (1952). Field tests of insecticides for the control of the salt-marsh sand fly in Florida. *Rep. 22nd.* and *23rd. Ann. Meeting, Florida Anti-Mosquito Assoc.* **Hertig, M.** (1950). Control of *Phlebotomus* sandflies in Peru. *Bull. World Health Org.*, 2 : 621-628. **Hill, M. A.** and **Roberts, E. W.** (1947). An investigation into the effects of gammexane on larvae, pupae and adults of *Culicoides impunctatus*. *Ann. Trop. Med. Parasitol.*, 41 : 143-163. **Hocking, B.** (1950). Further tests of insecticides against black flies (Diptera : Simuliidae), and a control procedure. *Scientific Ag.iculture* 30 : 489-508. **Hocking, B.** and **Richards, W. R.** (1952). Biology and control of Labrador black flies (Diptera : Simuliidae). *Bull: Ent. Res.*, 43 : 237-257. **Jamnback, H.** (1952). The importance of correct timing of larval treatments to control specific black flies. *Mosquito New'* 12 : 78-79. **Kettle, D. S.** (1949). An attempt to control *Culicoides impunctatus* in Scotland by barrier spraying. *Ann. Trop. Med. Parasitol*, 43 : 284-296. **Kindler, J. B.** and **Regan, F. R.** (1949). Larvicide tests on blackflies in New Hampshire. *Mosquito News* 9 : 108-112. **Travis, B. V.** (1949). Studies of mosquito and other biting insect problems in Alaska. *J. Econ. Ent.* 42 : 451-457. **Travis, B. V.** *et al.* (1951). Strip spraying by helicopter to control blackfly larvae. *Mosquito News* 11 : 95-98. **Travis, B. V., Smith, A. L.** and **Madden, A. H.** (1952). Effectiveness of insect repellents against balckflies. *J. Econ. Ent.*, 44 : 813-814. **Wanson, M., Courtois. L.** and **Lebied, B.** (1949). L'eradication du *Simulium damnosum* Theo. a Leopoldville. *Ann. Soc. Belge de Med. Trop.*, 29 : 373-403. **Wanson, M., Courtois, L.** and **Bervoets, W.** (1950). *Ann. Soc. Belge. de Med. Trop.*, 30 : 629-697.

HOUSE-FLY CONTROL

Larvicides. Water-soluble inorganic chemicals such as powdered borax, sodium fluosilicate, or sodium arsenite have been used for the control of house-fly larvae in latrines and manure piles. For example, borax is recommended for application to sewage sludge, at the rate of 2 ounces per thousand gallons. Diesel oil may also be applied at a rate of 1 gallon to 200 squre feet. To control fly breeding in compost heaps of faecal and waste matter, a 1-inch surface layer of 10 per cent borax in earth has proved highly effective in India.* Among the water-insoluble organic insecticides, PDB (a solid) and ODB (a liquid) have been effective; they are also ovicidal. Of the new chlorinated hydrocarbons, lindane, chlordane, aldrin and dieldrin are all approximately 100 times as larvicidal as borax or PDB. Even DDT is larvicidal, but it is usually applied to the surface of the breeding medium in order to kill the newly emerged flies. For example, a standard latrine box hole is treated with 2 ounces of 5 per cent residual spray solution or with 1 ounce of 10 per cent DDT dust. Open breeding places are treated at the rate of 100 mgm. of DDT per square foot with residual spray, or 200 mgm. per square foot with DDT dust.

* If treated manure is spread over fields, 20 lb. of contained borax per acre may damage crops.

Space Sprays. The control of adult house-flies was first accomplished by solutions of pyrethrum in kerosene (*Flit*) applied as a finely atomized space spray from a hand atomizer (Flit-gun) in a closed room. Later the thiocyanates such as *Lethanes* and *Thanite* were developed as synthetic substitutes to give the quick knockdown and kill characteristic of pyrethrum. Then certain synergists such as sesame oil, IN-930, n-propyl isome or DHS activator were added to extend the toxicity of the pyrethrins. When DDT appeared, it was found to be not only a residual poison but also an excellent insecticide for incorporation in space sprays.

DDT is highly toxic to house-flies, but it requires at least 10 minutes to knock them down. Its great value lies in being combined with pyrethrum to kill those flies which would have recovered from the fast initial knockdown caused by pyrethrum alone. Where 0.07 per cent pyrethrins were formerly required for knockdown and kill, the addition of 0.1 per cent DDT reduces the pyrethrin requirement to 0.03 per cent. Or 0.1 per cent DDT may be combined with 2 per cent *Lethanes* or *Thanite*. These formulations, dissolved in odourless kerosene, give complete control of flies in space sprays when emitted at the rate of 1 cc. per cubic metre from air atomizers. Aerosol bombs are now frequently used, in which the propellent is liquefied *Freon*; this liquid contains 10 to 20 per cent of a non-volatile oil, and also 0.3 per cent pyrethrins and 3 per cent DDT which have been dissolved in that oil. The solvent may consist of a mixture of cyclohexanone with sesame oil or lubricating oil, or certain of the aromatic oils (*e.g. Velsicol*); its function is to stabilize the droplets, produced by decompression of the *Freon*, at a suitable size for contacting the insects.

For direct contact killing of house-flies by space sprays, DDT is surpassed in toxicity by chlordane, heptachlor, aldrin, lindane and dieldrin, in ascending series. However, for use in domestic establishments, there is no reason to replace the DDT-pyrethrum aerosols already in wide usage. In barracks and dormitories heavy aerosol generators have been employed, and by this method chlordane has proved superior to lindane, which surpasses DDT and methoxychlor; a light deposit of approximately 5 mgms. per square foot on the walls affords a continuous fumigant effect which keeps flies away for 2 months. However, chlordane vapour is not entirely free of toxic hazard. In warehouses and storage rooms, lindane has come to be the favoured insecticide. It is applied at fairly frequent intervals by fine air atomizers; there are stationary installations provided with time switches and solenoid valves which will spray daily. There are also continuously operated vaporizers electrically heated with a light bulb or special element, in which BHC or lindane, mixed with an inert base containing such materials as glycols, carbowax and bentonite, is maintained at a temperature of 113°-116° C. Another method is to add an acetone solution of lindane to a Fiberglas filter inserted in air-conditioning systems. All these devices produce a concentrate of lindane in the air of between 5 and 20 mgms. per 1,000 cubic feet, sufficient to keep the number of surviving flies in rooms and barns at a low level. For safety, the daily output should not exceed 66 mgms. BHC per 1,000 cubic feet, and vaporizers should not be used in homes and sleeping quarters.

The combination of pyrethrins with the new synergists piperonylcyclonene and piperonyl butoxide (the *Pyrenone* formulae) has produced space sprays which are very effective and entirely free of any toxic hazard. Recently the synthetic analogue, allethrin, has been discovered to be as toxic to house-flies as the pyrethrins. Allethrin may be synergized by the parethrin synergists, such as piperonyl butoxide, propyl isome and octacide.

Residual Sprays. As with mosquitoes, the outstanding role of DDT is to provide effective residual sprays against house-flies. The procedure followed is essentially the same as in mosquito control. The walls and ceilings of rooms where food is prepared or consumed are uniformly sprayed, the food and utensils being suitably protected. Barns, stables and farmyard buildings are special candidates for treatment. On rough wood, a deposit of 50 mgms. of DDT per square foot may be expected to remain lethal to flies for 1 month, 100 mgms. per square foot for 2 months, and a deposit of 200 mgms. per square foot does not need renewal for 3 months. Sources of deterioration include the flies brushing the deposit off with their feet, or covering it with their excretion. If the wall surface is smooth, deposits may persist for a long time; deposits of 70 mgms. per square foot have remained insecticidal for 18 months on glass, wall-paper or planed wood. The painted fibre-board walls of a barn in Colorado, sprayed with DDT solution at 200 mgms. per square foot, remained toxic to flies for at least 7 years. The question of absorption and efflorescence of deposits has been covered at the end of Chapter III. In addition to window screens, the cords of electric lights and window-blinds should be treated, since they are favourite resting-places.

Strips of 16-mesh wire screening, 0.75 inches wide, may be dipped in DDT concentrate and tacked up horizontally on ceilings to serve as roosts; this stratagem may control flies in barns as a substitute for the residual spraying of walls. The application of residual wall-sprays in barns produces about 95 per cent reduction of the fly population; although the flies increase in numbers during the day as they enter from outside, they are killed overnight. The treatment of back-door areas and garbage cans in towns by 5 per cent DDT emulsions at frequent intervals can reduce the fly index by 70 to 85 per cent; weekly spraying of streets and alleys by 5 per cent DDT solution from mist blowers has reduced the fly index by 95 per cent in a Mississippi town. Aerial spraying with DDT at 0.5 lb. acre once a month in 1945 reduced the fly population of Manila to negligible proportions. The aerial spraying of Cairo in 1946 was effective in eliminating houseflies as an element in a typhoid epidemic.

Many of the other chlorinated hydrocarbons may be used for residual control, and aldrin, dieldrin, lindane, BHC, methoxychlor, DDD, chlordane and toxaphene have all given satisfactory control of non-resistant houseflies. For initial killing power combined with lasting toxicity, DDT is superior to BHC, chlordane, methoxychlor, toxaphene, DDT and DFDT, in that descending order. Methoxychlor has a lack of toxic hazard to recommend it for use in dairy barns, and weekly sprayings are very effective. Aldrin, dieldrin and lindane have given excellent residual control at half the dosage necessary with chlordane and DDT. Lindane is very rapid in action, flies contacting it do not leave without taking up a lethal dose, and it has a strong fumigant action; chlordane vapour is one-third as toxic as that of gamma-BHC. Lindane lasts considerably longer if suspensions are used. A mixture of 1 part gamma-BHC with 5 parts DDT has been found to have a slight synergistic effect on houseflies in India. Mixture of DDT and chlordane are more synergistic, and mixtures of 3 parts chlordane to 1 part DDT are used in central Italy. No synergism is exhibited between dieldrin and DDT. Dieldrin alone has the most lasting toxicity of all the insecticides and has quite fast initial killing power. Therefore, 0.5 to 1.25 per cent emulsions of dieldrin are now favoured for large-scale street and backyard spraying in Asiatic and American cities, and deposits of 50 mgms. of dieldrin per square foot are as effective as 200 mgms of DDT per square foot. However, in Egypt chlordane proved more effective than dieldrin. Strips of screening dipped in dieldrin concentrate have given good control of house-flies in dairy barns for 8 to 16 weeks.

The appearance of flies resistant to DDT, and also to BHC, chlordane and dieldrin has encouraged the search for residual spray insecticides outside the chlorinated hydrocarbon group. Pyrenone formulae containing pyrethrins and piperonyl butoxide have shown a residual toxicity of 11 weeks. Dilan has proved an effective substitute against DDT-resistant houseflies in California. Since parathion has as persistent a residual toxicity as DDT, other p-nitrophenyl thiophosphates, including Compound 3456, are being tested as residual sprays. The methyl analogue of parathion, which has a far less toxic hazard, is about two-thirds as insecticidal to the housefly. However, mixtures of parathion with DDT apppear to be far less dangerous than expected*. Malathon is being used in 1 per cent solution every 2 weeks in California.

The other alternative is to develop synergists for DDT. Curiously enough the pyrethrin synergists such as piperonylcyclonene, piperonyl butoxide and octacide have the power of inhibiting the detoxification of DDT by resistant flies, and thus acting synergistically. A group of chlorinated hydrocarbons, known previously as acaricides for tetranychid mites, are also synergistic with DDT against resistant flies. They include MR-60, MR-30 or Compound 876, and K-3926. But the most effective compound is di-p-chlorophenylmethylcarbinol, or DMC; this material is mixed in equal proportions with DDT for residual spraying. However, more success has been obtained with this mixture in space sprays. Other materials which are synergistic with DDT, in this case against normal as well as resistant flies, include DFDT, pyrolan (G-22008) and chlorinated chalcone.

Resistant Strains. The first DDT-resistant house-flies were observed in 1946 in the vicinity of Rome and Naples, Italy, where DDT had been applied since 1944. In the same year failure of DDT residual sprays was noted in Denmark, where DDT resistance became general in 1949. Resistance appeared in Sardinia in 1948, following the inception of DDT spraying in 1946. Operations in Sweden in 1946 uncovered, near the town of Arnas, a case of natural rather than acquired resistance.

In the U.S.A., DDT-resistant strains appeared in 1948 in New York, New Jersey, North Carolina, Geogia, Alabama, Florida and California. In the following year they were reported from Illinois and Oregon. In 1950 it was ubiquitous in Illinois, and by 1951 it was general in Florida. However, up to 1950 the nation as a whole was not involved, and in the northwestern states resistance was only occasional by 1951. In Canada, DDT-resistance appeared in Ontario and Quebec in 1949, and Alberta and British Columbia in 1950.

DDT-resistant strains became generally common in Mexico by 1950, evidence of them having first appeared in 1947. The end of the half-century also witnessed resistance in Trinidad, Puerto Rico and Panama. In Peru, Brazil and Argentina,

*Festooning barns with parathion-impregnated twine, and watering the floors with 0.05 per cent TEPP in dilute molasses, are ingenious new methods not yet approved.

house-flies in areas which had been treated with DDT residual sprays since 1946 or 1947 were significantly resistant by 1950. However, only slight resistance developed in similar areas in Venezuela. Resistance of *Musca domestica* to DDT was found in New South Wales (Australia) in 1950.

In India and Ceylon, where the house-flies belong to different species, no DDT resistance had appeared by 1951. An examination in 1949 of *M. vicina* from areas in Ceylon treated since 1947 yielded no evidence of resistance. An examination of *M. nebulo* from areas near Delhi, treated for 4 consecutive years previously, yielded no evidence of resistance.

Resistance to BHC was first discovered in 1949 in an Egyptian village, where BHC (but never DDT) had been applied monthly for a year previously. A BHC-resistant strain was discovered in 1950 in California, 5 months after its first use, which followed DDT treatments.

Resistance to chlordane was first encountered in 1949 in Sardinia, where chlordane had been added to the DDT spray schedule for 1 year. In 1950 it was reported from Italy, where chlordane has been used to combat DDT-resistance since 1948. In 1951 it was reported from an Egyptian village, where it had been used for 1 year, following both BHC and DDT.

Resistance to dieldrin first occurred in California in 1950, following resistance to DDT and BHC, and a more widespread case was encountered in Arizona in 1951. General resistance to chlorinated hydrocarbons is now shown by house-fly populations in Italy, Denmark, Sardinia, Florida, California, Panama and the Levant States.

Genetics of Resistance. DDT-resistant strains have been produced experimentally in the laboratory by taking the survivors of adulticide sprays performed on each generation. First reported in 1948 from Orlando, Florida, laboratories in Georgia, New York, New Jersey, Illinois and England have developed and published research on DDT-resistant strains of *M. domestica.* Although in some cases the resistance was found to build up gradually starting with the F_2 generation, in others the increase was delayed until the F_{12}, after which it rapidly built up to the F_{18} generation, where a constant level was reached. On the other hand, DDT pressure for 12 generations of *M. nebulo* failed to give a significant increase in resistance.

In most cases it has been found that stocks which are in process of developing resistance to DDT are becoming resistant to methoxychlor and DDD also. As the development of resistance proceeds, they come to show resistance to lindane, chlordane, dieldrin and toxaphene, and a single exposure to any one of these chlorinated hydrocarbons seems to result in greatly increased resistance in the progeny of the survivors. By this stage certain strains may show a certain amount of resistance to rotenone, pyrethrum, parathion and other insecticides, although this is not usual. There are some anomalies : a DDT-resistant strain from Italy that is resistant to pyrethrum but not to the other chlorinated hydrocarbons; a laboratory-induced pyrethrum-resistant strain that is also resistant to DDT; instances of methoxychlor-resistant flies that are not resistant to DDT. The general finding is that house-flies already resistant to one insecticide become resistant to other more readily than normal. The acquisition of resistance is accelarated where larval breeding sites become contaminated with DDT or other insecticides.

When DDT-resistant strains are bred in the laboratory in the absence of DDT, it may be found that the resistance is maintained at the same level for many generations, or that it fluctuates from generation to generation, or that the resistance gradually decreases (reversion). If a resistant fly is crossed with a normally susceptible fly, the F_1 progeny will be intermediate in resistance, and the F_2 obtained from inbreeding this generation will show a wide range of susceptibility and resistance exceeding either grandparent. There is no evidence of sex-linkage, as had been found in the case of HCN-resistance in the California red scale. All these results are consistent with the idea that resistance is inherited through multiple genes, whose separate effects "add up" to resistance. The cases of "reversion" are due to the stock being heterogenoeus. There is one example of resistance to knockdown by DDT of an Italian strain of house-flies being due to two alleles of a single gene.

Physiology of Resistance. The nature of the physiological changes involved in DDT-resistance has been the subject of considerable research. A change in behaviour has been noted in the U.S.A. to the extent that resistant flies may rest more on floors than on the walls and ceilings which are sprayed, that they may be less active, or that they may be affected soon after exposure and thus avoid picking up a lethal dose. In New York it was found that resistant strains had a longer larval period than the normal, and that even within a resistant population the late pupaters were the most resistant. Attempts to find morphological differences have been unsuccessful, with one exception : the naturally resistant house-flies from Arans (Sweden) had thicker pulvilli and intersegmental membranes, were more heavily pigmented, and were more hardy to extremes of temperature.

Evidently resistance is not referable to cuticular penetration, since resistant strains absorb DDT through the cuticle at the same rate as normal flies. But it has been found that resistant flies rapidly metabolize DDT by dehydrochlorinating it to DDE, and to other catabolites including DDA. Insects which are normally resistant to DDT such as the milkweed bug (*Oncopeltus fasciatus*), certain grasshoppers (*Melanoplus* spp.) and the Mexican bean beetle (*Epilachna varivestris*) also have this power. This ability is thermolabile, is lost at death of the fly, and shows the properties of enzymic processes. Certain compounds such as piperonylcyclonene and other pyrethrin synergists can inhibit this detoxification process, and thus act as adjuvants for DDT in the case of resistant flies. Other DDT synergists of this nature such as DMC, K-3926 and MR-60 act as decoys for the detoxification process and the DDT is left intact long enough to kill. But resistance to the DDT-synergist compounds may also build up. The insecticide Dilan (a DDT analogue with nitro groups instead of chlorine on the alkane nucleus) is effective against DDT-resistant flies because it is not detoxifiable by dehydrochlorination. Esters of iodoacetic and chloroacetic acids have been found to be powerful contact poisons for DDT-resistant house-flies.

It has also been found that DDT-resistant flies have an appreciably higher content of cytochrome oxidase than the normal. Flies resistant to DDT have a longer larval life and pupate later than susceptible flies. Although enzymic dehydrochlorination is a major factor, a cause of resistance may be also sought in the general vigour in oxidation-reduction processes.

REFERENCES

Baranyovits, F. (1951). *Nature* 168 : 940-961. (Effect of DDT and BHC on houseflies contacting it). **Barrett, W. L.** (1952). *J. Econ. Ent.*, 45 : 90-93. (Methoxychlor in dairy barns). **Bettini, S.**

and **Boccacci, M.** (1950). *Rivista di Parassitol.*, 13: 165-167. (Esters of chloroacetic acid for DDT-resistant flies). **Brown, A. W. A.** (1952). *Soap San. Chem.* 28(7): 135;137. (Resistance and DDT synergists : review). **Brown, A. W. A.** *et al.* (1951). 81*st. Ann. Rep. Ent. Soc. Ontario*, 1950, pp. 33-50. (Resistance of insects to insecticides : symposium). **Bruce, W. N.** (1949). *Bull. Illinois Nat. Hist. Survey* 25: 1-32. (Fumigant effect of BHC and chlordane). **Bruce, W. N.** and **Decker, G. C.** (1950). *Soap San. Chem.*, 26 (3): 122. (Cross-resistance to insecticides by houseflies). **Busvine, J. R.** (1949). *Ceylon J. Med. Sci.*, 6(4): 1-3. (No DDT resistance in *M. vicina*). **Chadwick, L. E.** (1952). *Am. J. Trop. Med. Hyg.*, 1: 404-411. (Physiology of DDT resistance : review). **DeLong, D. M., Boush, G. M.** and **Lea, A. O.** (1952). *J. Ecob. Ent.*, 45: 323-325. (Chlordane fogs for housefly control). **Dicke, R. J.** and **Paul, J. J.** (1951). *J. Econ. Ent.*, 44: 896-898. (Synergism of chlordane with DDT). **Fales, J. H.** *et al.* (1951). *J. Econ. Ent.*, 44: 23-28. (Studies on allethrin and trans-allethrin). **Fulton, R. A., Gelardo, R. P.** and **Sullivan, W. N.** (1952). *J. Econ. Ent.*, 45: 540-541. (Electric vaporizers for lindane). **Gersdorff, W. A., Nelson, R. H.** and **Mitlin, N.** (1951). *J. Econ. Ent.*, 44: 921-927. (Synergists for allethrin). **Harrison, C. M.** (1951). *Bull. Ent. Res.*, 42: 761-768. (Heterogeneity as a cause of reversion in resistant stocks). **Harrison, C. M.** (1952). *Trans. Roy. Soc. Trop. Med. Hyg.*, 46: 255-261. (Resistance of insects to insecticides). **Hess, A. D.** (1952). *Am. J. Trop. Med. Hyg.*, 1: 371-388. (Status of insecticide resistance in medical entomology : review). **Lal, B. N.** and **Srivastava, S. B.** (1950). *Indian J. Agr. Sci.*, 20: 239-250. (Borax to prevent fly breeding). **List, G. M.** (1952). *J. Econ. Ent.*, 45: 127-129. (DDT deposits effective after seven years). **Missiroli, A.** (1950). *Am. J. Trop. Med.* 30: 773-783. (Mixtures of chlordane and DDT). **Moore, S., Toczydlowski, A. H.** and **Sweetman, H. L.** (1951). *J. Econ. Ent.* 44: 731-733. (Lindane vaporizers in barns). **Pal, R.** (1951). *Trans. Roy. Soc. Trop. Med.*, 54: 125-126. (No DDT resistance in *M. nebulo*). **Pal, R.** (1951). *Nature* 167: 368. (Synergism of gamma-BHC with DDT). **Pimentel, D., Schwardt, H. H.** and **Norton, L. B.** (1951). *Soap San. Chem.*, 27(1): 102. (Screen-strip method for housefly control in barns). **Schoof, H. F., Siverly, R. E.** and **Coffey, J. H.** (1951). *J. Econ. Ent.* 44: 803-807. (Large-scale spraying of dieldrin, and development of resistance). **Sternburg, J.** and **Kearns, C. W.** (1952). *J. Econ. Ent.*, 45: 497-505. (Dehydrochlorinating ability of DDT-resistant insects). **Sullivan, W. N.** (1951). *J. Econ. Ent.*, 44: 125-126. (Lindane vaporized from plastic filters).

CONTROL OF BITING FLIES ON LIVESTOCK

Tabanid Flies. Cattle may be protected from the attentions of horse-flies by spraying them with a 10 per cent solution of butoxypropylene glycol 800 (*Crag* fly repellent) in deodorized kerosene, or with an emulsion containing 0.1 per cent pyrethrins and 1 per cent piperonyl butoxide. When this emulsion was sprayed on dairy animals in Illinois with a knapsack sprayer at the rate of 30 cc. per animal twice weekly, the attentions of *Tabanus sulcifrons* were reduced by 90 per cent; a similar result was obtained when 1 per cent pyrethrin and 10 per cent synergist were applied undiluted with a hand-sprayer at 4 cc. per animal twice weekly. On the other hand, 4 per cent DDT emulsions failed to protect cattle from the attentions of *Chrysops discalis*.

Tabanid larvae are susceptible to DDT, and the larvae of *Chrysops discalis* are killed within 2 days by a concentration of 0.2 parts per million of DDT in the muck in which it breeds. Aerial treatment with two applications of DDT at 1 lb. per acre has controlled *Tabanus nigrovittatus* in its shoreline breeding-grounds. This species may also be rapidly killed by soil application of 0.004 per cent pyrethrum emulsions.

Attempts have been made to control adult tabanids by area sprays from aircraft. When small plots of 10 acres each were treated with DDT, methoxychlor, toxaphene or chlordane at 2 lb. per acre, no appreciable reduction could be noted in the numbers of *T. sulcifrons* and *abactor*. When plots of 1 square mile were treated, significant reductions in the populations of *C. excitans*, *C. carbonaria* and *T. typhus* could be noted with DDT sprays; at 0.25 to 2 lb. per acre, and Complate control with lindane at 0.5 to 1 lb. per per acre. The application of oil to stagnant pools may kill the adults drinking from them and the young larvae as they hatch from the eggs laid on the vegetation above them.

Stable-flies and Horn-flies. The spraying of dairy cattle with certain modern insecticides has been found to protect them from the pestering attentions of *Stomoxys calcitrans* and *Siphona irritans*, and as a result to increase the milk yield by 25 to 35 per cent; beef cattle so protected may put on 20 per cent more wieght. The animals are usually sprayed every 3 weeks, using a power sprayer at 300 p.s.i. pressure, with 0.5 per cent suspensions of methoxychlor or DDD applied at the rate of 1 pint per head. If the cattle are not being used for milk, it is permissible to employ DDT, chlordane or toxaphene. With dairy cattle, a spray of 0.05 per cent pyrethrins and 0.5 per cent piperonyl butoxide (a "pyrenone" formula) is an entirely safe substitute. This type of spray, applied at 1 to 3 pints per head twice weekly, can give about 80 per cent reduction in the attentions of stable-flies. Of the known fly repellents, indalone, 2-phenylcyclohexanol and the pinene ether of glycol are effective against the stable-fly whereas DMP, Rutgers 612 and citronella are not*. *Crag* fly repellent (butoxypropylene glycol 800 in 10 per cent solution in refined kerosene) will keep these flies away, but *Crag* has proved not to be as effective as the "pyrenone" formula mentioned above. However, if the 10 per cent *Crag* is applied at the rate of 100 c.c. per cow every 2 to 3 days, it has proved better than the synergized pyrethrins. Proprietary emulsions are now available which contain both 0.06 per cent pyrethrins and 2 per cent of *Crag* repellent. The buffalo fly (*S. exigua*). of Australia has been controlled by 0.5 per cent DDT suspensions sprayed every 2 weeks.

The horn-fly develops in cattle dung. One method of control is to feed cattle daily doses of phenothiazine at the rate of 5 grams per 100 lb. body-weight, which has the effect of rendering their dung proof against larval breeding by this fly. The stable-fly develops in water-soaked decomposing vegetation; infestations in marine grass cast up on the beach in Florida could be cotrolled by applying 0.5 per cent DDT emulsion at the rate of 2 gal. per 100 square feet since the adult flies were killed as they emerged.

The stable-fly may also be reduced in numbers by the residual sprays applied in barns and houses for house-fly control. In this respect DDT and methoxychlor are superior to chlordane and toxaphene. Both horn-flies and stable-flies which attack cattle in open pasture may be combatted by installing cable-type back-rubbers, which are simply 4 strands of barbed wire, 16 feet long, suspended between two posts and wrapped in burlap sacking. The wrapping is soaked with 5 per cent DDT every 2 weeks; the cattle pass under the cable and rub the solution on to their back. The reduction in flies may reach 99 per cent, equivalent to the result of a 0.5 per cent spray applied every 3 days. Added to the sacking in furnace oil solution, DDT is better than methoxychlor, and both are superior to Dilan, toxaphene and lindane. A proprietary back-rubber (*Old Scratch*) consists of two 6-foot lengths of cable on which flat steel washers are strung; the insecticide is automatically dispensed, by the rubbing action, from a central pillar which contains 18 gallons of the solution. An

* Recently, dipropyl isocinchomeronate and butadiene-furfural copolymer have been found to be effective repellents for house-flies and stable-flies.

effective automatic sprayer consists of a treadle which releases a plunger pump to finely spray one-thirtieth of an ounce of a solution of 1 per cent pyrethrins and 10 per cent piperonyl butoxide; on the return trip through the chute the cattle similarly spray themselves on the other side.

Tsetse Flies. Species of the genus *Glossina*, carriers of the trypanosomes which cause sleeping-sickness and Nagana disease in Africa, are very susceptible to DDT poisoning by direct contact or by residual contact with the pulvilli. The adult fly is characterized more by its mobility than its abundance, and comparatively small populations per unit area are sufficient to cause concern. Emergence of the adult flies may continue for long periods from pupae present in the ground. Hessian cloth screens treated with DDT or BHC have been used as insecticidal traps; distributed at strategic points at the rate of about 10 per acre they have obtained up to 90 per cent control. Alternatively, the vegetation may be sprayed at points where the flies congregate; about 98 per cent control may be obtained by spraying the vegetation with 5 per cent solutions 4 times at weekly intervals, DDT being applied at 100 mg. per square foot, or lindane at 11 mg. per square foot. When 5 per cent DDT emulsions were used for this purpose, a formulation with fine DDT crystals in the aqueous phase (*Supona-D*) gave a 97 per cent reduction of flies which lasted for 36 weeks, whereas an ordinary emulsion gave 78 per cent reduction for 17 weeks.

The tsetse-fly problem has also been attacked on an area basis. DDT smokes from a burning mixture in canisters have proved unsatisfactory, the reduction not exceeding 63 per cent. DDT fogs applied by a thermal aerosol generator have given only about 50 per cent control. BHC smokes applied from 1-lb. canisters arranged so as to give a dosage of 1 oz. lindane per acre have given up to 90 per cent control; the range of the smoke from these canisters is about 60 yards in dense bush.

When applied from aircraft, DDT aerosols are very effective. Generated by injection of 10 per cent solutions into the exhaust stack of Anson aircraft, and applied eight successive times at 0.25 lb. per acre, they gave 99 per cent control of *G. morsitans* and *G. swynnertoni*. Similar applications of BHC resulted in 93 per cent control. Repeated treatment of an area in Zululand, with DDT aerosols succeeded in achieving almost complete extermination of *G. pallidipes*. When fine sprays of 10 per cent DDT or BHC solutions in oil were applied from a boom-and-nozzle assembly on the Anson aircraft in Tanganyika, 99 per cent control was obtained when the area was treated 0.25 lb. DDT per acre every 10 days for a period of three months.

The spraying of oxen with 9 per cent DDT solutions twice weekly has been found to kill 90 per cent of the female *G. pallidipes* feeding on them, and an 80 per cent reduction in the infestation level in the area. Cattle are treated with DDT or BHC dips against the tsetse-fly in Portuguese East Africa. It has been found possible to treat calves orally with BHC (containing 13 per cent of the gamma isomer) at 250 mg. per kilogram every week and so to render them toxic to any *Glossina* feeding on them.

REFERENCES

Blakeslee, E. B. (1945). *J. Econ. Ent.*, 38: 584-552. (DDT to control *Stomoxys* larvae). **Bruce, W. N.** (1952). Illinois Nat. Hist. Survcy, Biol. Note 27: 11 pp. (Automatic sprayer to control biting flies on cattle). **Bruce, W. N.** and **Decker, G. C.** (1951). *J. Econ. Ent.*, 44: 154-159. (Pyrethrins as control of *Tabanus sulcifrons* on cattle). **Colonial Insecticidal Committee** (1951). *3rd. Ann. Rep.*

Prog. Rep., 9, Documents DC 39918-1 and CIC 51 A/C, *etc.* Colonial Office, Great Smith St., London, S.W. 1. (Control of *Glossina* from aircraft). **Colonial Insecticides, Fungicides and Herbicides Committee** (1951). *Ann. Rep.* 1950-51, *Appendix* 1, *CMD* 8063. H. M. Stationery Office, London. (Control of *Glossina* from aircraft). **Du Toit, P. J.** (1948). *Vet. Med.*, 43 : 8. (Short account of DDT spraying of *Glossina*). **Gerry, B. I.** (1949). *J. Econ. Ent.* 42 : 888-890. (Area control of *T. nigrovittatus*). **Goodwin, W. J., Sloan, M. J.** and **Schwardt, H. H.** (1952). *J. Econ. Ent.* 45 : 121-122. (Pyrethrins to control biting flies on cattle). **Granett, P., Haynes, H. L.** and **Helm, R. W.** (1951). *J. Econ. Ent.*, 44 : 97-102. (Comparison of pyrethrins with Crag repellent). **Hixson, H.** and **Muma, M. H.** (1951). *J. Econ. Ent.*, 44 : 401-404. (Pyrethrins to control *Stomoxys* on cattle). **Lewis, E. A.** (1950). *Ann. Rep.* for 1949, E. African Tsetse and Trypan. Res. and Reclam. Org., p. 36. **Metcalf, C. L., Flint, W. P.** and **Metcalf, R. L.** (1951). Destructive and Useful Insects. McGraw-Hill Book Co., New York, 3rd. Ed. pp. 901-902. (Spraying cattle against biting flies). **Rogoff, W. M.** and **Moxon, A. I.** (1952). *J. Econ. Ent.*, 45 : 329-334. (Cable-type back-rubber for cattle). **Tiel, N. van.** (1952). *Trans. IXth. Int. Cong. Ent.*, Amsterdam, 1951. (Emulsions with DDT in aqueous phase for *Glossina*). **Toit, R. du, Kluge, E. B.** and **Fielder, O. G. H.** (1950). Tsetsefly and Trypan. Perman. Inter-African Bureau, Biblio. Bull., 7 : p. 249. **Travis, B. V.** and **Smith, A. L.** (1951). *J. Am. Vet. Med. Assoc.*, 119 : 214-216. (Repellents for *Stomoxys* on cattle). **Symes, C. B.** (1952). *J. Roy. Sanit. Inst.*, 72 : 498-512. (Insecticidal control of tsetse fly : review).

CONTROL OF FLIES PARASITIC ON LIVESTOCK

Attacks of the screw-worm fly (*Callitroga americana*) on wounds in cattle may be prevented by smearing them with pine-tar oil, or by applying finely powdered diphenylamine, diphenylene oxide or nitrophenetole. Once the wounds are infested, they may be treated with a mixture of 35 per cent diphenylamine, 32 per cent benzene, 10 per cent n-butyl alcohol, 21 per cent lamp-black, and 2 per cent *Triton X*-300 emulsifier. Diphenylamine is more toxic than DDT and its analogues to these larvae, but is surpassed in toxicity by chlordane, lindane and dieldrin. In the United Kingdom, BHC has been successfully employed to eliminate larvae of *C. hominivorax* without toxic hazard to the cattle. For the control of *C. macellaria* larvae breeding in carcasses, PDB, ODB or Thanite sprays have proved most suitable; DDT kills only the adult flies as they come to oviposit. Diphenylamine or biphenyl, added to benzene, may be used to protect sheep from myiasis by this species of fleeceworm. The application of area sprays of DDT against *Callitroga* spp. can obtain at best only 70 per cent population reduction even when 2 lb. are applied per acre.

Methods of control have been developed against the blowfly larvae which have become known as fleeceworms or wool maggots, notably *Lucilia sericata*, *L. cuprina* and *Phormia regina*. In Australia, *L. cuprina* was first combatted by jetting the breech of the sheep with calcium arsenite, to prevent oviposition or "strike." Maggots already established were eliminated by boric acid in bentonite, mixed either with creosote (the BTB dressing) or lysol and ODB (the BKB dressing). Now it has been found that 1 per cent BHC emulsion sprays can prevent strike and eliminate larvae of *L. cuprina*. In the United States, a mixture of biphenyl, benzene, n-butyl alcohol and emulsifier (formula M.S. 793 F) was applied to the site of entrance of *P. regina* and C. *macellaria*. But later it was discovered that 0.1 per cent lindane suspension sprays gave 95 per cent control of *L. sericata* and *P. regina*, and that BHC, chlordane, and especially toxaphene, were superior compounds to apply to infested wounds. In Britain, 0.5 per cent DDT sprays or dips have given effective protection of sheep against *L. sericata*. Like *Cal-*

litroga, the fleeceworm *Phormia regina* may not be controlled by area sprays of DDT at 2 lb. per acre, no more than 70 per cent reduction being obtained.

Cattle-grubs, which are larvae of the warble flies *Hypoderma bovis* and *H. lineatum*, have been controlled by the application of rotenone washes to the backs of the cattle, and of rotenone ointment to the larval cysts. Recent improvements consist in wholesale spraying of herds, with a view to generally reducing the local warble-fly population rather than preventing damage that has been already done; this is carried out with power sprayers operating at a pressure of 400 p.s.i. No insecticide has been found to be as effective as rotenone. BHC is effective only in salves, but not in suspension sprays; chlordane gives erratic results, toxaphene is inferior, and methoxychlor is ineffective. DDT can be induced to control warbles if throughly rubbed into the skin, to the detriment of the cow or her milk; as a spray it is ineffective. The control effected by rotenone may be increased by the addition of sulphur. A suspension containing 0.25 per cent rotenone and 10 per cent sulphur, applied as a spray at 2 quarts per head, may be expected to achieve 98 per cent control, and has proved superior to scrubbed hand-washes at 1 pint per head or 1 per cent rotenone dusts at 3 oz. per head.

Horse bots (*Gasterophilus* spp.) may be initially repelled by treating the fore quarters of the animal with a mixture of equal parts of pine tar and lard. The eggs, once laid, may be destroyed as they hatch by applying warm water washes containing 2 per cent phenol or creosote. Larvae that have reached the stomach may be eliminated by administration of toluene or carbon disulphide by a veterinarian using a stomach tube. The dose with toluene is 20 cc. per 100 lb. body-weight, and it gives excellent results. A more common remedy is carbon disulphide at 2 to 2.5 c.c. per 100 lb. body-weight; but it causes desquamation of the stomach lining. Another anthelmintic that should be effective is phenothiazine, at the recommended doses of 1 ounce per 1000 lb., or better 5 gm. per animal daily. Unfortunately BHC, which can be given orally with safety, has little effect on these larvae; lindane, chlordane, heptachlor, aldrin or dieldrin are equally ineffective.

The sheep ked (*Melophagus ovinus*) may be decisively controlled by the use of chlorinated hydrocarbons or rotenone. In Britain, 0.5 per cent DDT dips have been found to give complete control; in Australia, 0.2 per cent dips and even 0.1 per cent sprays do likewise. In the U.S.A. virtually complete control is obtained with emulsions of 0.2 per cent DDT or 0.1 per cent lindane. However, rotenone is more effective, cheaper and safer than DDT for use on sheep. When lambs were treated, 0.5 per cent rotenone dust gave better control than 5 per cent DDT dust. Other chlorinated hydrocarbons such as BHC, toxaphene, chordane or methoxychlor are superior to DDT or DDD, although 0.2 per cent suspension dips or sprays of all these insecticides afford almost complete control for at least 4 months. Rotenone is even better than the chlorinated hydrocarbons, since it can obtain complete control even with 0.005 per cent concentrations.

REFERENCES

Eddy, G. W. (1951). *J. Econ. Ent.*, 44: 254. (Larvicides for the screw-worm). **Errington, B. J.** and **Westerfield, C.** (1940). *Vet. Med.*, 35: 688-693. (Phenothiazine as anthelminthic for horses and mules). **Graham, O. H.** and **Alford, H. I.** (1951). *J. Econ. Ent.*, 44: 577-579. (Chlorinated hydrocarbon insecticides to remove horse bots). **Habermann, R. T.**, **Harwood, P. T.** and **Hunt,**

W. H. (1941). *North American Vet.*, 22 : 85-92. (Phenothiazine as anthelminthic for horses). **Howell, C. E.** and **Britton, J. W.** (1940). *Cornell Vet.*, 30 : 526-532. (Phenothiazine as an equine anthelminthic). **Reid, B. L.** (1947). *Agric. Gaz.* New South Wales 57 : 377-381. (Phenothiazine as anthelminthic for domestic animals). **Rude, C. S., Smith, C. L.** and **Parish, H. E.** (1945). *U. S. Dept. Agr. Bull. E*-633. 2 pp. (Biphenyl as remedy for fleeceworms). **Todd, A. C.** *et al.* (1950). *J. Am. Vet. Med. Assoc.*, 66 : 369. (Toluene or carbon disulphide to control horse bots). **United States Department of Agriculture.** (1926). Farmers' Bulletin 1503. (Control of horse bots).

CHAPTER V

INSECTICIDAL CONTROL OF OTHER SPECIES OF MEDICAL IMPORTANCE

CONTROL OF FLEAS

The control of fleas involves treating the house and surroundings, disinfesting any dogs or cats in the household, or eliminating rats after first disinfesting them. A 5 per cent spray in kerosene completely eliminates *Pulex irritans* and other fleas when applied to floors, rugs and the lower 2 feet of walls at a dosage of 1 gallon per 1200 square feet. The same result may be obtained with 5 or 10 per cent dusts applied at the rate of 0.5 lb. DDT per 1000 square feet. Lindane or dieldrin sprays are more effective than DDT.

The dog flea (*Ctenocephalides canis*) and the cat flea (*C. felis*) may be eliminated by dusting dogs with 10 per cent DDT at the rate of a quarter-teaspoonful for the average-sized dog; 5 per cent may be also used. Cases have been found where *C. felis* was resistant to 5 per cent DDT but not to 5 per cent chlordane dusts. A 5 per cent BHC dust will free a dog from fleas more quickly. DDT should not be used on cats, since they have the habit of licking themselves; rotenone or pyrethrum powders should be substituted on these animals.

The disinfesting of rats from their fleas should precede any programme of rodent control in plague areas. It has been done with 8 or 10 per cent DDT dust, applied as "tracking patches" measuring 6 by 8 inches and smoothed to a thickness of 1/32 inch or less, putting at least two patches on each rat-run; or the burrows and runs have been dusted with 10 per cent DDT. Approximately 90 per cent control of the rat flea (*Nosopsyllus fasciatus* and 80 per cent reduction of the oriental rat flea (*Xenopsylla cheopis*) may be expected. This method also halves the rat mite infestation, but only slightly reduces the rat lice. The sticktight flea (*Echidnophaga gallinacea*) is only 40 per cent reduced by this method; but it may be eliminated from dogs with 5 per cent, and from poultry with 10 per cent, DDT dusts.

The rats themselves as well as their fleas, may be eliminated by heavy doses of insecticidal dusts. A thick layer of 50 per cent DDT dust killed 90 per cent of the roof rats exposed in the Belgian Congo. BHC dusts containing 3 per cent of the gamma isomer have killed rats and their fleas in Kenya.

Formerly, pyrethrum and rotenone dusts or dips have been employed for removing fleas from man and animals, and sprays of *Lethane* thiocyanates, pyrethrum or rotenone have been applied in the house. The possibility that *X. cheopis* might develop resistance to DDT has been tested in the laboratory, but the results were negative. However, resistance to DDT has appeared in *Pulex irritans* in Greece and Peru.

REFERENCES

Hess, A. D. (1952). *Am. J. Trop. Med. Hyg.* 1 : p. 372. (DDT resistance in *Pulex*). **Kilpatrick, J. W.** and **Fay, R. W.** (1952). *J. Econ. Ent.*, 45 : 284-288. (DDT resistance in *Xenopsylla*). **Symes, C. B.** (1952). *J. Roy. Sanit. Inst.*, 72 : p. 509. (BHC as a raticide). **Wanson, M.** and **Camphyn, R.** (1949). *Ann. Soc. Belge Med. Trop.*, 29 : 549. (DDT as a raticide).

CONTROL OF LICE ON MAN AND ANIMALS

Lice on Humans. Since 1944, the standard lousicide has been a 10 per cent DDT dust in pyrophyllite. Blown into the clothing of the wearer, a dose of 1.5 ounces has given complete control of the body louse, *Pediculus humanus corporis.* Applied lightly and rubbed into the dry hair, it controls the head louse, *P. h. capitis.* Three weekly applications to the pubic hairs may be expected to eliminate the crab louse, *Phthirus pubis.* The body lice cease to feed 3 hours after treatment, are knocked down in 6 hours, and die in 20 hours; the human subject is protected against reinfestation for at least 3 weeks. The DDT powder fails to kill the eggs, but the young nymphs are killed as they hatch. Large numbers of people may be treated in a short time with 10 per cent DDT dust while wearing their ordinary clothing. A duster with a 6-inch extension tube, operated either by a hand plunger or with compressed air from a gasoline-driven compressor, is inserted into either sleeve, then the front of the collar, the front of the trousers, the back of the collar, and the back of the trousers; the subject's head and hat are also treated. The treatment of 1400 prisoners of war in North Africa, 77 per cent of whom were infested, achieved complete disinfestation of body lice. The mass treatment of about 2,500,000 individuals in Naples, during the louse-borne typhus outbreak of 1943-44, cut the daily incidence of new civilian cases from 60 down to 10, and prevented typhus appearing in any of the American forces there.

The handicap of DDT in not killing eggs may be overcome by the use of the NBIN emulsion, in which a concentrate, containing 6 per cent DDT, 12 per cent benzocaine and 14 per cent *Tween* 80 emulsifier dissolved in 68 per cent benzyl benzoate, is diluted with 5 volumes of water. The benzocaine (ethyl p-aminobenzoate) destroys the eggs, and thus the emulsion achieves control of all stages at once. Benzyl salicylate, or still better Lauseto-neu, may be substituted as the ovicidal agent. The benzyl benzoate used as the initial solvent also serves as a remedy for scabies. Being a liquid formulation, it is more acceptable for use in hair against head lice; moreover, it controls crab lice with a single application. The EEAY formula is a modification in which the 6 per cent DDT and 12 per cent benzocaine are simply dissolved in ethyl alcohol, which is then diluted with water.

DDT has been impregnated into underwear and has given complete protection to millions of troops without causing skin irritation. Cotton and wool garments are impregnated by dipping them into a solution of DDT in white spirit or other volatile solvent, so that the insecticide is taken up on the fibres to constitute 1 per cent of their weight. If further impregnated to the extent of 2 per cent, the DDT will still be effective after 6 to 8 washings of the garment. When necessary, reimpregnation may be effected by soaking in 2 per cent emulsions of DDT, or by dry-cleaning in a solvent which contains DDT.

The slowness of knockdown of lice by DDT has led to 0.2 per cent pyrethrins being added to the 10 per cent dust. Since pyrethrum reduces the "flowability" of the powder, 3 per cent diatomaceous earth (*Celite* 209) is admixed as a conditioner, and 1 per cent piperonyl butoxide synergist and 0.24 per cent isopropyl cresol anti-oxidant are also added. Alternatively, a 1 per cent lindane dust in pyrophyllite may be used to obtain a quick knockdown. British workers have found a 0.2 per cent liquid preparation of lindane to give excellent control of body lice and head lice.

The recent discovery of DDT-resistant body lice in Korea in 1951, and in Egypt in 1952, which can complete their life-cycle on cloth impregnated with double the normal dose of DDT, has once more focussed attention on pyrethrin lousicides. In 1941 the Americans introduced the MYL powder, which contained pyrethrum, N-isobytylundecylenamide synergist, phenol-S anti-oxidant, and 2,4-dinitroanisole as an ovicide. The modern "pyrethrum powder" contains allethrin with sulfoxide synergist and p-benzylaminophenol, diluted in pyrophyllite with *Celite* 209 conditioner. An equally good substitute for sulfoxide is triallyl a-carboxy-β-methyl-glutarate. Evidently allethrin is superior to the pyrethrins, at least as far as initial control is concerned.

Other formulations which have been developed for louse control in armies of the 20th century may be listed as follows :

NCI Powder :	naphthalene, creosote, iodoform.	British, 1914-18.
AL 63 „ :	naphthalene, tar acids, rotenone.	British, 1939.
Sherlice belt :	treated with lauryl thiocyanate.	British, 1940
Lauseto-Neu :	chloromethyl p-chlorophenyl sulphone, an excellent ovicide and lousicide.	German, 1939-45.
Diphenylamine louse powder, and bisethylxanthogen clothing impregnant.		Russian, 1939-45.

Salves containing pyrethrum or rotenone have proved highly effective in controlling head lice and crab lice. However, the 1 per cent rotenone cream fails to kill the eggs and may cause dermatitis. A 0.2 per cent solution of pyrethrum in refined kerosene, rubbed into the hair, kills both lice and nits. A solution of 25 per cent thiocyanoethyl laurate and 25 per cent n-butylcarbitol thiocyanate in refined paraffin, termed *Lethane* 384 *Special*, has given excellent results. Whereas this formulation may cause dermatitis, emulsions containing lauryl thiocyanate or isobornyl thiocyanate do not. When *Lethane* 384 *Special* is further diluted with an equal volume of white oil before application to the head, nothing worse than a very occasional mild case of erythema and desquamation of the scalp has ever appeared.

Another standard method of louse control is to supply personnel with a change of clothes and a bath at delousing and fumigation station. The bath is followed by a spray with the NBIN emulsion, since even carbolic soap does not kill the nits. The clothing is disinfested by autoclaving with steam, or by fumigation with hydrogen cyanide or methyl bromide; methyl allyl chloride has been used in Britain, trichloroacetonitrile in Germany, and methyl formate in Russia. Methyl bromide is the most ovicidal of the common fumigants; a dosage of 1 lb. per 1000 cubic feet is sufficient for a 30-minute exposure in ordinary chamber fumigation; and for only 15 minutes in fumigation with a 28-inch vacuum. The dosage of hydrogen cyanide in chamber fumigation is 1 to 1.5 lb. per 1000 cubic feet. Methyl bromide is also used in gas-tight delousing bags, in which clothing is placed along with a breakable ampoule of the fumigant to give a dosage equivalent to 9 lb. per 1000 cubic feet.

Lice on Animals. Most infestations of lice on farm livestock may be controlled by dipping them in 0.3 per cent emulsions of DDT. By this treatment, complete eradication may be obtained of the cattle biting louse (*Damalinia bovis**), the short-nosed sucking louse (*Haematopinus eurysternus*) on cattle, the hog louse (*H. adventicius*), the sheep biting louse (*D. ovis*) and goat lice (*D. hermsi, caprae* and *limbata*). The eggs are killed as they hatch, and protection may last for 4 months. In most of these cases there is evidence that concentrations down to 0.1 per cent DDT give effective control. However, the long-nosed sucking louse (*Linognathus vituli*) and the capillate louse (*Solenopotes capillatus*) on cattle require concentrations of 0.6 per cent DDT for complete mortality. The treatments employed for the hog louse (*Haematopinus suis*) are 0.5 per cent DDT emulsion sprays to kill the motile stages, or 0.75 per cent dips to kill the nits as they hatch. DDT is also effective against *H. asini* and *D. pilosa* on horses.

In dust form, 10 per cent DDT has given fairly satisfactory results against the cattle biting louse. When thoroughly rubbed into the skin at the rate of 3 ounces per animal, 5 per cent DDT may be expected to give good control of all lice on cattle. One or two treatments with 5 per cent DDT dust control the dog biting louse (*D. canis*) and the dog sucking louse (*H. piliferus*). However, the dusting of rat runways with 8 per cent DDT achieves only 25 per cent mortality of the rat louse (*Polyplax spinulosa*).

Both the biting lice (Mallophaga) and the sucking lice (Anoplura) are even more susceptible to BHC than to DDT. Dipped in suspensions containing 0.04 per cent or emulsions containing 0.02 per cent of the gamma isomer of BHC, cattle are completely disinfested of *Damalinia bovis*, *Haematopinus eurysternus*, and the refractory *Linognathus vituli* also. Emulsion sprays containing only 0.005 per cent gamma-BHC, or 1 per cent BHC dusts, are sufficient to control *D. bovis* completely. There is a report that emulsion dips containing 0.1 per cent gamma isomer kill the eggs of *H. eurysternus*, but this concentration would not be safe to use on young cattle. BHC lotions control the hog louse at concentrations of only 0.005 per cent gamma isomer. Baths containing only 0.0025 per cent gamma-BHC in emulsion completely eradicate the dog biting louse and the dog sucking louse. Of the other chlorinated hydrocarbons, chlordane has proved the equal of DDT for the hog sucking louse, and is superior for the horse biting louse (*Damalinia equi*). DDD is the equal of DDT for controlling the hog louse, and has no toxic hazard associated with its use.

Wettable powder suspensions are now recommended for both dips and sprays, since they offer the least risk of the livestock absorbing the insecticide through the skin. A choice of chlorinated hydrocarbons is now available, including DDT, DDD, methoxychlor, chlordane and toxaphene, to be used in 0.5 per cent concentrations for cattle and 0.2 per cent for hogs, sheep and goats. With lindane, only 0.025 per cent suspensions are required, but neither lindane nor toxaphene should be applied to calves. Only methoxychlor is permissible for dairy cattle, since the other insecticides contaminate the milk. When DDT, DDD or methoxychlor is used as against the hog louse, two treatments may be required; but with chlordane or toxaphene one should be sufficient.

Sprays containing 0.005 per cent pyrethrins with 0.05 per cent piperonyl butoxide synergist may be used as frequently as required without any associated toxic hazard to the cattle. Dips containing 1 per cent wettable sulphur with 0.005 per cent rotenone

---

* Also called *Bovicola bovis* and formerly *Trichodectes scalaris*.

added are effective treatments for cattle lice. A simple remedy for cattle and hogs is to brush raw linseed oil over the coat, uisng 1/4 pint per animal for cattle. Hogs may be disinfested by producing a wallow pool covered with a film of crude petroleum oil; this is offered for one day every week, the pool being filled with water alone on the intervening days. Cattle and hogs may be provided with rubbing posts which are wrapped in sacking impregnated with oil or insecticide, but such measures are useless when the lice infest parts of the body which cannot be rubbed.

Materials used in cattle dips for louse control formerly included 0.2 per cent As_2O_3 in the form of sodium arsenite, 0.05 per cent nicotine, and 0.3 per cent cresylic acid. Sodium fluoride dust has been used for biting lice on cattle and sheep. Derris dusts containing 0.5 per cent rotenone are still acceptable on cattle and sheep, but not on hogs.

Control of the various species of biting lice affecting poultry (*Lipeurus*, *Goniocetes*, *Menopon*, *Eomenacanthus*, etc.) may be achieved with 4 per cent DDT dusts or by spraying them with 4 per cent DDT wettable powder suspensions. This treatment is superior to the 33 per cent sodium fluoride dusts and 0.5 per cent nicotine dips formerly employed. However, *Falculifer rostratus* on pigeons is only partially controlled by 10 per cent DDT dust. The daily feeding of 250 mg./kg. DDT, 6 mg./kg. gamma-BHC or 5 mg./kg. pyrethrins to poultry has achieved reduction but not eradication of *Menopon* and *Goniocetes*.

REFERENCES

Blanton, F. S. (1951). *J. Econ. Ent.*, 44 : 814-815. (Bag method of fumigating clothing with methyl bromide). **Bushland, R. C.** *et al.* (1945). *J. Econ. Ent.*, 38 : 210-217. (DDT louse powder). **Buxton, P. A.** (1950). The Louse. Edward Arnold, London. 2nd Ed. 164 pp. **Department of the Army** (1947). *Tech. Manual TM*-632, Superintendent of Documents, Washington. pp. 45-50. (Methods of louse control with DDT). **Eddy, G. W., Carson, N. B.** and **Bushland, R. C.** (1947). *U. S. Dept. Agric. Bull. E*-733., 235 pp. (Chemical compounds tested as lousicides). **Hart, L. R.** (1946). *Min. Supply U. K. Res. Dev. Monograph* 9.602. 57 pp. (Impregnation of clothing with DDT). **Hurlbut, H. S., Altman, R. M.** and **Nibley, C.** (1952). *Science* 115 : 11. (DDT-resistant body lice in Korea). **Latta, R., Richardson, H. H.** and **Kindler, J. B.** (1946). *U. S. Dept. Agr. Circulat* 745. (Methyl bromide as a delousing agent). **Menon, P. B., Gupta, C. M. S.** and **Basu, B. C.** (1951). *Indian Vet. J.* 47 (6) :1-8. (Internal medication of poultry with insecticides against lice). **Roy, D. N.** and **Ghosh, S. M.** (1944). *Ind. Med. Gaz.*, 79: 589. **U. S. Department of Agriculture.** (1918). Farmers' Bulletin 909. (Control of lice on cattle). **U. S. Department of Agriculture.** (1920). Farmers' Bulletin 1085. (Control of lice on hogs). **U. S. Department of Agriculture** (1946). Farmers' Bulletin 1330. (Control of lice on sheep and goats). **Williams, R. W.** (1950). *Soap San. Chem.*, 26 (8) : 117-121. (The EEAY lousicide formula).

CONTROL OF BED-BUGS, INCLUDING FUMIGATION

Bed-bugs. An effective method of control of the bed-bug, *Cimex lectularius*, has been developed to take advantage of the outstanding qualities of DDT. Residual sprays of 5 per cent DDT in kerosene are applied to the lower two-thirds of the walls of the infested rooms, crevices being specially treated; and to the sides of mattresses, particularly the seams. Excellent control is obtained, and the surfaces remain residually toxic to bed-bugs for nearly 200 days. When 20 per cent DDT concentrates are used, the buildings are effectively protected for 12 months. Treatment of an entire town in Russia with 5 per cent emulsion (at 5 c.c. per square metre) achieved complete control, which persisted for at least 3 months. The use of DDT in the great cities of the United States has largely eliminated the bed-bug as a source of pest control operators' business, although DDT-resistant bed-bugs have been reported from Hawaii and the United States. In some cases bed-bugs have been combatted with thermally generated aerosols of DDT ducted from the TIFA machine through flexible pipe into houses; such a practice is effective but involves a fire hazard.

Of the other chlorinated hydrocarbons which have been tested, methoxychlor is as effective as DDT, and lindane more so. Technical BHC is a satisfactory substitute for DDT, but does not remain effective for so long a period. Although toxaphene is more toxic by direct contact, it is not as residually effective as DDT. In northern Germany, 0.05 per cent solutions of E-605 (probably methylparathion rather than parathion) have been employed, though they are not as reliable as DDT and involve a toxic hazard. Lauseto-Neu is more toxic than DDT to the bed-bug, but its long-term effectiveness has not been investigated. The insecticides which DDT superseded in sprays, including rotenone, pyrethrum, and certain thiocyanates, had no appreciably persistent residual properties.

The bed-bug formerly furnished one of the main reasons for the fumigation of house, the others being cockroaches and rats. Hydrogen cyanide was the most common fumigant, but ethylene oxide, chloropicrin and carbon bisulphide have been employed. Recently methyl bromide has become the preferred material; but in some countries the use of sulphur dioxide produced by burning sulphur is still common. The topic of household fumigation will be described in this section, notwithstanding the fact that it is now almost completely susperseded by residual spraying with DDT for bed-bug control.

Assassin Bugs. Although DDT is insecticidal both to *Triatoma* and *Rhodnius*, BHC has been favoured for control of these cone-nosed reduviid bugs. Residual wall-sprays of BHC suspensions heavily applied can eliminate *T. infestans*, vector of Chagas' disease, from Brazilian huts. Deposits of 500 mgms. BHC per square metre remain effective for about 3 months in Peru. The addition of pyrethrins has increased their toxicity. In Chile, heavy doses of dieldrin have given excellent kills of this species, and the deposits remained insecticidal for several months. But on the whole dieldrin, chlordane and DDT are less toxic to these insects than to most disease vectors.

Building Fumigation. It is possible to fumigate a building that is of tight construction, self-contained and detached, and has a cement basement or an impervious floor. The fumigation is performed by qualified operators, who work in teams of 2 or 3, and their first task is to seal the windows and doors with oiled paper and masking tape, leaving one exit open. They wear rubber gloves, suitable coveralls, and gas-masks equipped with the M2 canister for hydrogen cyanide or the M1 canister for methyl bromide. Starting with the rooms farthest from the exit, they commence releasing the gas and work through the building until they reach the exit, which they seal after leaving. The house is left locked and guarded for at least 6 hours; large warehouses may require a period of 72 hours. When HCN is employed, the minimum dosage is 6 ounces per 1000 cubic feet; warehouses containing stored products may require as much as 16 ounces per 1000 cubic feet. After fumigation the building is opened to the air for a period of 1 to 2 hours for small buildings, while large warehouses may require 12 to 72 hours.

One method of releasing the hydrogen cyanide was to add sodium to sulphuric acid. An earthenware crock of 50 per cent H_2SO_4 was set in each room, and for every 1000 cubic feet to be fumigated, 1 lb. of sodium cyanide in egg form was dropped into 1 gallon of the acid. A more up-to-date method is to use discoids; these are wafers of cellulose fibre, 6 inches in diameter, which have been impregnated with liquid HNC when stacked in cylindrical cans before sealing. The cans are set in position and opened by a special opener, and the discoids are shaken out broadcast over the floor. A third method, followed in the fumigation of citrus trees under tarpaulin tests, is to spray liquid hydrogen cyanide from a miniature pressure sprayer equipped with pump and nozzle. California pest control operators now fumigate small houses under immense tarpaulin covers.

The more volatile fumigants, such as methyl bromide or ethylene oxide, are applied from cylinders in which the gas has been compressed into a liquid at pressures up to 25 p.s.i. When fractional doses of fumigants are applied from cylinders, they are fitted with graduated glass applicators ranging in size from 100 c.c. to 5 lb., to measure off the liquid fumigant which flows in through a stand-pipe. Methyl bromide may be handled in 1-lb. cans; they are punctured by an auger attached to gasketed tubing which is used to carry the gas to the spots to be treated, such as stored products under individual tarpaulins.

REFERENCES

Babers, F. H. and **Pratt, J. J.** (1951). *U. S. Dept. Agr. Bull. E*-818., 45 pp. (Reference to DDT-resistant bedbugs in Hawaii). **Busvine, J. R.** (1946). *Nature* 158 : 22. (Comparative efficacy of insecticides for *Pediculus* and *Cimex*). **Buxton, P. A.** (1952). *Trans. Roy. Trop. Med. Hyg.*, 46 : p. 220. (Tabulated references to control of bedbugs). **Department of the Army** (1947). *Tech. Manual TM*-632, Superintendent of Documents, Washington. Chapters 5 and 16. (Methods of bedbug control and of fumigation). **Dias, E.** and **Pellegrino, J.** (1948). *Brasil-med.*, 62 : 185. (BHC and pyrethrins to control *Triatoma*). **Federal Security Agency** (1949). Rat-borne Disease Prevention and Control. U. S. Public Health Service, Atlanta. pp. 222-225. **Hess, A. D.** (1952). *Amer. J. Trop. Med. Hyg.*, 1 : 376. (Mention of resistance of *Triatoma* to chlorinated hydrocarbons). **Johson, M. S.** and **Hill, A. J.** (1948). Medical News Letter (Washington, U.S.A.)., 12 : 27-28. (DDT-resistant bedbugs in Hawaii). **Madden, A. H., Lindquist, A. W.** and **Knipling, E. F.** (1945). *J. Econ. Ent.*, 38 : 265-271. (DDT for control of bedbugs). **Neghme, A.** *et al.* (1949). *Bol. inform. parasit. Chilenas.*, 4 : 36. (Dieldrin for control of *Triatoma*). **Pirie, H.** (1951). *Chem.* and *Ind.* Feb. 10 : 1951, pp. 100-105. (BHC wallsprays to control *Triatoma*).

COCKROACH CONTROL

The cockroaches which commonly infest buildings—the German roach (*Blattella germanica*), the American roach (*Periplaneta americana*) and the oriental roach (*Blatta orientalis*), have formerly been combatted with roach powders containing sodium fluoride, sodium fluosilicate or pyrethrum. Stubborn infestations have eventually been destroyed by fumigation with hydrogen cyanide or other fumigant, but the building becomes open again to reinvasion.

The appearance of DDT did offer an improvement in control, particularly of the American roach, to which it is more toxic than to the German roach. A programme of cockroach control may be carried out by the residual application of 5 per cent DDT in kerosene, paying particular attention to the underside of tables, the rear of sinks and other protected places; supplemented with 5 per cent (or preferably 10 per cent) DDT dust blown into crevices of walls, base-boards and mouldings with a plunger, bulb or bellows type duster. This treatment produces its full effect within two days, and if it is repeated every month it is possible to obtain satisfactory and occasionally complete control. In cases where fumigation is carried out, it should be followed by a thorough application of DDT dust or spray.

Although pyrethrum roach powder is not as effective as DDT dust, pyrethrin sprays may give dramatic and immediate results. Application of 0.3 per cent pyrethrin spray will flush the roaches from their hiding places and rapidly knock them down, and as much as 98 per cent mortality may be obtained if the formulation contains 1.5 per cent of a piperonylcyclonene synergist. However, this treatment does not reach 'dead spots' in the building, and has little or no residual effect; even if control is complete, re-invasion may be expected in two months. In food establishments, the pyrethrins are preferred insecticides because of their freedom from health hazard, and they are used either as a 0.05 per cent spray plus 1 per cent piperonyl butoxide, or a 0.1 per cent dust plus 1 per cent piperonylcyclonene. A recent spray formula contains 0.06 per cent pyrethrins plus 0.5 per cent sulfoxide. Pyrethrins may be added to DDT formulations to hasten the knockdown of roaches; the Lethane thiocyanates have been used for the same purpose.

Although DDT is 10 times as toxic as sodium fluoride to the German roach, it is only one-third as toxic as toxaphene, and one-tenth as toxic as chlordane or lindane. Thus residual treatment of indoor surfaces with DDT at levels adequete for house-fly control are relatively ineffective against cockroaches. Moreover, laboratory experiments have shown that *Blattella* can develop resistance to DDT. Application of chlordane or lindane at 100 mgm. per square foot gives a high degree of control and prevents reinfestation for 10 weeks. The use of BHC and lindane is handicapped by their odour, while that of chlordane may involve a vapour hazard in poorly ventilated buildings. Nevertheless, 2 per cent chlordane sprays and dusts have been employed very effectively in many establishments without deleterious effects on human health. But definite evidence of acquired resistance of *Blattella* to chlordane has appeared in a city in Texas.

It is now possible to use cockroach repellents as treatments for boxes, cartons and other surfaces. Cumene isopropyl peroxide and butadiene-furfural copolymer have been found to be highly effective in this regard. A cement in which finely-divided copper powder is incorporated in magnesium oxychloride (Hubbellite) has proved to be repellent to cockroaches regardless of its colour.

REFERENCES

Gahan, J. B. *et al.* (1948). *J. Econ. Ent.*, 41 : 795-801. (Chlordane the best chlorinated hydrocarbon for cockroach control). **Goodhue, L. D.** and **Tissol, C. L.** (1952). *J. Econ. Ent.*, 45 : 133-134. (Cockroach repellents). **Greaves, T.** (1947). *J. Coun. Sci. Ind. Res. Australia* 20 : 425-433. (DDT for cockroach control). **Hazard, F. O.** (1945). *Soap San. Chem.*, 21 (4) : 129. (Hubbellite as a roach repellent surface). **Heal, R. E., Nash, K. B.** and **Williams, M.** (1953). *J. Econ. Ent.* 46 : 385-386. (Chlordane-resistant *Blattella* in Texas). **Hess, A. D.** (1952). *Amer. J. Trop. Med. Hyg.*,

1 : 380. (Laboratory-induced DDT-resistance in *Blattella*). **Mc-Govarn, E. R.** and **Piquett, P. G.** (1946). *Soap. San. Chem.*, 22(8) : 157, et seq. (Modern cockroach powders).

CONTROL OF ECTOPARASITIC MITES

Harvest Mites. The Trombiculid mites known as chiggers, red-bugs, harvest mites or velvet mites may be controlled in the brushy areas which they inhabit. Formerly sulphur dust was employed, but a dosage of 50 lb. per acre was required. Neither DDT at 10 lb. per acre nor benzyl benzoate at 5 lb. per acre give adequate control. The dinitro compounds DNOC and DNBP have given good control at 5 lb. per acre, while at the same time causing a certain amount of damage to the vegetation. Chlordane applied at 2 lb. per acre has given excellent control of chiggers (*Trombicula irritans* and *Acariscus masoni*) in the southern United States, as also has BHC at this dosage. BHC has also given good control of chiggers at 2 lb. per acre, while at a dosage of 1.5 lb. of the gamma isomer per acre it achieved over 95 per cent control of the grass mite (*Acomatacarus australiensis*) in New South Wales. Recently, lindane and heptachlor have proved effective for chigger control at 0.5 lb. per acre, and dieldrin has given excellent results at 0.2 lb. per acre.

The impregnation of clothing is another means of protecting personnel in areas infested by harvest mites. Benzyl benzoate was used during World War II, sometimes diluted 50 : 50 with dibutyl phthalate, to protect soldiers against the chiggers transmitting scrub typhus. Benzil and diphenyl carbonate proved superior to benzyl benzoate, particularly because of their fastness to laundering. Recently, clothing impregnants repellent to harvest mites, ticks and mosquitoes have been developed. One formulation is called M-1960, and it contains benzyl benzoate, 2-butyl-2-ethyl-1.,-3-propanediol and N-butylacetanilide. Since this has caused skin rash in a small proportion of individuals, it has been succeeded by M-2040, which contains 40 per cent N-butyl-4-cyclohexenol-1,2-dicarboximide, 30 per cent hendecenoic acid and 30 per cent N-butylacetanilide.

Scabies. The standard remedy for infestations of the itch mite (*Sarcoptes scabiei*) is benzyl benzoate, applied as a 25 per cent emulsion in alcohol with soft soap added. Two applications on consecutive nights may be expected to effect a cure; this treatment has eradicated infestations in both the American and British armed forces. The idea of incorporationg benzyl benzoate in an anti-scabies soap has been abandoned because it has caused dermatitis, especially in children. A similar alcoholic emulsion containing 20 per cent of turpentine has proved a promising scabicide, although this material may be irritating and involve a toxic hazard. Tetraethylthiuram monosulphide (TETMS or Tetmos) is effective against *Notoedres*, and under the name of *Tetmosol* it has been incorporated to the amount of 10 per cent in soaps effective against sarcoptic and notoedric scabies. Sulphur. mixed with vaseline or soft soap and applied as a salve will effect complete control, although the treatment is very unpleasant and may involve dermatitis. Of other ointments formerly used, those containing derris often cause severe dermatitis, and those containing pyrethrum do not kill the mites.

Livestock Mange. For the control of mange mites on livestock, BHC has proved to be excellent. BHC suspension sprays containing 0.035 per cent gamma isomer have completely controlled *Sarcoptes scabiei* on pigs, horses and cattle. No reinfestation occurs after *S. scabiei suis* has been eliminated by 0.05 per cent lindane suspension sprays. These now replace the standard formulation which until recently was 3 per cent lime-sulphur, made by dissolving 10 lb. of stone lime and 24 lb. of gallons of water. *Chorioptes bovis* is also completely controlled by lindane suspensions, although the subsequent protection against reinfestation does not extend to the 8-month period afforded by 3 per cent wettable sulphur or lime-sulphur sprays. When the animals are dipped, suspensions with as little as 0.015 gamma isomer have given an apparently complete cure of chorioptic mange on horses and cattle. A spray containing 1 per cent wettable sulphur and 0.15 per cent lindane in suspension, applied to cattle at 400 p.s.i. pressure, has given complete control of *C. bovis* for the entire winter.

DDT cannot be relied upon to control mange mites. It has proved inadequate against *Sarcoptes s. suis*; it has successfully controlled *Psoroptes bovis* in some cases, but not in others. It has been reported to be effective against *Psoroptes cuniculi* and *Notoedres cati*. When applied as an 8 per cent dust to rat runways, DDT achieves less than 50 per cent control of the tropical rat mite (*Liponyssus bacoti*); sulphur must be added to the dust for acaricidal action against this mite.

For the control of livestock mange, therefore, the lime-sulphur dips and sprays may be replaced by BHC sprays, which should be suspensions since emulsions containing 0.05 per cent of the gamma isomer may kill calves. This chlorinated hydrocarbon also supersedes creosote, along with other treatments such as rotenone, benzyl benzoate and nicotine applied in former days. In the control of the chicken mite (*Dermanyssus gallinae*) however, BHC sprays impart a flavour to the hens' eggs even if applied only to the henhouse, and thus is no better than the creosote, carbolineum or pentechlorophenol formerly employed. But the treatment of henhouses with 1 per cent DDT suspensions has given complete and long-lasting control of the chicken mites.

REFERENCES

Goodwin, W. J. and **Schwardt, H. H.** (1952). *J. Econ. Ent.*, 45 : 122-124. (Sulphur plus lindane for *Chorioptes bovis*). **Sweetman, H. L.** *et al.* (1951). *J. Econ. Ent.*, 44: 112. (Lindane for *S. scabiei suis*). **United States Department of Agriculture.** (1926). Farmers' Bulletin 1493. (Control of mange on livestock).

CONTROL OF TICKS AND OTHER ARTHROPODS

Ticks. The control of ticks by insecticides is divided into three phases : the area control of field populations, the treatment of buildings, and the disinfestation and protection of domestic animals. In addition, personal protection of men in infested areas may be achieved by toxic substances repellent to the young ticks.

Although not completely satisfactory, an area spray of DDT at 2 lb. per acre has given good control of the cattle tick (*Boophilus annulatus*) in grass-palmetto scrub, the American dog tick (*Dermacentor variabilis*) along road margins on Long Island, the spotted fever tick (*D. andersoni*) in sage-brush in the western U.S.A. and the gulf coast tick (*Amblyomma maculatum*) in the southern United States. A dosage of 1 lb. DDT per acre is sufficient to control the American dog tick, the lone star tick (*Am-*

blyomma americanum) and the black-legged tick (*Ixodes scapularis*). If 10 per cent DDT dust is used against the above-mentioned species, it should be applied at 40 lb. per acre. Lindane gives excellent control of the American dog tick at 1 lb. per acre; but BHC at 2 lb. per acre does not give lasting control of the lone star tick. Area sprays of chlordane or toxaphene at 2 lb. per acre, heptachlor at 1 lb. per acre, and parathion or aldrin at 0.5 lb. per acre, give good control of *A. americanum* and *D. variabilis* for about 2 months, although parathion is not sufficiently persistent. For the American dog tick, only dieldrin rivals DDT in effectiveness and persistence. The most effective treatment of areas for ticks and chiggers may be 10 to 20 gallons per acre of an emulsion made by diluting 3 gallons of an 18 per cent emulsifiable concentrate of dieldrin in 100 gallons of water. Nicotine sulphate and sodium arsenite have also been recommended for area control of ticks.

For personal protection, the stocking and clothing impregnants M-2040 and M-1960 are both effective against the lone star tick, because of the 30 per cent N-butylacetanilide that they contain. Of the two, M-2040 is superior. Dibutyl adipate and n-hexyl mandelate have been found to be excellent clothing impregnants for repelling *D. variabilis*, and dibutyl adipate has proved to be superior to N-butylacetanilide in this respect.

When larval ticks are found in houses, the walls and furniture should be treated with a residual spray of 5 per cent DDT in refined kerosene applied as for bed-bug control. If dogs have required disinfestation, residual sprays should be applied to the kennel and its immediate vicinity, especially in the case of the brown dog tick (*Rhipicephalus sanguineus*). The relapsing fever tick (*Ornithodorus moubata*) may be eliminated from the dirt floors of East African huts by two heavy sprayings with BHC at 1.25 gm. per square foot, or by BHC dusts containing 0.5 per cent of the gamma isomer. This tick is comparatively resistant to DDT. The floor tick of Kashmir (*O. crossi*) may be successfully attacked with DDT or BHC dusts, or better with DDT-BHC mixtures, applied every month.

For the disinfestation of dogs from the brown dog tick, the American dog tick and the lone star tick, 10 per cent DDT dusts have given complete control but are not always satisfactory. Emulsions containing 2 to 5 per cent DDT are effective, but are not entirely without health hazard. Cases resistant to 5 per cent DDT dust have been remedied with 2 per cent chlordane or 1 per gamma-BHC dusts. Dusts containing 5 per cent BHC completely paralyse lone star ticks, which must then be removed by hand. For safety, a bath containing 2 ounces of derris (or cube) powder and 1 ounce of neutral soap per gallon of lukewarm water is recommended.

The control of ticks infesting cattle is now most effectively carried out with suspensions or emulsions of BHC, or even better by its purified gamma isomer, lindane. Sprays containing 0.5 per cent BHC in suspension give complete control of all stages of the cattle tick and the lone star tick, and protect the animals for 7 to 10 days after treatment. Dips containing 0.5 per cent BHC, equivalent to 0.06 per cent gamma isomer, can prevent the eggs of the cattle tick from hatching. The blue tick (*Boophilus decoloratus*) of South America is even more susceptible, dips and sprays containing only 0.003 per cent gamma isomer giving complete control. Sprays containing 0.1 per cent gamma isomer protect cattle against the spotted fever tick for two weeks.

BHC is the best of the chlorinated hydrocarbons although both toxaphene and

DDT are highly effective. Toxaphene is widely used against *Boophilus annulatus* and *Amblyomma americanum* in the southwestern United States. The spraying of 2 per cent DDT in suspension gives excellent control of the gulf coast tick, while the cattle tick may be completely eliminated by 1 per cent DDT sprays. In Queensland, effective control of *B. australis* has been obtained by monthly dipping of cattle in 0.5 per cent DDT; however, this insecticide leaves some female cattle ticks still alive. In Texas, dipping or spraying with 0.5 per cent suspensions or emulsions of toxaphene or DDT has given excellent control of *B. annulatus*, as also has 0.03 per cent gamma-BHC. The blue tick is considerably less susceptible to DDT, and 1 per cent DDT dips achieve only 60 per cent control. With *Hyalomma mauritanicum* 5 per cent DDT has been applied twice weekly to protect calves from piroplasmosis, although the treatment cannot be recommended. However, 1 per cent DDT emulsion has become the standard wash for eliminating the winter tick (*Dermacentor albipictus*).

The arrival of the chlorinated hydrocarbons was the answer to the increasing resistance to sodium arsenite which the cattle tick had been developing in Brazil and Australia, and the blue tick had been showing on the east coast of South Africa. Arsenical dips also had the disadvantage of depressing the general condition of the cattle, which the organic insecticides do not. Concentrations used for the cattle tick were 0.25 per cent sodium arsenite containing 0.2 per cent As_2O_3, and the dips were repeated every 2 weeks. For the blue tick, a concentration of 0.8 per cent sodium arsenite (0.64 per cent As_2O_3) was necessary even before the resistant strain appeared. Rotenone sprays have also been used for the cattle tick, and nicotine as well as rotenone for the blue tick, before the advent of the chlorinated hydrocarbons. However, resistance to DDT is now shown by the cattle tick in certain parts of Australia, and the blue tick is showing resistance to BHC in places in South Africa.

For ticks infesting the ears of cattle, BHC has again proved valuable. The spinose ear tick (*Ornithodorus megnini*) of Texas and Mexico is satisfactorily controlled by injecting 1/2 ounce of 5 per cent BHC, dissolved in a mixture of 1 part xylene to 9 parts pine oil, with a syringe into the ears; while the east coast fever tick (*Rhipicephalus appendiculatus*) of Uganda is controlled by 5 per cent BHC sprays applied twice weekly. DDT in 5 per cent sprays is also satisfactory for *Rhipicephalus*, but inadequate for *Ornithodorus*.

Protection of cattle against *R. appendiculatus* has been experimentally obtained by the oral administration of weekly doses of gamma-BHC at the level of 33 mgms. per kilo body-weight. Partial control of *Argas persicus* on poultry has been obtained by a single dose of gamma-BHC at 12 mg./kg. or pyrethrins at 40 mg./kg.

Spiders, Scorpions and Centipedes. Arachnida are comparatively resistant to DDT. In forest spraying, only the phalangiids succumb at a dosage of 1 lb. per acre, but at 5 lb. per acre all exposed spiders are affected or killed. The black widow spider (*Latrodectus mactans*) may be killed by DDT if applied heavily around her web; chlordane and lindane are effective to a less degree. It has been found that the scorpion *Centrurus* in the West Indies may be readily eliminated by residual deposits of DDT. Centipedes (*Chilopoda*) are more readily killed by DDT than are spiders.

REFERENCES

Cole, M. M. and **Lloyd, G. W.** (1951). *J. Econ. Ent.*, 44: 1025. (Dibutyl adipate superior to N-butylacetanilide). **Dicke, R. J.** and **Morgan, B. B.** (1951). *J. Econ. Ent.*, 44: 991. (Control of brown dog tick with chlordane or lindane). **Gouck, H. K.** and **Fluno, J. A.** (1950). *J. Econ. Ent.*, 43 : 698-701. (Control of American dog tick). **Granett, P.** and **French, C. F.** (1951). *J. Econ. Ent.*, 44 : 93-97. (Dibutyl adipate and n-hexyl mandelate). **Kalra, S. L., Jacob, V. P.** and **K. N. A.** (1950). *Indian J. Med. Res.*, 38 : 457-466. (Internal medication of poultry with insecticides against *Argas*). **Symes, C. B.** (1952). *J. Roy. Sanit. Inst.*, 72 : 498-512. (Control of *O. moubata* mentioned).

RODENT CONTROL

The control of rats and mice by poisoning achieves permanent results only if the buildings are rat-proofed. Essentially this means that there should be no openings larger than 1/2 inch to allow entrance of the rats. Storage floors and basements should be of cement, all cracks in walls should be repaired and the surface made as smooth as possible, and rat guards should be placed over water pipes. The lower parts of screen doors, windows and chimneys should be covered with rat-wire of at least 2 mesh to the inch. Window screens should have metal frames, and doors should be provided with a metal threshold and an automatic door-closer. Rodent control is also futile unless the community has a programme of garbage removal.

A necessary preliminary of rat poisoning is to assess the population by means of tracking patches. Powdered talc (or flour) is shaken on the floor in patches measuring about 6 by 18 inches, and smoothed to a thin coating with a straightedge. At least two patches should be laid on each rat run. If the infestation is in a plague or typhus area, the tracking powder should contain 10 per cent of DDT. This eliminates the rat fleas and partially controls rat lice; it also kills mice which poison themselves orally by licking it off. Heavy applications of 50 per cent DDT dust have killed roof rats themselves in the Belgian Congo, and BHC dusts containing 3 per cent gamma isomer have proved raticidal in Kenya.

The control programme then depends on the location of the rat infestation. If it is in outdoor burrows or small enclosed spaces, it may be treated by placing granules of calcium cyanide in rat-holes, or by blowing calcium cyanide dust into them with a hand duster or foot pump. Each burrow should receive about 1/2 ounce of dust, and the burrow openings should then be plugged. Sulphur mixed with nitre and tallow may be burned on a rag at the end of a stick thrust in the rat-hole. If the infestation is in an air-tight self-contained building with a good basement floor, it may be fumigated with hydrogen cyanide, methyl bromide, chloropicrin, carbon disulphide or sulphur dioxide. The rat-killing dosage of hydrogen cyanide is 2 ounces per 1000 cubic feet, whereas the flea-killing dosage is 5 ounces per 1000 cubic feet, for a 5-hour exposure. If the infestation is in a high-class building, it may well be better to attempt saturation trapping rather than risk the odours from hidden poisoned carcasses.

In most cases the rat control will be accomplished by laying down baits, which are usually the size of a level teaspoonful, moistened to a ball or cube about 1/2 inch in diameter. Alternatively the bait may be placed on a 5-inch square of waxed paper which is twisted to form a "torpedo." A business establishment might call for the placing of 100 baits along the sides of rat-runs. The choice of material for the bait is best decided by test-baiting for one night with a variety of foods, without poison. Acceptable food materials fall into three classes : (*i*) meat and fish, (*ii*) fats and (*iii*) fruit and vegetables. The Norway rat (*Rattus norvegicus*, also called brown or wharf rat) prefers type *i*, and the roof rat (*Rattus rattus*, including the subspecies black rat,

Alexandrine rat and climbing rat) prefers type *iii*. The Norway rat is characterized by short ears and tail, the roof rat by long ears and tail.

All baits are made up with cereal as a binder; the meat is ground up and mixed with an equal quantity of rolled oats and a little wheat flour; the fat is added to 4 times its weight in rolled oats with a little fish oil; the fruit is chopped up into cubes and a little flour is sprinkled over it. Almost any food has an equal chance of proving the favourite bait in any given situation, and it is a moot point whether baits should have one flavour or a variety. At least it is agreed that the type of bait should be changed between successive baitings. After test-baiting, many workers then pre-bait for 2 more days with the favourite food before adding the poison. With all poisons except red squill, uneaten baits should be removed not later than 2 days after being put down, and should be burned or buried. Dead rats should be similarly disposed of daily. In plague areas the carcasses should be placed, with precautions against direct handling, into a tight tin or bag containing calcium cyanide.

Baits may be placed in special containers to prevent children, livestock and pets from gaining access to them. A 5-inch board leaning horizontally against the base of a wall gives useful protection indoors. A 3-inch sewer pipe with its entrances partially buried is useful outdoors. Bait boxes may be constructed of wood or sheet metal, provided at each end with an entrance hole measuring 2 by 2.5 inches across. These boxes may be padlocked after baiting and moved from one bait station to another. The baits themselves are often put in saucers, which may have a double rim to make them spill-proof.

Red squill is the safest rodenticide to use, since it is not highly poisonous and is itself a sufficiently strong emetic to prevent any animal except rats (which cannot vomit) from ingesting a toxic dose. It is obtainable as a powder or a liquid extract, derived from the bulb of *Urginea* (*Scilla*) *maritima*, which grows on the Mediterranean coast, particularly in Italy. The toxicity of red squill is due to several glucosides; it also contains calcium oxalate, which is urticaceous to the skin and gives a sharp unpleasant taste. To be satisfactory, the material should show a median lethal dose for rats of 600 mgms. per kilo or less. This is mixed with the food to give 10 per cent baits. Now fortified red squill is available with an m.l.d. of 200 to 250 mg./kg., and this is toxic enough to allow 5 per cent baits to be lethal. It is blended with the flour, or with the oil in type *ii* baits; it does not mix well with type *iii* baits. It is the logical first choice for a poisoning programme, and may be safely used outdoors and in residential areas. When used in type *ii* baits it may be left exposed for several days. Red squill is the only rat poison that is safe for use in food-processing establishments.

Compound 1080 is the name given to sodium fluoroacetate, which is an extremely powerful poison. It is crystalline, odourless and tasteless. The median lethal dosage for rats is approximately 4 mg./kg. Since no emetic is fast enough to prevent a lethal dose being ingested, and since there is no known antidote, 1080 must never be used in baits unless enclosed in boxes. When it is handled, a gauze face-mask must be worn. Its only application is in poisoned water, with 12 grams 1080 per gallon (approx. 0.26 per cent) for Norway rats, and 7 grams per gallon for roof rats and mice. For safety the water is dyed black with nigrosine. It is exposed in 1/2-ounce waterproofed paper cups, half filled, stamped with a skull-and-crossbones and dyed brown; the water is transferred with the aid of 12-ounce rubber bulbs. This method is used only in buildings which can be left unoccupied and sealed overnight. The 1080 kills fast by causing ventricular fibrillation of the heart. The dead rats are found, 1 to 8 hours after baiting, lying within 10 feet of the container. They should be immediately put into incinerators or deep burial pits, since dogs happening to eat them may also be poisoned. The poisoned water may be put in a shallow pan or chick fountain and enclosed in a locked bait box. Compound 1080 may be expected to kill over 90 per cent of the rats in an operation, but it should be used only when other methods have failed.

ANTU, or alpha-naphthylthiourea, is a unique rodenticide in that it is highly poisonous to the Norway rat but not to the roof rat or house mouse. The median lethal dose for adult Norway rats is 7 mg./kg., but young rats are considerably more resistant. ANTU resembles thiourea in inhibiting the respiratory enzyme, cytochrome oxidase; the thyroid is the main site of action, and administration of iodine or KI has a protective effect. Death is due to pulmonary oedema, and takes up to 48 hours. Rats taking sublethal doses quickly build up a tolerance, even to 50 times the normal lethal dose. The toxicity of ANTU to humans is apparently very low, but it is very poisonous to cats, dogs, chicks and pigs. It should therefore not be used on farms, and in cities it should be used only in cereal baits, which are not eaten by cats and dogs. The baits should contain 2 or 3 per cent ANTU, and an equal amount of tartar emetic should be added to prevent accidental ingestion. The safest way to use ANTU is as a 20 per cent dust in pyrophyllite applied in tracking patches about 3 feet long. On the whole, ANTU is becoming less favoured as a rodenticide.

Zinc Phosphide (ZnP) is a rodenticide of many years standing. It is a black powder with an odour which is disagreeable to man but possibly attractive to rats. The toxicity is about 40 mg.-kg. for rats, and probably also for man. The antidote is 1-4 gm. of copper sulphate, administered in water every 10 minutes until vomiting occurs, accompanied by a cathartic of magnesium sulphate. The baits should contain 1 per cent zinc phosphide, and this small amount requires thorough mixing; 0.375 per cent tartar emetic should be added. Zinc phosphide is particularly suited to coating cubes of type *iii* baits for roof rat control; the cubes are swirled in a bucket until all the powder has been taken upon the surfaces. For both species of rat, zinc phosphide is favoured to alternate with red squill baits. For treating garbage dumps, a bait containing 2 parts zinc phosphide, 1 part magnesium carbonate, 16 parts rolled oats and 48 parts ground meat or fish is recommended, since it decomposes to a non-toxic residue after exposure to several rains.

Thallium sulphate (Tl_2SO_4) has also been used as a rodenticide for many years. It is a heavy white crystalline powder, odourless and tasteless, and rather insoluble. Thallium sulphate is a slow-acting cumulative poison, with a median lethal dose of 15 mg./kg. for Norway rats. Its toxicity to man is similar, and it can be readily absorbed through the skin; rubber gloves should therefore be worn when it is handled. Although expensive, only 0.5 per cent of thallium sulphate is necessary in baits; 3 per cent tartar emetic should be added. Similarly, 0.5 per cent of this salt may be used in poisoned water, and the same precautions should be taken as for Compound 1080. Reliance may be placed on thallium sulphate baits to give results.

Arsenious oxide (As_2O_3), also known as arsenic trioxide or white arsenic, has a toxicity of 15 mg./kg. to rats if it is finely ground. Although slow acting, this poison is cheap. The recommended strength in baits is 3 per cent, with 1.125 per cent tartar emetic added. Sodium arsenite is more water soluble and is used in concentrations not exceeding 3 per cent. Either arsenious oxide or sodium arsenite may be used to the extent of 2.4 per cent in poisoned water, with the customary precautions.

Strychnine has seen extensive use in baits for control of house mice and field rodents; it is relatively ineffective against rats. It is available either as the alkaloid, a chloroform-soluble powder; or as the sulphate, a water-soluble powder. Strychnine is very toxic and rapid in action. Although extremely bitter, enough may be ingested accidentally to poison the nervous system.

Warfarin (W.A.R.F. Compound 42) is an unusual rodenticide which since its appearance in 1950 has shown itself to give the most thorough control of rats (except *R. hawaiiensis*). It is 20 to 30 times more toxic than dicumarol, the first coagulant to be tested as a rodenticide; the chemical nature of warfarin is 3(a-acetonylbenzyl)-4-hydroxycoumarin. Its effect is to inhibit the production of prothrombin by the liver of the rat, and thus it leads to a deterioration of the walls of blood vessels. A single dose of 50 mg./kg. is without effect, but 5 daily doses of 1 mg./kg. are lethal; the rats may be seen walking in the open with a slow, measured gait, and a painless death ensues from lung haemorrhage. Warfarin is thus unusually suitable because there is no "bait shyness," since all rats continue to take the bait avidly until their death. Accidental ingestion of a single dose of warfarin, on the other hand, will not kill man or domestic animals. An attempted suicide by eating 567 mgms. over a period of 6 days met with failure. There is virtually no danger to chickens and other domestic animals. The antidote for warfarin is Vitamin K.

Warfarin is used in concentrations of 0.025 per cent in baits. The commercial material is usually in the form of a 0.5 per cent dilution in corn starch, which may be dyed green. This is diluted with 19 parts of bait; ground corn meal is recommended, which may contain also one-fourth of rolled oats and 5 per cent of powdered sugar, or alternatively 20 per cent of cheese. Since the baits are to be left exposed for some time, their water content should not exceed 10 per cent, and they may be "moistened" with 5 per cent mineral oil or corn oil; 0.25 per cent p-nitrophenol may be added as a mould deterrent. A dollop of 2 ounces to 1 lb. is put in each place, preferably in a bait station, or at least in a shallow pan and covered with a board roof. The rats start to die in 3 days, and deaths continue for a week after; they apparently die in their holes, since the odour problem does not measure up to expectations. Water poisoned with 0.075 per cent warfarin may also be used. The only shortcoming of warfarin —its slowness in action—may be remedied by first using zinc phosphide where an immediate kill is also required. It is necessary to prebait when warfarin is used, since there is hesitation in taking these baits as soon as they are put down.

Other Rat Poisons. *Barium carbonate* is a cheap, reasonably effective, and fairly safe rodenticide; it is often used for treatment of garbage dumps. With a median lethal dose of approximately 100 mg./kg., it engenders a thirst which makes the rats run about and die in the open. It is used at concentrations between 5 and 20 per cent in baits. *Phosphorus* paste is largely obsolete, since it is too hazardous for use in dwellings, and it presents a fire hazard. *Lye* and *plaster of Paris* have been used as cheap makeshift rodenticides. Other more modern rodenticides include *Muritan* (sodium N-p-chlorophenyl diazothiourea), *Castrix* (m.l.d. 7 mg./kg., used in 0.05

per cent baits), *Tomorin* (a German anticoagulant material), *Actone* (m.l.d. 3 mg./kg.) and *GDD* 1189 (m.l.d. 250 mg.-kg.).

Package Deterrents. It is possible to impregnate cardboard and fibreboard containers of edible materials with repellent chemicals to prevent attack by rats. The list of effective materials includes sodium silicate, pentachlorophenol, trinitrobenzene, diazoaminobenzene and actidione. Another good impregnant, the cyclohexylamine complex of zinc dimethyldithiocarbamate, has been developed as a general rodent repellent under the name of ZAC; when mixed with a polyethylene polysulphide (p.e.p.s.) sticker it is marketed under the name of ZIP.

Field Rodents. In addition to species of the genus *Rattus*, certain other wild rodents in the field may be vectors of plague and therefore require control, such as wood rats (*Neotoma*), ground-squirrels (*Citellus*), prairie dogs (*Cynomys*) and marmots (*Marmota*). Their burrows may be fumigated with hydrogen cyanide dust applied with a foot pump at 10 to 20 grams per burrow; or with carbon disulphide (inflammable) at 15 c.c. per burrow, applied in a ball of sacking or with a pump; or with methyl bromide at 10 c.c. per burrow. The following bait materials have been placed at the openings of burrows : strychnine (0.2 per cent), thallium sulphate (1.5 per cent) and zinc phosphide (1 per cent). Compound 1080 (0.05 per cent in rolledo ats) has also been employed, but the bait should be coloured with 0.03 per cent brilliant yellow S.P. dye to prevent its being taken by birds.

REFERENCES

Department of the Army (1947). Insect and Rodent Control. Tech. Manual TM-632, Superintendent of Documents, Washington. 156 pp. **Federal Security Agency** (1949). Rat-borne disease prevention and control. U.S. Public Health Service, Atlanta. 292 pp. **Grosss, B., Baker, R. H.** and **Bonnet, D. D.** (1951). U.S. Public Health Reports 66 : 1727-1733. (Warfarin to control Hawaiian rat). **Hayes, W. J.** and **Gaines, T. B.** (1950). Control of Norway rats with residual rodenticide warfarin. U.S. Public Health Reports 65 : 1537-1555. **Hayne, D. W.** (1949). Michigan *Agric. Exp. Sta. Bull.*, 32 : 313. (The rat repellents ZIP and ZAC). **Herms, W. B.** (1939). Medical Entomology. Macmillan Company, New York. 3rd Ed. pp. 400-409. **Holmes, R. W.** and **Love, J.** (1925). Suicide attempt with warfarin. *J. Amer. Med. Assoc.*, 148 : 935-937. **Mallis, A.** (1945). Handbook of Pest Control. *McNair-Dorland*, New York. 554 pp. **O·Connor, J. A.** (1948). Use of blood anticoagulants for rodent control. *Rescarch* 1 : 334-336. **Schein, M. W.** (1950). Field tests of efficiency of rodenticide Compound W.A.R.F. 42. U.S. Public Health Reports 65 : 368-372. **Storer, T. I.** (1948). Controlling rats and mice. California *Agr. Exp. Sta.* Extension Circular 410., 36 pp. **Symes, C. B.** (1952). *J. Roy. Sanit. Inst.* 72 : pp. 508-509. (DDT and BHC as rodenticides).

INDEX

A

Acalyptrata Muscoidea, 143.
Acanthocheilonema perstans, insect carrier, 119.
Acanthoconops, 118.
Acarina, characters, 228; life history, 232; classification, 234.
Acrostichal bristles, in flies, 134.
Acylonitrile, 313.
Aedes, characters, 92; species, 93; aegypti, 94; diseases, 98, 99, 100; *Plasmodium gallinaceum*, 98; biological testing of antimalarial drugs, 98; breeding, 98; control, 101.
Aerial application of insecticides, 355.
Aerosol, apparatus for producing, 351, 357; bomb, 353; smoke, 354; for disinfesting aircrafts, 367; in mosquito control, 371,
Aeschna, 285.
Affinin, 310.
AL 63, as pediculicide, 392.
Aldrin, chemistry, 303; estimation, 303; mode of action, 328.
Aleurobius, 267.
Allantonema, parasites of flies, 140.
Allesandrini, method of estimation of DDT, 301.
Allethrin, 308; estimation, 308; in mosquito control, 367; against body lice, 308; control of lice, 392.
Allodermanyssus sanguinensis, 265.
Allyl chloride, 392.
Aluminium silicate, 316; oxide, 328.
Amblyomma, 242; carrier of *Rickettsia ruminantium*, 242.
Ametabolous, metamorphosis, 7.
Anabasine, 307.
Anaplasma, ticks as carriers, 229.
Annelida, 3.
Animal, classification, 1.
Anopheles, life history, 35; parasites, 40; races, 41; classification, 52; species, 42; malaria, 47; uses in therapeutics, 47; geographical distribution of malaria-carriers, 48; susceptibility of different species to malaria, 49; collection of larva, 50; of adults, 51; dissection, 51, 54; artificial infection with malaria, 55; identification of adults, 61; of larva, 70; filarial infection, 33, 90, 92; malaria control, 56, 360, 364.
Anophelini, characters, 20; classification, 33; difference from Culicini, 33.
Anoplura, order, 214; classification, 214; life history, 216.
Anthomyidae, 148, 163.
Antibiotics of animal origin, 310.
Apiochaeta species, 144; myiasis, 175.
Aponomma, 242.
Arachnida, characters, classification, 228.
Aramite, 304.
Araneida, 273.
Argas, difference from *Ornithodorus*, 249; *persicus*, 249.
Argasidae, 234; difference from Ixodidae, 235.
Armigeres, 87.
Armillifer, 276.
Arochlor, 304.
Arosol 151 B, 315.
Arsenical compounds as insecticides, 311; toxicity, 344; mode of action, 328.
Arthropoda, characters, classification, 1, 11; diseases, 11, 12, 13.
Aschiza, characters, 18, 128.
Asilidae, 123.
Aspirator for catching midges, 116.
Astigmata, 234.
Attaclay, 317.
Auchmeromyia luteola, 154.

B

Babesia, ticks as carriers, 241, 247.
Barraud's cage, 288.
Bed bugs, habits, 207; control, 212, 394.
Beetles, intermediate hosts of parasites, 204; infesting stored food, 205; vesicating, 203; scarabiasis, 203.
Bentonite, 317.
Benzene, used against infestation by screw-worm fly larvae, 387.
Benzocaine, for controlling *Pediculus*, 225, 391.
Benzyl benzoate, in scabies, 171, 270; in *Phthirus* infestation, 225; as clothing impregnant, 307; repellent to *Aedes*, 374. salicilate, in control of lice, 391, 393.
BHC (Benzene hexachloride), chemistry, 302; estimation, 303; mode of action, 333; toxic hazards, 342; smoke, 354; control of mosquitoes, 59, 84, 87, 362, 369; of flies, 142, 379; of tsetse fly, 386; against parasites of livestock, 239, 259, 388, 393, 398; control of fleas, 390; control of relapsing fever ticks, 259; in scabies, 271; against *D. folliculorum*, 272; against bed bugs, 395; resistance in flies, 381;
Berles's fluid, 287.
Berles's organs in Cimex, 208.
Bhatia's medium, 287.
Biological warfare, 17.
Biothermic method of fly control, 142.
Biting flies, 163.
Black spore, 48.
Blattella species, 283; life history, 283; control, 396.
Blattidae, 283.
Bles's medium, 287.
Blister beetles, 202.
Blowers, see spraying apparatus.
Bombs, aerosol, 353; in mosquito control, 367, 371.
Bombyliidae, 123.
Boophilus, 240; Texas fever, 240, 241; *annulatus*, as carriers of murine typhus, 186.

Borax, as insecticide, 312.
Boric acid, 312.
Botanical principles in insecticides, 310, 328, toxicity, 344.
Bovicola, 225; control, 393.
Brachycera, 18; families, 123.
Brazil spotted fever, carried by ticks, 246.
Brevirostrata, 235.
Bullis fever, carried by *Amblyomma*, 246.
Buthus, 275.
Butoxypropylene in *Tabanus* control, 384.

C

Caccobius, 203.
Calandra, 205.
Calcium caseinate, 316; oxide, 316; cyanide, 197; other compounds, 317.
Calliphora, 151.
Calliphoridae, 148, 151; control, 154, 387.
Callitroga species, control, 387.
Calyptrata Muscoidea, families, 147.
Canine leishmaniasis, 111.
Cantharidae, 203.
Carbon disulphide, in control of horse bots, 346, 388.
Carbon tetrachloride, 313.
Carboxide, 312.
Carpoglyphus, 267.
Celite, 316.
Centipede, 282.
Centrurinae, 275.
Ceratophyllus, 191.
Ceratopsyllidae, 188.
Chagas' disease, 212.
Chalcidoidea, 140.
Chaoborinae, 19.
Cheese mites, 267.
Chelicerata, 1.
Chelifer, 277.
Chiastopsylla, 191.
Chiggerosis, 199.
Chilopoda, 282.
Chironomidae, 117.
Chlordane, chemistry, 303; estimation, 303; mode of action, 320; concentrate, 318; toxic hazards, 342; in fly control, 379; resistance to, 382; control of screw-worm fly, 387; of sheep ked, 388; of animal lice 393; of bed bugs, 212.
Chlorinated acaricides, 304.
„ biphenyls, 303.
„ camphene, 303.
„ hydrocarbons, 300; mode of toxic hazards, 342; against sheep ked, 388.
Chlorophenyl sulphone, 304.
Chloropicrin, 198, 213, 313.
Chloropidae, 145.
Chordata, 1.
Chrysenthemum cinerariafolium, 308.
Chrysomyia species, 152, 153; myiasis, 153,242; control, 387.
Chrysops, 123; life history, 126; disease, 126; species, 127; control, 384.
Cimex, life history, 207; species, 210; control, 212; by methyl parathion, 305; by meta-cide, 305; by DDT, 394.
Cimicidae, 207.
Citronella oil, repellent to leech, 4; to mosquitoes, 373; to stable-flies, 385.
Cockroach, anatomy, 13; species, 283; control, 396.
Coelomyces, 40.
Coleoptera, 201; disease, 203.
Collembola, 284.
Composting in fly control, 112.
Congo floor-maggot, 154.
Conjunctivitis, carried by eye-flies, 146.
Conorhinus, see *Triatoma*.
Contact poisons, susceptibility to, 326; effect of temperature, 327.
Copra itch, 267.
Cordylobia, 155.
Corredentia, 284.
Correthrella, 20.
Coxal glands in ticks, 231.
Crag fly repellent, 384.
Creeping eruption, 161.
Crustacea, 1.
Ctenocephalides, 190.
Ctenopsylla, 191.
Ctenopthalmus, 191.
Ctenopsylla, 191.
Culex, characters, 21, 80; species, 81; *fatigans*, 82, 83, 84; control, 91; *pipiens*, 84; filarial infection, 88; yellow fever, 100.
Culicine, classification, 21; difference from Anophelini, 33.
Culicoides, 118; carriers of parasites, 119, 120; parasites of *Anopheles*, 41; control, 377.
Cuticle, permeability, 321.
Cylops, 280; carrier of guinea-worm, 281; control, 282
Cyclorrapha, 18, 132.

D

Damalinia, control, 393.
Danish-Lidov coloremetric method for estimation of Aldrin, 303.
DDD, chemistry, 301; mode of action, 332; toxic hazards, 341; use against mosquitoes, 369; in Simulim control, 376; in fly control, 380; in control of animal lice, 293.
DDT, chemistry, 300; estimation, 58, 301; synergists, 304; spray concentrates, 318; smoke, 354; mode of action, 322; 332; residual spray, 357; in mosquito control, 57, 360, 367, 370; resistance in insects, 371; repellent to mosquitoes, 369; in sand-fly control, 379; in Simulium control, 379; against Culicoides, 377; against fly, 142, 379; with Metacide against resistant house-fly, 306; concentrates, 318; against parasites of livestock, 239; against sheep ked, 388; in control of lice, 233, 391; of fleas, 198, 300; resistance in lice, 392; against animal lice, 393; against Trombicula, 265.
DFDT, 302.
Daphnia, 282.
Deo-Base, 314.
Demodicidae, 272.
Dermacentor, 237; immunity against, 237;

disease carrier, 237; causing tick paralysis, 238.
Dermacentroxenus rickettsi, 245.
Dermanyssus, 272.
Dermatobia, 162; *hominis*, dissemination, 163.
Dermatophyllus, 199.
Derris elliptica, 309.
Detergents, 314, 316.
Diatomite, 316.
Dibutyl phthalate, as repellent to insects, 116.
Dieldrin, chemistry, 303; mode of action, 320; toxic hazards, 343; in fly control, 380; Simulium control, 376; resistance to, 382; against screw-worm fly, 387.
Dilan, 302; in fly control, 381.
Dimethyl phthalate, repellent to leech, 4; as clothing impregnant, 265; against Notoedres, 271; vapour repellent, 116, 373, 377, 385.
Dinitro compounds, 306.
Dinobdella, 4.
Dinopsyllus, 191.
Diphyllobothrium latum, 281.
Diplopoda, 282.
Diphenylamine, in control of screw-worm fly, 387; of lice, 392.
Dipylidium caninum, developing in fleas, 186.
Diptera, classification, 18.
Dirofilaria species, carried by mosquitoes, 33.
Dissecting fluids, 290.
Dixinae, 19.
Dog ticks, control, 239.
Dolicopodidae, 123.
D. M. C., 304.
Dorsocentral bristles in flies, 134.
Dracontiasis, carrier, 281.
Dracunculus, 281.
Dragon flies, 284.
Dreft, 316.
Drosophila melanogaster, 143.
Drug prophylaxis in malaria, 59.
Dung beetles, 203.
Dusts, as diluents of insecticides, 316; in formulation of insecticides, 317; mode of action, 323, 328.

E

Echidnophaga, 199.
Encephalitis, carried by mosquitoes, 33; carried by ticks, 248.
Endrin, 304.
Entry of insecticides, modes of, 320, 322; tracheal, 325; into eggs, 304.
Ephemerida, 284.
Epicauta, 203.
EPN, 305.
Eristalis, see Tubifera.
Espundia, 112.
Ethylene dichloride, 313.
Ethylene oxide, 312, 356; fumigation against bed-bugs, 396; dichloride, 313, 346.
Eurypterida, 1.
Eye-flies, transmitting yaws, 146; conjunctivitis, 146.

F

Fannia, species, 163; myiasis, 163.
Fievre Boutonneuse, carried by ticks, 245.
Filaria, and mosquitoes, 88; collection of blood for filaria, 90; control, 91; Hetrazan in filaria control, 94.
Flea, anatomy, 180; life history, 182; diseases, 185; classification, 188; control, 196, 390.
Flesh flies, see *Sarcophaga*.
Fluorine compounds, as insecticides, 311, 312.
Fly, anatomy, 132; life history, 136; methods of examination, 147; catching, 291; diseases, 140; non-biting, 147; blood-sucking, 163; resistant strain, 381; control, 141, 378. Control of biting, 384.
Fumigants, 312; toxic hazards, 345.
Fumigation of buildings, 395.

G

Gamasidae, 272.
Gammexane, see Benzene hydrochloride.
Gametocytes, influence on mosquito infection, 55.
Gasterophilus, species, 161; control, 388.
Gene's organs in ticks, 231.
Genetics of resistance to insecticides, 382.
Genitol, 304.
Gix, 302.
Glossina, anatomy, 166; life history, 168; species, 170; carrier of trypanosomes, 172, 173; control, 173, 386.
Glycyphagus, 267.
Gnathospastoidis rouxi, 203.
Gnathostoma spinigerum, carrier, 281.
Grocer's itch, 267.
Guinea-worm disease, transmission, 281; prophylaxis, 282.

H

Habronema, carried by flies, 140; 165.
Haemadipsa species, 3.
Haemalum solution, Mayer's 290; acid solution, 290.
Haemaphysalis, 236.
Haematopinidae, 215.
Haematopinus, 225; *eurysternus*, 225; control, 393; control of other species, 393.
Haematopota, 123; life history, 127; disease, 126.
Haematoxylin, Ehrlich's, 290; Delafield's, 290.
Haemoglobinuria in animals, 248.
Haemoproteus columbae, transmission, 178.
Haller's organs in ticks, 230.
Hazards, toxic, of insecticides, 341.
Heart-water, carried by ticks, 248.
Heliopsis longipes, 310; *scabra*, 310.
Hellebore root, 310.
Hemimetabolous metamorphosis, 7.
Hemiptera, 207.
Hendecenoic acid, as insect repellent, 374; action compared to chlordane, 303.
Henig's solution, 290.
Heptachlor in fly control, 379.
Heterostigmata, 234.
HETP, 305.

Hetrazan, 91.
Hexachlor, 303.
Hippelates, 145; disease, 146.
Hippobosca, 177; species, 178.
Hirudinea, 3, 4.
Hoplopsyllus, 189.
House-flies, see Musca.
Hyalomma, 251; control, 399.
Hydrogen cyanide, 313; toxicity, 346; control of lice, 392; of fleas, 197; fumigation of building, 395; control of ticks, 159.
Hymenolepis carioca, 165; *diminuta*, fleas as carriers, 186.
Hymenoptera, 201, 202.
Hypoderma, 158; species, 159; life history, 160; control, 388.
Hypopus stage in the life of mite, 266.
Hystrichopsylla, 191.

I

Immunity, metazoal, 155; against ticks, 237.
Indalone, as clothing impregnant, 374; repellent to biting flies, 385.
Insecta, characters, 2, 4; classification, 11.
Insects, anatomy, 5; classification, 11; collection, preservation and killing, 286.
Interbifid spaces, 132.
Isodon torosus, 246.
Isodrin, 303.
Isoptera, 284.
Ixodes, 236; carriers of Q. fever, 246; causing tick paralysis, 243.
Ixodidae, 229; life history, 232.
Ixodoidea, 229; carriers of parasites, 229; classification, 234.

J

Janthinosoma, carriers of eggs of Dermatobia, 162.

K

Kala-azar, carried by *Phlebotomus*, 110; control, 116, 375.

L

Lactophenol, 287.
Laelaps, 272.
Langeron's fluid, 287.
Larvicides, chemical, 365.
Lashiohelea, 118.
Latrodectus, 273, 274.
Lauryl thiocyanate, 343.
Lauseto-neu, a louse ovicide, 391; against bed-bugs, 395.
Leech, anti-leech remedies, 4.
Leishmaniasis, human, 110; canine, 111.
Lemon grass oil, repellent to mosquitoes, 373.
Lethane, 343; action against mosquitoes, 367; against flies, 306; against fleas, 392; against lice, 392.
Lepidoptera, 204; vesicating, 205.
Leptoconops, 118.
Leptopsylla, 191.
Lethane, in fly control, 142; in flea control, 390; in louse control, 392.
Leucocytozon, 248.
Lice, head and body, 214; pubic, 224; carriers of diseases, 221; control, 223, 391; control, 394. of animals, 225; control, 392; biting, 226; control, 394.
Lime sulphur, 312.
Lindane, method of action, 320, 333; in fly control, 379; against screw-worm fly, 387; against sheep ked, 388; against lice, 391, 392; against bed-bugs, 395.
Linguatula, 276.
Linognathus, 215, 225; *vituli*, control, 393.
Liponyssus, 272; carrier of endemic typhus, 272.
Lissapol, 316.
Loa loa, transmission by Chrysops, 126.
Lonchocarpus, 309.
Longirostrata, 235.
Lubricating oils, 314.
Lucex, 302.
Lucilia, 151; myiasis, 175; control, 387.
Lutzia, 88.
Lymnatis nilotica, 3.
Lynchia maura, 178.
Lyperosia, 174; control, 385.
Lyponyssus bacoti, as carriers of murine typhus, 186.
Lytta, 203.

M

Maggot, therapy, 176; rearing adults, 176; identification, 177.
Mandibulata, 1.
Malaria and Anopheles, 47; mosquitoes, in India, 42; their distribution, 48; survey, 49; dissection of mosquitoes for parasites, 51, 54; detection of infection in mosquitoes, 52, 53; control, 56, 360, 364.
Malathon, 305; in fly control, 381.
Mallophaga, 226; control, 394.
Mange of livestock, caused by mites, 267; control, 398.
Mansonia, 84.
Mansonioides, species, 85; control, 87; filarial infection carried by, 92.
Margaropus, 242.
Mecoptera, 285.
Megarhinini, 21.
Megarhinus species, 21.
Meloidae, 203.
Melophagus ovinus, 178; control, 388.
Mermithides, parasites of Anopheles larvae, 40.
Mesocyclops, carrier of dracontiasis, 281.
Mesostigmata, 234.
Metacide, against bed-bugs and house-flies, 305.
Metamorphosis, different types, 7.
Metastigmata, 234.
Methoxychlor, chemistry, 301; estimation, 301; control of insects, 376, 379, 388, 393, 395; resistance to, 382.
Methyl bromide, 312; toxicity, 346; naph

thalenes, 315; used with parathion against bed-bugs, 213, 305; in control of lice, 392; for fumigation against bed-bugs, 395; against rodents, 198; 402; fleas, 198.
„ naphthalenes, 315.
„ parathion, 305.
Milletia, 309.
Mites, parasites of mosquitoes, 40; classification, 228; control, 398.
Mochlonynx, 32.
Mosquito, anatomy, 22; life history, 28; diseases carried by, 33; collection, 50; dissection, 51; preservation, 286; repellents, 373; control, 364.
Mounting media, 287.
Mundulea, 309.
Musca, anatomy, 132; life history, 138; diseases carried by, 140; characters of the genus, 148; species, 148; biting, 151; control, 142, 378.
Muscoidea, 132; classification, 143; Acalyptrata, 142; Calyptrata, 147.
Muscidae, 147; blood-sucking, 163.
Myiasis, classification, 175; subdermal, 163; urinary, 128, 143, 162, 163, 175.
MYL, in control of lice, 392.
Mylabris, 203.
Myriapoda, 1, 282.

N

Nacconole, 316.
NBIN, emulsion in control of human lice, 391.
NCI, as pediculicide, 392.
Nematocera, 18.
Neopsylla, 191.
Neutroleum alpha, 314.
Nicotine, 307; estimation, 307; modes of entry, 321, 322; mode of action, 331; toxicity, 344; bentonite, 307.
Nicteribiidae, 177.
Nosopsyllus, 189; as carriers of plague, 187; of murine typhus, 185.
Notoedres, 267.

O

Ocellus, 17.
Octacide, 308, 309.
Octochlor, used against Anopheles, 304.
Odonata, 284.
Oestridae, 148, 157.
Oestrus ovis, 158.
Oils as insecticides, 314, 315; mode of action, 328
Oligochaeta, 3.
Ommatidia, 16.
Onchocerciasis, human, 121; animal, 122; control, 375.
Onthophagus, 204.
Onychophora, 2.
Oocysts, 52, 53.
Ookinete, 53.
Ornithodorus, characters, 249, 251; difference from *Argas*, 249; *savignyi*, 251; *moubata*, 252; other species, 253, 257; relapsing fever carried by, 255; control, 259.
Oroya fever, carried by sand-flies; 112.
Orthoptera, 282.
Orthorrapha, 18, 282.
Oscinis pallipes, 146.
Otodectes cynotis, 267.

P

Pachydrilus lineata, 3.
Pachyrhizus, 309.
PDB, 313.
Paederus, 202.
Pangonia, 128.
Parameles, 246.
Parasitic index, 54.
Parathion, 305; mode of action, 333; for insect control, 305, 376, 381.
Paris green, 311, 341.
Pasteurella avicida, carried by ticks, 250.
Pattonella, 248.
Paurometabolous metamorphosis, 7.
Pedicinus, 215.
Pediculidae, 214.
Pediculoides, 266.
Pediculus, 215; species, 220; life history, 216; disease, 221; control, 223, 391.
Pedipalpi, 277.
Pentachlorophenol, 307.
Pentastomida, 276.
Pentrol, 316.
Periplaneta, 13, 283.
Peritrophic membrane, 15, 168.
Phenothiazine, 306; estimation, 306; toxic hazards, 343; in control of bot-flies, 388.
Phlebotomus, 104; life history, 108; diseases, 110; repellents, 116, 375; control, 116, 375.
Phenyl cellosolve, 223, 270.
Phoridae, 144.
Phormia regina, 175.
Phosphates, organic as insecticides, 305.
Phthirus pubis, 224; control, 225, 391.
Physiology of resistance to insecticides, 383.
Picrasma excelsa, 310.
Pinus, 205.
Piophila casei, 143.
Piperonyl butoxide, mode of action, 333; in insect control, 308, 379, 384.
„ cyclohexanol, 309.
„ cyclonene in fly control, 379; in Tabanid control, 384; mode of action, 333.
Piroplasmosis, ticks as carriers, 247.
Plague, mechanism of transmission, 186; fleas concerned, 186; rats, 186; control, 196, 390, 405.
Plasmodium, developing in mosquitoes, 98.
Polychaeta, 3.
Polyplax, 225; *spinulosus*, as carriers of murine typhus, 186, 215; control, 393.
Pulicidae, classification, 188; characters, 189.
Pupipara, 177.
Preipcitin test, 56.
Preservation of insects and larvae, 287.
Prestomal teeth, 167.
Presyn, 308.
Prostigmata, 234.
Psychoda, 103, control, 375.
Psychodidae, 103.

Psoroptes, 267.
Pupipara, 177.
Pygiopsylla, 190.
Pyrethrins, 307; estimation, 308; synergists, 309; mode of action, 328, 330.
Pyrethrum, 344, 373; in control of insects, 142, 379, 384, 390, 391.
Pyrolan, 306, 307.

Q

Q. fever, ticks as transmitters, 246.

R

Rats, plague, 186, 194; identification, 195; ectoparasites, 196; murine typhus, 185; control, 196, 402; poisons, 197, 405.
Relapsing fever ticks, 257.
Repellents to mosquitoes, 369, 373; to sand-flies, 374; to ticks, 374; to *Simulium*, 374.
Resistance, genetics, 382; physiology, 383.
Resistant strains of insects, 371, 381; control, 305.
Rhipicephalus, 238; carrier of typhus, 239; of Texas fever, 239; of equine piroplasmosis, 239; of *T. parva*, 239.
Rhipicentor, 242.
Rhizoglyphus parasiticus, 267.
Rhizopertha dominica, 205.
Rhodnius, 212; disease, 212; control, 213.
Rhothane, 302.
Ricinin, 310.
Rickettsia, different types and their carriers, 218, 242, 244.
Rickettsialpox, carried by a mite, 265.
Rocky Mountain spotted fever, carried by ticks, 244.
Rodents, field control, 198, 406.
Rotenone, sources of, 309; estimation, 310; mode of action, 331; toxicity, 344; in control of warble-flies, 388; of sheep ked, 388; of fleas, 390; of dog ticks, 239.
Ryania speciosa, 310.
Ryanodine, 310.

S

Sabadilla, 310.
Sand-flies, see Phlebotomus.
Sentocol, 316.
Sampling method of testing insecticides, 336.
Sao Paulo fever, carried by ticks, 246.
Saponair officianilis, 316.
Sarcophaga, 157.
Sarcophagidae, 156.
Sarcopsyllidae, 188; genera, 199.
Sarcoptes, life history, 267; remedies against infestation, 270.
Scabies, prevention, 398.
Scabrin, 310.
Scarabiasis, 203.
Scatophagidae, 147.
Schecter-Haller method of estimation of DDT, 301.
Schizophora, 18.
Schongastia, 261.
Schradan, 305.
Scorpionidea, 275; families, 275.
Screw-worm fly, control, 387.
Scrub typhus, 264.
Section cutting, 289.
Sherlice belt, 392.
Silvanus surinamensis, 205.
Simulium, control, 375.
Smoke, insecticidal, 354.
Soaps, 315.
Sodium arsenite, 311.
fluoride, 311.
fluosilicate, 311.
Solonopotes capillatus, 215, 393.
Solifugae, 277.
Spider bite, 274; treatment, 274.
Spirochaeta (*Treponema*) *novyi*, carried by ticks, 256.
Spirochaeta (*Treponema*) *recurrentis*, carried by lice, 221.
Spraying apparatus, 347.
„ residual, 367.
Stay-way lotion, 373.
Stigmatomyces baeri, parasite of flies, 140.
Stivalius, 187.
Stomoxys, 163; carrier of trypanosome of surra 165; of Habronema, 165; control, 385.
Store boxes, preparation, 287.
Stylosome, 261.
Sulfoxide, 392.
Sulphur compounds, 306, 312; dioxide, 312; as diluents, 317; toxicity, 346.
Supercel, 316.
Surra, carried by Tabanus, 125; by Stomoxys, 126, 165; by Glossina, 165; by other blood-sucking flies, 126.
Sylvatic plague, 198.
Sylvius, 128.
Synergists, for DDT, 304; 306; for pyrethrum, 309, 330; for allethrin, 308; mode of action, 330.
Syrphus, 128.

T

Tabanidae, 123.
Tabanus, 123; life history, 125; diseases, 125; species, 126; control, 384.
Talc, 317.
Tachinidae, 148.
Tarsonemidae, 266.
Tar distillate, 315.
Tartar emetic, 312.
Teepol K, 316.
Tephrosia, 309.
TEPP, 305.
Testing insecticides, 334; laboratory methods, 334; field methods, 336; sampling method, 337.
Tetmosol soap, 306.
Tetraethylthiurum monosulphide, 270, 306.
Texas fever, carried by ticks, 240; 241.
Thanite, 343.
Theileria, ticks as carriers, 244.
Thiocyanates, 306, 343, 390.
Thysanura, 284.

Ticks, responsible for different types of typhus, 246; human diseases carried by, 243; animal diseases carried by, 247; causing paralysis, 243; control, 399.
Tityus, 275.
Toxaphene, 304; estimation, 304; concentrate, 318; in insect control, 377; used against parasites of live stock, 388; control of lice, 393.
Tracheal penetration of insecticides, 325.
Trench fever, 223.
Tribolium castaneum, 205.
Trichodectes, 226.
Trichopsylla, 191.
Trichloroacetonitrile, 313.
Trichoptera, 285.
Triton X-100, 316.
Trogoderma, 205.
Trombicula, 260; life history, 263; diseases, 264; control, 265, 398.
Trypanosoma brucei, *gambiense* and *rhodesiense*, carried by *Glossina*, 166, 170;
cruzi, developing in *Triatoma*, 212;
evansi, carried by *Tabanus*, 125; other blood-sucking flies, 126, 165.
Tsetse-fly control, 386.
Tubifera, 128.
Tularaemia, ticks responsible, 244.
Tunga, 119.
Typhus, different types, 185, 221, 222, 244, 246.
Tyroglyphus, 267.
Turkey Red oil, 316.
Tween 80, 316, 376.

V

Valone, 307.
Vermiformia, 234.

W

Warble-fly, 378.
Wohlfahrtia, 157.

X

Xenopsylla, species, 186, 187, 192; 198; identification, 193.
Xiphosura, 277.

Y

Yellow fever, carried by *Aedes* and other mosquitoes, 99; control, 101, 102.
Yaws, transmitted by *H. pallipes*, 146.